International Rare
Book Prices

SCIENCE &
MEDICINE

1990

International Rare Book Prices

SCIENCE & MEDICINE

Series Editor: Michael Cole

1990

The Clique

International Rare Book Prices – Science & Medicine

ISBN 1 870773 17 9

North America
Spoon River Press, P.O. Box 3676
Peoria, Illinois 61614, U.S.A.

Typesetting by Maxiprint, York, England
Printed and bound by Unwin Brothers Ltd., Woking, England

Contents

Introduction and Notes

Science & Medicine is the fourth title in the annual series *International Rare Book Prices*. The other titles in the series are *The Arts & Architecture, Early Printed Books, Modern First Editions, Voyages, Travel & Exploration, 19th Century Literature.*

The series, generally referred to as *IRBP*, provides annual records of the pricing levels of out-of-print, rare or antiquarian books within a number of specialty subject areas and gives likely sources and suppliers for such books in Britain and the United States of America. It is intended to be used by both the experienced bookman and the newcomer to book-collecting.

Sources of information:

The books recorded each year in the various subject volumes of *IRBP* have been selected from catalogues of books for sale issued during the previous year by numerous bookselling firms in Britain and the United States. These firms, listed at the end of this volume, range in nature from the highly specialized, handling books solely with closely defined subject areas, through to large concerns with expertise across a broad spectrum of interests.

Extent of coverage:

IRBP concentrates exclusively on books published in the English language and, throughout the series as a whole, encompasses books published between the 16th century and the 1970s.

The 30,000 or so separate titles recorded in the annual volumes of *IRBP* vary greatly from year to year although naturally there is a degree of overlap, particularly of the more frequently found titles. Consecutive annual volumes do not, therefore, merely update pricings from earlier years; they give substantially different listings of books on each occasion. The value of the *IRBP* volumes lies in providing records of an ever-increasing range of individual titles which have appeared for sale on the antiquarian or rare book market.

Emphasis is placed throughout on books falling within the lower to middle range of the pricing scale (£10 - £250; $20 - $500) rather than restricting selection to the unusually fine or expensive. In so doing, *IRBP* provides a realistic overview of the norm, rather than the exception, within the booktrade.

Authorship and cross-references:

Authors are listed alphabetically by surname.

Whenever possible, the works of each author are grouped together under a single form of name irrespective of the various combinations of initials, forenames and surnames by which the author is known.

Works published anonymously, or where the name of the author is not recorded on the title-page, are suitably cross-referenced by providing the main entry under the name of the author (when mentioned by the bookseller) with a corresponding entry under the first appropriate word of the title. In cases of unknown, or unmentioned, authorship, entry is made solely under the title.

Full-titles:

Editorial policy is to eschew, whenever possible, short-title records in favour of full-, or at least more complete and explanatory, titles. Short-title listings do little to convey the flavour, or even the content, of many books - particularly those published prior to the nineteenth century.

Descriptions:

Books are listed alphabetically, using the first word of the title ignoring, for alphabetical purposes, the definite and indefinite articles *the, a* and *an.* Within this alphabetical grouping of titles, variant editions are not necessarily arranged in chronological order, i.e., a 2nd, 3rd or 4th edition might well be listed prior to an earlier edition.

Subject to restrictions of space and to the provisos set out below, the substance of each catalogue entry giving details of the particular copy offered for sale has been recorded in full.

The listings have been made so as to conform to a uniform order of presentation, viz: Title; place of publication; publisher or printer; date; edition; size; collation; elements of content worthy of note; description of contents including faults, if any; description and condition of binding; bookseller; price; approximate price conversion from dollars to sterling or vice versa.

Abbreviations of description customary within the booktrade have generally been used. A list of these abbreviations will be found on page *x.*

Collations:

Collations, when provided by the bookseller, are repeated in toto although it should be borne in mind that booksellers employ differing practices in this respect; some by providing complete collations and others by indicating merely the number of pages in the main body of the work concerned. The same edition of the same title catalogued by two booksellers could therefore have two apparently different collations and care should be taken not to regard any collation recorded in *IRBP* as being a definitive or absolute record of total content.

Currency conversion:

IRBP lists books offered for sale priced in either pounds sterling (£) or United States dollars ($). For the benefit of readers unaccustomed to one or other of these currencies, an approximate conversion figure in the alternative currency has been provided in parentheses after each entry, as, for example, "**£100 [** ≃ **$160]**", or, "**$60 [** ≃ **£37]**". The conversion is based upon an exchange rate of £1 sterling ≃ US $1.60 (US $1 ≃ £0.625 sterling), the approximate rate applicable at the date of going to press.

It must be stressed that the conversion figures in parentheses are provided merely as an indication of the approximate pricing level in the currency with which the reader may be most familiar and that fluctuations in exchange rates will make these approximations inaccurate to a greater or lesser degree.

Acknowledgements:

We are indebted to those booksellers who have provided their catalogues during 1989 for the purposes of *IRBP*. A list of the contributing booksellers forms an appendix at the rear of this volume.

This appendix forms a handy reference of contacts in Britain and the United States with proven experience of handling books within the individual specialist fields encompassed by the series. The booksellers listed therein are able, between them, to offer advice on any aspect of the rare and antiquarian booktrade.

Many of the listed books will still, at the time of publication, be available for purchase. Readers with a possible interest in acquiring any of the items may well find it worth their while communicating with the booksellers concerned to obtain further and complete details.

Caveat:

Whilst the greatest care has been taken in transcribing entries from catalogues, it should be understood that it is inevitable that an occasional error will have passed unnoticed. Obvious mistakes, usually typographical in nature, observed in catalogues have been corrected. We have not questioned the accuracy in bibliographical matters of the cataloguers concerned.

The Clique

Abbreviations

advt(s)	advertisement(s)	iss	issue
addtn(s)	addition(s)	jnt(s)	joint(s)
a.e.g.	all edges gilt	lge	large
ALS	autograph letter signed	lea	leather
altrtns	alterations	lib	library
Amer	American	ltd	limited
bibliog(s)	bibliography(ies)	litho(s)	lithograph(s)
b/w	black & white	marg(s)	margin(s)
bndg	binding	ms(s)	manuscript(s)
bd(s)	board(s)	mrbld	marbled
b'plate	bookplate	mod	modern
ctlg(s)	catalogue(s)	mor	morocco
chromolitho(s)	chromo-lithograph(s)	mtd	mounted
ca	circa	n.d.	no date
cold	coloured	n.p.	no place
coll	collected	num	numerous
contemp	contemporary	obl	oblong
crnr(s)	corner(s)	occas	occasional(ly)
crrctd	corrected	orig	original
cvr(s)	cover(s)	p (pp)	page(s)
dec	decorated	perf	perforated
detchd	detached	pict	pictorial
diag(s)	diagram(s)	port(s)	portrait(s)
dw(s)	dust wrapper(s)	pres	presentation
edn(s)	edition(s)	ptd	printed
elab	elaborate	qtr	quarter
engv(s)	engraving(s)	rebnd	rebind/rebound
engvd	engraved	rec	recent
enlgd	enlarged	repr(d)	repair(ed)
esp	especially	rvsd	revised
ex lib	ex library	roy	royal
f (ff)	leaf(ves)	sep	separate
facs	facsimile	sev	several
fig(s)	figure(s)	sgnd	signed
fldg	folding	sgntr	signature
ft	foot	sl	slight/slightly
frontis	frontispiece	sm	small
hand-cold	hand-coloured	t.e.g.	top edge gilt
hd	head	TLS	typed letter signed
ill(s)	illustration(s)	unif	uniform
illust	illustrated	v	very
imp	impression	vell	vellum
imprvd	improved	vol(s)	volume(s)
inc	including	w'engvd	wood-engraved
inscrbd	inscribed	w'cut(s)	woodcut(s)
inscrptn	inscription	wrap(s)	wrapper(s)
intl	initial		

Science & Medicine
1989 Catalogue Prices

Abady, J.
- Gas Analyst's Manual. London: Spon, 1902.
561 pp. Fldg plates, ills. Orig bndg.
(Book House) £20 [≃ $32]

Abbott, Charles, M.D.
- Upland and Meadow. A Poaetguissings
Chronicle. New York: Harper, 1886. 1st edn.
Sm 8vo. 397 pp. Orig bndg.
(Xerxes) $95 [≃ £59]

Abbott, Maude E. (ed.)
- Appreciations and Reminiscences: Sir William
Osler Memorial Number. Montreal: Intl Assn
of Medical Museums, Bulletin IX, (1927). 1st
edn. 2nd imp, with addtns. One of 1500.
xxxix, 634 pp. Frontis port, over 75 plates,
ports, facs. Orig cloth. *(Elgen)* $150 [≃ £93]

Abercrombie, John, physician
- Pathological and Practical Researches on
Diseases of the Brain and Spinal Cord. New
Edition, enlarged by the Author.
Philadelphia: 1843. 8vo. 324 pp. Occas
foxing. Upper hinge weak.
(Rittenhouse) $200 [≃ £125]

Abercrombie, John, writer on horticulture
- The British Fruit-Gardener; and Art of
Pruning ... Dublin: Whitestone, 1781. 1st
Irish edn. 12mo. Contemp calf, rebacked.
(Falkner) £65 [≃ $104]
- The Gardener's Pocket Journal. London:
1807. 10th edn. 12mo. 324 pp. Half calf.
(Wheldon & Wesley) £30 [≃ $48]
- The Gardener's Pocket Journal, and Daily
Assistant. London: 1821. 16th edn. 12mo.
324 pp. Frontis. Half calf gilt.
(Wheldon & Wesley) £20 [≃ $32]
- Abercrombie's Practical Gardener, or
Improved System of Modern Horticulture.
London: 1817. 2nd edn. 8vo. xxiv,680 pp.
Trifle foxed. Half mor.
(Wheldon & Wesley) £40 [≃ $64]

Abernethy, John
- Surgical Observations. On the Constitutional
Origin and Treatment of Local Diseases; and
on Aneurisms ... Philadelphia: Dobson, 1811.
1st Amer edn. 8vo. Half- titles, ix,[2],325 pp.
Usual browning. Contemp tree calf, jnts open
but attached, rubbed. *(Elgen)* $250 [≃ £156]

Abraham, H.
- Ashphalts and Allied Substances. Their
Occurrence, Modes of Production, Uses in
the Arts and Methods of Testing. London:
Crosby Lockwood, 1929. 3rd edn. 891 pp.
Ills. Orig bndg. *(Book House)* £25 [≃ $40]

Abraham, James Johnston
- Lettsom. His Life, Times, Friends and
Descendants. London: 1933. 1st edn. Lge
8vo. Num ills. Orig cloth, sl marked, t.e.g.
(Robertshaw) £35 [≃ $56]

Accum, Friedrich Christian (Frederick)
- Chemical Amusement, comprising a Series of
Curious and Instructive Experiments in
Chemistry ... Fourth Edition, with Plates and
considerably enlarged. London: Thomas
Boys, 1819. 12mo. xlviii,49-430 pp. Plates.
Orig polished calf backed mrbld bds.
(Charles B. Wood) $85 [≃ £53]
- Guide to the Chalybeate Spring of Thetford,
exhibiting the General and Primary Effects of
the Thetford Spa ... [London]: sold by T.
Boys; at the Spa ..., 1819. 1st edn. 12mo.
xiv,159,[i] pp. Fldg hand cold frontis, fldg
plate. Orig bds, uncut, jnts rubbed.
(Finch) £275 [≃ $440]
- A Treatise on the Art of making Good and
Wholesome Bread of Wheat, Oats, Rye and
Barley and other Farinaceous Grain. London:
Thomas Boys, 1821. 1st edn. 8vo. [ii],iv, 160,
xxiv pp. Hand cold aquatint title vignette.
Mod cloth, ex-lib but unmarked.
(Charles B. Wood) $275 [≃ £171]

Acheta Domestica (pseud.)
- See Budgen, M.L.

Acts of Parliament
- An Act to Prevent the Marriage of Lunaticks. London: Thomas Baskett & Robert Baskett, 1742. Folio. 15 & 16 Geo II, pp 705-708. Sl stained. Disbound.   *(Jarndyce)* **£40 [≈ $64]**

Acworth, W.M.
- The Railways of England. London: Murray, 1889. 1st edn. 8vo. 427 pp. 56 ills. Orig bndg, sl rubbed, shaken. *(Book House)* **£45 [≈ $72]**

Adami, J. George & McCrae, John
- A Text-Book of Pathology. Philadelphia: [1912]. 1st edn. 8vo. 759 pp. 11 cold plates, 304 text ills. Orig cloth, ex-lib, spine ends & crnrs frayed.   *(Elgen)* **$50 [≈ £31]**

Adams, E.D.
- Niagara Power. History of the Niagara Falls Power Co 1886-1918. Evolution of its Central Power Station and Alternating Current System. New York: 1927. 2 vols. 4to. 455; 504 pp. Maps, ills, diags. Orig bndg.
(Book House) **£120 [≈ $192]**

Adams, Henry Gardiner
- Beautiful Shells ... London: Groombridge, 1871. 3rd edn. 8vo. [iv],156 pp. 8 cold plates by Benjamin Fawcett, text engvs. 1 section springing. Orig pict brown cloth gilt, a.e.g., somewhat cockled & rubbed.
(de Beaumont) **£30 [≈ $48]**
- Humming Birds Described and Illustrated. London: n.d. 8vo. 8 hand cold plates. Orig dec cloth, a.e.g., recased.
(Henly) **£95 [≈ $152]**
- Nests and Eggs of Familiar Birds. London: 1871. 8vo. 238 pp. 16 cold plates. Orig cloth, trifle used and loose.
(Wheldon & Wesley) **£20 [≈ $32]**

Adams, Joseph, 1756-1818
- Memoirs of the Life and Doctrines of the late John Hunter ... Second Edition, corrected by the Author. London: 1818. 8vo. [iv], 262 pp, advt leaf. Silhouette port. Orig bds, uncut, new paper spine. *(Bickersteth)* **£110 [≈ $176]**
- Observations on Morbid Poisons, Phagedaena, and Cancer ... Laws of the Venereal Virus ... London: for J. Johnson, 1795. 1st edn. 8vo. [4],iv,328,[2] pp. Lib withdrawal stamp. Contemp half mor.
(Rootenberg) **$500 [≈ £312]**

Adams, Lionel E.
- The Collector's Manual of British Land and Freshwater Shells ... Second Edition. Leeds: 1896. Cr 8vo. [6],ii,214,xii advt pp. 11 plates (9 cold). Orig cloth gilt.
(Fenning) **£28.50 [≈ $46]**

Adams, W.H. Davenport
- Lighthouses and Lightships: a Descriptive and Historical Account of their Mode of Construction and Organization. London: Nelson, 1871. 322,index pp. Frontis, plates, ills. Foxed. Orig gilt dec cloth.
(Wreden) **$45 [≈ £28]**

Adams, W.H. Davenport & Giacomelli, H.
- The Bird World. London: Nelson, 1880. Roy 8vo. 464 pp. Num ills. Orig cloth gilt, a.e.g.
(Egglishaw) **£18 [≈ $28]**

Adanson, M.
- A Voyage to Senegal, the Isle of Goree and the River Gambia. Translated from the French. London: 1759. 8vo. [xiv],337,[1] pp. Margs of title browned. Calf, jnts cracked.
(Wheldon & Wesley) **£375 [≈ $600]**

Adrian, E.D.
- The Mechanism of Nervous Action; Electrical Studies of the Neurone. Philadelphia: 1935. 1st edn, 2nd printing. 8vo. 103 pp. Ills. Orig cloth.
(Elgen) **$65 [≈ £40]**

The Aero Manual ...
- The Aero Manual. A Manual of Mechanically- Propelled Human Flight ... Compiled by the Staff of "The Motor". London: Temple Press, 1909. 1st edn, 1st printing. Post 8vo. [xvi], 157,[3 advt] pp. Frontis, 161 ills. Crnr of 7 ff sl creased. Orig cloth.   *(Duck)* **£150 [≈ $240]**

Agassiz, Alexander
- Letters and Recollections of Alexander Agassiz with a Sketch of his Life and Work. Edited by G.R. Agassiz. Boston: Houghton, Mifflin, 1913. 8vo. xii,454 pp. Frontis port, 17 plates, fldg maps in pockets. Orig cloth gilt, uncut & unopened.
(Berkelouw) **$100 [≈ £62]**

Agassiz, Louis
- Bibliographia Zoologiae et Geologiae: a General Catalogue of all Books ... on Zoology and Geology, Enlarged, and Edited by H.E. Strickland. London: Ray Society, 1848-54. 4 vols. 8vo. Sm lib marks on titles. Some foxing. rec qtr mor.

(Bow Windows) **£195 [≈ $312]**
- The Classification of Insects from Embryological Data. Washington: Smithsonian Contribs to Knowledge, 1850. 4to. 28 pp. Plate. Plain wraps.
(Wheldon & Wesley) **£25 [≈ $40]**
- An Essay on Classification. London: 1859. 8vo. viii,381 pp. Ex lib with a few stamps. Orig cloth. *(Baldwin)* **£30 [≈ $48]**
- Lake Superior: its Physical Character, Vegetation, and Animals, compared with those of other ... Regions. With a Narrative of the Tour by J. Elliot Cabot. Boston: 1850. Roy 8vo. x,[ii],9-428 pp. Frontis, map, 15 plates. Title sl damaged. Orig cloth.
(Wheldon & Wesley) **£130 [≈ $208]**
- Louis Agassiz, his Life and Correspondence. Edited by E.C. Agassiz. Boston: 1886. 4th edn. 2 vols. 8vo. 12 ills. Cloth.
(Wheldon & Wesley) **£40 [≈ $64]**

Agassiz, Louis & Gould, A.A.
- Outlines of Comparative Physiology, touching the Structure and Development of the Races of Animals, Living and Extinct, edited from the Revised Edition. London: Bohn, 1851. 8vo. xxiv,442 pp. Cold frontis, 391 figs. Orig cloth.
(Wheldon & Wesley) **£30 [≈ $48]**

Agassiz, Louis & Hartt, F.
- Scientific Results of a Journey in Brazil. Boston: Fields, Osgood, 1870. 8vo. xxiii,620 pp. Num text figs. Orig cloth, waterstained.
(Gemmary) **$60 [≈ £37]**

Agnew, D. Hayes
- The Principles and Practice of Surgery ... Philadelphia: 1878-1883. 1st edn. 3 vols. Thick 8vo. Num ills. Orig sheep, rubbed, lacks vol 3 labels, vol 3 rear cvr bumped.
(Elgen) **$150 [≈ £93]**

Agricola, Georgius
- De Natura Fossilium (Textbook of Mineralogy). Translated by M.C. and J.A. Bandy. New York: Geological Society of America, Special Paper 63, 1955. 8vo. xii,240 pp. Orig cloth. *(Gemmary)* **$75 [≈ £46]**
- De Re Metallica. Translated from the First Latin Edition of 1556 ... By Herbert Clark Hoover and Lou Henry Hoover. London: Mining Magazine, 1912. 1st edn in English. Folio. Num text ills. Orig vellum over bds, largely unopened. 2 stray marks on spine. Sgnd by H.C. Hoover. *(Reese)* **$750 [≈ £468]**

Agrippa, Henry Cornelius
- The Vanity of Arts and Sciences. London: R. Everingham for R. Bentley ..., 1694. 3rd edn in English. [xx],368,[2] pp. Frontis. Contemp calf, front bd detached, jnts taped.
(Gach) **$285 [≈ £178]**

Aikin, Arthur
- The Natural History of the Year; being an Enlargement of Dr. Aikin's Calendar of Nature. London: for J. Johnson, 1798. 1st edn. 12mo. vi,[ii],195,[1] pp. Fldg plate. Contemp half calf. *(Burmester)* **£100 [≈ $160]**

Aikin, J.
- The Woodland Companion or a Brief Description of British Trees. London: 1802. 1st edn. 8vo. iv,92 pp. 28 dble-page plates. Half calf, rebacked. *(Henly)* **£45 [≈ $72]**

Aikin, John
- Biographical Memoirs of Medicine in Great Britain from the Revival of Literature to the Time of Harvey. London: for Joseph Johnson, 1780. Sole edn. Final advt leaf. Speckled sheep, hinges cracked, spine sl worn. *(Jarndyce)* **£220 [≈ $352]**

Ainslie, John
- Comprehensive Treatise on Land Surveying ... Edinburgh: Doig, 1812. 4to. 40 plates (many cold). Contemp calf.
(Emerald Isle) **£200 [≈ $320]**
- The Gentleman and Farmer's Pocket Companion and Assistant: consisting of Tables for finding the Contents of any Piece of Land ... Edinburgh: Constable, 1812. Sq 12mo. xx, 218 pp. Contemp sheep, rebacked.
(Blackwell's) **£75 [≈ $120]**

Ainsworth-Davis, J.R.
- The Natural History of Animals. London: 1903-04. 4 vols in 8. Roy 8vo. Cold & plain plates, text ills. Sl foxing. Orig cloth gilt, spines faded. *(Henly)* **£35 [≈ $56]**

Airy, George Biddell
- Mathematical Tracts on Lunar and Planetary Theories ... for the Use of Students in the University. Third Edition, Corrected ... Cambridge: 1842. 8vo. [viii],390 pp. 5 fldg plates. Contemp half calf, rubbed.
(Bickersteth) **£25 [≈ $40]**

Aitchison, L.
- A History of Metals. London: Macdonald, 1960. 2 vols. 4to. 647 pp. Ills. Ex-lib but good. Orig bndg. *(Book House)* **£35 [≈ $56]**

Aitken, R.B.
- Great Game Animals. New York: 1968. 192 pp. Num ills. Orig bndg. Dw.
(Trophy Room Books) **$125 [≈ £78]**
- Great Game Animals of the World. London: 1969. Lge 4to. Cold & b/w ills. Orig bndg. Dw.
(Grayling) **£60 [≈ $96]**

Aiton, William
- A Treatise on the Origin, Qualities and Cultivation of Moss-Earth, with Directions for Converting it to Manure ... Air: ptd by Wilson & Paul, & sold for the author by them ..., 1811. Only edn. 8vo. xxxix,357 pp. Orig bds, uncut, jnts weak but holding.
(Young's) **£75 [≈ $120]**

Albee, Fred H.
- Orthopedic and Reconstruction Surgery. Philadelphia: 1919. 1st edn. 8vo. 1138 pp. Num ills (inc cold). Orig cloth.
(Elgen) **$195 [≈ £121]**

Alder, J. & Hancock, A.
- The British Tunicata. London: Ray Society, 1905-12. 3 vols. 8vo. 3 ports, 66 plates, many cold. Endpapers & half-titles foxed. Orig cloth, v sl worming to cvrs of vol 2, back cvr vol 3 spotted.
(Wheldon & Wesley) **£75 [≈ $120]**

Alderson, John
- An Essay on Apparitions, in which their Appearance is Accounted for by Causes wholly Independent of Preternatural Agency. London: for Longman ..., 1823. New edn, rvsd & crrctd. 8vo. 53 pp. Half-title. Disbound, uncut.   *(Young's)* **£75 [≈ $120]**

Aldis, W.S.
- An Elementary Treatise on Solid Geometry. Cambridge: 1865. viii,188 pp. Orig cloth, spine relaid.   *(Whitehart)* **£38 [≈ $60]**

Alexander, William, M.D.
- Plain and Easy Directions For the Use of Harrowgate Waters ... Edinburgh: for A. Kincaid & W. Creech, & J. Dickson, 1773. 1st edn. 8vo. iv,92 pp. Stitched as issued, uncut, a little dusty externally. *(Finch)* **£65 [≈ $104]**
- Plain and Easy Directions For the Use of Harrowgate Waters ... Edinburgh: for A. Kincaid & W. Creech, 1773. 1st edn. 8vo. iv, 92 pp. Outer ff a bit soiled. Disbound.
(Burmester) **£45 [≈ $72]**

Algarotti, Francesco
- The Philosophy of Sir Isaac Newton

explained, in Six Dialogues, on Light and Colours, between a Lady and the Author. Glasgow: Urie,   1765.   Fcap   8vo. [2],xiv,17-280 pp, advt leaf. Errata slip. Lacks intl blank. Contemp calf.
(Spelman) **£220 [≈ $352]**
- Sir Isaac Newton's Theory of Light and Colours ... made familiar to the Ladies in Several Entertainments [Translated from the Italian by Elizabeth Carter] ... London: G. Hawkins, 1742. 2nd edn. 2 vols. 12mo. [iv], 211; [iv], 224 pp. Contemp sheep, rebacked.
(Burmester) **£250 [≈ $400]**

Alglave, Em. & Boulard, J.
- The Electric Light: Its History, Production, and Applications. Translated from the French by T. O'Conor Sloane ... New York: Appleton, 1884. 1st Amer edn. Roy 8vo. xviii, 458,advt pp. Frontis, plates, ills. Orig green cloth, sl rubbed & dull, minor soiling.
(Duck) **£150 [≈ $240]**

Ali, S.
- The Birds of Travancore and Cochin. 1953. Roy 8vo. xx,436 pp. 16 cold & 6 plain plates, 32 text figs. Orig cloth, sm holes in spine.
(Wheldon & Wesley) **£40 [≈ $64]**
- Indian Hill Birds. 1949. Cr 8vo. 64 cold & 8 plain plates. Orig cloth. Dw.
(Wheldon & Wesley) **£18 [≈ $28]**

Allen, Harrison
- A System of Human Anatomy, including its Medical and Surgical Relations. With a Section on Histology by E.O. Shakespeare. Philadelphia: Lea, 1884. 1st edn. Lge 4to. 812 pp. Errata leaf. 109 litho plates (many cold), 241 text ills. Three qtr leather, scuffed.
(Elgen) **$150 [≈ £93]**

Allman, G.J.
- A Monograph of the Fresh-Water Polyzoa. London: Ray Society, 1856. Folio. viii,119 pp. 11 plates (10 cold). Lib stamps. Half calf.
(Baldwin) **£50 [≈ $80]**

Alpheraky, Sergius
- The Geese of Europe and Asia. London: 1905. 4to. 25 cold plates. 4 plates sl foxed. Contemp half mor. *(Grayling)* **£400 [≈ $640]**

Alt, Adolf
- Lectures on the Human Eye in its Normal and Pathological Conditions. London: Putnam, 1884. 8vo. xvi,208 pp. 95 text ills. Sl marg browning. Orig cloth, spine ends & crnrs frayed.   *(Elgen)* **$40 [≈ £25]**

An Amateur Flagellant ...
- Experiences of Flagellation, a Series of Remarkable Instances of Whipping Inflicted on both Sexes. London: privately ptd, 1885. 8vo. 80 pp. Bookplate half torn out.
(Xerxes) **$45 [≈ £28]**

Amherst, Alicia
- See Cecil, The Hon Mrs Evelyn.

Amory, Robert
- A Treatise on Electrolysis and its Applications to Therapeutical and Surgical Treatment in Disease. New York: Wood, [1886]. 1st edn. 8vo. 307 pp. Ills. Orig cloth, inner hinges starting. *(Elgen)* **$65 [≈ £40]**

Amos, William
- Minutes in Agriculture and Planting ... Illustrated with Specimens of ... Natural Grasses ... Boston, Lincs.: J. Hellaby, 1804. 1st edn. 4to. [2],viii,92 pp. 3 plates of actual grass samples, 2 hand cold & 7 other plates. Contemp half calf, sl worn.
(Burmester) **£500 [≈ $800]**

Amphlet, J. & Rea, C.
- The Botany of Worcestershire. Mosses and Hepatics contributed by J.E. Bagnall. Birmingham: 1909. 8vo. xxiv,654 pp. Map. Orig cloth, jnts sl loose.
(Wheldon & Wesley) **£36 [≈ $57]**

Amuchastegui, Axel
- Some Birds and Mammals of North America. London: 1971. One of 505. Folio. 16 cold plates. Qtr mor. Slipcase.
(Wheldon & Wesley) **£450 [≈ $720]**
- Studies of Birds and Mammals in South Africa ... London: Tryon Gallery & George Rainbird, [1967]. Sm folio. 63 pp. Cold litho frontis, 24 cold plates. Half mor.
(Bickersteth) **£48 [≈ $76]**

An Anatomical Treatise of the Liver ...
- See Wainewright, Jeremiah.

Anderson, J.
- Dura Den. A Monograph of the Yellow Sandstone & its remarkable Fossil Remains. London: 1859. 8vo. 96 pp. Cold frontis, 6 hand cold plates. Mod cloth, lib b'plate.
(Baldwin) **£55 [≈ $88]**

Anderson, James
- Essays Relating to Agriculture, and Rural Affairs ... By a Farmer. Edinburgh: for T. Cadell, London ..., 1775. 1st edn. 2 parts in one vol. 8vo. xxxiii,[i],196; [v],200-472 pp. 24 ills on 3 fldg plates. Contemp calf, minor split at hd of spine. *(Burmester)* **£240 [≈ $384]**

Anderson, R.
- Lightning Conductors. Their History, Nature, & Mode of Application. London: Spon, 1879. 4to. 256 pp. Ills. New cloth.
(Book House) **£70 [≈ $112]**

Anderson, T.
- Volcanic Studies in Many Lands. Second Series. London: 1917. 8vo. xv,88 pp. 81 plates. Orig cloth. *(Baldwin)* **£25 [≈ $40]**

Anderson, William
- System of Surgical Anatomy. Part First [all published] ... New York: J.V. Seaman, 1822. 1st edn. 4to. 199 pp, errata leaf, 12 pp. 9 engvs. Text v foxed, plates with offsetting. Lib stamp on title. Rebound in paper bds using orig bds, uncut. *(Elgen)* **$400 [≈ £250]**

Andrade, C.S.
- Love Life of the Birds. Translated by H.M. Clark. Buenos Aires: 1952. 4to. 24 cold plates by A. Amuchastegui. Orig cloth.
(Wheldon & Wesley) **£30 [≈ $48]**

Andress, James
- The Parterre: or Beauties of Flora ... With Poetical Illustrations. London: Tilt & Bogue, 1842. 1st edn. Folio. 12 hand cold plates. Orig dec pressed green cloth gilt, upper jnts v sl worn. *(Gough)* **£1,750 [≈ $2,800]**

Andrews, Thomas
- The Scientific Papers of the Late Thomas Andrews, M.D., V.P. and Professor of Chemistry, Queen's College, Belfast. London: Macmillan, 1889. 8vo. 514 pp. Orig bndg. *(Emerald Isle)* **£50 [≈ $80]**

Annals of British Geology ...
- Annals of British Geology. A Critical Digest of all the Publications and Account of Papers read during the Year. Edited by J.F. Blake. For the Years 1890-93. London: 1891-95. 4 vols. All published. 8vo. Orig cloth, faded.
(Wheldon & Wesley) **£20 [≈ $32]**

Ansell, W.F.H.
- Mammals of Northern Rhodesia ... Lusaka: Government Printer, 1960. 1st edn. 8vo. 155 pp. 7 cold plates, maps, figs. Orig cloth. Dw.
(Terramedia) **$75 [≈ £46]**

Ansted, David Thomas
- Geology, Introductory, Descriptive, & Practical. London: 1844. 1st edn. 2 vols. 8vo. xxiv,506; xvi,572 pp. Lacks half-titles. Num engvd text diags. Light marg stain on 2 ff. Contemp half calf, rubbed, mor labels chipped.   *(Bow Windows)* **£95** **[≈ $152]**
- Scenery, Science and Art; being Extracts from the Note-Book of a Geologist and Mining Engineer. London: Van Voorst, 1854. 1st edn. 8vo. 4 tinted plates, text ills. Orig embossed cloth, lower jnt splitting.
  (Hannas) **£85** **[≈ $136]**

Anstie, Francis
- Stimulants and Narcotics, their Mutual Relation, with Special Researches on the Action of Alcohol, Ether, and Chloroform. London: Macmillan, 1864. 8vo. 489 pp. Orig bndg, hinges broken, needs new backstrip.
  (Xerxes) **$75** **[≈ £46]**

Antes, John
- Observations on the Manners and Customs of the Egyptians, the Overflowing of the Nile; with Remarks on the Plague ... London: Stockdale, 1800. 4to. 139,5 pp. Fldg map. Sl foxing. Orig bds, rubbed & soiled, lacks backstrip, untrimmed.
  (Worldwide Antiqu'n) **$175** **[≈ £109]**

Aplin, O.
- Birds of Oxfordshire. OUP: 1889. 1st edn. 8vo. viii,217 pp. Hand cold frontis, fldg map. Sl spotting. Orig cloth gilt, spine reprd.
  (Gough) **£30** **[≈ $48]**

Appleton, D., publisher
- Appleton's Dictionary of Machines, Mechanics, Enginework and Engineering. Illustrated with Four Thousand Engravings on Wood. New York: D. Appleton & Co, 1852. 1st edn. 2 vols. Thick 4to. 960; 960 pp. 4088 ills. Orig pict gilt half mor, cvrs sl damp spotted.   *(Charles B. Wood)* **$200** **[≈ £125]**
- Appleton's Dictionary of Machines, Mechanics Engine-Work and Engineering. New York: D. Appleton & Co, 1865. 2nd edn, rvsd. 2 vols. Thick 4to. 960; 960 pp. 3960 w'engvd ills. Old lib stamps on titles. Contemp half mor, hinges a bit rubbed.
  (Charles B. Wood) **$150** **[≈ £93]**

Arago, D. Francois J.
- Popular Astronomy. Translated from the Original and Edited by Admiral W.H. Smyth ... and Robert Grant. London: Longman ..., 1855-58. 1st edn of this trans. 2 vols. 8vo. xlviii, 707; xxxii,846 pp. 25 plates, 358 other

ills. Contemp calf, sides strengthened with cloth.   *(Fenning)* **£65** **[≈ $104]**

Arber, Agnes
- Herbals, their Origin and Evolution, a Chapter in the History of Botany 1470-1670. New Edition, rewritten and enlarged. Cambridge: 1938. 8vo. xxiv,326 pp. 27 plates, 131 text figs. Orig cloth.
  (Wheldon & Wesley) **£60** **[≈ $96]**
- Herbals, their Origin and Evolution ... New York: 1970. Reprint of Cambridge 1938 edn. 8vo. xxiv,326 pp. 27 plates. Orig cloth.
  (Wheldon & Wesley) **£20** **[≈ $32]**
- The Mind and the Eye. A Study of the Biologist's Standpoint. Cambridge: 1954. 8vo. xi, 146 pp. Orig cloth. Dw.
  (Bickersteth) **£20** **[≈ $32]**
- The Natural Philosophy of Plant Form. Cambridge: 1950. 8vo. xiv,247 pp. 47 ills. Orig cloth. *(Wheldon & Wesley)* **£30** **[≈ $48]**

Archer, G. & Godman, E.M.
- The Birds of British Somaliland and the Gulf of Aden. Vols 3 and 4. London: 1961. 2 vols. Imperial 8vo. 1008 pp. 2 maps, 14 cold plates by Thorburn & Gronvold. Orig cloth.
  (Wheldon & Wesley) **£120** **[≈ $192]**

Aristotle
- Insigne Artificium Aristotelis: or Aristotle's Compleat Master-Piece. In Two Parts ... added, Hippocrates his Treasure ... London: the booksellers, 1702. 12mo. Title, iv, 118,[2], [ii],24 pp. Frontis, fldg plate, text w'cuts. Some margs creased. Orig sheep sl worn.   *(Bickersteth)* **£110** **[≈ $176]**
- Aristotle's Masterpiece, Completed in Two Parts. The First containing the Secrets of Generation ... The Second Part being a Private Looking Glass for the Female Sex. New York: the Flying Stationers, 1807. 16mo. 137 pp. 4 text ills. V little foxing. Leather, worn.   *(Xerxes)* **$250** **[≈ £156]**

Arkell, W.J.
- The English Bathonian Ammonites. London: Pal. Soc. Monograph, 1951-59. 8 parts, complete. viii,264 pp. 33 plates. 1-5 disbound, 6-8 as issued.
  (Baldwin) **£75** **[≈ $120]**
- The Geology of Oxford. London: 1947. 8vo. viii,267 pp. 6 plates, 49 figs. Orig cloth. Dw.
  (Baldwin) **£36** **[≈ $57]**
- Oxford Stone. London: 1947. 1st edn. 8vo. 185 pp. 37 plates, 27 figs. Orig cloth. Dw.
  (Baldwin) **£25** **[≈ $40]**

Armstrong, E.A.

- The Folklore of Birds. London: Collins, New Naturalist Series, 1958. 1st edn. Cold frontis, 32 plates, 85 text figs. Orig cloth. Dw.
  (Egglishaw) **£75 [≈ $120]**
- A Study of Bird Song. London: 1963. 8vo. xv, 335 pp. 16 photo plates, 43 figs, 14 tables. Orig cloth. *(Wheldon & Wesley)* **£18 [≈ $28]**
- The Wren. London: New Naturalist, 1955. 1st edn. 8vo. viii,312 pp. 20 photos, 41 other ills. Orig cloth, b'plate & label on endpaper.
  (Wheldon & Wesley) **£60 [≈ $96]**

Armstrong, G.E.

- Torpedoes and Torpedo-Vessels. London: Charles Bell, 1896. 1st edn. Post 8vo. xvi, 287 pp. Frontis, 4 fldg plates, ca 50 photos & line ills. 1 leaf dog-eared. Orig navy cloth, spine sl dull. *(Duck)* **£65 [≈ $104]**

Armstrong, John

- The Art of Preserving Health: A Poem. London: for A. Millar, 1754. 1st edn. 4to. [ii], 134 pp. Disbound. *(Burmester)* **£30 [≈ $48]**
- The Art of Preserving Health: A Poem. New Edition. London: for A. Millar, 1757. 8vo. 97, [i] pp. Crnrs somewhat rounded. Orig wraps, edges worn & chipped.
  (Hollett) **£35 [≈ $56]**

Arnold, E.C.

- British Waders. Illustrated in Water-Colour, with Descriptive Notes. Cambridge: 1924. 4to. vii,102 pp. 51 cold plates. Orig cloth.
  (Wheldon & Wesley) **£45 [≈ $72]**

Arnold, F.H.

- Flora of Sussex ... London: Hamilton, Adams, 1887. 1st edn. 8vo. xxiv,118 pp. Map. A few pencil notes. Ink spot affecting 1st few ff. Orig green cloth, rubbed but sound. *(Claude Cox)* **£18 [≈ $28]**
- Flora of Sussex ... New Edition, with Numerous Additions. London: Simpkin Marshall ..., 1907. Sm 8vo. xxi,154 pp. Cold frontis, map. Orig cloth gilt, sl marked & faded. *(Hollett)* **£30 [≈ $48]**

Art ...

- Art of Preserving the Sight ... see Beer, Georg Joseph.
- The Art of Pruning Fruit-Trees ... see Venette, Nicolas.

Artis, E.T.

- Antediluvian Phytology, illustrated by a Collection of the Fossil Remains of Plants peculiar to the Coal Formations of Great

Britain. London: 1825. 1st edn. 4to. xiii,[i] pp. 24 plates (some v sl foxed & offset on text). A few pp taped in. Mod bds, sl worn.
  (Wheldon & Wesley) **£170 [≈ $272]**
- Antediluvian Phytology illustrated by a Collection of the Fossil Remains of Plants, Peculiar to the Coal Formations of Great Britain. London: 1838. 2nd edn. 4to. xiii,24 pp. 24 plates. Orig cloth.
  (Baldwin) **£95 [≈ $152]**

Artman, William

- Beauties and Achievements of the Blind. Rochester, New York: for the author, 1882. 1st edn. 387 pp. Raised letter frontis. Orig bndg. *(Xerxes)* **$45 [≈ £28]**

Aschoff, Ludwig

- Lectures on Pathology. Delivered in the United States, 1924. New York: 1924. 1st edn in English. 8vo. 365 pp. 35 text ills. Orig cloth. *(Elgen)* **$40 [≈ £25]**

Ashenhurst, T.R.

- Lectures on Practical Weaving: The Power Loom and Cloth Dissecting. Huddersfield: Broadbent, 1895. 8vo. 616 pp. Ills. Well used. Orig bndg. *(Book House)* **£28 [≈ $44]**

Ashley, Alfred

- The Art of Etching on Copper. London: John & Daniel A. Darling, [1849]. 1st edn. Oblong 4to. v,18,[10] pp. Etched title, 14 etched plates. Occas foxing. Orig chromolitho glazed bds, spine broken, disbound (gutta percha).
  (Charles B. Wood) **$85 [≈ £53]**

Ashmole, Elias

- Elias Ashmole (1617-1692). His Autobiographical and Historical Notes, His Correspondence ... Edited ... by C.H. Josten. Oxford: 1966. 1st edn. 5 vols. Plates, ports, fldg chart. Dws. *(Elgen)* **$200 [≈ £125]**
- See also Burman, Charles (ed.).

Ashton, J.

- Curious Creatures in Zoology. London: 1890. 8vo. xi,348 pp. 130 ills. Orig cloth.
  (Wheldon & Wesley) **£25 [≈ $40]**

Aston, F.W.

- Mass Spectra and Isotopes. London: 1933. 1st edn. xii,248 pp. 8 plates, diags. Orig bndg. *(Elgen)* **$125 [≈ £78]**

Astruc, Jean

- A General and Compleat Treatise on all Diseases incident to Children from their Birth to the Age of Fifteen ... London: John Nourse, 1746. 1st edn. 8vo. x,229,advt pp. Contemp calf, rebacked.   *(Rootenberg)* **$750 [≈£468]**

Atcheley, S.C.
- Wild Flowers of Attica. Edited by W.B. Turrill. Oxford: 1938. 4to. xix,60 pp. 22 cold plates. Orig cloth.
          (Wheldon & Wesley) **£30 [≈$48]**

Atkinson, J.
- A Compendium of the Ornithology of Great Britain ... London: Hurst, Robinson, 1820. 1st edn. 8vo. xii,232 pp. Contemp half calf, mrbld bds.   *(Gough)* **£30 [≈$48]**

Atkinson, John C.
- British Birds' Eggs and Nests. London: 1867. New edn. Cr 8vo. 12 cold plates. Orig cloth gilt, trifle used & loose.
          (Wheldon & Wesley) **£15 [≈$24]**
- British Birds' Eggs and Nests, Popularly Described. New Edition. London: Routledge, 1870. Lge 12mo. 182,[i advt] pp. 12 cold plates by W.S. Coleman, lge fldg table in pocket. Orig cloth gilt, a.e.g., spine faded, inner hinges cracked but sound.
          (Fenning) **£24.50 [≈$40]**

Atkinson-Willes, G.L. & Scott, Peter
- Wildfowl in Great Britain. A Survey of the Winter Distribution of the Anatidae and their Conservation. London: 1964. 8vo. Cold & other plates. Orig cloth. Dw.
          (Grayling) **£30 [≈$48]**

Audubon, John James
- The Birds of America ... New York & Philadelphia: 1840-44. Orig 8vo edn. 7 vols. Roy 8vo. 500 cold plates. 3 sm marg tears in text reprd. Some tissue foxing transferred to text. Contemp red mor gilt, fine.
          (Wheldon & Wesley) **£15,000 [≈$24,000]**
- The Birds of America. New York: 1937. 4to. Port, facs title, 500 cold plates. Orig cloth.
          (Henly) **£42 [≈$67]**
- The Birds of America. With an Introduction and Descriptive Text by W. Vogt. London: 1937. Roy 4to. xxvi pp. Port, facs of orig title-page, 500 cold plates. Orig buckram.
          (Wheldon & Wesley) **£35 [≈$56]**
- The Birds of America. New York: 1946. 4to. Port, facs title, 435 cold plates. Orig cloth.
          (Henly) **£28 [≈$44]**
- Letters 1826-1840. Edited by Howard Corning. Boston: The Club of Odd Volumes, 1930. One of 225. 2 vols. Lge 8vo. Orig linen backed bds, paper labels, unopened. Orig dec

slipcase.   *(Heritage)* **$250 [≈£156]**
- The Original Bird Paintings of John James Audubon. With a Preface by M.B. Davidson. London: 1966. 2 vols. 4to. 433 cold & 100 plain ills. Orig cloth. Slipcase.
          (Wheldon & Wesley) **£70 [≈$112]**

Audubon, John James & Bachman, J.
- The Quadrupeds of North America. New York: 1851-51-54. 2nd iss vol 1, 1st iss vols 2 & 3. 3 vols. Roy 8vo. 155 hand cold litho plates. Mod qtr mor.
          (Wheldon & Wesley) **£2,750 [≈$6,000]**

Avebury, Lord
- See Lubbock, Sir John, Baron Avebury.

Aveling, J.H.
- The Influence of Posture on Women in Gynecic and Obstetric Practice. Philadelphia: Lindsay & Blakiston, 1879. 1st Amer edn. 8vo. xv, 182 pp. text w'cuts. Orig cloth, spine ends frayed.   *(Elgen)* **$95 [≈£59]**

Averill, Charles
- A Short Treatise on Operative Surgery ... Designed for the Use of Students in Operations on the Dead Body. 1st American Edition, edited by John Bell. Philadelphia: 1823. 8vo. xi,229 pp. Top cvr detached.
          (Rittenhouse) **$50 [≈£31]**

Awsiter, John
- An Essay on the Effects of Opium considered as a Poison. London: G. Kearsly, 1763. 1st edn. 8vo. viii,70 pp. Uncut & sewn as issued. Cloth box.   *(Rootenberg)* **$650 [≈£406]**

Axe, J. Wortley (ed.)
- The Horse: its Treatment in Health & Disease. London: 1906. 9 vols. 4to. Num cold plates, many ills. Orig pict green cloth.
          (Argosy) **$500 [≈£312]**

Ayscough, James, optician
- A Short Account of the Eye and Nature of Vision. Chiefly designed to illustrate the Use and Advantage of Spectacles ... Second Edition. London: E. Say, for A. Strahan, 1752. 26 pp, advt leaf. Half-title. Fldg frontis. Orig stabbed pamphlet, uncut as issued.
          (Jarndyce) **£800 [≈$1,280]**

Babbage, Charles
- The Ninth Bridgewater Treatise. A Fragment ... London: Murray, 1837. 1st edn. 8vo. xxii, 23-240 pp. W'engvd text ills. Mod qtr calf.
          (Pickering) **$950 [≈£593]**

- On the Economy of Machinery and Manufactures ... Fourth Edition Enlarged. London: Charles Knight ..., 1835. 32mo in 8s. Steel-engvd title, iii-xii,[ii], xiii-xxiv, 408 pp. Orig brown cloth, a little worn.
(Pickering) **$700 [≈£437]**
- Passages from the Life of a Philosopher. London: Longman ..., 1854. 1st edn. 8vo. xii, 496 pp. 24 advt pp. Frontis. Orig cloth.
(Rootenberg) **$950 [≈£593]**
- Reflections on the Decline of Science in England and on Some of Its Causes. London: B. Fellowes, 1830. 1st edn. xvi,228,[4 advt] pp. Sl foxed. Polished prize calf gilt, front jnt tender. *(Elgen)* **$700 [≈£437]**
- Table of Logarithms of the Natural Numbers from 1 to 108000 ... Stereotyped. - Fourth Impression. London: [no publisher] 1841. Large Paper. 4to. xx,201,[1],[2] pp. Title inner marg strengthened. Lib stamp on title verso. Title sl soiled. Contemp calf, rebacked.
(Pickering) **$600 [≈£375]**

Babbitt, Edwin D.
- The Principles of Light and Color: including among other things the Harmonic Laws of the Universe ... East Orange, NJ: The College of Fine Forces, [1896]. 2nd edn. viii, 560,5 advt pp. Cold plates, ills. Orig dec cloth, spine cracked at ft. *(Wreden)* **$95 [≈£59]**

Badger, C.M.
- Floral Belles from the Green-House and Garden, Painted from Nature. New York: 1867. Folio. 66 pp. 16 hand cold plates. Some sl marg staining. Brown mor gilt, sl rubbed, rebacked.
(Wheldon & Wesley) **£900 [≈$1,440]**

Badham, Charles David
- A Treatise on the Esculent Funguses of England ... London: Reeve Bros, 1847. 1st edn. Roy 8vo. x,138,[ii],[14 advt] pp. Cold litho frontis, 20 plates (16 hand cold). Orig cloth gilt, rebacked, new endpapers.
(Lamb) **£80 [≈$128]**

Bagnall, J.E.
- The Flora of Warwickshire. The Flowering Plants, Ferns, Mosses and Lichens, and Fungi. London: 1891. One of 500, signed. 8vo. xxxiv, 579 pp. Map in pocket. Orig cloth, reprd.
(Wheldon & Wesley) **£50 [≈$80]**

Bailey, John & Culley, George
- General View of the Agriculture of the County of Cumberland ... London: C.

Macrae, 1794. 4to. 51 pp. Edge of final leaf reprd. Half-title laid down. A little used. Mod half calf gilt. *(Hollett)* **£50 [≈$80]**

Baily, William Hellier
- Figures of Characteristic British Fossils: with Descriptive Remarks. Volume 1 Palaeozoic. London: Van Voorst, 1875. 1st edn. lxxx,126 pp. 42 litho plates, 18 w'cuts. Binder's cloth.
(Hollett) **£50 [≈$80]**

Bain, Alexander
- The Emotions and the Will. London: John W. Parker, 1859. 1st edn. 8vo. xxviii,649,[11] pp. Orig cloth, lightly chipped.
(Gach) **$175 [≈£109]**
- Mental and Moral Science: a Compendium of Psychology and Ethics. London: Longmans, Green, 1868. 8vo. Contemp half calf, rather rubbed, front jnt cracked.
(Waterfield's) **£60 [≈$96]**
- The Senses and the Intellect. London: Longmans, Green, 1868. Lacks endpapers. Inner rear bd & last few ff of text dampstained. Text detaching from case.
(Gach) **$40 [≈£25]**

Baird, S.F., Cassin, John, & Lawrence, G.N.
- The Birds of North America ... Philadelphia: Lippincott, 1860. 1st edn. 2 vols. 4to. 100 cold litho plates. Orig green cloth, sl rubbed.
(W. Thomas Taylor) **$2,500 [≈£1,562]**

Baird, William
- The Natural History of the British Entomostraca. London: Ray Society, 1850. 8vo. viii, 364 pp. 36 plates (17 cold). Some sl foxing. Orig cloth, rebacked.
(Wheldon & Wesley) **£35 [≈$56]**

Baker, E.C.S.
- Cuckoo Problems. London: 1942. 8vo. xvi, 207 pp. 8 cold & 4 plain plates. Orig cloth.
(Wheldon & Wesley) **£25 [≈$40]**
- The Game Birds of India, Burma and Ceylon. London: 1921-30. 3 vols. Roy 8vo. 2 maps, 60 cold & 18 plain plates. Orig half mor (sl used, refixed) & (vol 3) orig cloth.
(Wheldon & Wesley) **£280 [≈$448]**
- The Game Birds of India, Burma and Ceylon. Vol 2 Snipe, Bustards and Sand-Grouse. London: 1921. Roy 8vo. xvi,328 pp. 2 cold maps, 19 cold & 6 plain plates. Tear in 1 plate reprd. Orig half mor, refixed & reprd.
(Wheldon & Wesley) **£50 [≈$80]**
- The Indian Ducks and their Allies. London: 1908. Roy 8vo. xi,292 pp. 30 cold plates by

Gronvold, Lodge & Keulemans. Some v sl foxing. Orig half mor, somewhat rubbed.
(Wheldon & Wesley) **£300 [≃ $480]**

Baker, H.R. & Inglis, C.M.
- The Birds of Southern India. Madras: 1930. 8vo. xxxiii,504 pp. 22 cold plates. V sl wormed. Orig cloth.   *(Henly)* **£68 [≃ $108]**

Baker, Henry
- Employment for the Microscope, In Two Parts ... London: 1753. 1st edn. 8vo. xiv, 442,[10] pp. 17 fldg plates. Contemp calf, rebacked. *(Wheldon & Wesley)* **£200 [≃ $320]**
- The Microscope Made Easy ... Fourth Edition. London: Dodsley, 1754. 8vo. ii,xvi, 311,[xiii] pp. 14 plates. Orig calf, rebacked.
(Charles B. Wood) **$295 [≃ £184]**
- The Microscope Made Easy ... Fourth Edition. London: Dodsley, 1754. 8vo. xvi,311, [13] pp. Fldg table, 15 plates (plate 13 defective). Contemp calf, hd of spine worn, jnts beginning to crack.
(Egglishaw) **£130 [≃ $208]**
- Of Microscopes, and the Discoveries made thereby. In Two Volumes: I. The Microscope Made Easy. II. Employment for the Microscope. London: Dodsley, 1785. 2 vols. 8vo. xxiii, 324; xxx,442,[x] pp. 15 (numbered 1-3,3*, 4-14) + 17 plates. Contemp calf, hinges cracked.
(Charles B. Wood) **$395 [≃ £246]**

Baker, Humphrey
- The Well-Spring of Sciences; teaching the Perfect Worke and Practise of Arithmetick ... London: 1650. Sm 8vo. [4],312,[21] pp. Lacks last leaf of index. Sm crnr torn from 1 leaf sl affecting text. Later half leather, sl worn.   *(Whitehart)* **£135 [≃ $216]**

Baker, J.A.
- The Peregrine. London: 1967. 8vo. 191 pp. Review pasted in front. Orig bndg. Dw.
(Wheldon & Wesley) **£20 [≃ $32]**

Baker, J.H.
- The Flora of the Lake District. London: George Bell, 1885. Only edn. 8vo. viii,262 pp, advt leaf. Orig cloth gilt.
(Hollett) **£45 [≃ $72]**

Baker, Thomas, 1625?-1689
- The Geometrical Key: or the Gate of Equations Unlock'd ... London: J. Playford, for R. Clavel ..., 1684. 1st edn. 4to. 19 ff, 7,[1] pp, 118 ff. Fldg table, 10 fldg plates. Contemp calf, rebacked, crnrs worn. Wing

B.517 (& B.516B).
(Pickering) **$2,000 [≃ £1,250]**

Bakewell, R.
- An Introduction to Geology. New Haven: B. & W. Noyes, 1839. 5th edn. 8vo. xxviii,596 pp. 2 cold & 6 b/w fldg plates, 16 text w'cuts. Orig cloth, cvr loose.
(Gemmary) **$100 [≃ £62]**

Baldwin, F.G.C.
- The History of the Telephone in the United Kingdom ... London: Chapman & Hall, 1925. Roy 8vo. xxvi,728 pp. Frontis, fldg table, 186 ills. Orig cloth, sl rubbed & bumped. Author's pres copy.   *(Duck)* **£185 [≃ $296]**

Baldwin, J.H.
- The Large and Small Game of Bengal and the North Western Provinces of India. London: 1876. 8vo. Ills. Half mor.
(Grayling) **£85 [≃ $136]**

Baldwin, James Mark
- Handbook of Psychology: Feeling and Will. New York: Henry Holt, 1891. 1st edn. 8vo. [ii], [xii],394,8 pp. Orig brown cloth, lower front & rear cvr stained. *(Gach)* **$65 [≃ £40]**
- Handbook of Psychology: Senses and Intellect. New York: Henry Holt, 1890. 2nd edn, rvsd. 8vo. [xvi],[344],8 pp. Orig brown cloth, mildly stained.   *(Gach)* **$65 [≃ £40]**
- History of Psychology: A Sketch and an Interpretation. London: Watts, 1913. 2 vols. 12mo. Orig cloth, cvrs faded.
(Gach) **$75 [≃ £46]**

Baldwin, James Mark (ed.)
- Dictionary of Philosophy and Psychology ... New Edition with Corrections. Gloucester, MA: Peter Smith, 1960. Reprint edn. Facs of the crrctd 1925 edn. 3 vols in 4. 4to. Ex-lib. Orig buckram.   *(Gach)* **$135 [≃ £84]**

Balfour, J.H.
- The Plants of the Bible. London: 1885. New edn. 8vo. 249 pp. Num ills. Orig cloth, trifle worn.   *(Wheldon & Wesley)* **£25 [≃ $40]**

Balfour-Browne, F.
- British Water-Beetles. London: Ray Society, 1940-58. 3 vols. 8vo. 6 plates, 67 maps, 266 text figs. Orig cloth, vol 1 spine faded.
(Egglishaw) **£65 [≃ $104]**
- British Water-Beetles. London: Ray Society, [1940]-58. Vols 1 & 2 the reprint. 3 vols. 8vo. 5 plates, 67 maps, 266 text figs. Orig cloth.
(Wheldon & Wesley) **£80 [≃ $128]**

Ball, Isaac
- An Analytical View of the Animal Economy. New York: for the author by G.J. Hunt, 1808. 1st edn. 12mo. '90' [=88] pp. 9 pp subscribers. Hand cold frontis, w'cut vignette. Sl foxing. Orig half leather, paper cvrd bds (v worn). *(Elgen)* **$250 [≈£156]**

Ball, John
- A New Compendious Dispensatory ... London: T. Cadell, 1769. 1st edn. 12mo. Advt leaf at end. Contemp mottled calf, rebacked *(Ximenes)* **$300 [≈£187]**

Ball, Sir Robert
- The Story of the Sun ... London: Cassell, 1893. 8vo. 11 plates, num ills. Orig gilt dec black cloth. *(Waterfield's)* **£25 [≈$40]**

Ball, W.W. Rouse
- A History of the Study of Mathematics at Cambridge. Cambridge: 1889. 1st edn. xvii, 264, 28 ctlg pp. Perf lib stamp on title. Orig cloth. *(Elgen)* **$60 [≈£37]**

Baltet, Charles
- The Art of Grafting and Budding. London: William Robinson, 1873. 1st English edn. Cr 8vo. [8],230,[2 advt] pp. Frontis, text ills. Orig cloth. *(Fenning)* **£24.50 [≈$40]**

Bancroft, Edward
- Experimental Researches concerning the Philosophy of Permanent Colours ... London: Cadell & Davies, 1813. 2nd, enlgd, edn. 2 vols. 8vo. lxi,542; 518 pp. Half calf, rubbed, spine ends sl defective, lacks labels. *(Hollett)* **£350 [≈$560]**

Banks, John
- A Treatise on Mills in Four Parts ... London: for W. Richardson, 1795. 1st edn. 8vo. xxiv,172,[4] pp. Half-title. Subscribers. Old bds, rebacked in cloth. *(Hollett)* **£450 [≈$720]**

Banks, Joseph
- Joseph Banks in Newfoundland and Labrador, 1766. His Diary, Manuscripts and Collections. Edited by A.M. Lysaght. London: 1971. 4to. xxviii,458 pp. Frontis, 8 cold maps, 12 cold plates, 96 pp of half-tone ills, 6 line drawings, 6 text figs. Orig cloth. Dw. *(Wheldon & Wesley)* **£50 [≈$80]**

Bannerman, David A.
- The Birds of Tropical West Africa ... London: 1930-51. 8 vols. Imperial 8vo. 7

maps, 83 cold & 3 plain plates, 695 text figs. Orig cloth, trifle used. *(Wheldon & Wesley)* **£500 [≈$800]**
- The Birds of Tropical West Africa, Vol 4, Eurylaemidae to Turdidae. London: 1936. Roy 8vo. xl,459 pp. Cold fldg map, 14 cold plates, 117 text figs. Orig buckram. Dw. *(Wheldon & Wesley)* **£60 [≈$96]**
- The Birds of Tropical West Africa, Vol 5 (Sylviidae to Zosteropidae). London: 1939. Roy 8vo. xliii,485 pp. 2 fldg map, 9 cold plates, text figs. Orig buckram. *(Wheldon & Wesley)* **£60 [≈$96]**
- The Birds of Tropical West Africa, Vol 6 Paridae-Fringillidae. London: 1948. Roy 8vo. xxxix,364 pp. 12 cold & 2 plain plates. Text figs. Orig buckram. *(Wheldon & Wesley)* **£60 [≈$96]**
- The Birds of Tropical West Africa, Vol 7 Weaver Birds. London: 1949. 4to. xxxv,413 pp. 14 cold plates by G. Lodge. Orig cloth. *(Wheldon & Wesley)* **£75 [≈$120]**
- The Birds of Tropical West Africa, Vol 8, Supplement and General Index. London: 1951. 4to. xxiv,552 pp. Orig cloth. *(Wheldon & Wesley)* **£40 [≈$64]**
- The Birds of West and Equatorial Africa. London: 1953. 2 vols. 8vo. 1526 pp. 30 cold & 24 plain plates, 433 text figs. Orig cloth. *(Wheldon & Wesley)* **£80 [≈$128]**
- The Canary Islands: Their History, Natural History and Scenery. London: Gurney & Jackson, 1922. 1st edn. 8vo. xv,365 pp. 4 maps, cold frontis by Roland Green, 2 cold plates by Gronvold, 81 b/w ills. Orig cloth, faded, sm 'splash' on spine. His 1st book. *(Gough)* **£95 [≈$152]**

Bannerman, David A. & W.M.
- Birds of Cyprus. London: 1958. Imperial 8vo. lxix,384 pp. Map, 16 cold & 15 half-tone plates, 2 line-drawings. Orig cloth. *(Wheldon & Wesley)* **£115 [≈$184]**
- Birds of Cyprus. London: Oliver & Boyd, 1958. Thick 4to. lxix,384 pp. 29 cold plates. Orig cloth gilt. Dw. *(Hollett)* **£150 [≈$240]**

Bannerman, David A. & Lodge, G.E.
- The Birds of the British Islands. London: 1953-63. 12 vols. Imperial 8vo. 386 cold & 2 plain plates. Orig cloth. *(Henly)* **£320 [≈$512]**

Banting, William
- Letter on Corpulence addressed to the Public. Philadelphia: Lippincott, 1863. 3rd edn. 16mo. 51 pp. Orig wraps. *(Xerxes)* **$45 [≈£28]**

Banyer, Henry
- Pharmacopoeia Pauperum: or, The Hospital Dispensatory. Containing the Medicines used in the Hospitals of London ... London: T. Warner, 1718. 1st edn. 12mo. 108 pp. Later plain wraps, upper wrapper detached, lower with early ink notes.
(Bickersteth) **£380 [≈ $608]**

Barbour, Thomas
- The Birds of Cuba. Massachusetts: Memoirs of the Nuttall Ornithological Club No. VI, June, 1923. 4to. 141 pp. 4 plates. Orig cloth.
(Bickersteth) **£55 [≈ $88]**

Barcroft, Sir Joseph
- Researches on Pre-Natal Life. Volume I [all published]. Springfield: 1947. 1st Amer edn. Lge 8vo. 292 pp. Ills. Orig cloth.
(Elgen) **$45 [≈ £28]**

Barcsay, Jeno
- Anatomy for the Architect. Budapest: Corvina, 1956. 1st edn. One of 1200. Folio. 308 pp. 142 plates. Orig cloth. Dw sl frayed.
(Claude Cox) **£25 [≈ $40]**

Bardswell, Frances A.
- The Herb Garden. Painted by the Hon. Florence Amherst and Isabelle Forrest. London: A. & C. Black Colour Book, 1930. 2nd edn. 16 cold plates. Orig cloth.
(Old Cathay) **£33 [≈ $52]**

Barker, John
- An Inquiry into the Nature, Cause, and Cure of the Present Epidemic Fever. Together with some General Observations ... London: for T. Astley ..., 1742. Sole edn. 8vo. Title, 128 pp. Title marg sl soiled. Orig grey wraps, torn at hd of spine.
(Bickersteth) **£190 [≈ $304]**

Barker, T.J.
- The Beauty of Flowers in Field and Wood, containing the Natural Orders of Families of British Wild Plants with their Moral Teachings. Bath, [1852]. 8vo. xii,228 pp. 10 plates (4 hand cold, others colour ptd & finished by hand). Orig cloth gilt spine relaid.
(Henly) **£45 [≈ $72]**

Barlow, P.
- An Elementary Investigation of the Theory of Numbers ... London: 1811. 1st edn. xiv, [i], 507 pp. 1st few pp with contemp underlinings & comments, sl stain on edge of last few pp, occas sl foxing. Lacks half- title. Later cloth.
(Whitehart) **£95 [≈ $152]**

Barnard, K.H.
- A Monograph of the Marine Fishes of South Africa Part 2 (Teleostei - Discocephali to end, Appendix). South Africa: 1925-27. Roy 8vo. 647 pp. 37 plates. Lib stamp on endpaper. Orig cloth. *(Wheldon & Wesley)* **£45 [≈ $72]**

Barnes, Eleanor C., Lady Yarrow
- Alfred Yarrow. His Life and Work. London: Edward Arnold, 1923. 1st edn. xvi,328 pp. 9 cold & 68 other plates, text ills. Orig navy cloth, uncut, edges v sl rubbed.
(Duck) **£45 [≈ $72]**

Barnes, Thomas
- A New Method of Propagating Fruit-Trees, and Flowering Shrubs ... London: for R. Baldwin, & J. Jackson, 1759. 1st edn. 8vo. 2 fldg plates. Later wraps. "Thomas Barnes" possibly a pseudonym for "Sir" John Hill.
(Ximenes) **$300 [≈ £187]**

Barr, P.
- Ye Narcissus or Daffodyl Flower, containing hys historie and culture ... with a Compleat Liste of all the Species and Varieties known to Englyshe Amateurs. London: 1884. 8vo. 48 pp. Ills. Orig wraps.
(Wheldon & Wesley) **£45 [≈ $72]**

Barrett, C.R.B.
- The History of the Society of Apothecaries of London. London: 1905. 1st edn. Lge 8vo. xl, 310 pp. Frontis & other ills. Random foxing & other marks. Orig cloth, cvrs marked & a little bumped. *(Bow Windows)* **£80 [≈ $128]**

Barrett, Sir William & Besterman, Theodore
- The Divining Rod. An Experimental and Psychological Investigation. London: Methuen, 1926. 1st edn. 8vo. xxiii,336 pp. 12 plates, 62 text ills. Orig cloth gilt, split at hd of spine. *(Minster Gate)* **£25 [≈ $40]**

Barrett-Hamilton, G.E.H.
- A History of British Mammals. Continued by M.A.C. Hinton. London: 1910-21. Parts 1-21, all published. Roy 8vo. 2 ports, 14 cold & 57 plain plates. Orig wraps.
(Wheldon & Wesley) **£75 [≈ $120]**

Barrow, John Henry
- See Landseer, Thomas.

Barrows, W.B.
- Michigan Bird Life ... with Special Reference

to its Relation to Agriculture. Michigan Agricultural College: Bulletin, 1912. Roy 8vo. xiv,822 pp. 222 ills. Orig cloth, spine trifle stained, inner jnts loose.
(Wheldon & Wesley) £28 [≈ $44]

Barry, Edward
- A Treatise on the Three Different Digestions, and Discharges of the Human Body ... London: A. Millar, 1759. 1st edn. 8vo. xvi, 434 pp. 1 text ill. Marg repr to lower edge of title. Qtr calf. *(Rootenberg)* $275 [≈ £171]

Barry, J.W.
- Railway Appliances. A Description of Details of Railway Construction (except Earthworks and Structures) ... and a Short Notice of Railway Rolling Stock. London: Longmans, 1890. 331 pp. Ills. Orig bndg, sl grubby.
(Book House) £30 [≈ $48]

Barth, J.B.P. & Roger, Henry
- A Practical Treatise on Auscultation. Translated with Notes by Patrick Newbigging. With an Appendix by L.M. Lawson. Lexington: 1847. 1st Amer edn. 8vo. 312 pp. Occas sl foxing, mild dampstaining. Rec buckram. *(Elgen)* $100 [≈ £62]

Bartholomew, J.G., et al.
- Atlas of Zoogeography ... Edinburgh: 1911. Folio. x,67,xi pp. 36 dble plates. Cloth.
(Wheldon & Wesley) £100 [≈ $160]

Bartlet, J.
- The Gentleman's Farriery: or, a Practical Treatise on the Diseases of Horses ... London: John Nourse, 1753. 1st edn. xx,294 pp. Frontis, 2 fldg plates. Sm stain affecting top of some pp. Old polished calf gilt, a little rubbed & defective. *(Hollett)* £150 [≈ $240]
- The Gentleman's Farriery: or, a Practical Treatise on the Diseases of Horses ... Sixth Edition, Revised. London: J. Nourse ..., 1767. 12mo. xxviii,370,[x] pp. Prelims misbound but present. 6 plates. Some worming, mostly marginal. Contemp calf, sl rubbed. *(Blackwell's)* £60 [≈ $96]

Bartlett, Sir Frederick
- Remembering. A Study in Experimental and Social Psychology. Cambridge: UP, 1950. 2nd edn. Roy 8vo. 317 pp. Orig bndg.
(Xerxes) $40 [≈ £25]

Barton, B.H. & Castle, T.
- The British Flora Medica, a History of the Medical Plants of Great Britain. New Edition

by J.R. Jackson. London: 1877. 8vo. xxii,447 pp. 48 cold plates. Orig cloth, trifle worn, spine faded.
(Wheldon & Wesley) £120 [≈ $192]

Barton, Benjamin Smith
- Elements of Botany: or Outlines of the Natural History of Vegetables. Philadelphia: for the author, 1803. 12,2,302, 32,168,38 pp. 30 plates, some fldg (frontis minor tear, old tape stain). Three qtr calf & bds, rubbed. *(Reese)* $850 [≈ £531]

Barton, J.
- A Lecture on the Geography of Plants. London: 1827. Post 8vo. [4],94 pp. 4 fldg maps. Orig bds, uncut, rebacked.
(Wheldon & Wesley) £30 [≈ $48]

Barzelius, J.J.
- The Use of the Blowpipe in Chemistry and Mineralogy. Translated by J.D. Whitney. Boston: Ticknor, 1845. 4th edn. 8vo. xv,237 pp. 3 fldg plates. Some browning. Orig cloth. Sgntr of A. Winchell dated 1854.
(Gemmary) $150 [≈ £93]

Baskin, Leonard
- Ars Anatomica. A Medical Fantasia. Thirteen Drawings. New York: Medicina Rara, 1972. One of 2500, signed. Elephant folio. 13 plates. Orig cloth, slipcase.
(Terramedia) $175 [≈ £109]

Basset, A.B.
- A Treatise on Hydrodynamics with Numerous Examples. London: 1888. 2 vols. xii,264; xv, 328 pp. A few text diags. Orig cloth, spines sl marked, lib stamps on front inner cvrs. *(Whitehart)* £25 [≈ $40]

Bate, George
- Pharmacopoeia Bateana: or, Bate's Dispensatory. Translated from the last Latin Edition ... Compleated ... Second Edition, with Emendations & Enlarged by William Salmon. London: 1700. [16],747,[12] pp, advt leaf. Sl browning. Orig calf, rebacked. Wing B.1089. *(Elgen)* $250 [≈ £156]

Bateman, J.
- A Monograph of Odontoglossum. London: [1864-] 1874. Imperial folio. 30 hand cold plates by W.H. Fitch. Sl marg defects & stains on some ff of text, creases in half-title, last leaf strengthened with document tape. 3 plates somewhat foxed. Green half mor.
(Wheldon & Wesley) £3,000 [≈ $4,800]

Bates, Ernest Sutherland & Dittemore, John V.
- Mary Baker Eddy: The Truth and the Tradition. New York: Knopf, 1932. 1st edn. 8vo. 476,xxxiv pp. Ills. Orig brown cloth. Dw sl chipped. This biography was suppressed after a short period of sale.
(Karmiole) **$100** [≃ £62]

Bates, G.L.
- Handbook of the Birds of West Africa. London: 1930. 8vo. xxiii,572 pp. Map, frontis, text figs. Orig cloth, trifle used.
(Wheldon & Wesley) **£50** [≃ $80]

Bates, H.W.
- The Naturalist on the River Amazon. London: 1863. 1st edn. 2 vols. 8vo. Map, ills. Orig cloth.
(Wheldon & Wesley) **£220** [≃ $352]

Bates, M.
- The Natural History of Mosquitoes. New York: 1949. 8vo. xv,379 pp. 16 plates. Orig cloth. *(Wheldon & Wesley)* **£20** [≃ $32]

Bateson, William
- Mendel's Principles of Heredity. A Defence. Cambridge: UP, 1902. 1st edn. Sm 8vo. xiv,[2],212 pp. Half-title. Frontis port. Orig cloth. *(Rootenberg)* **$350** [≃ £218]
- Mendel's Principles of Heredity. Cambridge: UP, 1909. 8vo. xiv,[2],396 pp. Half-title. 3 photo ports, 1 full-page & 5 dble-page cold plates, num text figs. Orig cloth, spine reprd.
(Rootenberg) **$250** [≃ £156]
- Mendel's Principles of Heredity. Cambridge: (1909) 1913. 3rd imp, with addtns. Roy 8vo. xiv,396 pp. 3 ports, 6 cold plates, 38 text figs. Orig cloth. *(Wheldon & Wesley)* **£35** [≃ $56]

Batten, Loring
- Relief of Pain by Mental Suggestion. New York: Moffat, 1917. 1st edn. 8vo. 157 pp. Orig bndg. *(Xerxes)* **$45** [≃ £28]

Bauer, Max
- Precious Stones. Translated from the German with Additions by L.J. Spencer. Rutland, VT: Tuttle, 1978. Cr 4to. xxvii,647 pp. 20 plates, 95 figs. Orig cloth. Dw.
(Gemmary) **$60** [≃ £37]

Baxandall, D.
- Catalogue of the Collections in the Science Museum ... Mathematics: 1. Calculating Machines and Instruments. London: HMSO,

1926. 85 pp. 13 plates. Orig ptd wraps.
(Whitehart) **£25** [≃ $40]

Baxter, Andrew
- Matho: or the Cosmotheoria Puerilis, a Dialogue ... First Principles of Philosophy and Astronomy are accomodated to the Capacity of Young Persons ... London: A. Millar, 1740. 1st edn. 2 vols. 8vo. [12],432; [6],395,[1] pp. Contemp calf, sl worn. John Mills b'plate. *(Rootenberg)* **$350** [≃ £218]
- Matho: or the Cosmotheoria Puerilis, a Dialogue ... First Principles of Philosophy and Astronomy are accommodated to the Capacity of Young Persons ... London: 1740. 1st edn in English. 2 vols. 8vo. Lacks leaf before title in vol 1 (blank?). Mod calf.
op(Hannas) **£75** [≃ $120]

Baxter, E.V. & Rintoul, L.J.
- The Birds of Scotland. London: Oliver & Boyd, 1953. 1st edn. 2 vols. 8vo. 2 cold frontis by Lodge, 24 photo plates. Orig cloth. Dws. *(Gough)* **£95** [≃ $152]
- The Birds of Scotland. Edinburgh: Oliver & Boyd, 1953. 2 vols. Roy 8vo. 2 cold frontis, 24 plates. Orig cloth.
(Egglishaw) **£72** [≃ $116]

Baxter, James Phinney, the 3rd
- The Introduction of the Ironclad Battleship. Cambridge, Mass.: Harvard UP, 1933. 1st edn. Super roy 8vo. x,[vi],398,[2 blank] pp. 7 plates inc frontis. Orig blue cloth, t.e.g., sl marked & soiled. *(Duck)* **£110** [≃ $176]

Bayley, Harold
- The Lost Language of Symbolism. An Inquiry into the Origin of Certain Letters, Words, Names, Fairy-Tales, Folklore, and Mythologies. London: Williams & Norgate, 1912. 2 vols. 8vo. x,376; viii,388 pp. Over 1400 text diags. Orig green cloth gilt.
(Karmiole) **$100** [≃ £62]

Bayma, Joseph
- The Elements of Molecular Mechanics. London: Macmillan, 1866. 1st edn. 8vo. [2 advt], xviii,266,[2 advt] pp. 3 plates. Orig cloth. *(Fenning)* **£65** [≃ $104]

Bazin, G.A.
- The Natural History of Bees ... Translated from the French. London: for J. & P. Knapton, 1744. 1st English edn. 8vo. xvi,452,16 pp. 12 fldg plates. Contemp calf, sl rubbed. *(Young's)* **£325** [≃ $520]
- The Natural History of Bees. Translated

from the French. London: 1744. 8vo. [xvi], 452, [16] pp. 12 fldg plates. Contemp calf, rebacked. Adapted by Bazin from Reamur's 'Memoires'.
(Wheldon & Wesley) **£160 [≈ $256]**

Beach, Rex
- Hands Off Dr. Locke. New York: Farrar & Reinhardt, 1934. Sm 8vo. 56 pp. Photo ills. Orig bndg. Dw. *(Xerxes)* **$50 [≈ £31]**

Beach, W. & Sheppard, W.D.
- Reformed Medical Journal. New York: 1832. Vol 1, nos 1-12, Jan-Dec 1832, all published. 8vo. Orig stiff wraps, cloth backstrip worn, bds stained. *(Xerxes)* **$350 [≈ £218]**

Beach, Wooster
- An Improved System of Midwifery, adapted to the Reformed Practice of Medicine ... New York: Jas. McAlister, 1847. 1st edn. 4to. 272 pp, 4 advt ff. 52 plates, mostly hand cold lithos. Waterstain in lower crnr, some foxing & browning. Rebound in calf backed bds.
(Pickering) **$1,800 [≈ £1,125]**

Beale, Lionel S.
- How to Work the Microscope. Third Edition. London: Harrison, 1865. xvi,272,[2 advt] pp. Mtd photo frontis, 56 plates, 2 ills. Occas sl spotting & browning. 1 sm marg tear. Orig green cloth, uncut, sl marked & soiled.
(Duck) **£30 [≈ $48]**
- The Microscope in its Application to Practical Medicine. London: Churchill, 1867. 3rd edn. 8vo. xxiii,320,5 ctlg pp. Plates, text figs. Orig cloth, rear hinge cracked, spine faded, crnrs sl worn.
(Savona) **£30 [≈ $48]**

Beamish, Richard
- Memoir of the Life of Sir Marc Isambard Brunel. Second Edition, revised and corrected. London: Longman ..., 1862. 8vo. xviii, [1],357,[2 advt] pp. Port, 8 plates, 8 other ills. Orig cloth gilt, inside jnts reprd.
(Fenning) **£85 [≈ $136]**

Bean, Percy & Scarisbrick, F.
- The Chemistry and Practice of Sizing. A Practical Treatise on the Sizing of Cotton Yarns, Practical Size Mixing, Tape Sizing, Ball or Warp Sizing, Hank Sizing ... Manchester: Kirkham, 1910. 8vo. 654,advt pp. Ills. Orig bndg. *(Book House)* **£20 [≈ $32]**

Bean, Percy & McCleary, William
- The Chemistry and Practice of Finishing. A Practical Treatise on Bleaching and the Finishing of White, Dyed and Printed Cotton Goods. Manchester & Boston, 1912. Vol 1 3rd edn, vol 2 2nd edn. 2 vols. 8vo. 22 plates, 96 mtd fabric samples. Orig cloth.
(Charles B. Wood) **$100 [≈ £62]**

Beard, Peter
- The End of the Game. London: 1965. 1st edn. Sm folio. Num cold & other ills. Orig bndg. Frayed dw. *(Grayling)* **£60 [≈ $96]**
- The End of the Game. London: 1978. 2nd edn. Lge 4to. Num ills. Orig bndg. Dw.
(Grayling) **£40 [≈ $64]**

Beasley, Henry
- The Book of Prescriptions, containing 3000 Prescriptions ... Second Edition. London: Churchill, 1859. 12mo. [2 advt],[2],xvi,548 pp. Lightly dusty in places. Rec paper bds.
(Fenning) **£32.50 [≈ $52]**
- The Druggist's General Receipt Book ... Third Edition. London: John Churchill, 1854. Sm 8vo. viii,483 pp. Contemp half calf, spine sl rubbed & chipped at hd.
(Spelman) **£50 [≈ $80]**

Beatson, Alexander
- A New System of Cultivation, without Lime, or Dung, or Summer Fallows, as practised at Knowle-Farm, in the County of Sussex ... London: Bulmer & Nicol, 1820. 1st edn. 8vo. [xvi], 163 pp. 4 plates. Sl browned. Orig bds, uncut, sl soiled, spine defective.
(Finch) **£68 [≈ $108]**

Beaufoy, Henry
- Scloppetaria: Or Considerations on the Nature and Use of Rifled Barrel Guns ... By a Corporal of Riflemen. London: C. Roworth for T. Egerton, 1812. 2nd edn. xxiv,252,[6] pp, errata leaf. Frontis, 12 plates. Occas marking. Early 19th c mor elab gilt by Bedford. *(Duck)* **£685 [≈ $1,096]**

Beaufoy, Mark
- Nautical and Hydraulic Experiments, with numerous Scientific Miscellanies. London: ptd at the Private Press of Henry Beaufoy ..., 1834. Only edn. 4to. cxix,688 pp. Engvd silhouette, 2 vignettes, 16 plates. Occas spotting. Orig cloth, rehinged.
(Young's) **£220 [≈ $352]**

Beaumont, William
- Experiments and Observations on the Gastric Juice. And the Physiology of Digestion. Boston: Lilly, Wait, 1834. 1st edn, 2nd iss. Narrow 4to. 280 pp. Occas minor foxing & sl staining. Orig muslin backed paper bds,

remains of label, spine cracking slightly.
(Reese) **$850 [≈ £531]**

- The Physiology of Digestion, with Experiments on the Gastric Juice. Burlington: 1847. 2nd edn. Last 2 ff v sl foxed. Orig bndg, minor wear at crnrs & spine ends. *(Rittenhouse)* **$400 [≈ £250]**

Bechstein, J.M.
- The Natural History of Cage Birds. London: 1888. Cr 8vo. vi,311 pp. Hand cold frontis, num text figs. Orig cloth gilt, a.e.g.
(Henly) **£25 [≈ $40]**

Beck, Bodog F.
- Bee Venom Therapy, Bee Venom, Its Nature and Its Effect on Arthritic and Rheumatoid Condition. New York: Appleton-Century, 1935. 1st edn. Sm 4to. 238 pp. Orig bndg.
(Xerxes) **$75 [≈ £46]**

Beck, Carl
- Roentgen Ray Diagnosis and Therapy. New York: 1904. 1st edn. xix,460 pp. 322 ills. Orig cloth, inner hinges cracked, shaken, hd of spine torn. *(Elgen)* **$150 [≈ £93]**

Beck, T. Romeyn
- On the Utility of Country Medical Institutions ... Albany: 1825. 8vo. 20 pp. Faint lib stamp on title. Title & last leaf foxed. Mod wraps. *(Hemlock)* **$175 [≈ £109]**

Beckmann, Johann
- A History of Inventions and Discoveries. Translated from the German by William Johnston. Second Edition, carefully corrected and enlarged by a Fourth Volume. London: 1814. 4 vols. 8vo. xvi,488; iv,419; iv,461; iv,682 pp. Rec paper bds.
(Fenning) **£245 [≈ $392]**
- A History of Inventions, Discoveries, and Origins. Translated from the German by W. Johnston. Fourth Edition, carefully revised and enlarged by W. Francis ... and J. W. Griffith. London: Bohn, 1846. 2 vols. Sm 8vo. 2 ports. Orig cloth.
(Fenning) **£35 [≈ $56]**
- A History of Inventions ... Translated from the German by William Johnston. Fourth Edition, revised and enlarged. London: Bohn, 1846. 2 vols. 8vo. xxiii,518; xii,548 pp. Orig cloth, ex-lib but nice copies.
(Charles B. Wood) **$110 [≈ £68]**
- A History of Inventions ... London: 1846. 4th edn. 2 vols. 12mo. Frontis ports. Orig dec cloth, sl worn. *(Elgen)* **$85 [≈ £53]**

Beddome, R.H.
- The Ferns of Southern India, being Descriptions and Plates of the Ferns of the Madras Presidency. Madras: 1863. 1st edn. 4to. xv,88,vii [of xv] pp. 272 plates (numbered 1-271, 28a). Front wrapper of orig part 20 bound in instead of title-page. Buckram. *(Wheldon & Wesley)* **£90 [≈ $144]**
- The Ferns of Southern India. Madras: 1873. 2nd edn. 4to. xv,88,xv pp. 272 plates (numbered 1-271, 28a). Margs of 2 ff defective. Buckram, bndg rather used.
(Wheldon & Wesley) **£110 [≈ $176]**
- The Ferns of British India (exclusive of those figured in 'The Ferns of Southern India'). Madras: 1865-70. 4to. In the orig 23 parts. 345 litho plates. Marg tears in 3 parts, marg stains in 4 parts. Rather used. Wraps. Without the 1876 Supplement.
(Wheldon & Wesley) **£175 [≈ $280]**
- The Ferns of Southern India, being Descriptions and Plates of the Ferns of the Madras Presidency. Madras: 1863. 4to. xv,88, vii pp. 272 plates (numbered 1-271, 28a). Hole in 1 or 2 ff reprd, some sm wormholes, sl foxing. Cloth.
(Wheldon & Wesley) **£120 [≈ $192]**
- Handbook to the Ferns of British India, Ceylon and the Malay Peninsula. Calcutta: 1883. 8vo. xiv,500 pp. 300 ills. Orig cloth, refixed. *(Wheldon & Wesley)* **£25 [≈ $40]**

Beer, Georg Joseph
- Art of Preserving the Sight unimpaired into extreme Old Age ... By an Experienced Oculist. The Fifth Edition, considerably augmented ... London: Colburn, 1822. 12mo. xi, 259 pp,2 advt ff. 1 plate. Rec cloth & bds, uncut. *(Bickersteth)* **£130 [≈ $208]**

Beet, G.
- The Grand Old Days of the Diamond Fields. Cape Town: Maskew Miller, 1888. 8vo. xix,192 pp. Photo ills. Orig cloth.
(Gemmary) **$75 [≈ £46]**

Beeton, Isabella
- The Book of Household Management. London: S.O. Beeton, 1864. 65th thousand. 8vo. xxxix, 1112 pp. 12 chromolitho plates, num w'cut ills. Rebound in contemp style half calf, gilt spine, mor label.
(Frew Mackenzie) **£140 [≈ $224]**
- The Book of Household Management. London: Ward, Lock, [ca 1880]. New edn, rvsd & crrctd, with addtns. 313th thousand. Cold ills. Orig dark green bds, red roan spine sl chipped. *(Box of Delights)* **£30 [≈ $48]**

- Beeton's Every-Day Cookery and Housekeeping Book. London: Ward, Lock, Bowden, 1891. New edn, enlgd. 1st edn thus. 8vo. lxxxii,568,[i] advt pp. Cold fldg frontis, 7 cold & 27 b/w plates, text ills. Prelims spotted. Inscrptn on endpaper. Orig pict grey cloth. *(de Beaumont)* **£98 [≈ $156]**

Begbie, J. Warburton
- Selections from the Works of the late J. Warburton Begbie. Edited by Dyce Duckworth. London: New Sydenham Society, 1882. 8vo. xxiv, 422 pp. Port. Orig cloth. *(Bickersteth)* **£24 [≈ $38]**

Begin, L.J.
- Application of the Physiological Doctrine to Surgery. Translated from the French by Wm. Sims Reynolds. Charleston [SC]: E.J. Van Brunt, 1835. 1st Amer edn. 8vo. 227 pp. Some browning. Lib stamp on title. Orig cloth backed bds, v worn. *(Elgen)* **$300 [≈ £187]**

Beirne, B.P.
- British Pyralid and Plume Moths ... London: Warne, Wayside and Woodland Series, 1952. 8vo. 208 pp. 16 cold plates, 189 text figs. Orig cloth. Dw. *(Egglishaw)* **£35 [≈ $56]**

Belkin, J.N.
- The Mosquitoes of the South Pacific (Diptera, Culicidae). Berkeley, CA: 1962. 2 vols. Cr 4to. 1036 pp. Maps, charts, ills. Orig cloth. *(Wheldon & Wesley)* **£40 [≈ $64]**

Bell, A.M.
- Locomotives. Their Construction, Maintenance, and Operation. London: Virtue, 1948. 6th rvsd edn. 2 vols. 4to. 452 pp. Ills. Orig bndg. *(Book House)* **£30 [≈ $48]**

Bell, A.R., et al.
- Railway Mechanical Engineering. A Practical Treatise by Engineering Experts. London: Gresham, 1923. 2 vols. 4to. 290; 284 pp. Fldg plates, ills. Orig bndg. *(Book House)* **£30 [≈ $48]**

Bell, Alexander Graham
- Memoir upon the Formation of a Deaf Variety of the Human Race. Washington: National Academy of Sciences, 1884. Folio. 86 pp. Fldg chart. Orig cloth gilt. *(Jenkins)* **$200 [≈ £125]**

Bell, Alexander Melville
- Visible Speech: The Science of Universal Alphabetics ... London: Simpkin, Marshall,

1867. Inaugural Edition. 4to. 126,[4] pp. 16 litho plates, num text ills. Orig dark blue-green cloth gilt, extremities sl rubbed. *(Karmiole)* **$350 [≈ £218]**

Bell, Benjamin
- A Treatise on the Theory and Management of Ulcers ... New Edition. Edinburgh: Charles Elliot, 1789. 8vo. 486,2 advt pp. Crnr of a few ff sl dampstained. Sgntr on title & p.17. Period mottled calf gilt, rebacked. *(Rankin)* **£75 [≈ $120]**
- A Treatise on the Theory and Management of Ulcers ... Boston: Thomas & Andrews, 1791. 1st Amer edn. 8vo. 295 pp. Frontis (sm tear not affecting image). Occas sl foxing. Contemp tree calf, rubbed, front jnt tender. *(Elgen)* **$275 [≈ £171]**

Bell, Sir Charles
- Essays on the Anatomy of Expression in Painting. London: for Longman, Hurst, Rees & Orme, 1806. 1st edn. 4to. xii,186 pp. 6 plates, 26 text engvs. Lacks half-title & advt leaf. Contemp mor, gilt spine, a.e.g., sl rubbed & scratched at edges, spine ends sl worn. *(Bickersteth)* **£250 [≈ $400]**
- The Hand Its Mechanism and Vital Endowments as Evincing Design ... London: William Pickering, 1833. 1st edn. 8vo. xvi, 288 pp. 8 pp inserted advts at front. W'engvd text ills. Contemp polished calf, rubbed, headcap chipped. *(Pickering)* **£850 [≈ $531]**
- The Hand: Its Mechanism and Vital Endowments, as Evincing Design. Philadelphia: Carey, Lea & Blanchard, 1833. 1st Amer edn. 12mo. 213,[3],advt pp. Contemp mauve cloth, jnts worn, front bd re-attached. *(Gach)* **$285 [≈ £178]**
- The Hand, its Mechanism and Vital Endowments as evincing Design. London: William Pickering, 1837. 8vo. Contemp straight grained mor, lacks top panel of backstrip, prize stamp on cvrs. *(Waterfield's)* **£45 [≈ $72]**
- Illustrations of the Great Operations of Surgery ... London: Longman, Hurst ..., 1821. 1st edn. Oblong folio. viii,134 pp. 20 hand cold engvd plates (17 fully cold). Light browning. Ownership inscrptn on title. Orig bds, uncut, a little worn. *(Rootenberg)* **$1,850 [≈ £1,156]**

Bell, F.D. & Young, Frederick
- Reasons for Promoting and Cultivating New Zealand Flax. London: Smith Elder ..., 1842. 1st edn. 8vo. 34,4 advt pp. Stitched as issued. *(Young's)* **£40 [≈ $64]**

Bell, Sir Isaac Lowthian

- Chemical Phenomena of Iron Smelting ... London: Routledge & Spon, 1872. 1st edn. xxiv, 435 pp. 3 fldg tables. Spotting. Marg notes. Orig cloth, gilt spine. Family association copy.   *(Duck)* £75 [≈ $120]
- Principles of the Manufacture of Iron and Steel ... London: Routledge, 1884. 1st edn. 8vo. xx,744 pp. 10 cold charts. Orig green cloth gilt, extremities sl rubbed.
                    (Karmiole) $60 [≈ £37]

Bell, Jacob & Redwood, Theophilus

- Historical Sketch of the Progress of Pharmacy in Great Britain. London: 1880. 1st edn. 8vo. Orig cloth.   *(Robertshaw)* £25 [≈ $40]

Bell, John

- Engravings of the Bones, Muscles and Joints. Part First, containing Engravings of the Bone. First American from the 2nd London Edition. Philadelphia: A. Finley, 1816. 4to. 108 pp. 12 plates, 3 outline plates (offset). Foxing. Minor ex-lib. Contemp sheep, v worn.   *(Elgen)* $175 [≈ £109]

Bell, John & Charles

- The Anatomy and Physiology of the Human Body ... The Third American, from the fourth English Edition. New York: Collins & Co, 1817. 3 vols. 8vo. xl,402; xxvi,[2 advt],420; [x], 357,[3 advt] pp. 34 plates, text ills. Foxed. Lib marks. Contemp calf, rebacked.   *(Heritage)* $500 [≈ £312]

Bell, Thomas, 1792-1880

- The Anatomy, Physiology, and Diseases of the Teeth. London: S. Highley, 1829. 1st edn. 8vo. xiii,[2],329 pp, advt leaf. 11 plates. Mod half calf, uncut.
                 (Rootenberg) $850 [≈ £531]
- A History of British Quadrupeds, including the Cetacae. London: 1874. 2nd edn. 4to. xviii, 474,2 pp. 160 w'engvs. Orig cloth, spine faded.   *(Henly)* £30 [≈ $48]
- A History of British Reptiles. London: 1849. 2nd edn. 8vo. xxiv,159,i pp. 50 w'engvs. Orig cloth, spine faded.   *(Henly)* £30 [≈ $48]
- A History of British Reptiles. London: 1849. 2nd edn. 8vo. xxiv,[ii],159 pp. 50 w'engvs. Orig cloth. *(Wheldon & Wesley)* £25 [≈ $40]
- A History of the British Stalk-Eyed Crustacea. London: 1853. 8vo. lxv,386,4 advt pp. 174 w'engvs. Calf, mrbld edges & endpapers.   *(Henly)* £44 [≈ $70]
- A History of the British Stalk-Eyed Crustacea. London: 1853. 8vo. lxv,386 pp. 174 w'engvs. Orig cloth.

(Wheldon & Wesley) £48 [≈ $76]
- A History of the British Stalk-Eyed Crustacea. London: 1853. 8vo. lxv,386,4 advt pp. 174 w'engvs. Sl foxing at ends. Orig cloth, spine sl faded.   *(Henly)* £38 [≈ $60]
- A History of the British Stalk-Eyed Crustacea. London: Van Voorst, 1853. 1st edn. 8vo. lxv,386,4 advt pp. 174 ills. Orig cloth.   *(Fenning)* £28.50 [≈ $46]

Bell, Walter G.

- The Great Plague in London in 1655. London: 1924. 374 pp. 30 ills. Orig bndg.
                 (Rittenhouse) $50 [≈ £31]

Belling, J.

- The Use of the Microscope. New York: McGraw-Hill, 1930. 1st edn, 2nd imp. 8vo. xi, 315 pp. Text ills. Orig cloth.
                    (Savona) £18 [≈ $28]

Belo, Jane

- Trance in Bali. New York: 1960. xiii,284 pp. 108 ills. Orig silver dec red cloth. Dw.
                    (Lyon) £65 [≈ $104]

Belt, Thomas

- The Naturalist in Nicaragua: A Narrative ... London: Murray, 1874. Sm 8vo. 403,32 ctlg pp. Fldg map, 26 ills. Orig dec cloth, spine ends sl worn, ex-lib, new endpapers.
                    (Schoyer) $65 [≈ £40]

Bemiss, Elijah

- The Dyer's Companion; in Two Parts. New-London: Cady & Eells, [1806]. 1st edn. 12mo. 118,75,[viii] pp. Old writing on endpapers. Orig sheep, rubbed.
              (Charles B. Wood) $450 [≈ £281]

Benedikt, R.

- The Chemistry of the Coal-Tar Colours. Translated ... and edited, with additions, by E. Knecht. Second Edition, revised and enlarged. London: George Bell & Sons, 1889. 8vo. xii,333,23 pp. Orig cloth.
               (Charles B. Wood) $95 [≈ £59]

Benjamin, E.H., secretary

- California Mines and Minerals. San Francisco: California Miner's Assoc, 1899. Souvenir Edition for the California Meeting of the A.I.M.E. 8vo. 450 pp. Maps, plates, text figs. Half leather, mrbld bds.
                    (Gemmary) $100 [≈ £62]

Bennet, J. Henry

- On   the   Treatment   of   Pulmonary

Consumption by Hygiene, Climate, and Medicine. London: 1866. 8vo. 56 pp. Disbound. *(Bickersteth)* **£30 [≈ $48]**

Bennett, A.W.
- The Flora of the Alps. London: Nimmo, 1896. 1st edn. 2 vols. 8vo. 120 cold plates. Orig cloth gilt, t.e.g.
(Egglishaw) **£85 [≈ $136]**
- The Flora of the Alps. London: 1897. 2 vols. 8vo. 120 cold plates. Orig cloth, sl used.
(Wheldon & Wesley) **£60 [≈ $96]**
- The Flora of the Alps. London: Nimmo, 1897. 2 vols. 8vo. 120 cold plates. Orig cloth gilt. *(Hollett)* **£85 [≈ $136]**

Bennett, Edward Turner
- The Gardens and Menagerie of the Zoological Society Delineated ... Chiswick: 1830-31. 1st edn. 2 vols. 8vo. Num engvs by W. Harvey & others. Half mor gilt.
(Egglishaw) **£95 [≈ $152]**
- The Gardens and Menagerie of the Zoological Society Delineated. London: 1831. 2 vols. 8vo. Num engvs by W. Harvey. Calf.
(Wheldon & Wesley) **£60 [≈ $96]**
- The Gardens and Menagerie of the Zoological Society Delineated. Vol I Quadrupeds. Chiswick: 1831. 8vo. xii,308 pp. Num w'cuts. Orig cloth, rebacked.
(Wheldon & Wesley) **£25 [≈ $40]**
- The Tower Menagerie ... London: Robert Jennings, 1829. 1st edn. Tall 8vo. [xviii], 241 pp. Name on fly, clipping on front pastedown. Orig leather spine, bds, bds soiled. *(Hartfield)* **$175 [≈ £109]**

Bennett, James Risdon
- Cancerous and other Inter-Thoracic Growths, their Natural History and Diagnosis ... London: 1872. 8vo. viii,189 pp. 5 plates. Cancelled lib stamps & numbers on title. Orig cloth, spotted. Author's pres inscrptn. *(Bickersteth)* **£20 [≈ $32]**

Bennett, John Hughes
- Researches into the Action of Mercury, Podophyllin, and Taraxacum on the Biliary Secretion; being the Report of the Edinburgh Committee of the BMA. Second Edition, with Appendix. Chicago: 1874. 1st Amer edn. 8vo. 80 pp. Some marg staining. Orig limp cloth, worn *(Elgen)* **$175 [≈ £109]**

Bennett, Sanford
- Old Age. Its Cause and Prevention. New York: Physical Culture, 1912. 1st edn. 8vo. 394 pp. Orig bndg. *(Xerxes)* **$25 [≈ £15]**

Benson, C.W. & White, C.M.N.
- Checklist of the Birds of Northern Rhodesia. Lusaka: Government Printer, 1957. 166 pp. Fldg cold map, cold frontis, 7 cold & 20 other plates. Orig cloth. Dw.
(Terramedia) **$75 [≈ £46]**

Bentham, G.
- Handbook of the British Flora, together with Illustrations of the British Flora and Further Illustrations of British Plants. London: 1946-49. 3 vols. Cr 8vo. Orig cloth.
(Henly) **£30 [≈ $48]**

Bentham, T.
- Asiatic Horns and Antlers. Calcutta: 1908. Hardcvr edn. 96 pp. Num ills.
(Trophy Room Books) **$400 [≈ £250]**
- Asiatic Horns and Antlers. Calcutta: 1908. Softcvr edn. 96 pp. Num ills.
(Trophy Room Books) **$200 [≈ £125]**

Bentley, P.
- Colne Valley Cloth from the Earliest Times to the Present Day. Huddersfield: Woollen Export Group, 1947. 8vo. 71 pp. Fldg map, ills by Harold Blackburn. Orig bndg. dw.
(Book House) **£20 [≈ $32]**

Bequaert, J.
- A Revision of the Vespidae of the Belgian Congo based on the Collection of the American Museum Congo Expedition ... New York: Bulletin of the Amer Mus Nat Hist, 1918. 8vo. 384 pp. 6 plates (2 cold), 267 text figs. Buckram, orig wraps bound in. Author's inscrptn. *(Bickersteth)* **£60 [≈ $96]**

Bergson, H.
- Creative Evolution. Translated by A. Mitchell. London: 1911. 1st edn. 8vo. xv,425 pp. A few pencil notes. Orig cloth.
(Wheldon & Wesley) **£20 [≈ $32]**

Beringer, B.
- Underground Practice in Mining. London: Mining Publ, 1928. 8vo. 255 pp. Ills. Orig bndg. *(Book House)* **£18 [≈ $28]**

Berjeau, J. Ph.
- Homeopathic Treatment of Syphilis, Gonorrhea, Spermatorrhoea and Urinary Diseases. Revised with Numerous Additions by J.H.P. Frost. Philadelphia: Bericke & Tafel, [ca 1870]. 12mo. 256 pp. Orig bndg.
(Xerxes) **$90 [≈ £56]**

Berkeley, George
- Alciphron, or The Minute Philosopher, in Seven Dialogues. London: Beecroft, 1767. 4th edn. Sm crnr replaced at hd of title. Contemp calf, rebacked.
(Emerald Isle) **£150 [≃ $240]**
- Alciphron, or The Minute Philosopher: in Seven Dialogues. New Haven (Conn.): for Increase Cooke & Co, 1803. 1st Amer edn. 8vo. Mod leather backed mrbld bds.
(Gach) **$125 [≃ £78]**
- The Medicinal Virtues of Tar Water fully explained ... London: Dublin ptd, London reptd, for the proprietors of the Tar-Water Warehouse ..., 1744. 1st London edn. 8vo. 32 pp. Clean tears in 2 ff without loss. Outer ff sl soiled. Stitched as issued, uncut.
(Burmester) **£130 [≃ $208]**
- Siris, A Chain of Philosophical Reflexions and Inquiries Concerning the Virtues of Tar Water and Divers Other Subjects. London: W. Innys, 1744. 2nd edn. Title-page mtd. Later half calf. *(Emerald Isle)* **£85 [≃ $136]**
- The Works. Oxford: Clarendon Press, 1901. Rvsd & enlgd edn, 1st printing. 4 vols. 12mo. 2 plates in vol 1. Orig blue cloth.
(Gach) **$135 [≃ £84]**

Berkeley, M.J.
- Handbook of British Mosses ... London: Reeve, 1863. 1st edn. 8vo. xxxvi,324 pp. 1 plain & 23 hand cold plates. Some marg notes. Orig cloth. *(Egglishaw)* **£45 [≃ $72]**
- Handbook of British Mosses ... London: Reeve, 1895. 2nd edn. 8vo. 324 pp. 1 plain & 23 hand cold plates. Orig cloth.
(Egglishaw) **£35 [≃ $56]**
- Introduction to Cryptogamic Botany. London: 1857. 8vo. ix,604 pp. 127 text figs. Orig cloth, trifle used & faded.
(Wheldon & Wesley) **£30 [≃ $48]**
- Outlines of British Fungology. London: 1860. 8vo. xvii,442 pp. 1 plain & 23 hand cold plates by W. Fitch. Orig cloth, sl worn.
(Wheldon & Wesley) **£60 [≃ $96]**
- Outlines of British Fungology. London: 1860. 8vo. xvii,442,16 pp. 23 hand cold & 1 plain plates. Orig cloth, spine faded.
(Henly) **£60 [≃ $96]**

Bernhard, O.
- Light Treatment in Surgery. Translated by R. King Brown. London: 1926. 1st edn in English. 8vo. 317 pp. Ills. Orig cloth.
(Elgen) **$65 [≃ £40]**

Bernheim, Bertram M.
- Blood Transfusion, Hemorrhage and the

Anemias. Philadelphia: [1917]. 1st edn. 8vo. 259 pp. Ills. Occas foxing. Orig cloth, sl shelfwear. *(Elgen)* **$100 [≃ £62]**

Bernheim, H.
- Suggestive Therapeutics, a Treatise on the Nature and Uses of Hypnotism. Translated from the Second and Revised French Edition by C.A. Herter. New York: Putnam, 1895. Roy 8vo. 420 pp. Orig bndg.
(Xerxes) **$75 [≃ £46]**

Bernutz, Gustave & Goupil, Ernest
- Clinical Memoirs of the Diseases of Women. Translated and edited by Alfred Meadows. London: New Sydenham Soc, 1866-67. 1st edn in English. 2 vols. 8vo. Ills. Lib stamps on titles. Orig cloth, sl shelf wear, front edge faded. *(Elgen)* **$100 [≃ £62]**

Berry, Katherine Fiske
- A Pioneer Doctor in Old Japan. The Story of John C. Berry, M.D. New York: Revell, 1940. 1st edn. 8vo. 247 pp. Photo plates. Orig bndg. *(Xerxes)* **$40 [≃ £25]**

Berthollet, Claude Louis
- Essay on the New Method of Bleaching, by means of Oxygenated Muriatic Acid ... From the French ... by Robert Kerr. Edinburgh: Creech; London: Robinson, 1790. 1st English edn, & 1st sep edn. 8vo. xxviii,139 pp. Fldg plate. Orig mrbld bds, rebacked with calf spine. *(Charles B. Wood)* **$650 [≃ £406]**

Berthollet, Claude Louis & A.B.
- Elements of the Art of Dying and Bleaching. Translated from the French ... by Andrew Ure. London: Thomas Tegg, 1841. New edn, rvsd. 8vo. xvi,540 pp. 7 plates on 4. Orig cloth, dull, front inner hinge cracked.
(Charles B. Wood) **$150 [≃ £93]**

Berzelius, J.J.
- The Use of the Blowpipe in Chemical Analysis and in the Examination of Minerals. London: 1822. 8vo. xxxix,344 pp. Lge fldg table, 3 plates. Half calf, rebacked.
(Henly) **£120 [≃ $192]**

Bessemer, Sir Henry
- Sir Henry Bessemer, F.R.S. An Autobiography. With a Concluding Chapter. London: Offices of "Engineering", 1905. 1st edn. 4to. xvi,380 pp. Port frontis, 50 plates, text ills. Orig blue cloth, sl marked & soiled, hinges cracked. *(Duck)* **£120 [≃ $192]**

Bethell, H.A.
- Modern Guns and Gunnery. A Practical Manual ... Woolwich: 1910. 3rd edn, rewritten. Imperial 8vo. xvi,443 pp. 48 photo & 3 line plates, 3 fldg tables, 150 text ills. Orig cloth, sl soiled & worn.
(Duck) £75 [≈ $120]

Bettany, G.T.
- Life of Charles Darwin. London: 1887. 1st edn. 8vo. 175,31 pp. Orig cloth, t.e.g.
(Elgen) $65 [≈ £40]

Bewick, Thomas
- Bewick Gleanings: being Impressions from Copper Plates and Wood Blocks engraved in the Bewick Workshop, edited with Notes, by Julia Boyd. Newcastle: 1886. Large Paper, numbered & signed by the editor. 4to. xxiv,108,104 pp. 54 plates, num ills. Half mor, t.e.g.
(Wheldon & Wesley) £300 [≈ $480]
- Birds, being Impressions from Original Wood-Blocks. Newcastle: David Esslemont, 1984. One of 147. 8vo. xvii,31,[i] pp. 24 plates, 5 tail pieces. Orig qtr cloth gilt. Dw.
(Blackwell's) £50 [≈ $80]
- A General History of Quadrupeds. London: 1791. 2nd edn. Roy 8vo. x,483 pp. 212 w'engvd figs, 107 vignettes & tail-pieces. Title a little browned. Calf, rebacked, new endpapers. Roscoe 2a, variant A.
(Henly) £150 [≈ $240]
- A General History of Quadrupeds. Newcastle: 1807. 5th edn. Large Paper (145 x 233 mm). Roy 8vo. x,525 pp. Num w'cuts. Contemp half calf, trifle rubbed.
(Egglishaw) £110 [≈ $176]
- A General History of Quadrupeds. Newcastle upon Tyne: S. Hodgson ..., 1790. 1st edn. Lge 4to. W'engvs. V sl spotting, not affecting ills. Rebound in qtr mor.
(Hartfield) $295 [≈ £184]
- A History of British Birds. Newcastle: Hodgson, 1797-1804. 2nd & 1st edns. 2 vols. Sm 8vo. 448 w'engvs. Dusty. Occas foxing. Later half mor, edges rubbed.
(Marlborough) £150 [≈ $240]
- A History of British Birds. Newcastle: 1821. 2 vols. 8vo. Num w'cuts. Mod qtr leather, press marks on spines. Roscoe 24-25c.
(Wheldon & Wesley) £140 [≈ $224]
- A History of British Birds. Newcastle: Edw. Walker, 1826. 2 vols. 8vo. xliv,382; xxii,482 pp. W'engvs. Light foxing. Mod calf, a.e.g., by Bayntun.
(Bromer) $350 [≈ £218]
- A History of British Birds. Newcastle: ptd by Charles Henry Cook, for R.E. Bewick, 1832.

2 vols. 8vo. xl,386,xxii; 414 pp. Text ills & vignettes. 19th c calf, mrbld panels inlaid on cvrs, rubbed, spines faded, some reprs to vol 1 spine.
(Heritage) $250 [≈ £156]
- A History of British Birds. Newcastle: 1847. 2 vols. 8vo. Num w'engvs. Orig cloth.
(Henly) £150 [≈ $240]
- Works. London: Bernard Quaritch, 1885. Memorial Edition. One of 750. 5 vols. Num ills. Contemp brown half mor gilt, t.e.g.
(John Smith) £400 [≈ $640]

Beyschlag, F., et al.
- The Deposits of the Useful Minerals & Rocks: Their Origin, Form, and Content. London: Macmillan, 1914. 2 vols (all published). 8vo. xxviii,514; xxi,515-1262 pp. 467 ills. Orig cloth, hinges cracking.
(Gemmary) $90 [≈ £56]

Bichat, Xavier
- Pathological Anatomy. The Last Course of Xavier Bichat, from an Autographic Manuscript of P.A. Beclard ... Translated from the French ... Philadelphia: Grigg, 1827. 1st edn in English. 8vo. 232 pp. Some foxing, damp staining. Leather, worn, front hinge cracked.
(Elgen) $150 [≈ £93]
- A Treatise on the Membranes in General, and on Different Membranes in Particular. New Edition, enlarged by ... Notice of the Life and Writings of the Author ... Boston: 1813. 1st edn in English. 8vo. 259 pp, errata leaf. Marg worm in 30 pp. Foxing. Calf, worn.
(Elgen) $350 [≈ £218]

Bickerton, Thomas H.
- A Medical History of Liverpool from the Earliest Days to the Year 1920. London: 1936. 1st edn. 4to. xx,313 pp. Port frontis, 2 fldg maps, num plates & ills. Orig cloth. Dw.
(Elgen) $45 [≈ £28]

Bicknell, W.I.
- The Natural History of the Sacred Scriptures and a Guide to General Zoology. London: [?1850-51]. 2 vols. 12mo. Pict title & 317 cold plates. Without the fldg frontis sometimes found. Contemp half calf gilt.
(Wheldon & Wesley) £60 [≈ $96]

Biden, C.L.
- Sea-Angling Fishes of the Cape (South Africa). Oxford: 1930. 8vo. xii,304 pp. 2 maps, 48 plates. Orig cloth.
(Wheldon & Wesley) £20 [≈ $32]

Bidwell, Shelford
- Curiosities of Light and Sight. London: Swan

Sonnenschein, 1899. xii,226 pp. Orig cloth, spine faded.   *(Wreden)* **$45 [≈ £28]**

Bierman, William
- The Medical Application of the Short Wave Current. With a Chapter on Physical and Technical Aspects by Myron Schwarzschild. Baltimore: William & Wilkins, 1942. 2nd edn. 8vo. 344 pp. Ills. Orig bndg.
(Xerxes) **$100 [≈ £62]**

Bigelow, Henry Jacob
- Surgical Anaesthesia: Addresses and Other Papers. Boston: 1900. 1st coll edn. 8vo. 378 pp. Occas sl foxing. Orig cloth, t.e.g.
(Elgen) **$125 [≈ £78]**

Bigelow, Horatio R. (ed.)
- An International System of Electro-Therapeutics. Philadelphia: Davis, 1894. 1st edn. 8vo. xxxii,[1147],32 ctlg pp. Num ills. Orig cloth, front hinge cracked, spine ends worn.   *(Elgen)* **$95 [≈ £59]**

Billings, John Shaw (ed.)
- The National Medical Dictionary. Philadelphia: Lea, 1890. 1st edn. 2 vols. Thick 4to. Orig green cloth, spine ends frayed, gouge on foredge of 1 cvr.
(Elgen) **$125 [≈ £78]**

Bingham, C.
- The Manufacture of Carbide of Calcium. London: Raggett, 1916. 219 pp. Ills. Orig bndg.   *(Book House)* **£15 [≈ $24]**

Bingley, William
- Animal Biography ... London: 1824. 6th edn. 4 vols. 20 plates. Vol 1 prelims v sl wormed. Calf, worn, 1 bd detached.
(Egglishaw) **£28 [≈ $44]**
- Animal Biography ... London: Rivington ..., 1824. 4 vols. 12mo. xliv,307; 367; 324; 356 pp. 20 plates. Contemp tree calf, mor labels, SPCK blind stamps, short splits in sides.
(Claude Cox) **£35 [≈ $56]**

Binns, Edward
- The Anatomy of Sleep ... London: John Churchill, 1842. 1st edn. 8vo. x,394 pp. Chromolitho frontis, 4 plates. Orig cloth, 2 old clippings pasted to endpapers.
(Charles B. Wood) **$1,500 [≈ £937]**

Binns, William
- An Elementary Treatise on Orthographic Projection ... Tenth Edition. London: Spon, 1882. 8vo. [22],138 pp. 23 fldg plates, ca 60

other ills. Orig cloth gilt.
(Fenning) **£18.50 [≈ $30]**

Bion, Nicolas
- The Construction and Principal Uses of Mathematical Instruments. Translated from the French ... London: 1723. 1st edn in English. Folio. vii,264 pp. 26 plates. Sev ff browned. Mod calf antique. 1758 supplement bound at end, 60 pp, 4 plates.
(W. Thomas Taylor) **$3,500 [≈ £2,187]**

Birch, John
- An Essay on the Medical Application of Electricity. London: J. Johnson ..., 1803. 1st edn (?). 4to. iv,57,[3] pp. Sl marg damp stain to last gathering. Contemp wraps, paper label on front cvr, uncut, spine worn & chipped.
(Rootenberg) **$300 [≈ £187]**

Birch, Thomas
- The Life of the Honourable Robert Boyle. London: Millar, 1744. 1st sep edn. 8vo. Advts. Lib stamp on title & at end. Cloth, lib b'plate.   *(Rostenberg & Stern)* **$150 [≈ £93]**

Bird, F.J.
- The American Practical Dyer's Companion ... Philadelphia: Henry Carey Baird, 1882. 1st edn. 8vo. xxxvii,17-388,31 pp. 172 mtd samples (2 more than called for on title). Orig dec cloth. *(Charles B. Wood)* **$500 [≈ £312]**
- The Dyer's Hand-Book ... Manchester & London: 1875. 1st edn. 8vo. viii,9-104 pp. 21 mtd samples of cloth. Pattern no. 22 not present; it never was pasted in. Orig cloth, rebacked. *(Charles B. Wood)* **$260 [≈ £162]**

Biringuccio, Vannoccio
- The Pirotechnia. New York: (1942) 1959. 4to. 477 pp. Ills. Orig bndg. Slipcase sl rubbed.   *(Book House)* **£40 [≈ $64]**
- The Pirotechnia ... Translated by Cyril Stanley Smith and Martha Teach Gnudi. New York: Basic Books, 1959. Roy 8vo. xxvi,477 pp. Text w'cuts. Slipcase.
(Gemmary) **$60 [≈ £37]**

Bischof, G.
- Elements of Chemical and Physical Geology. Translated by Benjamin H. Paul and J. Drummond. London: Cavendish Society, 1854 - 55- 59. 3 vols. 8vo. xxiii,455; xxiii,523; xvii, 565 pp. Num tables. Fine ex-lib. Orig dec cloth, worn. *(Gemmary)* **$675 [≈ £421]**

Bishop, Frederick
- The Illustrated London Cookery Book ...

London: 1852. 1st edn. xxxi,460 pp. Engvd frontis, addtnl pict title, fldg plate, num ills. Orig pict gilt dec dark green cloth, fine copy. *(Box of Delights)* **£120 [≈ $192]**

- The Wife's Own Book of Cookery, containing upwards of 1500 Original Receipts ... London: Ward, Lock, [ca 1856]. xvi,398,advt pp. Engvd frontis & half-title, fldg plate, 250 text engvs. Orig pict gilt cloth, gilt a little dulled, sm split at headband. *(Box of Delights)* **£75 [≈ $120]**

Bishop, Louis F. & Neilson, John, Jr.

- History of Cardiology. With an Introduction by Victor Robinson. New York: 1927. 1st edn. 8vo. 71 pp. 11 ports, 1 facs. Orig cloth. Pres note sgnd by Bishop laid in.
(Elgen) **$35 [≈ £21]**

Bjorling, P.R.

- Water or Hydraulic Motors. London: Spon, 1903. 8vo. 287 pp. Ills. Orig bndg.
(Book House) **£20 [≈ $32]**

Black, G.V.

- Descriptive Anatomy of the Human Teeth. Philadelphia: White Dental Manufacturing Co, 1902. 4th edn. 8vo. 169 pp. Ills. Pencil notes on endpapers. Orig bndg.
(Xerxes) **$75 [≈ £46]**

Blackall, John

- Observations on the Nature and Cure of Dropsies ... added, an Appendix ... 1st American, from the 3rd London Edition. Philadelphia: Webster, 1820. 8vo. xxii,274 pp. Some foxing & browning. Contemp calf, scuffed & worn, hinges cracked.
(Elgen) **$175 [≈ £109]**

Blackmore, Sir Richard

- A Treatise of Consumptions and Other Distempers belonging to the Breast and Lungs. London: for John Pemberton ..., 1724. 1st edn. Sm 8vo. xxxvii,[3],223,[1] pp. Orig panelled calf, sl rubbed, hd of spine worn.
(Elgen) **$350 [≈ £218]**

Blackwall, J.

- A History of the Spiders of Great Britain and Ireland. London: Ray Society, 1861-64. Folio. vi,384 pp. 29 hand cold plates (a few foxed, blind stamps). Orig bds, rather used.
(Wheldon & Wesley) **£240 [≈ $384]**

- A History of the Spiders of Great Britain and Ireland. London: Ray Society, 1861-64. 2 vols in one. Folio. 29 hand cold plates. Upper blank marg of plate 12 waterstained, margs of 2 plates reprd. Buckram.
(Wheldon & Wesley) **£300 [≈ $480]**

Blagrave, Joseph

- Blagrave's Supplement; or, Enlargement, to Mr. Nich. Culpeper's English Physician ... The Second Impression ... London: for Obadiah Blagrave, 1677. Sole edn. 8vo. [iv],237, [xiv], 47,20 ctlg pp. Few old ink notes. Endpapers creased. Orig calf. Wing B.3112. *(Bickersteth)* **£180 [≈ $288]**

Blair, David

- The Universal Preceptor; being a General Grammar of the Arts, Sciences, and Useful Knowledge. London: Richard Phillips, 1816. 8th edn, improved. 12mo. 317 pp. Calf.
(Moon) **£150 [≈ $240]**

Blake, Robert

- An Essay on the Structure and Formation of the Teeth in Man and Various Animals. Dublin: William Porter, 1801. 1st edn in English. 8vo. [10],xii,240,[4] pp. Errata & advt ff. 9 plates. Orig half calf, elab gilt, red & green labels. *(Rootenberg)* **$850 [≈ £531]**

Blanchan, Neltje

- Birds that Hunt and are Hunted. New York: 1904. Sm 4to. xi,359,[6 advt] pp. 24 cold plates. Orig cloth, sl worn.
(Hollett) **£35 [≈ $56]**

Blankaart, Steven

- The Physical Dictionary. Wherein the Terms of Anatomy, the Nature and Causes of Diseases, Chyrurgical Instruments, and their Use, are accurately described ... Fifth Edition ... London: 1708. 8vo. [iv],318,[2 advt] pp. Dble page plate. Sl foxing. Rec qtr calf.
(Burmester) **£85 [≈ $136]**

Blegny, Nicholas de

- The Art of Curing Venereal Diseases, done into English by J.H., M.D. London: Brown & Strahan, 1707. 1st English edn. 8vo. Some light browning. Rec mor.
(David White) **£95 [≈ $152]**

Blew, William C.A.

- Brighton and Its Coaches. A History of the London and Brighton Road. With Some Account of the Provincial Coaches ... London: Nimmo, 1894. Imperial 8vo. xxii,354 pp. 20 hand cold plates. Orig pict dec blue cloth gilt, t.e.g., spine sl marked. *(Duck)* **£195 [≈ $312]**

Blith, W.

- The English Improver Improved, or the Survey of Husbandry Surveyed. Third Impression much augmented. London: 1652. Sm 4to. [liv],262,[20] pp. Engvd title & 4 plates. Extra leaf bound before Kk & replacement leaf bound in at L12. 19th c roan. *(Wheldon & Wesley)* **£300 [≈$480]**

Bloch, Iwan
- Ethnological and Cultural Studies of the Sex Life in England as Revealed by Its Erotic and Obscene Literature and Art ... New York: Falstaff Press, [1934]. 1st edn in English. 8vo. 434 pp, in 48 pp of ills. Orig parchment backed dec bds.   *(Gach)* **$40 [≈£25]**

Blodget, Lorin
- Climatology of the United States ... Philadelphia: 1857. 1st edn. Roy 8vo. 536 pp. Subscribers' list. Fldg charts. Orig cloth, unopened, sl worn.
        (M & S Rare Books) **$175 [≈£109]**
- Climatology of the United States ... Philadelphia: 1857. 1st edn. Roy 8vo. 536 pp. Subscribers' list. Fldg charts. Orig cloth, faded, tear in upper spine.
        (M & S Rare Books) **$125 [≈£78]**

Blumenbach, J. Fred.
- The Institutions of Physiology. Translated from the Latin of the Third and Last Edition ... by John Elliotson, M.D. Second Edition. London: Bensley, 1817. 8vo. xvi,426 pp. Orig half calf, rebacked.
        (Charles B. Wood) **$5,000 [≈£3,125]**

Blunt, Wilfrid
- The Art of Botanical Illustration. London: 1950. 1st edn. 8vo. xxxi,304 pp. 46 cold & num other ills. Orig cloth trifle faded. Sl frayed dw. *(Wheldon & Wesley)* **£48 [≈$76]**
- The Art of Botanical Illustration. London: 1950. 1st edn. 8vo. xxxi,304 pp. 79 plates (32 cold), text figs. Orig cloth, spine faded.
        (Henly) **£25 [≈$40]**

Board of Agriculture ...
- Communications to the Board of Agriculture; on Subjects relative to the Husbandry, and Internal Improvement of the Country. London: Bulmer, 1797-1800. 2 vols. 4to. lxxxii,412,[ix]; viii,501,[i] pp. 77 plates (2 cold). Contemp half calf gilt, spines sl rubbed.   *(Blackwell's)* **£400 [≈$640]**

Bodenheimer, F.S.
- Animal Life in Palestine. An Introduction to the Problems of Animal Ecology and Zoogeography. Jerusalem: 1935. Roy 8vo. 506 pp. 70 plates, 77 text figs. Orig cloth, trifle worn, good ex-lib.
        (Wheldon & Wesley) **£35 [≈$56]**

Boenninghausen, Clemens M.F. von
- Boenninghausen's Essay on the Homeopathic Treatment of Intermittent Fever. By Charles Julius Hempel, MD. New York: Radde, 1845. 1st edn in English. 1st US edn. 8vo. 56 pp. New wraps.   *(Xerxes)* **$125 [≈£78]**

Boerhaave, Hermann
- A New Method of Chemistry ... Translated from the Original Latin ... Added Notes and an Appendix ... by Peter Shaw. Third Edition, corrected. London: Longman, 1753. 2 vols. xxx, 593,1; ii,410,[37] pp. 17 plates. Some foxing. New calf spines, mrbld bds, orig labels. *(Charles B. Wood)* **$650 [≈£406]**

Bolton, J.
- Filices Britannicae; and History of the British Proper Ferns. Leeds: 1785. Large Paper (240 x 310 mm). 4to. xvi,59,[5] pp. 31 hand col engvd plates. Some sl marg water staining. Mod qtr calf, uncut. A 2nd vol was published in 1790. *(Wheldon & Wesley)* **£320 [≈$512]**

Bolton, John
- Geological Fragments collected principally from Rambles among the Rocks of Furness and Cartmel. Ulverston: 1869. 1st edn. 8vo. 264 pp. 5 plates. Facs letter at front. Orig cloth gilt.   *(Hollett)* **£95 [≈$152]**
- Geological Fragments collected principally from Rambles among the Rocks of Furness and Cartmel. Ulverston: 1869. 1st edn. 8vo. 264 pp. 5 plates. Orig cloth gilt.
        (Hollett) **£85 [≈$136]**

Bong Ham, Kim
- On the Kyungrak System. Pyongyang: Foreign Languages Publishing House, 1964. 4to. 41 pp. English text. 38 cold plates. Orig bndg.   *(Xerxes)* **$60 [≈£37]**

Bonhote, J. Lewis
- Birds of Britain. London: A. & C. Black, 1907. Thick 8vo. xii,405 pp. 100 cold ills by Dresser. Orig brown cloth, blocked & lettered in gilt & colours, rear hinge sl strained.
        (Blackwell's) **£35 [≈$56]**
- Birds of Britain. London: A. & C. Black, 1917. Thick 8vo. 100 cold ills by Dresser. Edges & prelims sl spotted. Orig dec cloth gilt.   *(Hollett)* **£45 [≈$72]**
- Birds of Britain. London: A. & C. Black, (1907) 1917. 8vo. x,405 pp. 100 cold ills by Dresser. Orig dec cloth, sl rubbed.
        (Egglishaw) **£14 [≈$22]**

Bonner, James

- A New Plan for Speedily Increasing the Number of Bee-Hives in Scotland ... Edinburgh: J. Moir, 1795. 1st edn. 8vo. [2], xx, 258 pp, advt & errata leaf. Half-title. Intl & final blanks. Contemp tree calf gilt, gilt spine, yellow edges. Subscriber's copy.
(Spelman) **£220 [≈ $352]**
- A New Plan for Speedily Increasing the Number of Bee-Hives in Scotland. Edinburgh: 1795. 8vo. xx,260 pp. Bds, rather crudely rebacked. *(Wheldon & Wesley)* **£80 [≈ $128]**

Bonney, T.G.

- Annals of the Philosophical Club of the Royal Society written from its Minute Books. London: 1919. 1st edn. x,286 pp. Orig bndg, unopened. *(Elgen)* **£80 [≈ £50]**
- The Story of Our Planet. London: 1893. 1st edn. 8vo. xvi,592,18 advt pp. 6 cold plates, 170 text figs. Sm lib stamp on title verso. Orig pict cloth. *(Henly)* **£18 [≈ $28]**

Bonney, T.G. & Walton, E.

- Flowers from the Upper Alps, with Glimpses of their Homes. London: 1869. 1st edn. Roy 4to. 12 chromolithos by E. Walton, text by Bonney. The plates are mounted & the mounts are foxed. Orig cloth gilt, trifle worn.
(Wheldon & Wesley) **£100 [≈ $160]**

Bonnycastle, John

- An Introduction to Mensuration and Practical Geometry. Eleventh Edition corrected and improved. London: for J. Johnson, 1812. 8vo. xii,276 pp. Title vignette. Lacks half-title. Orig sheep, hinges cracked. *(Claude Cox)* **£20 [≈ $32]**

Boole, George

- A Treatise on the Calculus of Finite Differences. Cambridge: Macmillan, 1860. 1st edn. 8vo. [6],248 pp. Orig green cloth.
(Rootenberg) **$600 [≈ £375]**
- A Treatise on Differential Equations. Cambridge: 1859. 1st edn. xv,485 pp. Fldg plate. Orig cloth, faded & dust stained.
(Whitehart) **£25 [≈ $40]**

Borden, W.C.

- The Use of the Roentgen Ray by the Medical Department of the United States Army in the War with Spain (1898) ... Washington: 1900. 1st edn. 4to. 98 pp. 38 plates. Orig bndg, some shelf wear. *(Elgen)* **$225 [≈ £140]**

Boreman, T.

- A Description of above Three Hundred

Animals ... Edinburgh: 1787. 12mo. 4,[4],207 pp. Frontis, 98 engvd plates. Sl browning. 1 tear reprd without loss. Edges sl trimmed. Mod qtr calf.
(Wheldon & Wesley) **£100 [≈ $160]**

Borg, J.

- Descriptive Flora of the Maltese Islands, including the Ferns and Flowering Plants. Malta: 1927. 8vo. 846 pp. A few marg tears. Orig buckram, trifle faded, good ex-lib.
(Wheldon & Wesley) **£40 [≈ $64]**

Born, Max

- Atomic Physics. London: 1935. 1st English edn. xii,352 pp. A few text figs. Pres copy sgnd by Born. *(Whitehart)* **£40 [≈ $64]**
- The Constitution of Matter. Translated by E.W. Blair and T.S. Wheeler. New York: [1923]. 1st English edn. vii,80 pp. 37 diags. Orig cloth, dull, remains of sm label on front cvr. *(Whitehart)* **£18 [≈ $28]**
- Experiment and Theory in Physics. Cambridge: 1943. 1st edn. [iv],44 pp. Orig ptd wraps. Sgnd pres copy from Born.
(Whitehart) **£25 [≈ $40]**
- Physics and Politics. Edinburgh: 1962. 1st English edn. vii,86 pp. Dw.
(Whitehart) **£15 [≈ $24]**
- Physics in My Generation. London: [1956]. viii,232 pp. 15 text figs. Sm lib stamp on title verso & inside front cvr. Orig cloth. Dw. Pres copy sgnd by Born to Sir H. Dale.
(Whitehart) **£25 [≈ $40]**

Bossu, Jean Bernard

- Travels through that Part of North America formerly called Louisiana ... Added by the Translator a Systematic Account of all the Known Plants of English North America ... London: 1771. 2 vols. [8],407; [4],432 pp. Old marks removed. Later half calf, mrbld bds. *(Reese)* **$1,500 [≈ £937]**

The Botanist's Calendar ...

- The Botanist's Calendar, and Pocket Flora: arranged according to the Linnaean System ... London: T. Bensley for B. & J. White, 1797. 1st edn. 2 vols. Sm 8vo. Half-titles. Hd of titles reprd. 19th c half cloth, spine hds v sl frayed. *(Burmester)* **£90 [≈ $144]**

Bouillon-Lagrange, Edme Jean Baptiste

- A Manual of a Course of Chemistry ... London: G. Auld, for J. Cuthell, & Vernor & Hood, 1800. 1st edn in English. 2 vols. 8vo. 17 plates. Contemp tree calf, gilt spines, contrasting labels (sl rubbed).

(Ximenes) **$500 [≈ £312]**

Boulenger, G.A.
- Catalogue of the Lizards in the British Museum. London: 1885-87. 2nd edn. 3 vols. 8vo. 96 litho plates. Lib stamp on titles. Orig cloth, unopened.   *(Egglishaw)* **£210 [≈ $336]**
- The Tailless Batrachians of Europe. London: Ray Society, 1897-98. 2 vols. 8vo. iii, 376, subscribers, advt pp. 6 maps, 24 plates (16 cold), 124 text figs. Orig cloth gilt, t.e.g.
        (Egglishaw) **£130 [≈ $208]**

Boulger, G.S.
- Familiar Trees ... New Edition, revised and enlarged. London: Cassell, 1906-07. 3 vols in 2. 8vo. Num photo ills. Buckram.
        (Egglishaw) **£20 [≈ $32]**

Boulton, W.S. (ed.)
- Practical Coal-Mining ... London: Gresham, 1909. 3 vols. 4to. Plates, diags, text ills. Cloth backed bds, mor spine, bds sl scuffed.
        (Stewart) **£45 [≈ $72]**

Bourne, John
- A Treatise on the Steam Engine in its Application to Mines, Mills, Steam Navigation, and Railways. By the Artizan Club. London: 1853. New (4th?) edn. 4to. 258 pp. 33 plates, 349 w'engvs. Mod cloth bds. Largely written by Bourne.
        (Book House) **£140 [≈ $224]**
- A Treatise on the Steam Engine in its Various Applications to Mines, Mills, Steam Navigation, Railways, and Agriculture ... By the Artizan Club. Sixth Edition. London: [ca 1862]. Lge 4to. 495 pp. 37 plates, ills. Mod cloth. Ex libris C.A. Parsons.
        (Book House) **£140 [≈ $224]**

Boutcher, William
- A Treatise on Forest-Trees ... Second Edition. Edinburgh: for the author by J. Murray, 1778. 4to. xlviii,259,[5] pp. Calf backed mrbld bds. *(Karmiole)* **$200 [≈ £125]**

Bowditch, Henry I.
- The Young Stethoscopist; or, the Student's Aid to Auscultation ... New York: S.S. & W. Wood, 1848. 2nd edn. 8vo. [9],304 pp. Ills. Minimal foxing. Contemp calf, front hinge cracked, hd of spine defective.
        (Hemlock) **$200 [≈ £125]**

Bowerbank, J.S.
- A Monograph of the British Spongiadae. Vols 1 & 2 [of 4]. London: Ray Society, 1864-66.

2 vols. 8vo. 37 plates. Orig cloth.
        (Henly) **£35 [≈ $56]**
- A Monograph of the British Spongiadae. London: Ray Society, 1864-82. 4 vols. 8vo. 146 plates. Lib stamps on reverse of plates, crnrs of last 2 plates waterstained. Orig cloth, trifle worn, jnts of 2 vols beginning to crack.
        (Wheldon & Wesley) **£85 [≈ $136]**

Bowles, E.A.
- A Handbook of Crocus and Colchicum for Gardeners. London: 1952. Rvsd edn. 8vo. 222 pp. 32 plates (12 cold). Orig cloth.
        (Wheldon & Wesley) **£25 [≈ $40]**

Bowring, Sir John
- The Decimal System in Numbers, Coins, and Accounts: Especially with reference to the Decimalisation of the Currency and Accountancy of the United Kingdom. London: 1854. 1st edn. 8vo. 245 pp. 7 port plates, 120 engvs of coins. Orig purple cloth, sl rubbed.   *(Young's)* **£54 [≈ $86]**

Boyle, F.
- The Woodland Orchids. London: Macmillan, 1901. 1st edn. 4to. 274 pp. 16 chromolitho plates. Some, mostly light, spotting, affecting 1 plate. Orig blue cloth, water stained but sound.   *(Gough)* **£40 [≈ $64]**

Boyle, Robert
- A Disquisition about the Final Causes of Natural Things ... London: H.C. for John Taylor, 1688. 1st edn, 2nd iss. 8vo. 8 ff, 274 pp, 3 ff. Contemp calf, gilt lines on sides, rebacked. Wing B.3946.
        (Offenbacher) **$1,850 [≈ £1,156]**
- Experiments and Considerations about the Porosity of Bodies, in Two Essays ... London: for Sam. Smith ..., 1684. 1st edn. 8vo. [iv], 145 pp. Contemp calf, gilt spine, hd of spine chipped, surface of the leather damaged on both bds. Wing B.3966.
        (Pickering) **$2,500 [≈ £1,562]**
- Medicina Hydrostatica: or Hydrostaticks applyed to the Materia Medica ... London: for Samuel Smith ..., 1690. 1st edn. 8vo. [xiv], 217, [6 table],[1 Postscript],[ii],14 ctlg pp. Engvd frontis. Occas sl foxing. Early ptd booklabel on half-title. Contemp sheep rebacked.   *(Pickering)* **$3,500 [≈ £2,187]**
- Some Considerations about the Reconcileableness of Reason and Religion. By T.E. a Lay-man ... annex'd ... a Discourse on Mr. Boyle ... London: 1675. 1st edn. 12mo. [iv], xviii,[ii errata],126,[2 blank; vi,[ii errata], 39 pp. Contemp calf, spine reprd. Wing B.4024.   *(Burmester)* **£350 [≈ $560]**

- Some Considerations touching the Usefulnesse of Experimental Natural Philosophy, propos'd in a Familiar Discourses to a Friend ... Oxford: 1663. 1st edn. 2 parts in one vol. 4to. Collation as in Fulton. Later period-style calf. Wing B.4029. *(Hemlock)* **$3,200 [≈ £2,000]**
- Some Considerations touching the Usefulnesse of Experimental Natural Philosophy. Propos'd in a Familiar Discourse to a Friend ... Oxford: 1664. 2nd edn, 2nd iss. 4to. [xvi],416,[18] pp. Collation as in Fulton. Edges trifle browned. Mod qtr lea. *(Gach)* **$500 [≈ £312]**
- The Theological Works ... Epitomiz'd. In Three Volumes ... By Richard Boulton. London: W. Taylor, 1715. 1st edn. 3 vols. 8vo. [xiv], xxv,[i], 432,[viii]; [xxiv],440; [viii], xxi,[xi], 464 pp. Port. Some spots. Contemp calf, extremities worn, rubbed, sm splits in jnts. *(Clark)* **£150 [≈ $240]**
- Tracts Consisting of Observations about the Saltness of the Sea ... London: E. Flesher for R. Davis, Oxford, 1674. 1st edn. 8vo. Collation as in Fulton 113. Contemp sheep, orig unlettered spine laid down. Wing B.4053. *(Pickering)* **$2,000 [≈ £1,250]**

Brabourne, Lord & Chubb, C.
- The Birds of South America. Vol 1. Checklist. London: 1912. Imperial 8vo. xix, 504 pp. Cold map. Binder's cloth, spine faded. *(Wheldon & Wesley)* **£50 [≈ $80]**

Bracken, Henry
- Farriery Improv'd: or, a Compleat Treatise upon the Art of Farriery ... London: for J. Clarke, & J. Shuckburgh, 1738. 12mo. Contemp calf, gilt spine (sl wear). *(Ximenes)* **$250 [≈ £156]**
- The Traveller's Pocket-Farrier: or a Treatise upon the Distempers and Common Incidents happening to Horses upon a Journey ... London: B. Dod, 1743. 1st edn. 12mo. [6], 151,[9] pp. Red & black title. Contemp calf, cvrs loose. *(Rootenberg)* **$175 [≈ £109]**

Bradbury, Fred
- Jacquard Mechanism and Harness Mounting. Belfast & Manchester: 1912. 8vo. 355,12 advt pp. 356 figs. Orig maroon cloth gilt. *(Moon)* **£35 [≈ $56]**

Braddon, W. Leonard
- Cause and Prevention of Beri-Beri. London: Rebman, 1907. 1st edn. Sm 4to. 544 pp. Charts (1 fldg). Orig bndg. *(Xerxes)* **$150 [≈ £93]**

Bradley, Richard
- New Improvements of Planting and Gardening, both Philosophical and Practical in Three Parts ... Fourth Edition. London: W. Mears, 1724. 8vo. xvi,63, viii,[65]-183, viii, [185]-435, [i],104, [i],36,[xvi] pp. 21 plates. Some marking. Contemp calf, rebacked. *(Blackwell's)* **£120 [≈ $192]**
- A Philosophical Account of the Works of Nature ... To which is added an Account of the State of Gardening as it now is in Great Britain. London: 1721. 4to. [xix],194 pp. 28 hand cold plates. Lacks last unnumbered leaf. Trifle foxed. Mod half calf antique style. *(Wheldon & Wesley)* **£250 [≈ $400]**

Brady, G.S.
- A Monograph of the Free and Semi-Parasitic Copepoda of the British Islands. London: Ray Society, 1878-80. 3 vols. 8vo. 96 litho plates, some partially hand cold. Orig cloth, sl splits in backstrip. *(Egglishaw)* **£44 [≈ $70]**

Bragg, W.H. & W.L.
- X Rays and Crystal Structure. London: G. Bell & Sons, 1916. 2nd edn. 8vo. viii,229 pp. 4 plates, 75 text figs. Orig cloth. *(Gemmary)* **$90 [≈ £56]**

Braid, James
- Neurypnology; or, the Rationale of Nervous Sleep, considered in relation with Animal Magnetism. London: John Churchill, 1843. 1st edn. 8vo. xxii,265,[3] pp, inc half-title, errata & 2 pp advts. Orig cloth, upper hinge reprd. Author's pres inscrptn. *(Rootenberg)* **$2,000 [≈ £1,250]**

Bramwell, John Milne
- Hypnotism: Its History, Practice and Theory. London: Grant Richards, 1903. 1st edn, 1st printing. 8vo. [ii],xiv,478,[2] pp. Orig green cloth. *(Gach)* **$60 [≈ £37]**

Brande, W.T.
- Outlines of Geology; being the Substance of a Course of Lectures. London: 1817. 1st edn. 8vo. viii,144 pp. Fldg cold plate. Orig bds, uncut, somewhat soiled. *(Wheldon & Wesley)* **£60 [≈ $96]**

Brander, A.A. Dunbar
- Wild Animals in Central India. London: 1923. 8vo. xv,296 pp. 19 ills. Orig bndg. *(Wheldon & Wesley)* **£40 [≈ $64]**
- Wild Animals in Central India. London: Arnold, 1927. 2nd edn. 8vo. xv,296,16 advt pp. Frontis, photo ills, text diags. Orig

maroon cloth gilt.
(Bates & Hindmarch) **£75 [≈ $120]**

Brassey, Thomas
- Unarmoured Ships. London: Longmans, Green, 1875. 37 pp. Half-title. Disbound.
(Jarndyce) **£40 [≈ $64]**

Brees, S.C.
- A Glossary of Civil Engineering ... London: Bohn, 1844. 8vo. 310 pp. Num w'engvd text ills. Rec cloth.
(Charles B. Wood) **$100 [≈ £62]**

The Brewer ...
- See Loftus, William.

Brewer, J.A.
- Flora of Surrey ... from the MSS. of the late J.D. Salmon and from other Sources. Compiled for the Holmesdale Natural History Club, Reigate. London: 1863. 12mo. xxiv,367 pp. 2 cold geological maps in pockets. Orig cloth, reprd.
(Wheldon & Wesley) **£48 [≈ $76]**
- A New Flora of the Neighbourhood of Reigate, Surrey, with Lists of the Fauna. London: 1856. Post 8vo. ix,194 pp. Fldg map. Orig cloth.
(Wheldon & Wesley) **£30 [≈ $48]**

Brewster, Sir David
- Letters on Natural Magic addressed to Sir Walter Scott. London: 1832. 1st edn. Sm 8vo. W'cut text ills (1 with overlay). Later half calf.
(Robertshaw) **£35 [≈ $56]**
- Letters on Natural Magic addressed to Sir Walter Scott. London: 1832. 1st edn. 351 pp. Ills. Half calf. *(Moorhead)* **£40 [≈ $64]**
- Letters on Natural Magic, addressed to Sir Walter Scott. London: Murray, Family Library, 1834. viii,351 pp. Text ills, inc 2 with hinged flaps. Orig fawn cloth, upper cvr ptd as title, a little worn, splits in jnts.
(Box of Delights) **£30 [≈ $48]**
- The Life of Sir Isaac Newton. London: Murray's Family Library, 1831. 1st edn. Sm 8vo. xv,366 pp. Port, title vignette, text figs. Mod bds. *(Bickersteth)* **£75 [≈ $120]**
- The Life of Sir Isaac Newton. London: John Murray, 1831. 1st edn. 8vo. xv,[1],366 pp. Port, w'cut text ills. Port & title foxed & offset. Orig red sand-grained cloth (probably a re-issue bndg), gilt spine lettering, blind stamped sides. *(Pickering)* **$200 [≈ £125]**
- The Stereoscope: Its History, Theory and Construction ... London: Murray, 1856. 1st edn. 8vo. iv,235,[1 blank],[4 advt],32 ctlg pp.

Num text figs. Occas sl foxing. Orig brown cloth, v sl shelfwear.
(Heritage) **$750 [≈ £468]**
- A Treatise on Optics. A New Edition. London: Lardner's Cabinet Cyclopaedia, 1833. 2nd edn. Sm 8vo. Orig cloth, ptd paper label. *(Fenning)* **£32.50 [≈ $52]**

Bridge, J.H.
- The Inside History of the Carnegie Steel Company. New York: Aldine, 1903. 8vo. 369 pp. Ills. Orig bndg, cvrs marked.
(Book House) **£18 [≈ $28]**

A Brief Letter from a Young Oxonian ...
- See Wallis, John.

Briggs, B.
- Trees of Britain. Their Form and Character. London: 1936. 8vo. 430 pp. 125 plates. Orig green cloth. *(Henly)* **£35 [≈ $56]**

Briggs, Richard
- The New Art of Cookery, According to the Present Practice ... Philadelphia: 1792. 1st Amer edn. 16mo. xiii, xii-xvi,557,[1] pp. Sm piece chipped from lower crnr of title with loss of a few letters. Some browning & loose pp. Contemp calf, cvrs detached.
(M & S Rare Books) **$750 [≈ £468]**

Bright, Richard
- Clinical Memoirs on Abdominal Tumours and Intumescence. Edited by G. Hilaro Barlow. London: New Sydenham Soc, 1860. 8vo. xviii, 326 pp. 79 w'cuts. Sm lib stamps on title. Orig cloth, front cvr faded at ft.
(Elgen) **$200 [≈ £125]**

Brindley, James
- The History of Inland Navigations. Particularly those of the Duke of Bridgewater, in Lancashire and Cheshire ... London: 1766. 1st edn. 8vo. [vi],88 pp. 2 fldg plates (1 stained across 1 crnr). New calf backed bds. A 2nd, sep, part was issued the same year.
(Young's) **£120 [≈ $192]**

Bristowe, W.S.
- The Comity of Spiders. London: Ray Society, 1939-41. 2 vols. 8vo. 22 plates, 96 text figs. Orig cloth, trifle faded, very good ex-lib.
(Wheldon & Wesley) **£85 [≈ $136]**
- The World of Spiders. London: New Naturalist, 1958. 1st edn. 8vo. 4 cold & 32 other plates. Orig cloth. Dw.
(Wheldon & Wesley) **£30 [≈ $48]**

British Birds ...
- British Birds. An Illustrated Magazine devoted chiefly to Birds of the British List. Vols 1-76. London: 1907-83. Complete set, with all the Supplements & special indexes. 8vo. Publisher's cloth.
(Wheldon & Wesley) **£900 [≈$1,440]**

British Optical Manufacturers' Association
- Dictionary of British Scientific Instruments. London: Constable, 1921. 334 pp. Ills. Ca 70 pp affected by damp at outer edges. Orig bndg. *(Book House)* **£20 [≈$32]**

British Pharmacopoeia ...
- See under Pharmacopoeia.

Britten, F.J.
- The Watch and Clock Makers' Handbook, Dictionary and Guide. 14th edition, revised by J.W. Player. London: Spon, 1938. Thick 8vo. vi,547 pp. Ills. Orig cloth gilt.
(Hollett) **£40 [≈$64]**

Britten, James
- European Ferns. London: [1881]. 1st edn. 4to. vii,xliv,196 pp. 30 cold litho plates, num text figs. Orig maroon half calf, gilt spine, a.e.g. *(Bickersteth)* **£160 [≈$256]**
- European Ferns. London: [1881]. 4to. xliv, 196 pp. 30 cold plates. Orig dec cloth gilt, trifle used. *(Wheldon & Wesley)* **£35 [≈$56]**

Britten, James & Holland, R.
- A Dictionary of English Plant Names. London: English Dialect Society, [1878]-1886. 8vo. xxviii,618 pp. A little foxing. Orig buckram, spines faded.
(Wheldon & Wesley) **£75 [≈$120]**

Britton, N.L. & Rose, N.J.
- The Cactaceae. Descriptions and Illustrations of Plants of the Cactus Family. [Washington: 1919-23], facs reprint (but with plain plates) 1964. 4 vols in 2. 4to. 1068 pp. 137 plates, 1120 ills. Orig cloth.
(Wheldon & Wesley) **£52 [≈$83]**

Brocklehurst, H.
- Game Animals of the Sudan. Their Habits and Distribution. A Handbook for Hunters. London: 1931. 170 pp. Fldg map, 57 cold ills. Orig pict gilt cloth.
(Trophy Room Books) **$325 [≈£203]**

Brodie, Sir Benjamin C.
- Pathological and Surgical Observations on the Diseases of the Joints. Third Edition,

with alterations and improvements. London: 1834. 8vo. vii,344 pp. Orig cloth backed mrbld bds, uncut, hd of spine reprd.
(Bickersteth) **£90 [≈$144]**

Brook, Richard
- Cyclopaedia of Botany ... New Family Herbal. London: W.M. Clark & R. Brook, n.d. Thick 8vo. lix,711 pp. 96 hand cold plates. Occas sl spotting or fingering. Old half calf gilt, rather chipped & scraped, hinges splitting. *(Hollett)* **£75 [≈$120]**
- New Cyclopaedia of Botany and Complete Book of Herbs ... London: W.M. Clark, & Richard Brook, Huddersfield, [1853]. 2 vols. 8vo. lix,733,[i Directions] pp. 100 hand cold plates. Mod half mor gilt.
(Hollett) **£120 [≈$192]**

Brooke, J.
- The Wild Orchids of Britain. London: 1950. One of 1140. Folio. 139 pp. Frontis, 40 cold plates. Orig cloth.
(Wheldon & Wesley) **£110 [≈$176]**

Brookes, Richard, M.D.
- The Natural History of Fishes and Serpents ... Third Edition, corrected, to which is added An Appendix containing the Whole Art of Float and Fly Fishing ... London: for F. Power, Brookes's Nat Hist Vol 3, 1790. 12mo. [xii], 258,[6] pp. W'cut plates. Contemp calf. *(Egglishaw)* **£50 [≈$80]**
- The Natural History of Insects, with their Properties and Uses in Medicine. Third Edition. London: for F. Power, Brookes's Natural History Vol 4, 1790. 12mo. xx,21-333 pp. W'cut plates. Contemp tree calf, red label. *(Egglishaw)* **£52 [≈$83]**
- A New and Accurate System of Natural History. The Second Edition, Corrected. London: for Carnan & Newbery, 1772. 6 vols. 12mo. 147 plates. Top edge of vols 1 & 6 v sl dampstained. Contemp leather, gilt spines, vol 1 spine extremities sl worn.
(Schoyer) **$300 [≈£187]**

Brookes, S.
- An Introduction to the Study of Conchology. London: 1815. 4to. vii,160 pp. 9 hand cold & 2 plain plates. Some ltd foxing of the text & sm blind stamps to plates. New half calf.
(Wheldon & Wesley) **£225 [≈$360]**

Brooks, C. Harry
- Practice of Autosuggestion by the Method of Emile Coue. New York: Dodd Mead, 1922. 7th edn. 8vo. 119 pp. Orig bndg.
(Xerxes) **$40 [≈£25]**

Broster, L.R.
- Endocrine Man. A Study in the Surgery of Sex ... London: 1944. 1st edn. 8vo. 144 pp. Orig cloth, spine sunned. *(Elgen)* **$35 [≈ £21]**

Brothers, A.
- Photography: its History, Processes, Apparatus, and Materials ... London: Charles Griffin, 1892. 8vo. xv,364,[xiv],64 pp. 24 plates in different processes, 122 w'engvd ills. Orig cloth. *(Charles B. Wood)* **$250 [≈ £156]**

Brougham, Henry Peter Brougham, 1st Baron
- Dialogues on Instinct; with Analytical View of the Researches on Fossil Osteology. London: 1844. 1st edn. 12mo. 272 pp. Half calf, rubbed. *(Elgen)* **$35 [≈ £21]**

Brousseau, Kate
- Mongolism. A Study of the Physical and Mental Characteristics of Mongolian Imbeciles. Baltimore: Williams & Wilkins, 1928. 1st edn. Roy 8vo. 210 pp. Photo ills. Lacks front free endpaper. Orig bndg, front hinge cracking. *(Xerxes)* **$55 [≈ £34]**

Browinowski, Gracius J.
- The Birds of Australia ... Melbourne: Charles Stuart & Co, 1890. 1st edn. 6 vols. Lge 4to. 301 chromolitho plates. Lib stamps on titles only. Contemp maroon half mor, gilt dec spines. *(Gough)* **£3,650 [≈ $5,840]**

Brown, Alfred
- Old Masterpieces in Surgery. Omaha: privately ptd, 1928. 1st edn. 8vo. 263 pp. 56 plates. Orig cloth backed bds, spine ends sl frayed, minor ex-lib markings. *(Elgen)* **$150 [≈ £93]**

Brown, E.O.F.
- Vertical Shaft Sinking. London: Benn, 1927. 4to. 432 pp. Ills. Mod cloth. *(Book House)* **£35 [≈ $56]**

Brown, J. Campbell
- A History of Chemistry from the Earliest Times. Second Edition, edited by Henry Hilton Brown. Philadelphia: 1920. xxix,[2],543 pp. Port frontis, 106 ills. Sm lib stamp. Orig cloth, spotted, inner hinge starting. *(Elgen)* **$85 [≈ £53]**

Brown, James
- The Forester. A Practical Treatise on the Planting, Rearing, and General Management of Forest Trees ... Second Edition, enlarged.

London: Blackwood, 1851. 8vo. xiv,526 pp. Fldg plate, text figs. Orig green cloth gilt, faded. *(Blackwell's)* **£45 [≈ $72]**
- The Forester. A Practical Treatise on the Planting, Rearing and General Management of Forest Trees. London: 1851. 2nd edn. 8vo. xiv, 526,16 pp. Text figs. Orig cloth. *(Henly)* **£40 [≈ $64]**
- The Forester. A Practical Treatise on the Planting, Rearing and General Management of Forest Trees. London: 1861. 3rd edn. Roy 8vo. x, 700,16 pp. Diag, charts, 149 text figs. Half calf, spine relaid. *(Henly)* **£48 [≈ $76]**

Brown, Percy
- American Martyrs to Science through the Roentgen Rays. Springfield: [1936]. 1st edn. 8vo. 276 pp. Ports. Orig cloth. Dw. *(Elgen)* **$50 [≈ £31]**

Brown, R.
- The Peoples of the World: being a Popular Description of the Characteristics, Condition, and Customs of the Human Family. London: Cassell ..., 1890-94. 6 vols. Lge 8vo. 60 plates, num other ills. Some minor foxing. Orig dec cloth gilt, immaculate. *(Bow Windows)* **£150 [≈ $240]**

Brown, Richard
- A History of Accounting and Accountants. Edinburgh & London: T.C. & E.C. Jack, 1905. 1st edn. 8vo. xvi,459 pp. Frontis, 22 ills. Orig dark green cloth, uncut, bds sunned. *(Pickering)* **$250 [≈ £156]**
- A Letter from a Physician in London to his Friend in the Country; Giving an Account of the Montpellier Practice in Curing the Venereal Disease ... London: for J. Roberts, 1730. Sole edn. 4to. 24 pp. Title dusty & margs sl creased & chipped. Sewed as issued. *(Bickersteth)* **£45 [≈ $72]**

Brown, Robert, of the British Museum
- Miscellaneous Botanical Works. London: Ray Society, 1866-68. 2 vols. 8vo. Orig cloth. With an Atlas of 38 plates, folio, orig bds. Together 3 vols. *(Wheldon & Wesley)* **£145 [≈ $232]**

Brown, Samuel
- Lectures on the Atomic Theory and Essays Scientific and Literary. Edinburgh: Constable; London: Hamilton, Adams, 1858. 1st edn. 2 vols. 8vo. x,[ii],357; [iv],384 pp. Orig brown cloth. Family inscrptn. *(Burmester)* **£120 [≈ $192]**

Brown, Thomas
- A Manual of Modern Farriery ... Diseases ... Racing, Hunting, Coursing, Shooting, Fishing, and Field-Sports generally ... London: Virtue, [1846]. 8vo. viii,920 pp. Engvd & ptd titles, frontis, 18 plates. Contemp half calf, rebacked, bds rubbed.
(Blackwell's) **£125 [≈ $200]**

Brown, Capt. Thomas
- An Atlas of the Fossil Conchology of Great Britain and Ireland ... London: 1889. 4to. 115 plain plates. MS Index. Some pencil marg notes. Foredge of title strengthened. Qtr cloth, rebacked. *(Henly)* **£150 [≈ $240]**
- The Book of Butterflies, Sphinxes and Moths. London & Edinburgh: 1832-34. 1st edn. 3 vols. Sm 8vo. 144 hand cold plates. A little minor foxing, sl offsetting of text on some plates in vols 1 & 2. Orig cloth, vol 1 jnts reprd, trifle used.
(Wheldon & Wesley) **£170 [≈ $272]**
- The Book of Butterflies, Sphinxes and Moths. London & Edinburgh: Whittaker, Constable's Miscellany, 1832-34. 1st edn. 3 vols. 12mo. 3 vignettes, 144 hand cold plates. Orig cloth, paper labels, vol 1 sl loose & cloth rippled, vol 3 rubbed.
(Egglishaw) **£130 [≈ $208]**
- The Conchologist's Text-Book, embracing the Arrangements of Lamarck and Linnaeus. Glasgow: 1833. 1st edn. 180 pp. 19 steel engvd plates. Trifle foxed. Cloth, ex-lib.
(Wheldon & Wesley) **£30 [≈ $48]**
- The Taxidermist's Manual; or the Art of Collecting, Preparing, and Preserving Objects of Natural History ... London: Thomas C. Jack, 1885. Cr 8vo. xii,150 pp. 6 plates, 4 text ills. Orig green cloth.
(Blackwell's) **£65 [≈ $104]**

Brown, Thomas, 1778-1820
- Lectures on the Philosophy of the Human Mind. Edinburgh: James Ballantyne for W. & C. Tait ..., 1820. 1st edn, 1st printing. 4 vols. 8vo. Lacks half-titles. Lib stamps, pockets removed. Leather backed mrbld bds, spines & edges worn, vol 1 spine defective.
(Gach) **$225 [≈ £140]**
- Observations on the Zoonomia of Erasmus Darwin, M.D. Edinburgh: for Mundell & Son ..., 1798. 1st edn. 8vo. xxiv,560 pp, 8 pp inserted front advts, 1 leaf. Lightly foxed. Drab bds, untrimmed, mod paper backstrip.
(Gach) **$375 [≈ £234]**

Browne, Charles A.
- A Source Book of Agricultural Chemistry.

Waltham: Chronica Botanica, Vol 8, No 1, 1944. Roy 8vo. x,290 pp. Frontis, ills. Orig ptd stiff paper cvrs. *(Elgen)* **$45 [≈ £28]**

Browne, M.
- Practical Taxidermy. London: Upcott Gill, (1884). 2nd edn, rvsd & enlgd. 8vo. viii, 354,advt pp. Plates, 57 text ills. Orig dec cloth. *(Egglishaw)* **£22 [≈ $35]**
- Practical Taxidermy. London: 1922. 3rd edn. 8vo. xiv,281 pp. Ills. Stamps on title. Orig cloth. *(Wheldon & Wesley)* **£20 [≈ $32]**

Browne, Sir Thomas
- Posthumous Works ... Printed from his Original Manuscripts. London: for E. Curll & R. Gosling, 1712. 1st edn. 8vo. Port, 20 plates (of 22, lacks Memorial & School plates). Contemp calf, rebacked. Thos. Fountayne's b'plate. *(Hannas)* **£65 [≈ $104]**
- Pseudodoxia Epidemica: or, Enquiries into Very Many received Tenents ... London: 1650. 2nd edn, crrctd & enlgd. Folio. 8 ff,339 pp. Title & text within dble rules. Calf, rebacked. Wing B.5159.
(Argosy) **$400 [≈ £250]**
- Religio Medici. The fourth Edition, Corrected and amended. With Annotations, never before published ... London: 1656. 8vo. [xvi],174,[x], 285 [≈ 185]-297,[v] pp. 4 advt pp. Sep title to Observations. Lacks A1 (blank?). Sl wear & foxing. 19th c cloth. Wing B.5172. *(Clark)* **£65 [≈ $104]**
- Religio Medici. The seventh Edition, Corrected and Amended ... London: Andrew Crook, 1672. Large Paper. 3 parts in one vol. Lge 4to. [6],144 pp. Few ff sl spotted. Early 18th c calf backed bds, worn, front hinge loose. *(Hemlock)* **$500 [≈ £312]**
- The Works ... London: for Tho. Bassett ..., 1686. 1st edn. Folio. Red & black title. Frontis port (mtd). Title-page somewhat dusty. Rebound in period style calf, orig label laid down. Wing B.5150.
(Hartfield) **$850 [≈ £531]**
- Works. Including his Life and Correspondence. Edited by Simon Wilkin. London & Norwich: William Pickering, Josiah Fletcher, 1836-35. 4 vols. 8vo. Port, fldg facs, 3 other plates, 3 pedigrees. Orig cloth, 2 spines relaid, sl worn & rubbed.
(Clark) **£100 [≈ $160]**

Browning, John
- How to Work with the Spectroscope ... London: J. Browning, [late 1800s]. 8vo. 48 pp. Over 30 ills. Orig cloth bds, front bd sl worn & discold. *(Savona)* **£40 [≈ $64]**

Bruff, Peter
- A Treatise on Engineering Field-Work ... Second Edition, Corrected & Enlarged. London: 1840. 8vo. viii,176 pp. Lacks plate 6. Orig black cloth, hd of spine roughly reprd.
(Lamb) **£20 [≃ $32]**

Brunker, J.P.
- Flora of the County of Wicklow. Dundalk, 1950. 8vo. xii,310 pp. Map. Orig cloth.
(Wheldon & Wesley) **£30 [≃ $48]**

Brunton, T. Lauder
- Lectures on the Action of Medicines. Being the Course of Lectures on Pharmacology and Therapeutics. New York: Macmillan, 1898. 1st edn. Roy 8vo. 673 pp. Ills. Orig bndg, front hinge cracked.
(Xerxes) **$65 [≃ £40]**

Buchan, Alexander
- A Handy Book of Meteorology. Edinburgh & London: Blackwood, 1867. 1st edn. Post 8vo. [vi], 204,advt pp. 5 chart plates, 53 text ills. Occas spotting. Orig maroon cloth, uncut, spine faded.
(Duck) **£40 [≃ $64]**

Buchan, William
- Domestic Medicine ... Second Edition, with considerable Additions. London: Strahan ..., 1772. 8vo. 2 pp ctlg. B4-5 soiled. New cloth.
(Stewart) **£150 [≃ $240]**
- Domestic Medicine ... New and Enlarged Edition ... Edinburgh: 1820. 8vo. xx,523 pp. 1 section rather loose. Some spotting. Old cloth gilt, rubbed.
(Hollett) **£30 [≃ $48]**

Buchanan, J.F.
- Practical Alloying. A Compendium of Alloys and Processes for Brass Founders, Metal Workers and Engineers. Cleveland, Ohio: Penton, 1910. 8vo. 205 pp. Ills. Orig bndg.
(Book House) **£15 [≃ $24]**

Buc'hoz, Pierre Joseph
- The Toilet of Flora; or, a Collection of the most Simple and Approved Methods of Preparing Baths, Essences, Pomatums, Powders, Perfumes ... For the Use of Ladies ... London: Murray, 1779. New edn. 8vo. [xii],252 pp. Integral engvd frontis. Orig sheep, sl worn.
(Claude Cox) **£120 [≃ $192]**

Buck, Albert H.
- Diagnosis and Treatment of Ear Diseases. New York: Wood, 1880. 1st edn. 8vo. vii,411 pp. 28 w'cuts. Orig cloth, mottled.
(Elgen) **$50 [≃ £31]**

Buck, George W.
- A Practical and Theoretical Essay on Oblique Bridges. London: 1839. 1st edn. 4to. v, [3],43,[1 advt] pp. 12 plates (6 fldg). Sm piece of title marg cut away. Orig cloth.
(Fenning) **£110 [≃ $176]**

Buckland, Frank
- Curiosities of Animal Life. London: 1878-79. Popular Edition. 4 vols. 12mo. Polished calf, fine.
(Wheldon & Wesley) **£28 [≃ $44]**

Buckland, William
- Geology and Mineralogy considered with reference to Natural Theology. London: 1837. 2nd edn. 2 vols. Fldg cold cross-section, 69 plates. Half calf, mrbld bds.
(Baldwin) **£100 [≃ $160]**
- Geology and Mineralogy considered with reference to Natural Theology. London: 1837. 2nd edn. 2 vols in one. 8vo. 88 plates (inc lge fldg hand cold section). Half calf, sl rubbed.
(Henly) **£90 [≃ $144]**
- Geology and Mineralogy considered with reference to Natural Theology. London: 1837. 2nd edn. 2 vols. Fldg cold cross-section, 69 plates. Orig cloth, hinges little weak.
(Baldwin) **£85 [≃ $136]**
- Geology and Mineralogy considered with reference to Natural Theology. London: William Pickering, Bridgewater Treatise, 1837. 2 vols. 8vo. xv,468; vii,131 pp. Hand cold fldg plate, 87 b/w plates. Orig calf.
(Gemmary) **$225 [≃ £140]**
- Geology and Mineralogy considered with reference to Natural Theology. London: 1858. 3rd edn. 2 vols. 8vo. Frontis, 90 plates. Sm lib stamps on title versos. Some plates in vol 2 sl foxed. Tree calf, spines relaid.
(Henly) **£72 [≃ $115]**
- Geology and Mineralogy as exhibiting the Power, Wisdom and Goodness of God. London: 1870. 4th edn. 2 vols. 8vo & oblong 4to. 90 plates, some fldg (plate 1 is a lge hand cold section of the earth's crust). Sm lib stamps on title versos. Orig cloth, spines relaid.
(Henly) **£48 [≃ $76]**
- Reliquiae Diluvianae; or, Observations on the Organic Remains contained in Caves ... Other Geological Phenomena ... London: 1823. 1st edn. 4to. vii,303 pp. Table, 27 plates (sl foxed). New cloth. *(Baldwin)* **£200 [≃ $320]**
- Reliquiae Diluvianae; or, Observations on the Organic Remains contained in Caves ... Other Geological Phenomena ... London: 1824. 2nd edn. 4to. vii,303 pp. Fldg table, 27 plates (3 cold). Contemp bds, uncut, sl worn, new calf spine. *(Wheldon & Wesley)* **£180 [≃ $288]**

Buckler, W.

- The Larvae of British Butterflies and Moths. London: 1887-89. Vols 2 & 3 only. 8vo. 36 hand cold plates. Orig cloth, vol 3 a little faded.   *(Henly)* **£75 [≈ $120]**
- The Larvae of the British Butterflies and Moths. Vol 2. The Sphinges or Hawk-Moths and part of the Bombyces. London: Ray Society, 1887. 8vo. 18 hand cold plates. Orig cloth gilt.   *(Egglishaw)* **£52 [≈ $83]**

Buckley, Arabella

- The Fairy Land of Science. London: Stanford, 1879. 1st edn. 8vo. viii,244,32 advt pp. Num engvs. Orig dec cloth, moderate wear.   *(Bookmark)* **£15.50 [≈ $25]**

Bucknill, Sir John Alexander

- The Birds of Surrey. London: Porter, 1900. 1st edn. 8vo. lvi,374 pp. Fldg map, 6 photogravures, 13 b/w ills. Orig brown cloth gilt. Henry Feilden b'plate.
          (Gough) **£50 [≈ $80]**

Bucknill, Sir John Charles

- A Manual of Psychological Medicine ... Philadelphia: Lindsay & Blakiston, 1874. 3rd rvsd & enlgd edn. 8vo. Sgntr & stamp on title. Frontis supplied in facs. Mod cloth.
          (Gach) **$185 [≈ £115]**
- The Psychology of Shakespeare. London: Longman ..., 1859. 8vo. vii,264 pp. Victorian cloth, crown chipped, hinges cracked.
          (Gach) **$175 [≈ £109]**

Buckton, George Bowdler

- Monograph of the British Cicadae or Tettigae. London: Macmillan, 1890-91. 1st edn. 2 vols. 8vo. vi,[i],lxxviii, 133,[ii]; [iv],211 pp. 82 hand cold plates, each with leaf of ptd text. Contemp half calf. Author's inscrptn & ALS tipped-in.   *(Bickersteth)* **£130 [≈ $208]**

Budd, William

- Typhoid Fever, Its Nature, Mode of Spreading and Prevention. New York: 1931. reprint of 1874 edn. One of 800. Lge 8vo. 184 pp. Frontis, 4 plates (1 cold). Orig cloth, slipcase.   *(Elgen)* **$75 [≈ £46]**

Budge, E.A. Wallis

- The Divine Origin of the Craft of the Herbalist. London: Culpeper House, 1928. 8vo. xii, 96 pp. 13 ills. Orig green cloth. dw.   *(Blackwell's)* **£95 [≈ $152]**
- The Divine Origin of the Craft of the Herbalist. London: Society of Herbalists, 1928. 1st edn. 8vo. xi,96 pp. 13 ills. Orig

cloth gilt. Dw with sm tear on top edge.
          (Hollett) **£55 [≈ $88]**

Budgen, Miss M.L.

- Episodes of Insect Life, by Acheta Domestica M.E.S. London: Reeve & Benham, 1849-51. 1st edn. 3 vols. 8vo. 3 frontis, engvd chapter headings & tail-pieces. Orig dec cloth gilt, hd of 2 spines sl frayed.
          (Egglishaw) **£115 [≈ $184]**
- Episodes of Insect Life by Acheta Domestica M.E.S. London: 1849-51. 3 vols. 8vo. 3 hand cold frontis, num hand cold chapter headings, engvd tail-pieces. Sl foxing in the front of vols 2 & 3. Half mor gilt.
          (Wheldon & Wesley) **£200 [≈ $320]**
- Episodes of Insect Life by Acheta Domestica M.E.S. New York: 1851-52. 1st Amer edn. 3 vols. 8vo. 3 engvd frontis, num engvd chapter headings & tail-pieces. Orig pict cloth gilt, edges v sl worn.
          (Wheldon & Wesley) **£120 [≈ $192]**

Buffman, Herbert E., et al.

- The Household Physician. A Twentieth Century Medica. Boston: 1905. 1st edn. Lge thick 8vo. 1436 pp. 27 cold plates, 3 manikins.   *(Rittenhouse)* **$60 [≈ £37]**

Buffon, Georges-Louis Leclerc, Comte de

- Natural History. London: 1792. 2 vols. 8vo. Frontis, 2 engvd titles, 107 engvd plates. Sm waterstains to 4 plates. Calf backed bds, worn, jnts starting to crack.
          (Henly) **£45 [≈ $72]**
- Buffon's Natural History ... From the French. With Notes by the Translator. London: for the proprietor, 1797. 10 vols. 83 copper engvs. Contemp tree calf, gilt dec spines, dble green & black mor labels.
          (Hartfield) **$750 [≈ £468]**

Bull, Marcus

- Experiments to determine the Comparative Value of the Principal Varieties of Fuel used in the United States ... Apparatus used for their Combustion. Philadelphia, New York, Boston, London: 1827. 1st edn. Roy 8vo. Frontis. Browned. Orig bds, rebacked.
          (Duck) **£225 [≈ $360]**

Buller, W.L.

- A History of the Birds of New Zealand. London: [1872-] 1873. 1st edn. [One of 500]. Roy 4to. xxiii,384 pp. Frontis (inner marg reprd) & 35 hand cold plates plates by Keulemans. Lacks the 5 advt pp at end. Red half mor gilt, a.e.g., trifle rubbed.
          (Wheldon & Wesley) **£1,900 [≈ $3,040]**

- A History of the Birds of New Zealand. London: privately ptd, 1888. 2nd edn. One of 1000. 2 vols. Folio. 2 plain & 48 cold plates by Keulemans. Orig dec dark green mor gilt, a.e.g. *(Wheldon & Wesley)* **£1,800 [≈ $2,880]**
- Birds of New Zealand. Edited and brought up to date by E.G. Turbott. London: 1967. Folio. xviii,261,[1] pp. 48 cold plates. Orig cloth.   *(Wheldon & Wesley)* **£45 [≈ $72]**

Bulmer, John
- Anthropometamorphosis: Man Transformed: or, The Artificial Changling ... London: William Hunt, 1653. 2nd edn. 4to. l,559,[30] pp. W'cut ills. Engvd title & frontis cropped at edges. A few marg stains. Early 18th c mor gilt, contemp gauffered edges gilt.
(Gough) **£750 [≈ $1,200]**

Bunyard, G. & Thomas, O.
- The Fruit Garden. London: 1906. 2nd edn. Roy 8vo. xiii,507 pp. 120 plates, num text figs. Orig cloth, trifle used.
(Wheldon & Wesley) **£28 [≈ $44]**

Burbank, Luther
- Luther Burbank. His Methods and Discoveries and their Practical Application. Edited by J. Whitson, R. John and H.S. Williams. New York: 1914-15. 12 vols. 8vo. 1260 cold plates. Orig cloth, faded, vol 2 trifle worn & stained.
(Wheldon & Wesley) **£90 [≈ $144]**

Burbridge, F.W.
- Cool Orchids, and How to Grow Them ... London: 1874. 1st edn. Post 8vo. [vi],160 pp. 4 hand cold plates, text ills. Orig green cloth gilt, a.e.g.   *(Bow Windows)* **£60 [≈ $96]**

Burdell, Harvey & John
- Observations on the Structure, Physiology, Anatomy and Diseases of the Teeth. In Two Parts. New York: Gould & Newman, 1838. 1st edn. 8vo. 96 pp. Half-title. Text w'cuts. Foxing. Orig cloth, some spotting, spine ends frayed.   *(Elgen)* **$250 [≈ £156]**

Burdick, Gordon G.
- X-Ray and High Frequency in Medicine. Chicago: 1909. 318,[10] pp. Ills. Orig bndg.
(Elgen) **$75 [≈ £46]**

Burgess, N.G.
- The Photograph Manual; a Practical Treatise, containing the Cartes de Visite Process ... New York: Appleton, 1863. 8vo. 267, [iii] pp. Orig cloth, faded.
(Charles B. Wood) **$250 [≈ £156]**

Burleigh, T.D.
- Birds of Idaho. Caldwell, Idaho: 1972. Roy 8vo. xiii,467 pp. Map, 12 cold plates, ills. Orig bndg. *(Wheldon & Wesley)* **£20 [≈ $32]**
- Birds of Idaho. Idaho: Caldwell, 1972. 8vo. xiii,467 pp. 12 cold plates, text ills. Orig cloth. Dw.   *(Henly)* **£15 [≈ $24]**

Burn, Robert S. (ed.)
- The Illustrated Architectural, Engineering, & Mechanical Drawing-Book ... Third Edition, revised. London: Ward & Lock, [ca 1860]. 8vo. Frontis, ca 300 ills. Orig cloth gilt, sl signs of use.   *(Fenning)* **£14.50 [≈ $24]**
- The Illustrated London Practical Geometry and its Application to Architectural Drawing. Second Edition. London: Ingram, Cooke, [1853]. 8vo. 83,[12 advt] pp. Frontis, addtnl dec title, 284 ills. Orig cloth gilt.
(Fenning) **£12.50 [≈ $20]**

Burnley, James
- The History of Wool and Woolcombing. London: Sampson Low, 1889. 1st edn. Lge 8vo. 486 pp. Ills. Orig bndg.
(Book House) **£45 [≈ $72]**
- The History of Wool and Woolcombing. With an Appendix of Wool Combing Patents. London: Sampson Low, 1889, reprinted 1969. 487 pp. Ills. Orig bndg.
(Book House) **£18 [≈ $28]**

Burns, Allan
- Observations on some of the most frequent and important Diseases of the Heart ... Edinburgh: James Muirhead for Thomas Bryce, 1809. 1st edn. 8vo. iv,[2],322 pp. Contemp half calf. Steevens Lib stamp on title, sgntr of Dr. Terence East on flyleaf.
(Rootenberg) **$1,600 [≈ £1,000]**

Burns, John
- The Anatomy of the Gravid Uterus. With Practical Inferences relative to Pregnancy and Labour. Boston: Cushing & Appleton ..., 1808. 1st Amer edn. 8vo. 2 plates. Orig mrbld bds, drab paper backstrip, paper label, upper jnt cracked, hd of spine sl worn.
(Ximenes) **$175 [≈ £109]**

Burrows, E.I.
- Elements of Conchology according to the Linnean System. London: 1844. 8vo. 28 plates (25 hand cold). Orig cloth, spine relaid.   *(Henly)* **£98 [≈ $156]**

Burt, W.H.
- Tuberculosis or Pulmonary Consumption. Its

Prophylaxis and Cure by Suralimentation of Liquid Food. Chicago: Keener, 1890. 1st edn. 8vo. 233 pp. Orig bndg, cvr worn.
(Xerxes) **$45 [≈ £28]**

Burton, Robert

- The Anatomy of Melancholy ... By Democritus Junior ... The Fourth Edition, corrected and augmented by the Author. Oxford: for Henry Cripps, 1632. 4to. Engvd title. Early 19th c russia gilt, upper jnt cracked but firm. *(Spelman)* **£480 [≈ $768]**

- The Anatomy of Melancholy ... By Democritus Junior ... London: for Peter Parker, 1676. 8th edn. Folio. Dedic leaf, address to the reader, final advt leaf, half-title. Engvd title. Contemp calf, rebacked, orig label preserved, upper jnt cracked but firm.
(Heritage) **$500 [≈ £312]**

- The Anatomy of Melancholy ... To which is prefixed an Account of the Author. London: for Vernor, Hood ..., 1806. 11th edn, crrctd. 2 vols. 8vo. xxiv,461; 601,[12] pp. New half calf. *(Young's)* **£78 [≈ $124]**

- The Anatomy of Melancholy ... London: for Vernor, Hood & Sharpe, 1806. 2 vols. 8vo. xxiv, 461; 601,[10] pp. 2 frontis. Old panelled calf gilt, spines a little chipped & rubbed.
(Hollett) **£48 [≈ $76]**

- The Anatomy of Melancholy ... London: for B. Blake, 1836. 8vo. vi,744 pp. Frontis (rather offset). Old panelled diced calf gilt.
(Hollett) **£40 [≈ $64]**

- The Anatomy of Melancholy ... Illustrated by E. McKnight Kauffer. London: The Nonesuch Press, 1925. One of 750. 2 vols. Lge 8vo. Orig qtr parchment, ptd paper bds, sl soiled & rubbed.
(Thornton's) **£245 [≈ $392]**

Burton, W.K.

- Practical Guide to Photographic and Photo-Mechanical Printing. Second Edition, revised and enlarged. London: Marion & Co, 1892. 8vo. xvii,415,46 pp. Num illust advts. Orig cloth. *(Charles B. Wood)* **$125 [≈ £78]**

Bushnan, J.S.

- The Natural History of Fishes, particularly their Structure and Economical Uses. Edinburgh: Jardine's Naturalist's Library, Ichthyology Vol 2, 1840. 1st edn. Cr 8vo. 219 pp. Vignette title, port, 32 hand cold plates. Orig cloth. *(Egglishaw)* **£38 [≈ $60]**

Busk, Hans

- The Rifle: and how to use it ... Fourth Edition, considerably enlarged and improved.

London: Routledge ..., 1859. Sm 8vo. [2],225, [4],4 advt pp. 2 plates, num text ills. Orig roan backed cloth.
(Fenning) **£45 [≈ $72]**

Butcher, R.W.

- A New Illustrated British Flora. London: 1961. 2 vols. 8vo. 1825 ills. Orig cloth, trifle used. *(Wheldon & Wesley)* **£45 [≈ $72]**

Butler, Arthur Gardiner

- Birds of Great Britain and Ireland. London: Brumby & Clarke, [1908]. 1st edn. 2 vols. 4to. 107 chromolithos of birds by Gronvold & 8 of eggs. Orig blue cloth, sl faded.
(Gough) **£375 [≈ $600]**

- British Bird's Eggs: a Handbook of British Oology. London: Janson, 1886. 8vo. viii,219 pp. Frontis, 37 chromolitho plates. Orig cloth, unopened. *(Egglishaw)* **£28 [≈ $44]**

- Foreign Birds for Cage and Aviary. London: [1908-10]. 2 vols. Cr 4to. Num ills. Orig cloth. *(Wheldon & Wesley)* **£50 [≈ $80]**

- Foreign Finches in Captivity. London: 1894. 1st edn. Roy 4to. viii,332 pp. 60 hand cold plates by F.W. Frohawk. 1 plate trifle foxed. Orig dec cloth, trifle faded.
(Wheldon & Wesley) **£1,450 [≈ $2,320]**

- Foreign Finches in Captivity. London: 1899. 2nd edn. Roy 8vo. viii,317 pp. 60 colour ptd plates by F.W. Frohawk. New cloth.
(Wheldon & Wesley) **£200 [≈ $320]**

- Lepidoptera Exotica, or Descriptions and Illustrations of Exotic Lepidoptera. London: 1869-74. 4to. [xi],190 pp. 1 plain & 63 cold plates. Later cloth.
(Egglishaw) **£450 [≈ $720]**

Butler, Charles

- The Feminine Monarchy; or the History of Bees, written in Latin ... and now translated by W.S. London: 1704. 6th edn. 12mo. [iv],150 pp. Frontis. Contemp calf, rebacked.
(Wheldon & Wesley) **£140 [≈ $224]**

Butler, Edward A.

- Our Household Insects. An Account of Insect-Pests found in Dwelling-houses. London: Longmans Green, 1883. Sm 8vo. ix,244, [22 (of 24) ctlg] pp. 7 plates, 113 text figs. Orig blue cloth gilt.
(Blackwell's) **£21 [≈ $33]**

Butler, Samuel, 1835-1902

- Evolution, Old and New ... London: Hardwicke & Bogue, 1879. 1st edn. 8vo. xii, 384,[advts dated Feb.1879] pp. Orig ptd dec brown cloth, jnts & edges rubbed.

(Gach) **$150 [≃ £93]**
- Unconscious Memory: A Comparison between The Theory of Dr. Ewald Hering ... and "The Philosophy of the Unconscious" of Dr. Edward von Hartmann ... London: David Bogue, 1880. 1st edn. 8vo. viii,288,32 pp. Orig ptd brown cloth. *(Gach)* **$175 [≃ £109]**

Buxton, J.
- The Redstart. London: New Naturalist, 1950. 1st edn. 8vo. xii,180 pp. Cold frontis, 19 photos, 20 maps & diags, 2 text figs. Orig cloth. Dw. *(Wheldon & Wesley)* **£30 [≃ $48]**

Buxton, P.A.
- The Natural History of Tsetse Flies. London: 1955. Cr 4to. xviii,816 pp. 47 plates, text figs. Blind stamps at beginning. Orig cloth, spine spotted. *(Wheldon & Wesley)* **£40 [≃ $64]**

Buxton, R.
- A Botanical Guide to the Flowering Plants, etc. found Indigenous within 16 Miles of Manchester. London: 1849. Post 8vo. xxiv,168 pp. Orig cloth, trifle used, spine faded. *(Wheldon & Wesley)* **£35 [≃ $56]**

Bywater, John
- An Essay on the History, Practice, and Theory, of Electricity. London: for the author ..., 1810. Sole edn. 8vo. [2],iii, [2], 127 pp. 2 fldg plates. Orig bds, rebacked, orig label, uncut. *(Fenning)* **£110 [≃ $176]**
- An Essay on the History, Practice, and Theory, of Electricity. London: for the author ..., 1810. Sole edn. 8vo. [2],iii, [2], 127 pp. 2 fldg plates. Contemp half calf. *(Fenning)* **£85 [≃ $136]**

Cabot, Richard C.
- Facts on the Heart. Philadelphia: 1926. 1st edn. 8vo. 781 pp. Ills. Orig cloth, spine faded. *(Elgen)* **$75 [≃ £46]**

Cadbury, D.A., et al.
- A Computer Mapped Flora of Warwickshire. London: 1971. 4to. ix,768 pp. Cold geol map, 979 maps, 10 transparent overlays. Orig cloth. *(Wheldon & Wesley)* **£35 [≃ $56]**

Caelius Aurelianus
- On Acute Diseases and On Chronic Diseases. Edited and translated by I.E. Drabkin. [Chicago: 1950]. 8vo. xxvi,vii,1019 pp. Orig cloth. Dw. *(Elgen)* **$90 [≃ £56]**

Cahen, E. & Wooton, W.O.
- The Mineralogy of the Rarer Metals. A

Handbook for Prospectors. London: Charles Griffith, 1912. 12mo. xxviii,211 pp. Orig cloth. *(Gemmary)* **$35 [≃ £21]**

Cairnes, John Elliott
- Some Leading Principles of Political Economy newly expounded. London: Macmillan, 1874. 1st edn. 8vo. xix,[i],506,54 ctlg pp. Orig cloth, spine relaid. *(Claude Cox)* **£48 [≃ $76]**

Calderwood, Henry
- The Relations of Mind and Brain. London: Macmillan, 1884. 2nd edn. 8vo. xx,528 pp. 50 text ills. Orig pebbled tan cloth. *(Gach)* **$85 [≃ £53]**

Calderwood, W.L.
- The Life of the Salmon, with reference more especially to the Fish in Scotland. London: Edward Arnold, 1907. 8vo. xxiv,160 pp. 8 plates. Orig cloth gilt. *(Egglishaw)* **£16 [≃ $25]**
- The Salmon Rivers ... of Scotland. London: Edward Arnold, 1909. One of 250 Large Paper. 4to. x,442 pp. 4 cold & 34 other plates. Orig 2-tone cloth, t.e.g. *(Egglishaw)* **£150 [≃ $240]**

Calisch, E.
- Electric Traction. London: Loco Pub, 1913. Sm 8vo. 116 pp. Ills. Orig bndg. *(Book House)* **£18 [≃ $28]**

Calvert, Albert F.
- Daffodil Growing for Pleasure and Profit. London: Dulau, 1929. 8vo. xix,412 pp. Cold frontis, 16 ills of pests & diseases, 220 photos of varieties. Orig gilt dec cloth. *(Blackwell's)* **£135 [≃ $216]**

Cameron, A.G.
- The Wild Red Deer of Scotland. London: 1923. 8vo. Ills. Sl foxing. Orig bndg, sl faded. *(Grayling)* **£30 [≃ $48]**

Camp, W.M.
- Notes on Track Construction and Maintenance. Chicago: the author, 1903. 1st edn. Imperial 8vo. viii,1214,24 advt pp. Over 600 ills. Orig cloth, pict gilt dec spine, lib mark on spine, cvrs sl worn & soiled, hinges split, spine ends sl frayed. *(Duck)* **£65 [≃ $104]**

Campbell, Andrew
- Petrol Refining. London: Charles Griffin, 1922. 2nd edn, rvsd. xvi,298,299-308 illust

advt pp. 138 ills, inc 29 fldg. Orig maroon cloth. *(Duck)* **£30 [≈$48]**

Campbell, L. & Garnett, W.
- The Life of James Clerk Maxwell. London: 1882. xvi,662 pp. 8 plates, 12 figs. Orig cloth. *(Baldwin)* **£40 [≈$64]**

Campbell, M.S.
- The Flora of Uig (Lewis). A Botanical Exploration. Arbroath: 1945. 8vo. 63 pp. Map, 6 plates. Orig cloth.
(Wheldon & Wesley) **£20 [≈$32]**

Cannan, Edwin
- History of the Theories of Production and Distribution in English Political Economy from 1776 to 1848 ... London: Percival & Co, 1893. 1st edn. 8vo. xi,[1 blank],410,[1 advt], [1 blank],[31 advt] pp. Orig dark blue cloth gilt, a bit rubbed. *(Pickering)* **£220 [≈£137]**

Cannon, Walter Bradford
- Bodily Changes in Pain, Hunger, Fear and Rage: An Account of Recent Researches into the Function of Emotional Excitement. New York: Appleton, 1915. 1st edn. 8vo. [xiv], 311, [3] pp. Orig russet cloth. Defective dw. *(Gach)* **$125 [≈£78]**
- Bodily Changes in Pain, Hunger, Fear and Rage ... New York: Appleton, 1929. 2nd edn, 1st printing. Sm 8vo. [xviii],404,[2] pp. Orig russet cloth. Chipped dw. *(Gach)* **$75 [≈£46]**

Cantor, Alfred J.
- Handbook of Psychosomatic Medicine with particular reference to Intestinal Disorders. New York: Julian Messner, 1951. 1st edn. 8vo. 302 pp. Orig bndg. Pres copy.
(Xerxes) **$55 [≈£34]**

Capper, James
- Meteorological and Miscellaneous Tracts, applicable to Navigation, Gardening, and Farming ... Cardiff: J.D. Bird ..., [1809]. 1st edn. 8vo. xx,211,[1] pp. Half-title. 6 fldg tables. Rec bds, uncut. Author's pres copy.
(Burmester) **£120 [≈$192]**
- Observations on the Winds and Monsoons; illustrated with a Chart, and accompanied with Notes, Geographical and Meteorological. London: C. Whittingham, 1801. 1st (only?) edn. 4to. 234 pp. Lacks half-title. Lge fldg map. Sl foxing. Calf backed mrbld bds. *(Fenning)* **£110 [≈$176]**

Carey, George
- Biochemic System comprising the Theory,

Pathological Action, Therapeutical Application, Materia Medica, and Repertory of Schuessler's Twelve Tissue Remedies. St. Louis: Luyties, 1908. 8vo. 444 pp. Orig bndg, hinges taped. *(Xerxes)* **$90 [≈£56]**

Carmichael, Andrew
- A Memoir of the Life and Philosophy of Spurzheim. Boston: 1833. 1st Amer edn. Sm 8vo. 96 pp. Orig bndg. *(Elgen)* **$150 [≈£93]**

Carpenter, Sir H. & Robertson, J.M.
- Metals. OUP: 1939. 2 vols. Sm 4to. 1485 pp. Orig bndg. *(Book House)* **£20 [≈$32]**

Carpenter, William B.
- The Microscope and its Revelations. London: 1875. 5th edn. Sm 8vo. xxxii,848 pp. 25 plates, 449 text ills. Orig cloth.
(Elgen) **$95 [≈£59]**
- The Microscope and its Revelations ... Revised by W.H. Dallinger. London: Churchill, 1891. 7th edn. 8vo. xviii,1099,xvi pp. Frontis, 20 plates (inc cold), 756 text ills. Orig cloth, spine ends & bd edges worn, inner hinges cracked. *(Savona)* **£40 [≈$64]**

Carpue, Joseph Constantine
- An Introduction to Electricity and Galvanism; with Cases, shewing their Effects in the Cure of Diseases ... London: A. Phillips, 1803. 1st edn. 8vo. viii,112 pp. 3 fldg plates. 1st & last ff foxed. Advt tipped to last leaf. Contemp calf backed mrbld bds, worn. *(Rootenberg)* **$950 [≈£593]**

Carrel, Alexis & Lindbergh, Charles A.
- The Culture of Organs. New York: 1938. 1st edn. 8vo. 221 pp. Num ills. Orig cloth.
(Elgen) **$225 [≈£140]**

Carroll, Lewis
- Symbolic Logic. Part I. Elementary. London: 1897. 4th edn. xxxi,199 pp. Text diags. Sm lib stamp on title. Thick card cvrs.
(Whitehart) **£35 [≈$56]**

Carter, Susannah
- The Frugal Housewife: or, Complete Woman Cook ... Philadelphia: for Mathew Carey, 1802. 132 pp. 2 plates. Sm lib blindstamp on title. Contemp half calf, mrbld bds, worn & rubbed, front bd detached, lib b'plate.
(Reese) **$750 [≈£468]**

Carus, Paul
- The Soul of Man: An Investigation of the Facts of Physiological and Experimental

Psychology. Chicago: The Open Court Publ Co, 1891. 1st edn. Frontis. Orig dec blue cloth, crnrs bumped.   *(Gach)* **$50 [≈ £31]**

Cassell's Illustrated Exhibitor ...
- Cassell's Illustrated Exhibitor ... of all the Principal Objects in the International Exhibition of 1862. London: 1862. 4to. 272, advt pp. Ills. Orig bndg, cvrs faded & rubbed, shaken.   *(Book House)* **£35 [≈ $56]**

Cassella & Co, Leopold
- The Dyeing of Wool, including Wool-Printing, with the Dyestuffs. Frankfort: 1905. Lge 8vo. x,397 pp. 408 actual samples. Orig cloth gilt.   *(Hollett)* **£120 [≈ $192]**

Cassier's Magazine
- Electric Power Number. June, 1904. 4to. 256,178 advt pp. Ills. Orig bndg, cvrs a little soiled.   *(Book House)* **£35 [≈ $56]**
- Electric Railway Number. August, 1899. 4to. 540,228 advt pp. Orig bndg.
          (Book House) **£80 [≈ $128]**

Castaing, John
- An Interest-Book ... Exactly Examined ... London: for the author ..., 1720. 4th edn. 12mo. 88 ff. Signed by Castaing on 1st leaf of text. Contemp calf, gilt spine.
          (Burmester) **£38 [≈ $60]**
- An Interest-Book at 4,5,6,7,8 per C. From 1000 l. to 1 l. For 1 Day to 92 Days, and for 3,6,9,12 Months. Fourth Edition. London: H. Hills, 1724. 16mo. 74 pp. Half-title. Contemp calf.   *(Spelman)* **£75 [≈ $120]**

Castellani, A.
- Gems: Notes and Extracts. London: Bell & Daldy, 1871. 8vo. vi,241 pp. Orig cloth.
          (Gemmary) **$100 [≈ £62]**

Castro, D'Oliveira
- Elements of Therapeutics and Practice according to the Dosimetric System. New York: Appleton, 1888. 1st Amer edn. 8vo. xvi,492 pp. Orig brown cloth.
          (Karmiole) **$50 [≈ £31]**

A Catechism of Mineralogy ...
- See Pinnock, William.

Catlin, George
- Breath of Life or Mal-Respiration and its Effects upon the Enjoyments & Life of Man. New York: Wiley, (1861) 1869. 8vo. 76 pp. Ills. Orig bndg, worn. *(Xerxes)* **$175 [≈ £109]**
- The Lifted and Subsided Rocks of America ...

London: Trubner, 1870. Sm 8vo. xii,228 pp. Fldg map. Orig green cloth gilt.
          (Karmiole) **$100 [≈ £62]**

Catlow, Agnes
- Drops of Water; their Marvellous and Beautiful Inhabitants displayed by the Microscope. London: Reeve & Benham, 1851. Sq 12mo. xviii,194 pp. 4 hand cold plates. Orig cloth, spine ends worn.
          (Egglishaw) **£30 [≈ $48]**
- Popular Conchology; or the Shell Cabinet Arranged. London: 1843. Post 8vo. xx,300 pp. Text figs. Blind stamp on title. Orig cloth, rebacked.
          (Wheldon & Wesley) **£20 [≈ $32]**

Catlow, Joseph Peel
- On the Principles of Aesthetic Medicine ... London: Churchill; Birmingham: Hudson, 1867. 1st edn. 8vo. 325,[3] pp. Orig cloth, front jnt cracked, some wear to jnts, endleaves foxed.   *(Gach)* **$250 [≈ £156]**

Catlow, Maria E.
- Popular British Entomology. New Edition. London: Routledge, 1860. Sm 8vo. x,280 pp. 16 cold plates. Orig cloth gilt, sl loose.
          (Egglishaw) **£22 [≈ $35]**

Caton, John Dean
- The Antelope & The Deer. A Comprehensive Scientific Treatise ... including ... Capacity for Domestication of the Antilocapra & Cervidae of North America. New York: Forest & Stream, [1877]. 2nd edn. 8vo. xii,426 pp. Plates, ills. 1st sgntr reprd. Orig cloth.   *(Terramedia)* **$45 [≈ £28]**

Cavallo, Tiberius
- A Complete Treatise of Electricity in Theory and Practice with Original Experiments. London: Edward & Charles Dilly, 1777. 1st edn. 8vo. xvi,viii,412,[4] pp. Errata, advt leaf. 3 fldg plates (sm split in 1 fold). Half calf.
          (Rootenberg) **$550 [≈ £343]**
- The History and Practice of Aerostation. London: for the author ..., 1785. 1st edn. viii, 1-326,index [327]-[383] pp. 2 fldg plates. Occas sl spotting. Contemp speckled polished calf gilt, gilt spine & labels.
          (Duck) **£750 [≈ $1,200]**

Cavendish, Henry
- The Electrical Researches ... written between 1771 and 1781, edited from the Original Manuscripts ... by J. Clerk Maxwell. Cambridge: UP, 1879. 1st edn. Lge 8vo. lxvi, 454, 32 advt pp. Prelims sl foxed. Orig cloth.

(Rootenberg) **$450 [≈ £281]**
- Observations on Mr. Hutchins's Experiments for determining the Degree of Cold at which Quicksilver Freezes ... Read at the Royal Society, May 1, 1783. London: J. Nichols, 1784. Orig offprint. 4to. [ii],26 pp. Stab sewn as issued, uncut & unopened.
(Pickering) **$850 [≈ £531]**

Cayley, A.
- An Elementary Treatise on Elliptic Functions. London: 1895. 2nd edn. xiii,386 pp. Orig cloth, sl marked.
(Whitehart) **£35 [≈ $56]**

Cecil, The Hon Mrs Evelyn (Alicia Amherst)
- A History of Gardening in England. London: 1895. 1st edn. Roy 8vo. xvi,399 pp. 65 ills. Cloth gilt. *(Wheldon & Wesley)* **£50 [≈ $80]**
- A History of Gardening in England. London: 1896. 2nd edn. Roy 8vo. xiv,405 pp. 68 ills. Orig cloth. *(Wheldon & Wesley)* **£50 [≈ $80]**

The Certain Method to Know the Disease ...
- See Rudd, Sayer.

The 'Challenger' Reports
- See Sclater, P.L.

Challis, J.
- Notes on the Principles of Pure and Applied Calculation ... Cambridge: 1869. xxxi, lxiv,696 pp. 1 plate. Orig cloth, spine sl worn.
(Whitehart) **£25 [≈ $40]**

Chalmers, P.R.
- Birds Ashore and A-Foreshore. London: 1935. 4to. 180 pp. 16 cold plates, num ills, by Winifred Austen. Binder's cloth.
(Wheldon & Wesley) **£25 [≈ $40]**
- Mine Eyes to the Hills. An Anthology of the Highland Forest. London: A. & C. Black, 1931. 1st edn. Roy 8vo. xv,368 pp. 8 cold plates, text ills. Orig cloth.
(Egglishaw) **£36 [≈ $57]**

Chalmers, Thomas
- On the Wisdom and Goodness of God as Manifested in the Adaptation of External Nature to the Moral and Intellectual Constitution of Man. London: William Pickering, 1839. 1st edn. 2 vols. 8vo. 290; 302 pp. 19th c polished half calf gilt, sl rubbed.
(Gach) **$135 [≈ £84]**

Chambers, George F.
- Descriptive Astronomy. Oxford: 1867. xxxviii, 816 pp. Addenda leaf. Frontis, 37 plates, num text ills. Orig cloth, some wear esp at spine ends. *(Elgen)* **$60 [≈ £37]**

Chambers, Robert
- Ancient Sea-Margins, as Memorials of Changes in the Relative Level of Sea and Land. London: W.S. Orr, 1848. Roy 8vo. vi,337,[1] pp. Litho frontis, fldg map, text ills. Contemp polished calf gilt, a.e.g., jnts tender, ends cracking. *(Duck)* **£60 [≈ $96]**
- Ancient Sea-Margins, as Memorials of Change in the Relative Level of Sea and Land. London: 1848. 8vo. vi,337 pp. Frontis, cold map, text figs. Some foxing & old lib stamps. Orig cloth, worn.
(Wheldon & Wesley) **£60 [≈ $96]**
- Vestiges of the Natural History of Creation. London: 1844. 1st edn. Cr 8vo. vi, 390 pp. Orig red cloth.
(Wheldon & Wesley) **£200 [≈ $320]**

Chandler, John
- A Treatise of the Disease called a Cold; shewing its General Nature, and Causes ... London: for A. Millar, 1761. 1st edn. 8vo. [iv], 123 pp. Rec bds.
(Burmester) **£125 [≈ $200]**

Channing, Walter
- A Treatise on Etherization in Childbirth ... Boston: Ticknor, 1848. 1st edn. 8vo. viii, 400 pp. Ills. Lib stamp on 2 pp. Orig bds, t.e.g., spine relaid, new endpapers.
(Elgen) **$525 [≈ £328]**

Chanter, C.
- Ferny Combes. A Ramble after Ferns in the Glens and Valleys of Devonshire. London: 1856. 2nd edn. Sm 8vo. viii,118 pp. Map, 8 hand cold plates. Orig cloth, fine.
(Wheldon & Wesley) **£30 [≈ $48]**

Chapman, Abel
- Bird-Life of the Borders on Moorland and Sea. London: 1907. 2nd edn. 8vo. xii,458 pp. Cold map, num ills. Orig buckram.
(Wheldon & Wesley) **£45 [≈ $72]**
- First Lessons in the Art of Wildfowling. London: 1896. 8vo. Fldg frontis, other plates. Orig bndg. *(Grayling)* **£120 [≈ $192]**

Chapman, Henry T.
- Varicose Veins: Their Nature, Consequences, and Treatment, Palliative and Curative.

London: Churchill, 1856. 1st edn. Sm 8vo.
viii,98 pp. Orig cloth, sl shelf wear.
(Elgen) **$75 [≈ £46]**

Chapuis, Alfred & Jacquet, E.
- Technique and History of the Swiss Watch,
from its Beginnings to the Present Day ...
Olten, Switzerland: 1953. Sm folio. 278 pp.
42 cold & 190 other plates, text ills. Orig cloth
gilt, a name stamp on the foredge.
(Fenning) **£60 [≈ $96]**

Charsley, F.A.
- The Wild Flowers around Melbourne.
London: 1867. Folio. 13 cold lithographs.
Lower inner crnr of 1st 4 ff waterstained, 1
plate v sl foxed. Orig cloth, trifle rubbed, used
& loose.
(Wheldon & Wesley) **£800 [≈ $1,280]**

Chase, Revd H.
- Two Years and Four Months in a Lunatic
Asylum from Aug 20, 1863 to Dec 20, 1865.
Saratoga Springs: [no publisher named],
1868. 16mo. 184 pp. Testimonials at end.
Orig bndg, qtr inch at bottom crnr of cvrs &
text bitten away.   *(Xerxes)* **$200 [≈ £125]**

**Chatfield, Charles Hugh & Taylor,
Charles Fayette**
- The Airplane and its Engine. New York:
McGraw Hill, 1928. 1st edn. viii,329 pp.
Num ills. Orig blue cloth. *(Duck)* **£25 [≈ $40]**

Cheeseman, T.F.
- Manual of the New Zealand Flora. London:
1906. 8vo. xxxvi,1199 pp. [With] Illustrations
of the New Zealand Flora. London: 1914. 2
vols. Roy 8vo. 251 plates. Together 3 vols.
Orig cloth, v sl spotted.
(Wheldon & Wesley) **£200 [≈ $320]**

Chemistry ...
- Chemistry as applied to the Arts and
Manufactures. London: [ca 1880]. 8 vols. 4to.
Num ills. Orig gilt dec cloth.
(Baldwin) **£40 [≈ $64]**

Cheselden, William
- The Anatomy of the Human Body ... The
Vth Edition with Forty Copper Plates
Engrav'd by Ger: Vandergucht. London:
William Bowyer, 1750. 5th edn, but 1st with
these plates. 8vo. [xii],336 pp. Engvd frontis
& title, 40 plates (sl offsetting). Calf, gilt
spine.   *(Pickering)* **$1,800 [≈ £1,125]**

Cheshire, F.R.
- Bees and Bee-Keeping, Scientific and

Practical. London: [1886-88]. 2 vols. 8vo. 9
plates, 198 text figs. Sl foxing. New cloth.
(Wheldon & Wesley) **£75 [≈ $120]**

Cheshire, John
- A Treatise upon the Rheumatism ... London:
for C. Rivington, & W. Cantrell in Derby,
1723. 1st edn. 8vo. [xii],[44] pp. Pp 37-44
misnumbered 29-36. Title v sl dusty, crnrs
creased, old ink notes at end. Sewed as issued.
(Bickersteth) **£160 [≈ $256]**

Cheyne, George
- Dr. Cheyne's Own Account of Himself and of
his Writings ... London: J. Wilford, 1743. 1st
edn. 8vo. [iv],63 pp. 8 blank ff bound in at
end, 2 with MS notes. Orig grey upper
wrapper, lacks spine & lower wrap.
(Bickersteth) **£140 [≈ $224]**

Cheyne, W. Watson (ed.)
- Recent Essays by Various Authors on
Bacteria in Relation to Disease. London: New
Sydenham Soc, 1886. 8vo. xvi,650 pp. 8 litho
plates (6 cold). Orig cloth, spine ends frayed,
front inner hinge cracked.
(Elgen) **$125 [≈ £78]**

Chittenden, F.J. (ed.)
- Dictionary of Gardening ... see under Royal
Horticultural Society.

Chittenden, Russell
- Nutrition of Man. New York: Stokes, 1907.
1st edn. Sm 4to. 321 pp. Photo ills. Orig
bndg. Sgnd pres slip tipped-in.
(Xerxes) **$175 [≈ £109]**

Chomel, Noel
- Dictionnaire Oeconomique: or, The Family
Dictionary ... done into English ... Revised
and recommended by R. Bradley. London: D.
Midwinter, 1725. 1st English edn. 2 vols.
Folio. Red & black titles. Num w'cut ills.
Contemp panelled calf, hds of spines reprd.
(Spelman) **£800 [≈ $1,280]**
- Dictionnaire Oeconomique: or, The Family
Dictionary ... done into English ... Revised
and recommended by Mr. R. Bradley.
Dublin: J. Watts ..., 1727. 2 vols. Folio. Num
w'cut ills. Contemp calf, sm reprs.
(Jarndyce) **£950 [≈ $1,520]**

Chretien, Charles P.
- An Essay on Logical Method. Oxford &
London: 1848. 1st edn. 8vo. viii,220,16 advt
pp. Orig dark brown cloth, paper label.
(Pickering) **$400 [≈ £250]**

Christie, A.B.

- Infectious Diseases, with Chapters on Venereal Disease. Third Edition. London: 1956. 8vo. 344 pp. 9 plates. Orig cloth.
(Bickersteth) **£18 [≈ $28]**

Christie, H. Kenrick

- Technique and Results of Grafting Skin. New York: 1931. 1st Amer edn. 8vo. xii,67 pp. 31 ills. Orig cloth. *(Elgen)* **$50 [≈ £31]**

Church, A.H.

- The Chemistry of Paints and Painting. Third Edition, revised and enlarged. London: Seeley, 1901. Apparently deluxe edn. 8vo. xx, 355 pp. Hand cold frontis. Orig cloth, untrimmed. *(Charles B. Wood)* **$100 [≈ £62]**

Church, Archibald & Peterson, F.

- Nervous and Mental Diseases. Philadelphia: Saunders, 1901. 3rd edn. Thick 4to. 869 pp. 322 ills. Orig bndg, hinges cracked.
(Xerxes) **$75 [≈ £46]**

Church, William Conant

- The Life of John Ericsson. New and Cheaper Edition. London: Sampson Low, 1892. 2 vols. 2 port frontis, 2 other plates, num text ills. Orig cloth, sl marked, soiled & scratched, 1 hinge cracked but firm. *(Duck)* **£95 [≈ $152]**

Cicero

- M. Tullius Cicero Of the Nature of the Gods; in Three Books. With ... Notes ... added, An Enquiry into the Astronomy and Anatomy of the Ancients. London: for R. Francklin, 1741. 1st edn of this trans. 8vo. [iv], 283,[9] pp. Contemp calf, sl worn.
(Burmester) **£50 [≈ $80]**

The Circle of the Sciences ...

- See Wylde, James (ed.).

Clancey, P.A.

- Gamebirds of Southern Africa. Cape Town: 1967. 2nd edn. Cr 4to. xviii,224 pp. 12 cold plates, 10 maps, 35 figs. Art leather.
(Wheldon & Wesley) **£20 [≈ $32]**
- A Handlist of the Birds of Southern Mocambique. Lourenco Marques: 1971. Roy 8vo. 325 pp. 40 maps, 39 cold plates (numbered 1-40; plate 19 was not published), 6 photos. Inscrptn on title. Orig wraps.
(Wheldon & Wesley) **£40 [≈ $64]**

Clapham, A.R., Tutin, T.G. & Warburg, E.F.

- Flora of the British Isles. London: 1952. 1st

edn. Cr 8vo. liv,1591 pp. 79 figs. Orig green cloth. Dw. *(Henly)* **£15 [≈ $24]**
- Flora of the British Isles. CUP: 1962. 2nd edn. Thick 8vo. Orig cloth gilt. Dw.
(Hollett) **£30 [≈ $48]**

Clapperton, R.H.

- The Paper Making Machine; Its Invention, Evolution and Development. Oxford: Pergamon, 1967. 4to. 365 pp. 186 plates, some fldg. Orig cloth. Dw. *(Gough)* **£50 [≈ $80]**

Clark, Edwin

- The Britannia and Conway Tubular Bridges ... London: 1850. 2 vols 8vo plus folio Atlas of plates. 820 pp. 4 pp subscribers. Map. 2 plans, 6 tinted & 2 b/w lithos. Orig cloth bds. Atlas qtr leather & dec cloth. Atlas partly disbound, 3 or 4 plate edges partly frayed.
(Book House) **£1,100 [≈ $1,760]**

Clark, F. Le Gros

- Lectures on the Principles of Surgical Diagnosis; especially in relation to Shock and Visceral Lesions ... London: Churchill, 1870. 1st edn. 8vo. 345 pp. Text w'cuts. Orig cloth, shelfworn. *(Elgen)* **$75 [≈ £46]**

Clark, George L.

- Applied X-Rays. Third Edition. New York: McGraw-Hill, 1940. Roy 8vo. xviii,674 pp. 342 ills. Orig cloth. *(Duck)* **£20 [≈ $32]**

Clark, J.M. & Rudemann, R.

- The Eurypterida of New York. Albany, New York: New York State Museum Memoir 14, 1912. 2 vols. 4to. 1-439; 440-628 pp. Map, 1 cold & 88 b/w plates, 121 figs. Orig green cloth. *(Gemmary)* **$100 [≈ £62]**

Clark, J.W. & Hughes, T.M.

- Life and Letters of the Reverend Adam Sedgwick. London: 1890. 2 vols. 8vo. xiii, 539; vii,640 pp. 2 frontis. Orig cloth, 1 spine crudely reprd, 1 spine faded.
(Baldwin) **£45 [≈ $72]**

Clark, Samuel

- The British Gauger: or, Trader and Officer's Instructor ... Excise and Customs .... London: for J. Nourse, 1765. 2nd edn. Sm 8vo. [iv],v,[i], 432, *433-*448, 433,[1],x pp. 6 fldg plates (1 hand cold) (sl water stained). Last few ff partly browned. Rec qtr calf.
(Burmester) **£180 [≈ $288]**

Clark, W.E. le G.

- Early Forerunners of Man. A Morphological

Study of the Evolutionary Origin of the Primates. London: 1934. Roy 8vo. xvi,296 pp. 89 ills. Orig cloth, trifle used.
(Wheldon & Wesley) £30 [≃ $48]

Clarke, Sir Arthur
- A Code of Instructions for the Treatment of Sufferers from Railroad and Steam-Boat Accidents, Sudden Attacks of Illness ... Dublin: 1849. 1st edn. 8vo. xvi,176 pp. Orig cloth.   *(Spelman)* £120 [≃ $192]
- The Young Mother's Assistant; or, A Practical Guide for the prevention and treatment of the Diseases of Infants and Children. Second Edition. London: 1828. 12mo. xiii, [1], 176,[8],4 advt pp. Half calf, gilt spine.   *(Hemlock)* $150 [≃ £93]

Clarke, Edward Daniel
- The Life and Remains ... see Otter, W.

Clarke, Louisa Lane
- Objects of the Microscope. London: Groombridge, 1871. 4th edn. 8vo. viii,230,[2 advt] pp. 8 cold plates, text engvs. Orig pict cloth gilt, a.e.g., sm spots to cvrs.
(de Beaumont) £48 [≃ $76]

Clarke, R.Y.
- The Rail and the Electric Telegraph comprising a Brief History of Former Modes of Travelling and Telegraphic Communication. By Peter Progress. London: the author, 1847. Sm 8vo. 60,80 pp. Ills. Some foxing of plates. Orig dec bndg, rebacked.   *(Book House)* £20 [≃ $32]

Clarke, W.E.
- Studies in Bird Migration. London: Gurney & Jackson, 1912. 2 vols. 8vo. Maps, charts, ills. Orig buckram gilt.
(Egglishaw) £40 [≃ $64]

Classes ...
- The Classes and Orders of the Linnaean System of Botany ... see Duppa, R.

Clater, Francis
- Every Man his own Cattle Doctor ... Revised, with some new matter added by D. McTaggart. Halifax: Milner & Sowerby, 1861. Post 8vo. 216,[16 ctlg] pp. Half-title. Engvd frontis. Orig purple cloth gilt, jnts sl rubbed.   *(Blackwell's)* £30 [≃ $48]

Clay, Reginald S. & Court, Thomas H.
- The History of the Microscope. London: 1932. 1st edn. Sm 4to. xiv,266 pp. Errata slip.

164 ills. Orig bndg.   *(Elgen)* $275 [≃ £171]

Clayton, William
- The Theory of Emulsions and their Technical Treatment. Fourth Edition. London: Churchill, 1943. Super roy 8vo. viii,492 pp. 103 ills. Orig maroon cloth, gilt titling sl faded.   *(Duck)* £20 [≃ $32]

Clements, F.
- Blast Furnace Practice. London: Benn, 1929. 3 vols. 4to. 1447 pp. Fldg plates in pockets, ills. Rebound in cloth.
(Book House) £125 [≃ $200]

Clerk, D.
- The Gas, Petrol, and Oil Engine. London: Longmans, 1916-13. 2 vols. 402; 838 pp. Ills. Orig bndgs, not quite uniform.
(Book House) £30 [≃ $48]

Clifford, William Kingdon
- Mathematical Papers. Edited by Robert Tucker ... London: 1882. 1st edn. lxx,658,2 advt pp. 13 plates, litho page of MS. Orig bndg, inner hinges started, hd of rear jnts torn, crnrs worn.   *(Elgen)* $100 [≃ £62]

Clinton-Baker, H.
- Illustrations of Conifers. Hertford: 1909-13. One of 300. 3 vols. 4to. 2 frontis, 229 plates. Rec qtr mor, orig wraps bound in. [With] Illustrations of New Conifers, by Clinton-Baker & A.B. Jackson, 1935. Frontis & 95 plates. Uniformly bound.
(Henly) £600 [≃ $960]
- Illustrations of Conifers. Hertford: 1909-13. One of 300. 3 vols. 4to. 2 frontis, 229 plates from photos. A little minor foxing. Orig bds, trifle used.
(Wheldon & Wesley) £280 [≃ $448]

Clodd, Edward
- Pioneers of Evolution from Thales to Huxley ... London: Grant Richards, 1897. 1st edn. 8vo. [xii],250 pp, advt leaf. 4 ports. Orig cloth, t.e.g., cvrs discold.
(Bickersteth) £20 [≃ $32]

Clymer, R. Swinburne
- Medicines of Nature. The Thomsonian System. Quakertown, Penn.: Humanitarian Society, 1926. 8vo. 224 pp. Sl staining of top marg. Orig bndg, cvrs wrinkled due to water damage.   *(Xerxes)* $95 [≃ £59]

Coale, William Edward
- A Treatise on the Causes, Constitutional

Effects and Treatment of Uterine Displacements. Boston: David Clapp, 1852. 8vo. 52 pp. Some browning. New wraps.
(Xerxes) **$75 [≈ £46]**

Cobbett, William
- The American Gardener ... Stereotype Edition. London: C. Clements, 1821. 12mo. Unpaginated. 4 plates. Blank portion at hd of title torn off. Mrbld bds, rebacked in calf, rubbed. *(Karmiole)* **$85 [≈ £53]**
- Cottage Economy ... First American from the First London Edition. New York: Stephen Gould & Son ..., 1824. 1st Amer edn. Advt leaf after title, 4 pp ctlg at end. Frontis. Orig bds. *(Wreden)* **$65 [≈ £40]**
- Cottage Economy ... New Edition. London: William Cobbett, 1835. 12mo. 200,12 ctlg pp. Orig cloth backed bds, uncut, paper label (defective), soiled & rubbed.
(Claude Cox) **£38 [≈ $60]**
- Cottage Economy ... With a Preface by G.K. Chesterton. London: Peter Davies, 1926. 1st edn thus. x,220 pp. Engvd plate by Eric Gill. Orig cloth backed bds.
(Box of Delights) **£18 [≈ $28]**
- The English Gardener. London: 1829. 1st edn. 12mo. [500] pp. Text figs. Title margs strengthened, crnr of 1 leaf restored with sl loss of text. Sl foxed at beginning & end. Mod half calf, preserving old spine.
(Wheldon & Wesley) **£75 [≈ $120]**
- The Woodlands: or, a Treatise on the preparing of Ground for Planting ... London: William Cobbett, 1825 [-1828]. 1st edn. 8vo. Final advt leaf. A few ills. Half calf by Sangorski & Sutcliffe, uncut.
(Hannas) **£120 [≈ $192]**

Cocchi, Antonio
- The Pythagorean Diet of Vegetables Only, conducive to the Preservation of Health, and the Cure of Diseases ... London: for R. Dodsley, sold by M. Cooper, 1745. 1st edn in English. 8vo. iv,91 pp. Title sl soiled, outer margs v sl affected by damp. Rec bds.
(Burmester) **£275 [≈ $440]**

Cochran-Patrick, R.W.
- Early Records relating to Mining in Scotland. Edinburgh: Douglas, 1878. One of 350. 4to. 205 pp. Half leather, a little worn.
(Book House) **£165 [≈ $264]**

Cochrane, The Hon Basil
- An Improvement in the Mode of Administering the Vapour Bath, and in the Apparatus connected with it ... London: John

Booth, 1809. 4to. [vi],22 pp. 11 plates (5 fldg). Sl spotting or browning. Orig bds, uncut, paper spine renewed, sl wear to sides. Pres copy. *(Duck)* **£250 [≈ $400]**

Cockayne, T.O. (ed.)
- Leechdoms, Wortcunning and Starcraft of Early England ... London: Longman ..., 1864-66. 1st edn. 3 vols. Roy 8vo. 2 cold frontis. Sm lib stamp on titles. Orig half leather, inner hinges cracked, leather v worn esp at hd of spines, bds chipped.
(Elgen) **$300 [≈ £187]**
- Leechdoms, Wortcunning and Starcraft of Early England ... London: 1864-66, reprint 1961. 3 vols. 8vo. Orig cloth.
(Wheldon & Wesley) **£45 [≈ $72]**

Coetlogon, The Chevalier Dennis de
- Natural Sagacity the Principal Secret, if not the Whole in Physick; All Learning, without this, being in effect Nothing ... London: for T. Cooper, 1742. 1st edn. 8vo. iv, 44 pp. Title & last leaf sl dusty. Sewed as issued.
(Bickersteth) **£250 [≈ $400]**

Coffin, A.I.
- Botanic Guide to Health and the Natural Pathology of Disease. London: British Medico- Botanic Establishment, 1852. 25th edn. 8vo. xxiv, 384 pp. Mod half calf gilt. Cutting on pastedown. *(Hollett)* **£50 [≈ $80]**

Cohausen, Johann Heinrich
- Hermippus Redivivus: or, the Sage's Triumph over Old Age and the Grave. Second Edition, carefully corrected and much enlarged. London: J. Nourse, 1749. 8vo. [8], 248 pp. Minor discoloration front endpaper & ft of title. Rebound in half calf.
(Spelman) **£140 [≈ $224]**
- Hermippus Redivivus: or the Sage's Triumph over Old Age and the Grave ... Second Edition, carefully corrected and much enlarged. London: J. Nourse, 1749. 8vo. [8], 248 pp. Old repr to title. Contemp half calf, upper cvr held by cords, lower hinge cracked.
(Claude Cox) **£60 [≈ $96]**

Colbatch, J.
- A Dissertation concerning Misletoe, a most wonderful Specifick Remedy for the Cure of Convulsive Distempers. London: [ca 1730]. 6th edn. 8vo. 72 pp. Contemp calf.
(Wheldon & Wesley) **£60 [≈ $96]**

Cole, A.C.
- The Methods of Microscopical Research. A Practical Guide to Microscopical

Manipulation. London: Bailliere, Tindall & Cox, 1895. 2nd edn, enlgd & rewritten. 8vo. viii, 207,6 advt,41 ctlg pp. A few marg ink notes. Orig cloth, spine ends & crnrs worn.
(Savona) **£25 [≃ $40]**

Cole, G.A.J.
- Aids in Practical Geology. London: 1891. 1st edn. Cr 8vo. xiv,402,40 advt pp. 136 text figs. Orig cloth. *(Henly)* **£15 [≃ $24]**
- Open Air Studies. An Introduction to Geology Out-of-Doors. London: 1895. 1st edn. 8vo. xii,322,32 advt pp. 12 plates, 33 text ills. Lib stamp on title verso & at end. Orig cloth. *(Henly)* **£14 [≃ $22]**

Cole, R.J.
- A History of Comparative Anatomy, from Aristotle to the Eighteenth Century. London: 1944. 8vo. viii,524 pp. Num ills. Paper sl cockled. Orig cloth, sl used, inner jnt reprd with tape. *(Wheldon & Wesley)* **£35 [≃ $56]**

Cole, William Henry
- Light Railways at Home and Abroad. London: Charles Griffin: 1899. 1st edn. xii,399,[1] pp. 9 fldg plates, 2 fldg appendices, frontis, text ills. Near contemp prize half mor gilt, t.e.g., 2 jnt ends v sl rubbed. *(Duck)* **£85 [≃ $136]**

Coleman, W.S.
- British Butterflies. Figures and Descriptions of Every Native Species. London: Routledge, 1863. New edn. Sm 8vo. vii,179 pp. 14 cold & 2 plain plates. Orig cloth gilt.
(Egglishaw) **£14 [≃ $22]**

Collenette, C.L.
- Sea-Girt Jungles, Experiences of a Naturalist with the 'St. George' Expedition (Pacific). London: [1926]. 8vo. 275 pp. 36 ills. Cloth, used. *(Wheldon & Wesley)* **£30 [≃ $48]**

Collinge, W.E.
- Food of Some British Wild Birds. York: 1924-27. 2nd edn, rvsd & enlgd. 4to. xxii,427 pp. Port, 8 b/w plates, 47 diags. Half mor.
(Gough) **£40 [≃ $64]**

Collins, Joseph
- The Doctor Looks at Literature. Psychological Studies of Life and Letters. New York: Doran, 1923. 1st edn. 8vo. 317 pp. Orig bndg. *(Xerxes)* **$40 [≃ £25]**

Collins, Samuel
- Paradise Retriev'd: plainly and fully

demonstrating the ... Method of Managing and Improving Fruit-Trees ... With a Treatise on Mellons and Cucumbers. London: John Collins, Seedsman, 1717. 8vo. [2],v,[5],[5], 6-106 pp. 2 fldg plates. Contemp calf, rebacked. *(Spelman)* **£180 [≃ $288]**

Colloquia Chirurgica ...
- See Handley, James.

Colquhoun, Patrick
- A Treatise on Indigence; Exhibiting a General View of the National Resources for Productive Labour ... London: Hatchard, 1806. xii, 302,index pp, advt leaf. Fldg chart at p 23. Title v sl worn at ft. Occas sl foxing. Rebound in contemp style half leather.
(Moon) **£125 [≃ $200]**
- A Treatise on the Police of the Metropolis ... London: H. Baldwin ..., 1800. 6th edn. 8vo. [xvi],xvi,655,31 pp. Half-title. 2 fldg tables. Light tan mor, mrbld sides & edges.
(Young's) **£130 [≃ $208]**

Colson, Nathaniel
- The Mariners New Kalendar. Containing the Principles of Arithmetick and Geometry ... Directions for Sailing into Some Principal Harbours. London: Thomas Page, William & Fisher Mount, 1724. 132 pp. W'cuts, tables. Some stains. Contemp calf.
(Karmiole) **$300 [≃ £187]**

Columella, L.J.M.
- Of Husbandry, in Twelve Books: and his Book concerning Trees, translated into English, with several Illustrations from Pliny, Cato, Varro, Palladius, and other Antient and Modern Authors. London: 1745. 4to. xiv,[xiv], 600,[8] pp. Contemp calf, rebacked.
(Wheldon & Wesley) **£130 [≃ $208]**

Combe, Andrew
- The Physiology of Digestion considered with relation to the Principles of Dietics ... Edinburgh: MacLachlan & Stewart ..., 1836. 2nd edn, rvsd & enlgd. 8vo. xxviii, 350, [8] pp. Half-title. Orig cloth backed bds, label chipped. *(Young's)* **£180 [≃ $288]**

Combe, George
- Elements of Phrenology. Edinburgh: John Anderson, 1828. 3rd edn, enlgd. Cr 8vo. Fldg frontis (backed), 1 plate (browned). Contemp half calf. *(Stewart)* **£185 [≃ $296]**

Comer, George N.
- A Simple Method of Keeping Books, by Double-Entry, without the Formula or

Trouble of the Journal ... Sixth Edition. Boston: Tappan, Whittemore, & Mason, 1850. 8vo. 104, [2 advt],iv,[2 advt] pp. 4 pp prospectus on endpaper. Orig cloth, sl soiled.
(Pickering) **$600 [≈ £375]**

Commerell, Abbe de
- An Account of the Culture and Use of the Mangel Wurzel, or Root of Scarcity. Translated from the French. London: Dilly & Phillips, 1787. 3rd edn in English. 8vo. xxxix, [i],51 pp. Hand cold frontis (shaved at ft). Old wraps.   *(Burmester)* **£60 [≈ $96]**

Companion to the Botanical Magazine ...
- See Hooker, Sir William Jackson.

The Complete Brewer ...
- The Complete Brewer, or the Art and Mystery of Brewing Explained ... London: Coote, 1760. Sm 8vo. Soiled. Calf, dusty & worn, upper jnt split, hinges cracked.
(Marlborough) **£130 [≈ $208]**

The Complete Grazier ...
- The Complete Grazier, or, Gentleman and Farmer's Directory ... written by a Country Gentleman, and Originally Designed for Private Use. London: 1776. 4th edn. 8vo. viii, 252,xv pp. Contemp calf, jnts cracked, trifle worn.   *(Wheldon & Wesley)* **£50 [≈ $80]**

Comstock, J.L.
- An Introduction to Mineralogy. Hartford: H.F. Sumner & Co, 1833. 2nd edn. Cr 8vo. 339 pp. Ills. New cloth.
(Gemmary) **$150 [≈ £93]**
- An Introduction to Mineralogy. New York: Robinson, Pratt, & Co, 1836. 3rd edn. Cr 8vo. 339 pp. Ills. Leather, worn.
(Gemmary) **$150 [≈ £93]**
- An Introduction to Mineralogy. New York: Pratt, Woodford & Co, 1845. 12th edn. Cr 8vo. 369 pp. Ills. New cloth.
(Gemmary) **$150 [≈ £93]**
- The Young Botanist: being a Treatise on the Science, prepared for the use of Persons just commencing the Study of Plants. New York: 1835. 8vo. x,9-259 pp. Hand cold frontis, 191 text figs. Sl foxing. Orig cloth, worn, jnts split.   *(Wheldon & Wesley)* **£33 [≈ $52]**

Comstock, W.P.
- Butterflies of the American Tropics, the Genus Anea (Lepidoptera Nymphalidae). New York: Amer Museum Nat Hist, 1961. 4to. xiii, 214 pp. 30 cold plates. Buckram.
(Wheldon & Wesley) **£50 [≈ $80]**

Comte, A.
- The Book of Birds, edited and abridged from the Text of Buffon ... Translated by B. Clarke. London: 1841. Roy 8vo. xxxiv,292 pp. 38 hand cold plates. Some v sl foxing. Mor gilt. *(Wheldon & Wesley)* **£150 [≈ $240]**

Conant, Clarence
- An Obstetric Mentor. A Handbook of Homeopathic Treatment required during Pregnancy, Parturition and the Puerperal Season. New York: Chatterton, 1884. 12mo. 212 pp. Sl browning. Orig bndg, sl dampstained.   *(Xerxes)* **$135 [≈ £84]**

A Concise History ...
- A Concise History of Steam Carriages on Common Turnpike Roads, and the Progress of their Improvement. London: 1834. 33,[1] pp, advt leaf (creased). Some browning & spotting. Orig ptd yellow wraps with vignette, wraps soiled & worn. *(Duck)* **£225 [≈ $360]**

Conder, Josiah, architect
- The Flowers of Japan and the Art of Floral Arrangement. Tokyo: 1891. Folio. 14 cold & 40 plain plates, num text figs. Half calf, orig pict wraps bound in.
(Wheldon & Wesley) **£200 [≈ $320]**
- Landscape Gardening in Japan, and Supplement. Tokyo: 1893. 2 vols. Imperial 4to. 77 plates, 55 text ills. Perf lib stamp on Supplement title. Orig dec cloth, Supplement bndg somewhat spotted & reprd.
(Wheldon & Wesley) **£200 [≈ $320]**

Conversations ...
- Conversations on Botany ... see Fitton, E. & S.M.
- Conversations on Chemistry ... see Marcet, Jane.
- Conversations on Geology; comprising a Familiar Explanation of the Huttonian and Wernerian Systems ... London: for Samuel Maunder, 1828. 8vo. xxii,[ii],371 pp. 12 engvd or w'cut plates (4 hand cold). Orig cloth backed bds, sl worn, label rubbed away.
(Burmester) **£175 [≈ $280]**

Conway, Herbert & Stark, Richard B.
- Plastic Surgery at the New York Hospital One Hundred Years Ago, with Biographical Notes on Gurdon Buck ... New York: 1953. 1st edn. 8vo. 110 pp. Frontis, num ills. Orig cloth.   *(Elgen)* **$75 [≈ £46]**

The Cook's Oracle ...
- See Kitchiner, William.

Cook, Sir Edward
- The Life of Florence Nightingale. London: 1913. 2 vols. 8vo. 6 ports, facs. Random spots, hd of a few ff sl bumped. Orig cloth, t.e.g.   *(Bow Windows)* **£75 [≈ $120]**

Cooke, M.C.
- British Desmids, a Supplement to British Freshwater Algae. London: 1887. 8vo. xiv,205 pp. 66 cold plates. Orig cloth, trifle worn, rebacked. *(Wheldon & Wesley)* **£70 [≈ $112]**
- British Edible Fungi. London: 1891. 8vo. 237 pp. 12 cold plates. Orig cloth.
         (Wheldon & Wesley) **£18 [≈ $28]**
- British Fresh-Water Algae ... [with] British Desmids, a Supplement to British Fresh-Water Algae. London: 1882-84-87. Together 3 vols. 8vo. 130 + 66 cold plates. Half calf, t.e.g, & orig cloth, t.e.g.
         (Egglishaw) **£190 [≈ $304]**
- Fungoid Pests of Cultivated Plants. London: RHS, 1906. 8vo. xv,278 pp. 24 plates (22 cold). Buckram. Prof. J.H. Burnett's copy.
         (Egglishaw) **£18 [≈ $28]**
- Illustrations   of   British   Fungi (Hymenomycetes).   London: 1881-91. Incomplete set. 8 vols. 8vo. 846 (of 1198) cold plates. Vols 1 & 2 new cloth, the remainder unbound. Vols 1 & 2 are complete, while vols 6-8 lack only 49 plates.
         (Wheldon & Wesley) **£700 [≈ $1,120]**
- A Plain and Easy Account of British Fungi. London: Robert Hardwicke, 1862. Sm 8vo. vi, 148, [6 advt] pp. 24 hand cold plates. Text sl foxed. Orig dec cloth gilt.
         (Hollett) **£30 [≈ $48]**

Coolidge, J.L.
- A Treatise on the Circle and the Sphere. Oxford: 1916. 602 pp. A few figs. Orig cloth, sl worn.   *(Whitehart)* **£25 [≈ $40]**

Coolidge, Richard H.
- Statistical Report on the Sickness and Mortality in the Army of the United States, 1839-1855. Washington: 1856. Lge folio. 703 pp. Fldg map.   *(Jenkins)* **£485 [≈ $303]**

Cooper, A.
- The Complete Distiller ... London: for P. Vaillant ..., 1757. 1st edn. 8vo. [xiv],266, [14] pp. Fldg plate. Lib stamp on title verso. Old calf, front jnt partly cracked.
         (Young's) **£180 [≈ $288]**
- The Complete Distiller ... London: for J. Hamilton, 1797. 8vo. x,259,[12] pp. Fldg plate. Orig bds, uncut, spine paper defective, a little worn.   *(Hollett)* **£150 [≈ $240]**

Cooper, C.S. & Westell, W.P.
- Trees and Shrubs of the British Isles, Native and Acclimatised. London: 1909. 2 vols. 4to. 16 cold & 70 plain plates. Edges of cold plates sl foxed. Orig cloth gilt. *(Henly)* **£45 [≈ $72]**

Cope, E.D.
- The Crocodilians, Lizards, and Snakes of North America. Washington: 1900. 8vo. 1139 pp. 36 plates, 347 text figs. A few blind stamps. Cloth.
         (Wheldon & Wesley) **£70 [≈ $112]**
- The Vertebrata of the Cretaceous Formations of the West. Washington: 1875. 4to. iv,302 pp. 57 plates. Orig cloth, rebacked. Vol 2 of Hayden's Report.
         (Wheldon & Wesley) **£100 [≈ $160]**
- The Vertebrata of the Tertiary Formations of the West. Book 1. Washington: 1884. 4to. xxxiv, 1009 pp. Fldg table, 133 plates (of 134, lacks plate 72a). New cloth. Vol 3 of Hayden's Report. *(Wheldon & Wesley)* **£100 [≈ $160]**

Copley, Esther
- The Housekeeper's Guide, or, A Plain and Practical System of Domestic Cookery. London: Jackson & Walford, 1834. xi,[4 plates],407 pp. Engvd frontis, addtnl engvd title. New bds, buckram spine, old endpapers with MS receipts retained.
         (Box of Delights) **£100 [≈ $160]**

Cordell, Eugene Fauntleroy
- The Medical Annals of Maryland 1799-1899 ... Baltimore: Williams & Wilkins, 1903. 1st edn. 8vo. [iv],889,[5] pp. 32 plates. Orig ptd green cloth.   *(Gach)* **$50 [≈ £31]**

Coriat, Isador Henry
- The Meaning of Dreams. Boston: Little, Brown, 1915. 1st edn. Sm 8vo. [xiv],194,[6] pp. Orig ptd green cloth. Annotated by Boris Sidis.   *(Gach)* **$85 [≈ £53]**

Cornaro, Lewis
- Discourses on a Sober and Temperate Life. Wherein is Demonstrated by his own Example, the Method of Preserving Health to Extreme Old Age ... London: for Benjamin White, 1779. New edn. 8vo. xii,188 pp. Contemp calf, a little worn, front jnt weakening.   *(Young's)* **£30 [≈ $48]**
- Sure and Certain Methods of attaining a Long and Healthful Life: With Means of Correcting a Bad Constitution, &c. ... London: for Daniel Midwinter ..., 1727. 4th edn. 12mo. xl,120 pp. New calf.
         (Young's) **£65 [≈ $104]**

Cornish, C.J.
- Animals at Work and Play: their Activities and Emotions. London: 1896. 1st edn. 8vo. ix, [2], 323,16 advt pp. 12 plates. Orig cloth gilt.   *(Fenning)* **£24.50 [≈$40]**

Correvon, H.
- Rock Garden and Alpine Plants. Edited by H. Barron. New York: 1930. 8vo. xv,544 pp. 8 cold & 8 plain plates. Orig cloth.
           (Wheldon & Wesley) **£25 [≈$40]**

Correvon, H. & Robert, P.
- The Alpine Flora. Translated by E.W. Clayforth. Geneva: [1912]. 8vo. 436 pp. 1 plain & 100 cold plates. Orig cloth.
           (Wheldon & Wesley) **£25 [≈$40]**

Corson, Hiram
- Claims of Literary Culture. An Address Introductory to the Preliminary Course in the Hahnemann Medical College of Philadelphia. Philadelphia: Hahnemann College, 1875. 8vo. 48 pp. Cardboard cvr.
           (Xerxes) **$75 [≈£46]**

Costa, E.M. da
- Elements of Conchology: or an Introduction to the Knowledge of Shells. London: 1776. 8vo. viii,iii-vi,318,[1] pp. 7 hand cold plates. New half calf, antique style.
           (Wheldon & Wesley) **£150 [≈$240]**

Costard, George
- The History of Astronomy, with its Application to Geography, History and Chronology ... London: James Lister, 1767. 1st edn. 4to. xvi,308 pp, errata leaf. W'cut ills, overlays. Orig polished tree calf gilt, rear bd detached.   *(Elgen)* **£300 [≈£187]**

Cotes, Roger
- Hydrostatical and Pneumatical Lectures ... Published with Notes by his Successor Robert Smith ... London: for the editor, & S. Austen ..., 1738. 1st edn. 8vo. [xvi],243,[1 errata], [6],[4 advt] pp. 5 fldg plates. Contemp calf, rubbed, jnts sl worn.
           (Pickering) **$950 [≈£593]**

Cotton, Charles
- The Planters Manual ... [Translated by Charles Cotton]. London: Henry Brome, 1675. 1st English edn. Sm 8vo. [6],139,[1 blank],[4 ctlg] pp. Addtnl engvd title. Sl marg soiling. Early reprs to B4. Contemp sheep, rebacked, crnrs sl worn.
           (Spelman) **£600 [≈$960]**

- The Planters Manual ... [Translated from the French of R. Triquet by Charles Cotton]. London: Henry Brome, 1675. Sm 8vo. [vi], 139, [4 advt] pp. Addtnl engvd title (cropped at ft & foredge). 18th c half green sheep, rubbed, sl soiled. Wing C.6388.
           (Blackwell's) **£300 [≈$480]**

- The Planters Manual ... London: for Henry Brome, 1675. 1st edn. Sm 8vo. [vi],139,[1 blank], [4 advt] pp. Addtnl engvd title. Port bound in after title. 19th c polished calf, upper jnt cracked. Wing C.6388.
           (Pickering) **$1,000 [≈£625]**

Cotton, W.C.
- A Short and Simple Letter to Cottagers from a Conservative Bee-Keeper, with Other Letters. Penzance: 1838. 2nd edn. 8vo. 24 pp. 2 plates. Wraps.
           (Wheldon & Wesley) **£25 [≈$40]**

Couch, Jonathan
- A History of the Fishes of the British Islands. London: Groombridge, 1862-65. 4 vols. Roy 8vo. 252 cold plates. colour ptd by Fawcett of Driffield, finished by hand. Orig pict blue cloth gilt, back cvr vol 4 marked, hd of spines sl worn.   *(Egglishaw)* **£370 [≈$592]**
- A History of the Fishes of the British Islands. London: Groombridge & Sons, 1866-67. 2nd edn. Vol 2 1st edn, dated 1863. 4 vols. Lge 8vo. 252 hand cold plates, text ills. Insignificant foxing to a few plates. Half mor, t.e.g., barely rubbed. Slipcase.
           (Heritage) **$1,750 [≈£1,093]**
- A History of the Fishes of the British Isles. London: 1877. 4 vols. Roy 8vo. 252 cold plates. Vol 1 foxed at start. Orig cloth, refixed, trifle worn, vols 1 & 2 trifle worn.
           (Wheldon & Wesley) **£300 [≈$480]**
- A History of the Fishes of the British Isles. London: George Bell, 1877. 4 vols. Lge 8vo. Chromolitho plates by Fawcett of Driffield, finished by hand. Some spotting. Orig cloth gilt.   *(Hollett)* **£350 [≈$560]**
- Illustrations of Instinct deduced from the Habits of British Animals. London: 1847. 1st edn. 8vo. xii,343,4 advt pp. Prelims & endpapers sl browned. Orig cloth gilt.
           (Hollett) **£55 [≈$88]**

Coues, E.
- Birds of the Colorado Valley. Part 1, Passeres to Laniidae, with Bibliographical Appendix. Washington: 1878. All published. 8vo. xvi,807 pp. 70 figs. Orig cloth, rebacked.
           (Wheldon & Wesley) **£40 [≈$64]**
- Birds of the Northwest. A Handbook of the

Ornithology of the Region drained by the
Missouri River and its Tributaries.
Washington: 1874. 8vo. xi,791 pp. Orig
cloth, spine sl worn.
                    (Wheldon & Wesley) **£20 [≈ $32]**

Couper, Robert
- Speculations on the Mode and Appearances
of Impregnation in the Human Female ...
Edinburgh: C. Elliot; London: C. Elliot & T.
Kay, 1789. 1st edn. 8vo. 149,[1] pp. Qtr calf
over mrbld bds.   *(Rootenberg)* **$550 [≈ £343]**

Court, W.H.B.
- The Rise of the Midland Industries
1600-1838. OUP: 1953. 2nd edn. 271 pp. Ills.
A few marg notes. Orig bndg.
                    (Book House) **£25 [≈ $40]**
- The Rise of the Midland Industries
1600-1838. OUP: 1953. 2nd edn. 271 pp. Ills.
Ex-lib, lacks fly. Orig bndg. Dw.
                    (Book House) **£15 [≈ $24]**

Cousins, Frank
- Sundials. A Simplified Approach by Means of
the Equatorial Dial. Foreword by J.G. Porter.
London: John Barker, 1969. 8vo. 247 pp. Ills.
Orig cloth gilt.   *(Blackwell's)* **£55 [≈ $88]**

Coventry, B.O.
- Wild Flowers of Kashmir. London: 1923-30.
Series 1-3. 3 vols. Cr 8vo. 150 cold plates.
Orig pict cloth gilt, spines sunned.
                    (Farahar) **£75 [≈ $120]**

Coward, T.A.
- Birds of the British Isles and their Eggs.
Edited and Revised by J.A.G. Barnes.
London: 1969. 8vo. xvi,359 pp. 161 plates of
birds (many cold), 16 cold plates of eggs. Orig
cloth.   *(Wheldon & Wesley)* **£20 [≈ $32]**

Coward, T.A. & Oldham, C.
- The Vertebrate Fauna of Cheshire and
Liverpool Bay. London: Witherby, 1910. 1st
edn. 2 vols. 8vo. Fldg map, 55 ills. Orig cloth
gilt, spines browned. *(Hollett)* **£75 [≈ $120]**

Cowper, William, 1666-1709
- Myotomia Reformata: Or an Anatomical
Treatise on the Muscles of the Human Body
... London: for Robert Knaplock ..., 1724. 1st
folio edn. Folio. [10],lxxvii,194,[4] pp. Engvd
frontis & 66 plates, plus 1 in outline, text ills.
Contemp MS notes. Elab gilt dec calf.
                    (Rootenberg) **$5,500 [≈ £3,437]**

Cox, F. Augustus
- The Care of the Skin in Health and Disease.
London: Alexander & Shepherd, 1890. 56 pp.
Orig red cloth.   *(C.R. Johnson)* **£35 [≈ $56]**

Cox, Herbert E.
- Handbook of the Coleoptera or Beetles of
Great Britain and Ireland. London: 1874. 2
vols. 8vo. Orig cloth, trifle faded & used.
                    (Wheldon & Wesley) **£30 [≈ $48]**
- Handbook of the Coleoptera or Beetles of
Great Britain and Ireland. London: E.W.
Janson, 1874. 1st edn. 2 vols. 8vo. viii,527;
[2],366 pp. Orig cloth. *(Fenning)* **£35 [≈ $56]**

Coxe, John Redman
- The Philadelphia Medical Dictionary ...
Philadelphia: Thomas Dobson, 1808. 1st edn.
8vo. viii,[466] pp. Contemp calf, front bd
detached.   *(Gach)* **$75 [≈ £46]**
- The Philadelphia Medical Dictionary ...
Philadelphia: Dobson, 1817. 2nd edn. 8vo.
433, 2 advt pp. Sl stained throughout. Lacks
free fly. Leather.   *(Elgen)* **$50 [≈ £31]**

Crace-Calvert, Frederick
- Dyeing and Calico Printing ... Manchester,
London & New York: 1876. 1st edn. 8vo.
xxxii, 509,iii pp. Fldg table, 14 w'engvd ills,
56 mtd fabric samples. Orig cloth, inner
hinges cracked.
                    (Charles B. Wood) **$300 [≈ £187]**

Cramer, John Andrew
- Elements of the Art of Assaying Metals. In
Two Parts ... Translated from the Latin ...
added, Several Notes and Observations ...
London: 1741. 1st edn in English. 8vo. [xii],
1-204, half-title, 203-208, [205]-470,[viii] pp.
6 plates. Orig calf, sm crack in one jnt.
                    (Bickersteth) **£360 [≈ $576]**

Craufurd, George
- The Doctrine of Equivalents; or an
Explanation of the Nature, Value and Power
of Money ... in Two Parts. Rotterdam: Locke
& Co, 1803. 1st complete edn. 8vo. [viii],280
pp. Contemp half roan, jnts cracked, spine
rubbed & worn at ends, extremities worn.
                    (Pickering) **$1,250 [≈ £781]**

Crawfurd, O.
- A Year of Sport and Natural History.
Shooting, Hunting, Coursing, Falconry and
Fishing. London: 1895. Lge 4to. Num plates,
text ills. Orig cloth, sl rubbed.
                    (Grayling) **£25 [≈ $40]**

Crawshay, R.
- The Birds of Tierra del Fuego. London: 1907. One of 300. Imperial 8vo. xl,158 pp. Map, 21 hand cold plates by Keulemans, 23 photo plates. Half mor.
(Wheldon & Wesley) **£550 [≈ $880]**

Cresy, Edward
- An Encyclopaedia of Civil Engineering, Historical, Theoretical, and Practical. London: Longman ..., 1847. 1st edn. 2 vols. 8vo. 1655 pp. 3057 w'engvd ills. Orig polished half calf.
(Charles B. Wood) **$225 [≈ £140]**

Creyke, W.R.
- Book of Modern Receipts containing Full Instructions for producing all Kinds of Enamel, Underglaze, & Majolica Colours ... Hanley: J. Hitchings, 1884. Sm 8vo. 96 pp. Orig green cloth gilt. *(Spelman)* **£60 [≈ $96]**

Crile, George W.
- Diseases Peculiar to Civilized Man ... New York: 1934. 1st edn. 8vo. 427 pp. Fldg table, ills. Orig cloth, sl soiled. *(Elgen)* **$30 [≈ £18]**
- Hemorrhage and Transfusion. New York: 1909. 1st edn. 8vo. 560 pp. Ills. Orig cloth, inner hinges cracked. *(Elgen)* **$175 [≈ £109]**
- The Origin and Nature of the Emotions; Miscellaneous Papers ... Philadelphia: 1915. 1st edn. 8vo. 240 pp. Plates, ills. Title & last page foxed. Orig cloth. *(Elgen)* **$30 [≈ £18]**

Cripps-Day, F.H.
- The Manor Farm, to which are added The Boke of Husbandry and The Booke of Thrift. London: 1931. 4to. xxxviii,114,66 pp. Bds.
(Wheldon & Wesley) **£45 [≈ $72]**

Crisp, F.
- Mediaeval Gardens, "Flowery Medes" and other Arrangements of Herbs, Flowers, and Shrubs, grown in the Middle Ages ... London: 1924. One of 1000. 2 vols. 4to. 562 ills. Endpapers & prelims foxed. Orig white buckram (almost mint). Dws.
(Wheldon & Wesley) **£200 [≈ $320]**

Critchley, MacDonald
- The Parietal Lobes. London: E. Arnold, [1953]. 1st edn. 8vo. 480 pp. Cold frontis, num text ills. Orig cloth.
(Elgen) **$175 [≈ £109]**

Crocker, A.
- The Elements of Land-Surveying ... London: for Richard Phillips ..., 1806. Sm 8vo. xii,

279,[9 advt]' pp. Fldg hand cold frontis (sl torn), title vignette, 19 plates. Tree calf gilt, spine ends sl defective, lower hinge cracking.
(Hollett) **£120 [≈ $192]**

Crookes, W. & Rohrig, W.
- A Practical Treatise on Metallurgy. London: Longmans, Green, 1868. Vol 1 only. 8vo. xxviii,724 pp. 207 text figs. Orig cloth.
(Gemmary) **$125 [≈ £78]**

Crookes, William
- A Practical Handbook of Dyeing and Calico Printing. London: Longmans, Green, 1874. 1st edn. Thick 8vo. xvi,730,[1],28 pp. 11 plates, 36 w'engvd ills, 47 mtd samples. Orig qtr cloth, rebacked.
(Charles B. Wood) **$350 [≈ £218]**

Crothers, T.D.
- Morphinism and Narcomanias from Other Drugs. Their Etiology, Treatment and Medico-Legal Relations. Philadelphia: Saunders, 1902. 1st edn. 8vo. 351 pp. Orig bndg. *(Xerxes)* **$75 [≈ £46]**

Crouch, E.A.
- An Illustrated Introduction to Lamarck's Conchology; contained in his Histoire Naturelle des Animaux sans Vertebres. London: 1826. 4to. iv,47 pp. 22 hand cold plates. 1 plate marg stained, 1 sm marg tear reprd. Sm blind stamps on plates. Cloth.
(Wheldon & Wesley) **£220 [≈ $352]**
- An Illustrated Introduction to Lamarck's Conchology ... London: 1827. 4to. iv,47 pp. 22 litho plates (3 sl discold). Orig cloth, spine ends sl worn. *(Egglishaw)* **£36 [≈ $57]**

Crowder, W.
- A Naturalist at the Seashore. New York: 1928. 1st edn. Roy 8vo. 8 cold & num other plates. Orig cloth, spine faded.
(Henly) **£20 [≈ $32]**

Cudworth, William
- Life and Correspondence of Abraham Sharp, the Yorkshire Mathematician and Astronomer ... London: Sampson Low, 1889. xvi,342 pp. Fldg table, ills, facs. Newly bound, calf spine, gilt label, elephant hide vellum sides. *(Box of Delights)* **£48 [≈ $76]**
- Life and Correspondence of Abraham Sharp, the Yorkshire Mathematician and Astronomer, and Assistant of Flamsteed ... London: 1889. 1st edn. Large Paper. 8vo. xvi,342 pp. Frontis, 5 plates, 2 tables, text w'cuts. Three qtr vellum, t.e.g, some soiling & shelfwear. *(Elgen)* **$275 [≈ £171]**

Cuffe, Robert
- The Woodhall or Iodine Spa, Lincolnshire. Second Edition. London: 1868. 8vo. 27 pp. Frontis. Disbound. Author's pres inscrptn.
(Bickersteth) **£25 [≈ $40]**

Cullen, Thomas Stephen
- Cancer of the Uterus. Also, The Pathology of Diseases of the Endometrium. New York: 1900. 1st edn. Lge 8vo. 693 pp. 11 cold litho plates, ptd tissue guards, over 300 text ills. Orig cloth, worn, upper jnt torn, inner hinge cracked, ex-lib.
(Elgen) **$85 [≈ £53]**

Cullen, William
- Lectures on the Materia Medica ... Philadelphia: for the subscribers by Robert Bell, 1775. 1st Amer edn. Lge 4to. viii,512 pp. Some browning. Mod cloth.
(Hemlock) **$500 [≈ £312]**

Culpeper, Nicholas
- The Complete Herbal ... annexed, the English Physician Enlarged, and Key to Physic ... New Edition. London: Thomas Kelly, 1835. 4to. 398,[4] pp. Frontis port, 20 hand cold plates. Occas sl soiling & browning. Mod half calf gilt.
(Hollett) **£150 [≈ $240]**
- The Complete Herbal, a New Edition ... to which are now first annexed The English Physician and Key to Physic. Birmingham: Kynoch Press for Imperial Chemical (Pharmaceutical) Ltd, 1953. Roy 8vo. x,603 pp. Port, 16 cold plates. Half mor.
(Wheldon & Wesley) **£40 [≈ $64]**
- The English Physician Enlarged With Three Hundred and Sixty-Nine Medicines, made of English Herbs ... London: 1752. 12mo. 387 pp. Somewhat foxed. Mod wraps.
(Argosy) **$100 [≈ £62]**
- The English Physician Enlarged, with 369 Medicines made of English Herbs. London: 1788. 12mo. xii,371 pp. Antique style calf.
(Henly) **£90 [≈ $144]**
- Culpeper's English Physician and Complete Herbal ... Fourteenth Edition. Edited by Sibly. London: J. Adlard, 1810. 4to. xvi,400 pp.
(Gough) **£250 [≈ $400]**
- Culpeper's Last Legacy ... London: N. Brooke, 1657. 3 parts in one vol. 8vo. Port frontis. Contemp calf, inner rear hinge loose. Wing C.7519.
(Hemlock) **$350 [≈ £218]**
- Semeiotica Uranica: or, an Astrological Judgement of Diseases from the Decumbiture of the Sick, much enlarged ... Third Edition. London: Nath. Brooke, 1658. 8vo. [15],'224" [≈ 234],[12] pp. Sep title to'Urinalia". Port frontis. Text diags. Contemp calf. Wing

C748.   *(Hemlock)* **$500 [≈ £312]**

Cumberland, Richard
- A Philosophical Enquiry into the Laws of Nature ... Translated into English, with large Explanatory Notes, and an Appendix, by the Reverend John Towers. Dublin: 1750. 4to. lxxxvi, 597, [ii],134 pp. [7] pp subscribers after Preface. Orig calf, rebacked.
(Bickersteth) **£185 [≈ $296]**

Cumming, Alexander
- The Elements of Clock and Watch-Work adapted to Practice. In Two Essays ... London: for the author ..., 1766. 1st edn. 4to. [viii],192,[11 index],[1 errata],[1] pp. 16 fldg plates. Text sl foxed. Rec contemp style calf. Percy Webster's copy with note by him.
(Pickering) **$2,000 [≈ £1,250]**

Cunningham, Brysson
- A Treatise on the Principles and Practice of Harbour Engineering. London: Charles Griffin, 1908. 1st edn. Roy 8vo. xii,284 pp. Fldg maps, plates, ills. Orig gilt dec roy blue cloth. Publisher's pres copy.
(Duck) **£35 [≈ $56]**

Cunningham, J.T.
- The Natural History of the Marketable Marine Fishes of the British Islands. London: 1896. 8vo. xvi,375 pp. 2 maps, 159 text figs. Some sl stains in lower margs. New cloth.
(Wheldon & Wesley) **£18 [≈ $28]**

Curiosities of Ornithology ...
- Curiosities of Ornithology. With Beautifully Coloured Illustrations from Drawings by T.W. Wood and other Eminent Artists. L. Groombridge, [ca 1880]. 8vo. 64 pp. 10 chromolitho plates. Some foxing to outer pp. Orig cloth gilt, a.e.g., extremities worn, hinges tender.
(Bromer) **£125 [≈ $78]**

Curling, T.B.
- A Practical Treatise on the Diseases of the Testis, and of the Spermatic Cord and Scrotum. Philadelphia: Carey & Hart, 1843. 8vo. xxii,568 pp. Num ills. Occas sl spotting. Contemp calf.
(Hemlock) **$100 [≈ £62]**

Curr, John
- The Coal Viewer and Engine Builder's Companion. Sheffield: for the author, by John Northall, 1797. Sole edn. Cr 4to. 96 pp. 5 fldg plates. Marking & soiling. Probably contemp half calf, rebacked & recrnrd, rear endpapers soiled, back flyleaf partly wormed.
(Duck) **£825 [≈ $1,320]**

Curtis, C.H.

- Orchids, their Description and Cultivation. London: 1950. Cr 4to. 288 pp. 30 cold & 48 plain plates. Orig cloth, faded.
  (Wheldon & Wesley) **£25 [≈ $40]**
- Orchids, their Distribution and Cultivation. London: 1950. 1st edn. 4to. viii, 274 pp. 30 cold & 48 plain plates. Orig cloth, new endpapers. Dw.
  (Henly) **£28 [≈ $44]**

Curtis, J.T. & Lillie, J.

- Epitome of Homeopathic Practice. London: James Leath, 1850. New edn. 16mo. 215 pp. Orig bndg.
  (Xerxes) **$125 [≈ £78]**

Curtis, John

- British Entomology. Being Illustrations and Descriptions of the General of Insects found in Great Britain and Ireland. London: 1862. 8 vols, complete. Roy 8vo. 770 hand cold plates. Mod buckram, uncut.
  (Egglishaw) **£2,950 [≈ $4,720]**
- British Entomology ... Coleoptera. London: Lovell Reeve, 1862. 2 vols. Roy 8vo. 256 hand cold plates. Half leather.
  (Egglishaw) **£875 [≈ $1,400]**
- British Entomology ... Dictyoptera, Dermaptera ... London: Lovell Reeve, 1862. Roy 8vo. 93 hand cold plates. Half leather.
  (Egglishaw) **£360 [≈ $576]**
- Farm Insects: being the Natural History and Economy of the Insects Injurious to the Field Crops of Great Britain and Ireland. London: 1860. Roy 8vo. 528 pp. 16 hand cold plates, 69 text figs. Half calf, trifle worn.
  (Wheldon & Wesley) **£70 [≈ $112]**
- Farm Insects: being the Natural History and Economy of Insects Injurious to the Field Crops ... London: Van Voorst, 1883. Lge 8vo. [4], 528 pp. 16 hand cold plates, 69 w'engvd vignettes. Sl stain on endpaper & lower outer crnr of central section. Orig cloth.
  (Claude Cox) **£60 [≈ $96]**

Curtis, William

- Practical Observations on the British Grasses ... Fourth Edition with Additions ... To which is now added a Short Account of ... Blight ... by Sir Joseph Banks. London: Symonds, 1805. 8vo. [2],58,14,[2] pp. Fldg plate, 6 plates. Some browning. Contemp half calf, worn.
  (Claude Cox) **£40 [≈ $64]**
- Practical Observations on British Grasses ... London: 1812. 5th edn, with addtns. 8vo. 116 pp. 8 hand cold plates. New bds.
  (Wheldon & Wesley) **£25 [≈ $40]**

Curwen, J.C.

- Hints on Agricultural Subjects, and on the Best Means of Improving the Condition of the Labouring Classes. Second Edition, Improved and Enlarged. London: 1809. 8vo. xxiv,286,[2] pp. 4 plates (foxed), 4 fldg tables. Orig bds, partly unopened, rebacked, sl rubbed.
  (Blackwell's) **£105 [≈ $168]**

Cushing, Harvey

- The Doctor and His Books. Cleveland: privately ptd, [1927]. 1st edn. 8vo. 26 pp. Orig stiff wraps.
  (Elgen) **$50 [≈ £31]**
- The Life of Sir William Osler. Oxford: Clarendon Press, 1925. 1st edn, 1st printing. 2 vols. 8vo. xvi,686; xii,728 pp. Ills. Orig blue cloth, gilt spines.
  (Karmiole) **$150 [≈ £93]**
- The Life of Sir William Osler. Oxford: Clarendon Press, 1925. 1st edn, 1st iss. 2 vols. 8vo. Plates, ills. Occas sl foxing. Orig cloth.
  (Elgen) **$250 [≈ £156]**
- The Life of Sir William Osler. Second Impression. Oxford: 1925. 2 vols. 8vo. 2 frontis, over 40 other ills. Orig cloth, spine ends just a little fingered, sm mark on 1 cvr.
  (Bow Windows) **£95 [≈ $152]**
- The Life of Sir William Osler. Oxford: Clarendon Press, 1925. 1st edn, 3rd printing. 2 vols. 8vo. [xvi],685,[3]; xii,728 pp. 44 plates. Edges of a few ff chipped. Orig blue cloth.
  (Gach) **$75 [≈ £46]**
- The Life of Sir William Osler. Oxford: Clarendon Press, 1925. 3rd edn. 2 vols. 8vo. 685; 728 pp. Photo ills. Orig bndg.
  (Xerxes) **$60 [≈ £37]**
- The Meningiomas arising from the Olfactory Groove and their Removal by the Aid of Electro-Surgery. Glasgow: Jackson, Wylie, 1927. 1st edn. 8vo. 53 pp. 28 ills. Orig stiff wraps.
  (Elgen) **$200 [≈ £125]**
- Realignments in Greater Medicine: their Effect upon Surgery and the Influence of Surgery upon them. London: Frowde, 1914. 8vo. 27 pp.
  (Elgen) **$55 [≈ £34]**
- The Western Reserve and its Medical Traditions. Cleveland: privately ptd, [1924]. 8vo. 33 pp.
  (Elgen) **$85 [≈ £53]**

Cushing, John

- The Exotic Gardener; in which the Management of the Hot-House, Green-House, and Conservatory, is fully and clearly delineated according to Modern Practice ... London: A. MacPherson ..., 1812. 1st edn. 8vo. x,3 subscribers, 227 pp. Sl spotted. Contemp half mor. *(Young's)* **£105 [≈ $168]**
- The Exotic Gardener; in which the

Management of the Hot-House, Green-House, and Conservatory, is fully and clearly delineated ... Second Edition, much improved. London: Nicol, 1814. 8vo. xii,236 pp. Contemp tree calf, gilt spine, upper jnt reprd, new label.   *(Spelman)* **£65 [≈ $104]**

Cutter, Ephraim
- Therapeutical Drinking of Hot Water. New York: Kellogg, 1883. 1st edn. 8vo. 9 pp. Orig wraps.   *(Xerxes)* **$60 [≈ £37]**

Cuvier, G.L.
- The Animal Kingdom ... Translated from the last French Edition. London: [1833-] 1834-37. 8 vols in 4. 8vo. 7 ports, 738 plates (nearly all cold), complete. Contemp calf, rebacked. *(Wheldon & Wesley)* **£400 [≈ $640]**
- The Animal Kingdom. London: Henderson, 1834. 1st edn thus. 8 vols. 8vo. Ca 700 plates, mostly hand cold. Contemp half calf, elab gilt spines, 2-cold labels, by Thurnam of Carlisle.   *(Gough)* **£475 [≈ $760]**
- Essay on the Theory of the Earth. Translated from the French by R. Kerr. With Mineralogical Notes, etc. by R. Jameson. Edinburgh: 1813. 1st edn in English. 8vo. xiii, 265,[1] pp. 2 plates. Contemp diced calf gilt, rebacked, cvrs rather stained. Ducal b'plate.   *(Wheldon & Wesley)* **£100 [≈ $160]**

Cyclopaedia ...
- Cyclopaedia of Obstetrics and Gynecology ... see Grandin, Egbert H.

Daglish, E.F.
- Birds of the British Isles. London: 1948. One of 1500. Roy 8vo. xviii,222 pp. 48 plates (25 hand cold). Orig blue buckram, t.e.g.   *(Henly)* **£68 [≈ $108]**
- Birds of the British Isles. London: 1948. One of 1500. Roy 8vo. xviii,222 pp. 48 plates (25 hand cold). V sl foxing. Orig blue buckram, t.e.g.   *(Henly)* **£55 [≈ $88]**

Dale, G.E.
- A Familiar Essay on Electricity, with an Arrangement of Experiments ... Liverpool: M. Galway, 1812. Apparently sole edn. 4to. 45 pp. Contemp calf, orig plain brown upper wrapper bound in, rebacked, cvrs scratched. Author's inscrptn.
   (Bickersteth) **£125 [≈ $200]**

Dallas, W.S.
- List of the Specimens of Hemipterous Insects in the British Museum. London: 1851-52. 2 parts in one vol. 12mo. 590 pp. 15 litho plates. New cloth.   *(Egglishaw)* **£36 [≈ $57]**

Dallimore, W.
- Holly, Yew and Box. With Notes on Other Evergreens. London: 1908. 1st edn. 8vo. xiv, 284,4 pp. Num plates. Orig pict cloth.
   (Henly) **£38 [≈ $60]**

Dallmeyer, Thomas R.
- Telephotography. An Elementary Treatise on the Construction and Application of the Telephotographic Lens. London: 1899. 1st edn. 4to. 145 pp. 26 plates, some diags, some ff sl dusty. Pres blindstamp on title. Orig green cloth gilt.
   (Moon) **£80 [≈ $128]**

Dammermann, K.W.
- The Agricultural Zoology of the Malay Archipelago, the Animals injurious and beneficial to Agriculture, Horticulture and Forestry ... Amsterdam: 1929. Roy 8vo. xi,473 pp. Map, 40 plates (13 cold), 179 text figs. Orig cloth.
   (Wheldon & Wesley) **£45 [≈ $72]**

Damon, R.
- Geology of Weymouth, Portland, and the Coast of Dorsetshire. London: 1884. 8vo. xii, 250 pp. Cold map. Orig cloth.
   (Baldwin) **£35 [≈ $56]**

Dana, David D.
- The Fireman: The Fire Departments of the United States, with a Full Account of All Large Fires ... Boston: James French, 1858. 8vo. 368,12 advt pp. Frontis, num w'cut plates. Orig red cloth gilt, a bit soiled, spine extremities sl rubbed.
   (Karmiole) **$150 [≈ £93]**

Dana, Edward Salisbury
- The System of Mineralogy of James Dwight Dana 1837-1868. Descriptive Mineralogy. Sixth Edition, Fourth Thousand. Entirely rewritten and much enlarged ... New York: 1906. 4to. lxiii, 1134,x,75 pp. Num ills. Half mor, worn.   *(Elgen)* **$275 [≈ £171]**
- A Text-Book of Mineralogy with an Extended Treatise on Crystallography and Physical Mineralogy. New York: 1900. 8vo. vii,593 pp. Cold plate, num text figs. Orig cloth, spine relaid.   *(Henly)* **£18 [≈ $28]**

Dana, James Dwight
- Characteristics of Volcanoes, with Contributions of Facts and Principles from the Hawaiian Islands ... London: Sampson Low ..., [1890?]. Roy 8vo. xvi,399 pp. 16 plates, 59 text ills. A few marks. Orig cloth, uncut, edges sl worn, hinge cracked but

sound.   *(Duck)* £185 [≈ $296]
- Corals and Coral Islands. London: 1872. 1st
English edn. 8vo. 3 maps, cold frontis, 3
plates, num text ills. Calf.
                          (Henly) £140 [≈ $224]
- Corals and Coral Islands. New York: 1872.
398 pp. 2 fldg maps, cold litho frontis, plates,
ills. Three qtr polished calf, spine gilt extra,
shelf-rubbed.   *(Reese)* $200 [≈ £125]
- Descriptive Mineralogy. London & New
York: 1883. 5th edn, with the 3 appendixes.
8vo. Num text figs. Orig green cloth, spines
relaid.   *(Henly)* £68 [≈ $108]
- Manual of Geology. New York: 1875. 2nd
edn. 8vo. xvi,828 pp. Frontis, fldg cold map,
1122 text figs. Orig cloth, spine worn.
                          (Henly) £18 [≈ $28]
- A Manual of Mineralogy for the Use of
Students. 1851. 3rd edn. 12mo. 432 pp. Num
text figs. Orig cloth, spine faded, rebacked,
new endpapers.   *(Henly)* £15 [≈ $24]
- A Manual of Mineralogy, including
Observations on Mines, Rocks, Reductions of
Ores and Applications of the Science to the
Arts. 1854. 12mo. xii,432 pp. Num text figs.
Waterstained at end. Orig cloth, spine relaid.
                          (Henly) £15 [≈ $24]
- Manual of Mineralogy. London: 1857. 2nd
edn. 8vo. 456 pp. Text figs. Sl foxed. Orig
cloth, spine relaid.   *(Henly)* £18 [≈ $28]
- The System of Mineralogy ... see Dana,
Edward Salisbury.
- United States Exploring Expedition ... Atlas.
Zoophytes. Philadelphia: 1859. Lge folio. 12
pp. 61 plates, some tinted. Scattered foxing.
Orig half calf & cloth, front hinge broken.
                    (Reese) $3,000 [≈ £1,875]

Dance, S.P.
- The Art of Natural History: Animal
Illustrators and their Work. London: 1978.
Roy 4to. 224 pp. Num ills (many cold). Orig
cloth, slipcase.
                 (Wheldon & Wesley) £36 [≈ $57]
- The Encyclopaedia of Shells. London: 1974.
Roy 8vo. 288 pp. 1500 cold photos. Orig
cloth.   *(Wheldon & Wesley)* £35 [≈ $56]
- The Shell Collector's Guide. London: 1976.
Roy 8vo. 192 pp. Num cold & plain ills. Orig
cloth.   *(Wheldon & Wesley)* £25 [≈ $40]

Darling, F.F.
- Natural History in the Highlands and
Islands. London: Collins, New Naturalist
Series, 1947. 1st edn. 32 cold & 32 other
plates. Orig cloth. Dw.
                       (Egglishaw) £18 [≈ $28]

Darling, W. Fraser
- West Highland Survey. An Essay in Human
Ecology. OUP: 1955. 8vo. Frontis. Orig
bndg. Dw.   *(Grayling)* £60 [≈ $96]

Darwin, Charles
- Charles Darwin and the Voyage of the
"Beagle". Edited by N. Barlow. London:
1945. 8vo. 279 pp. Map, 15 plates. Orig
cloth.   *(Wheldon & Wesley)* £28 [≈ $44]
- The Descent of Man ... London: 1871. 1st
edn, 1st iss. 2 vols. 8vo. Sl foxing. Sm stain
on inner marg of a prelim leaf in each vol.
Orig green cloth, vol 2 a trifle worn & refixed.
Freeman 937.
              (Wheldon & Wesley) £220 [≈ $352]
- The Descent of Man ... London: 1871. 1st
edn. 1st iss, with advts at end of both vols
dated Jan. 1871. Vol 1 p 297 starts with
'transmitted'. Vol 2 has inserted postscript, &
errata on verso of title. 2 vols. 8vo. Orig green
cloth. Freeman 937.
              (Wheldon & Wesley) £385 [≈ $616]
- The Descent of Man, and Selection in
Relation to Sex. London: Murray, 1871. 1st
edn, 2nd iss. 2 vols. 8vo. 16 advt pp dated Jan
1871 in both vols. Foxing on end &
preliminary ff. Orig green cloth, sl worn,
partly unopened. Freeman 938.
                  (Rootenberg) $400 [≈ £250]
- The Descent of Man ... Second Edition,
revised and augmented ... Tenth Thousand.
London: Murray, 1874. 8vo. xvi,688 pp.
Lacks half-title & errata slip. 1 section sprung.
19th c half calf, mor label, W.H. Smith label,
rather rubbed.   *(Claude Cox)* £30 [≈ $48]
- The Descent of Man ... London: 1883. 2nd
edn, 17th thousand. 8vo. Orig cloth, ft of
spine trifle worn. Freeman 957.
              (Wheldon & Wesley) £40 [≈ $64]
- The Descent of Man. London: 1890. 2nd
edn. 27th thousand. 8vo. Orig cloth. Freeman
970.   *(Wheldon & Wesley)* £30 [≈ $48]
- The Descent of Man ... Second Edition,
revised and Augmented. Twenty-Ninth
Thousand. London: 1890. 8vo. xvi,693 pp.
Figs. Sgntr on endpaper. Orig cloth. Freeman
971.   *(Bow Windows)* £40 [≈ $64]
- The Descent of Man. London: (1874) 1896.
2nd edn. Cr 8vo. xvi,693 pp. 78 text figs.
Orig cloth. Freeman 979.
              (Wheldon & Wesley) £20 [≈ $32]
- The Descent of Man and Selection in
Relation to Sex. London: 1896. 2nd edn, rvsd
& augmented, 33rd thousand. 693 pp. Ills.
Orig cloth.   *(Moorhead)* £35 [≈ $56]
- The Descent of Man. London: Murray, 1922.
Popular Edition. 8vo. xix,1031,[4 advt] pp.

Text ills. Orig leaf green cloth, faded & soiled. Freeman 1027.   *(Blackwell's)* **£20 [≈ $32]**

- The Different Forms of Flowers on Plants of the same Species. London: Murray, 1877. 1st edn. 8vo. viii,352,[32 ctlg dated March 1877] pp. 15 text figs, 38 tables. Occas sl pencil underlinings & marg notes. Orig cloth. Freeman 1277.   *(Bickersteth)* **£220 [≈ $352]**

- The Different Forms of Flowers on Plants of the Same Species. London: Murray, 1877. 1st edn. 8vo. viii,352,[32 advt dated March 1877] pp. Text ills. Light browning on prelim & final ff. Orig green cloth, a few reprs. Freeman 1277.   *(Rootenberg)* **$475 [≈ £296]**

- The Different Forms of Flowers on Plants of the same Species. London: 1877. 1st edn. 8vo. viii,352,32 advt pp. Text figs. Sm lib stamp on front pastedown. Cloth, sm tear at hd of spine.   *(Henly)* **£85 [≈ $136]**

- The Different Forms of Flowers on Plants of the same Species. London: 1880. 2nd edn, 2nd iss. 8vo. xvi,352 pp. Ills. Minor foxing. Orig cloth. *(Wheldon & Wesley)* **£70 [≈ $112]**

- The Different Forms of Flowers on Plants of the same Species. London: 1884. 3rd thousand. 8vo. xxiv,352 pp. Text figs. Cloth. Freeman 1281.
          (Wheldon & Wesley) **£60 [≈ $96]**

- The Effects of Cross and Self Fertilisation in the Vegetable Kingdom. London: Murray, 1876. 1st edn. 8vo. viii,482 pp. Errata slip after p viii. Title sl spotted. Pale pencil notes in 1st 8 pp & at end. Orig cloth. Freeman 1249.   *(Bickersteth)* **£220 [≈ $352]**

- The Effects of Cross and Self Fertilisation in the Vegetable Kingdom. London: Murray, 1876. 1st edn, 1st iss. 8vo. viii, 482 pp. Errata slip. 109 tables, 1 diag. Orig green cloth. Freeman 1249.   *(Rootenberg)* **$450 [≈ £281]**

- The Effects of Cross and Self Fertilisation in the Vegetable Kingdom. London: 1876. 1st edn. Cr 8vo. viii,482 pp. Errata slip. A little foxing at the beginning. Orig cloth. Freeman 1249.   *(Wheldon & Wesley)* **£180 [≈ $288]**

- The Effects of Cross and Self Fertilisation in the Vegetable Kingdom. London: 1876. 1st edn. 8vo. viii,482 pp, errata slip. Sl foxing at ends. Sm lib stamp on front pastedown. Cloth, sl worn.   *(Henly)* **£120 [≈ $192]**

- The Effects of Cross and Self Fertilisation in the Vegetable Kingdom. London: 1888. 2nd edn. 8vo. viii,487 pp. Orig green cloth, with spelling 'Fertilization' on spine. Freeman 1254.   *(Wheldon & Wesley)* **£50 [≈ $80]**

- The Effects of Cross and Self Fertilisation in the Vegetable Kingdom. London: 1891. 3rd edn. 8vo. viii,487 pp. Orig cloth. Freeman 1256.   *(Wheldon & Wesley)* **£48 [≈ $76]**

- The Expression of the Emotions in Man and Animals. London: 1872. 1st edn, 2nd iss. Cr 8vo. vi,374 pp. Advts dated Nov 1872. 7 heliotype plates (3 fldg) numbered in Arabic. Orig cloth, trifle used & neatly refixed. Freeman 1142.
          (Wheldon & Wesley) **£100 [≈ $160]**

- The Expression of the Emotions in Man and Animals. London: Murray, 1872. 1st edn, 2nd iss. 8vo. vi,374,[4 advt] pp. Endpapers & title spotted. Orig cloth gilt, sl worn, sm piece chewed from top edge of upper bd. Freeman 1359.   *(Hollett)* **£85 [≈ $136]**

- The Expression of the Emotions in Man and Animals. London: 1872. 1st edn. Orig cloth, a few spots on cvrs.
          (Moorhead) **£200 [≈ $320]**

- The Expression of the Emotions in Man and Animals. London: 1873. 10th thousand. 8vo. vi, 374 pp. Advts dated Oct 1884. 7 plates. Orig cloth, neatly refixed. Freeman 1144.
          (Wheldon & Wesley) **£55 [≈ $88]**

- The Expression of the Emotions in Man and Animals. Tenth Thousand. London: 1873. 8vo. vi, 374 pp. Advts dated 1872 & 1873. 7 heliotype plates, other ills. Some spotting of title & a few other ff. Orig cloth, upper jnt a little torn at hd. Freeman 1144.
          (Bow Windows) **£75 [≈ $120]**

- The Expression of the Emotions in Man and Animals. Edited by Francis Darwin. London: Murray, 1904. Popular Edition. 8vo. viii, 397, [i advt] pp. Ills. Title browned. Orig leaf green cloth. Freeman 1157.
          (Blackwell's) **£21 [≈ $33]**

- The Formation of Vegetable Mould, through the Action of Worms. London: 1881. 1st edn. 8vo. vii,326 pp. 15 text figs. Orig cloth, foredge of bndg & last few margs somewhat stained. Freeman 1357.
          (Wheldon & Wesley) **£70 [≈ $112]**

- The Formation of Vegetable Mould. London: Murray, 1881. 3rd thousand. 8vo. vii,326,1 advt pp. Errata leaf. Text ills. Endpapers trifle browned. Orig green cloth gilt, few spots to upper bd, rear hinge cracking. Freeman 1359.   *(Hollett)* **£25 [≈ $40]**

- The Formation of Vegetable Mould, through the Action of Worms, with Observations on their Habits. London: Murray, 1883. 8th thousand, crrctd. Sm 8vo. vii,328,[i advt] pp. 15 text figs. Orig green cloth gilt, recased. Freeman 1366.   *(Blackwell's)* **£45 [≈ $72]**

- The Formation of Vegetable Mould through the Action of Worms, with Observations on their Habits. London: Murray, 1904. 13th thousand. 8vo. viii,298,[i advt] pp. 15 text figs. Orig light green cloth, pigment damaged

by insects, hinges started. Freeman 1389.
                              (Blackwell's) **£35 [≈ $56]**
- Formation of Vegetable Mould through the Action of Worms. London: Murray, 1904. 8vo. viii, 298 pp. 10 plates. Orig cloth.
                              (Wheldon & Wesley) **£15 [≈ $24]**
- The Foundations of the Origin of Species, Two Essays written in 1842 and 1844, edited by his son Francis Darwin. Cambridge: 1909. 8vo. xxx,263 pp. Port, plate. Orig buckram. Freeman 1556.
                              (Wheldon & Wesley) **£50 [≈ $80]**
- Geological Observations on the Volcanic Islands and Parts of South America visited during the Voyage of H.M.S. Beagle. London: 1876. 2nd edn. Cr 8vo. xiv,647 pp. 2 maps, cold plate of sections, 4 plates, 40 text figs. Orig brown cloth. Freeman 276.
                              (Wheldon & Wesley) **£135 [≈ $216]**
- Insectivorous Plants. London: 1875. 1st edn. 8vo. x,462 pp. 30 text figs. A little dust soiling, sm stamp on title & half-title. Orig cloth, sl used, new endpapers, jnts reprd. Freeman 1217.
                              (Wheldon & Wesley) **£120 [≈ $192]**
- Insectivorous Plants. London: 1875. 1st edn. 8vo. x,462 pp. 30 text figs. Orig cloth. Freeman 1217.
                              (Wheldon & Wesley) **£175 [≈ $280]**
- Journal of Researches into the Geology and Natural History of the Various Countries visited by H.M.S. Beagle ... London: 1840. 1st edn, 3rd iss. 8vo. [iv],[vii],xiv,629,[16 advt dated Aug 1839] pp. 2 fldg maps. Orig cloth, spine reprd. Freeman 12, bndg variant a.
                              (Rootenberg) **$1,500 [≈ £937]**
- Journal of Researches ... London: 1845. 2nd edn, crrctd with addtns. Cr 8vo. viii,519 pp. Orig red cloth, trifle used, inner jnt cracked. Freeman 14c.
                              (Wheldon & Wesley) **£100 [≈ $160]**
- Journal of Researches into the Natural History and Geology of the Countries Visited during the Voyage of the H.M.S. Beagle ... New York: Harper, 1846. 1st Amer edn. 2 vols. Scattered foxing. Ugly later cloth. Freeman 16.   *(Reese)* **$150 [≈ £93]**
- Journal of Researches ... London: 1860. 10th thousand. 8vo. xv,519 pp. Orig green cloth, spine reprd, inner jnts taped. Freeman 20.
                              (Wheldon & Wesley) **£85 [≈ $136]**
- A Naturalist's Voyage. Journal of Researches into the Natural History and Geology of the Countries visited during the Voyage of H.M.S. "Beagle" ... London: Murray, 1889. 8vo. x,519,[4 advt] pp. Port. Orig cloth gilt, rhea hunt on cvr, 3/6 on spine. Freeman 49.
                              (Blackwell's) **£45 [≈ $72]**

- A Naturalist's Voyage. Journal of Researches into the Natural History and Geology of the Countries visited during the Voyage of H.M.S. Beagle. New Edition. London: 1890. Cr 8vo. xi,500 pp. Port & title sl foxed. Orig cloth, sl worn. Freeman 58.
                              (Wheldon & Wesley) **£20 [≈ $32]**
- Journal of Researches ... during the Voyage of H.M.S. "Beagle". London: Minerva Library, 1894. 12th edn. 8vo. 492 pp. 16 plates & text figs. Orig cloth. Freeman 77.
                              (Wheldon & Wesley) **£25 [≈ $40]**
- Journal of Researches into the Natural History and Geology of the Countries visited during the Voyage of H.M.S. "Beagle" ... New Edition. London: Murray, 1901. 8vo. xvi,521, [7 advt] pp. 26 ills. Orig cloth. Freeman 97.   *(Blackwell's)* **£25 [≈ $40]**
- Journal of Researches ... London: 1913. New edn. 8vo. 2 maps, 105 ills. Orig cloth. Freeman 125.
                              (Wheldon & Wesley) **£35 [≈ $56]**
- Life and Letters ... see Darwin, Francis.
- A Manual of Scientific Enquiry ... see Herschel, Sir John Frederick William (ed.).
- The Movements and Habits of Climbing Plants. London: 1875. 2nd edn (1st edn in book form). 8vo. viii,208 pp. 13 figs. Orig cloth, lib b'plate. Freeman 836.
                              (Wheldon & Wesley) **£70 [≈ $112]**
- The Movement and Habits of Climbing Plants. Second Thousand. London: Murray, 1876. 8vo. viii,208,32 advt pp. 13 text figs. Prelims sl spotted. Orig green cloth.
                              (Gough) **£60 [≈ $96]**
- The Movements and Habits of Climbing Plants. Second Edition Revised. New York: 1876. 1st Amer edn. 8vo. viii,208 pp. 13 text figs. Orig brown cloth, good ex-lib. Freeman 838.   *(Wheldon & Wesley)* **£35 [≈ $56]**
- The Movements and Habits of Climbing Plants. London: 1891. 5th thousand. 8vo. x, 208 pp. Orig cloth. Freeman 846.
                              (Wheldon & Wesley) **£40 [≈ $64]**
- The Movements and Habits of Climbing Plants. London: Murray, 1906. Popular Edition. 8vo. x,208 pp. 13 text figs. Orig light green cloth, backstrip soiled. Freeman 852.
                              (Blackwell's) **£21 [≈ $33]**
- The Movements and Habits of Climbing Plants. London: (1906). 8vo. Some worming but sound. Cloth. Freeman 855.
                              (Wheldon & Wesley) **£15 [≈ $24]**
- On the Origin of Species by means of Natural Selection ... London: Murray, 1859. 1st edn. 8vo in 12s. ix,[1],502,[32 advt dated June 1859 (Freeman variant 3)] pp. 1 fldg plate. Orig cloth (Freeman variant a), uncut, minor

reprs. Qtr mor box. Freeman 373.
(Rootenberg) **$7,500 [≃ £4,687]**
- On the Origin of Species ... Fifth Thousand.
London: 1860. 2nd edn. 8vo. x,502 pp.
Murray advts not present. Orig cloth.
Freeman 376.
(Wheldon & Wesley) **£250 [≃ $400]**
- On the Origin of Species ... London: 1866.
4th edn (8th thousand). 8vo. xxi,593 pp. Fldg
diag. Orig green cloth, trifle used. Freeman
385.   *(Wheldon & Wesley)* **£185 [≃ $296]**
- The Origin of Species ... Fifth Edition, with
additions and corrections. New York:
Appleton, 1870. Apparently 5th Amer edn.
8vo. 447 pp. Orig brown cloth. Freeman 390.
(Wheldon & Wesley) **£80 [≃ $128]**
- The Origin of Species ... London: 1872. 6th
edn (11th thousand), with addtns & crrctns.
Cr 8vo. xxi,458 pp. Orig cloth. Freeman 391.
(Wheldon & Wesley) **£200 [≃ $320]**
- The Origin of Species ... London: 1873. 6th
edn (13th thousand), with addtns & crrctns.
8vo. xxi,458 pp. Fldg chart. Orig cloth,
refixed. Freeman 396.
(Wheldon & Wesley) **£100 [≃ $160]**
- The Origin of Species ... London: 1876. 6th
edn (18th thousand). 8vo. xxi,458 pp. Fldg
diag. Lib stamps on fly-leaf & half-title,
b'plate. Cloth. Freeman 401.
(Wheldon & Wesley) **£85 [≃ $136]**
- The Origin of Species ... London: 1884. 6th
edn (26th thousand). 8vo. xxi,458 pp. Fldg
diag. Orig cloth, trifle worn. Freeman 412.
(Wheldon & Wesley) **£30 [≃ $48]**
- The Origin of Species ... London: (1872)
1888. 33rd thousand. 8vo. xxi,432 pp. Fldg
diag. Orig cloth. Freeman 423.
(Wheldon & Wesley) **£35 [≃ $56]**
- The Origin of Species ... Sixth Edition, with
additions and corrections to 1872, (Thirty-
Third Thousand). London: Murray, 1888.
8vo. xxi,458,[32 ctlg] pp. Sl foxing. Orig
cloth. Freeman 423. *(Blackwell's)* **£40 [≃ $64]**
- The Origin of Species ... Sixth Edition, with
additions and corrections (Forty-Seventh
Thousand). London: Murray, 1888. 8vo.
xxi,432 pp. Orig cloth, hinges working loose.
Freeman 446.   *(Blackwell's)* **£35 [≃ $56]**
- The Origin of Species. London: 1890. 6th
edn, 39th thousand. 8vo. xxi,458 pp. Fldg
diag. Orig cloth, trifle rubbed, front jnt loose.
Freeman 436.
(Wheldon & Wesley) **£28 [≃ $44]**
- The Origin of Species. London: 1897. 6th
edn (49th thousand). 8vo. xxi,432 pp. Fldg
diag. Orig cloth. Freeman 451.
(Wheldon & Wesley) **£20 [≃ $32]**

- The Origin of Species. London: 1901. 8vo.
Port, diag. Orig cloth, sl stained. Freeman
474.   *(Wheldon & Wesley)* **£15 [≃ $24]**
- The Origin of Species ... London: Murray,
1906. Sm 8vo. xxxi,703,[i advt] pp. Orig
cloth. Freeman 499. *(Blackwell's)* **£25 [≃ $40]**
- The Origin of Species ... A Variorum Text,
edited by Morse Peckham. Philadelphia:
1959. 8vo. 816 pp. Orig cloth. Dw. Freeman
588.   *(Wheldon & Wesley)* **£35 [≃ $56]**
- The Power of Movement in Plants. London:
Murray, 1880. 1st edn. 8vo. x,592,[32
advt dated May 1878] pp. Text ills. Orig
green cloth, a few reprs. Freeman 1325.
(Rootenberg) **$475 [≃ £296]**
- The Power of Movement in Plants. London:
1880. 1st edn. 8vo. x,592,[32 advt dated May
1878] pp. 196 text figs. Orig cloth, unopened.
Freeman 1325.
(Wheldon & Wesley) **£150 [≃ $240]**
- The Power of Movement in Plants. Assisted
by Francis Darwin. Second Thousand.
London: 1880. 2nd edn. Cr 8vo. x,592 pp.
196 text figs. 1st 2 pp carelessly opened. Orig
cloth, label removed from front cvr.
(Wheldon & Wesley) **£50 [≃ $80]**
- The Structure and Distribution of Coral
Reefs. London: 1874. 2nd rvsd edn. 8vo. 3
fldg charts (2 cold), 14 text figs. Prize label on
flyleaf. Sm tear in 1 chart reprd. Orig green
cloth. Freeman 275.
(Wheldon & Wesley) **£100 [≃ $160]**
- The Structure and Distribution of Coral
Reefs. London: Minerva Library, 1891. 8vo.
Orig cloth.   *(Henly)* **£20 [≃ $32]**
- The Variation of Animals and Plants under
Domestication. London: 1868. 1st edn. 1st
issues. 2 vols. 8vo. 6 lines (Freeman in error)
of errata vol 1, 7 lines vol 2. Advts dated April
1867 vol 1, Feb 1868 vol 2. Half mor gilt.
Freeman 877.
(Wheldon & Wesley) **£180 [≃ $288]**
- The Variation of Animals and Plants under
Domestication. London: Murray, 1868. 1st
edn. 1st iss of text & bndg. 2 vols. Roy 8vo.
viii, 411,[1],[32 advt dated April 1867]; viii,
486,[2 advt dated Feb 1868] pp. Orig cloth,
Mudie's label on upper bds.
(Fenning) **£285 [≃ $456]**
- The Variation of Animals and Plants under
Domestication. London: 1875. 2nd edn, rvsd.
4th thousand. 1st issue of the rvsd edn. 2 vols.
8vo. 43 w'cuts. Endpapers rather foxed. Orig
cloth. Freeman 880.
(Wheldon & Wesley) **£75 [≃ $120]**
- The Variation of Animals and Plants under
Domestication. London: 1899. 2nd edn, 8th
imp. 2 vols. 8vo. Text figs. A little foxing.

Orig cloth. Freeman 898.
(Wheldon & Wesley) **£45 [≈ $72]**

- The Variation of Animals and Plants under Domestication. Authorized edition with Preface by Asa Gray. New York: [1868]. 2 vols. 8vo. Orig cloth, recased. Freeman 879.
(Wheldon & Wesley) **£48 [≈ $76]**

- The Variation of Animals and Plants under Domestication. New York: 1898. 2 vols. 8vo. Orig cloth, partly unopened. Freeman 897.
(Wheldon & Wesley) **£25 [≈ $40]**

- The Variation of Animals and Plants under Domestication. London: 1899. 8th impression of the 2nd edn, rvsd. 2 vols. 8vo. text figs. A little foxing. Orig cloth. Freeman 898.
(Wheldon & Wesley) **£45 [≈ $72]**

- On the Various Contrivances by which British and Foreign Orchids are fertilised by Insects ... London: Murray, 1862. 1st edn. 8vo in 12s. vi,365,[32 advt dated Dec 1861] pp. 1 fldg plate, 33 text w'cuts. Orig plum cloth with gilt orchid on cvr (Freeman 800a).
(Rootenberg) **£650 [≈ £406]**

- On the Various Contrivances by which British and Foreign Orchids are fertilised by Insects ... London: 1862. 1st edn. 8vo. vi, 365 pp. Fldg plate, 33 text figs. Orig plum cloth. Freeman 800a.
(Wheldon & Wesley) **£250 [≈ $400]**

- The Various Contrivances by which Orchids are fertilised by Insects ... London: 1890. 2nd edn, 5th thousand. edn. 8vo. xvi,300 pp. 38 ills. Orig cloth. Freeman 810.
(Wheldon & Wesley) **£40 [≈ $64]**

- The Various Contrivances by which Orchids are Fertilised by Insects. Second Edition, revised. Fifth Thousand. London: Murray, 1890. 8vo. xvi,300,[32 ctlg] pp. Num text ills. Orig cloth. Freeman 810.
(Blackwell's) **£30 [≈ $48]**

- The Various Contrivances by which Orchids are fertilised by Insects ... London: 1904. Popular edn. 8vo. xvi,300 pp. Ills. Orig cloth. Freeman 816.
(Wheldon & Wesley) **£30 [≈ $48]**

- See also Herschel, Sir John F.W. (ed.).

Darwin, Charles, Fitzroy, R. & King, P.P.
- Narrative of the Surveying Voyage of H.M. Ships Adventure and Beagle between the Years 1826 and 1836. London: 1839-40. 4 vols (inc vol 2 Appendix). 8vo. 12 maps, 44 plates. Vol 1 plates waterstained, vol 2 trifle foxed. Mod half mor. Freeman 10.
(Wheldon & Wesley) **£1,800 [≈ $2,880]**

Darwin, Erasmus
- The Botanic Garden; a Poem in Two Parts ... London: 1791-94. Vol 1 2nd edn, vol 2 4th edn. 2 vols in one. 4to. 2 frontis, 18 plates (1 by William Blake). Some foxing. Later half calf.
(Wheldon & Wesley) **£180 [≈ $288]**

- Phytologia; or the Philosophy of Agriculture and Gardening, with the Theory of Draining Morasses, and with an Improved Construction of the Drill Plough. London: J. Johnson, 1800. 4to. viii,612,[12] pp. 12 plates (foxed & stained). Contemp calf.
(Blackwell's) **£190 [≈ $304]**

- Phytologia; or the Philosophy of Agriculture and Gardening. London: 1800. 4to. viii, 612,[12] pp. 12 plates. Old diced calf, rebacked. *(Wheldon & Wesley)* **£160 [≈ $256]**

- Zoonomia; or, the Laws of Organic Life. London: 1801. 3rd edn. 4 vols. 8vo. 11 plates (6 cold). Half calf gilt.
(Wheldon & Wesley) **£120 [≈ $192]**

Darwin, Francis
- Charles Darwin: His Life told in an Autobiographical Chapter and in Selected Series of Letters. Edited by F. Darwin. London: 1892. 1st edn of this abridged version. 8vo. vi,348 pp. Port & facs. Mor, sl rubbed. Freeman 1461. *(Gough)* **£35 [≈ $56]**

- Charles Darwin: His Life told in an Autobiographical Chapter and in a Selected Series of his Published Letters. Edited by Francis Darwin. London: 1902. 6th thousand in this form. 8vo. viii,348 pp. Port & facs. Inscrptn on title. Orig cloth. Freeman 1473a.
(Blackwell's) **£30 [≈ $48]**

- The Life and Letters of Charles Darwin ... Fifth Thousand Revised. London: Murray, 1887. 3 vols. 8vo. 3 port frontis, 1 facs. Orig green cloth gilt, rather stained. Freeman 1453. *(Blackwell's)* **£75 [≈ $120]**

- The Life and Letters of Charles Darwin. New York: Appleton, 1887. 1st Amer edn, 1st printing. 2 vols. 8vo. [ii],viii,558,[4]; [iv], [vi],562,[4] pp. Orig brown cloth, sl worn. Freeman 1456. *(Gach)* **£65 [≈ £40]**

- Life and Letters of Charles Darwin. Seventh Thousand, Revised. London: Murray, 1888. 3 vols. 8vo. 3 frontis. Orig green cloth.
(Gough) **£60 [≈ $96]**

Daubeny, C.
- Lectures on Roman Husbandry delivered before the University of Oxford. Oxford: 1857. 8vo. xvi,328 pp. 12 plates (1 cold). Blind stamp on title. Orig cloth, reprd.
(Wheldon & Wesley) **£60 [≈ $96]**

Davidson, J. Brownlee & Chase, Leon Wilson
- Farm Machinery and Farm Motors. New York & London: 1920. Post 8vo. vi,[ii],513 pp. 377 ills, inc frontis. Sl marg dampstain last 2 ff. Orig cloth, hd of spine frayed.
(Duck) **£30 [≈ $48]**

Davidson, Richard Oglesby
- A Disclosure of the Discovery and Invention and a Description of the Plan of Construction and Mode of Operation of the Aerostat: or a New Mode of Aerostation. St. Louis (MO): 1840. 1st edn. 8vo. 1 ill. 2 old lib stamps. Old wraps, loose. *(Ximenes)* **$1,250 [≈ £781]**

Davie, O.
- Methods in the Art of Taxidermy. Philadelphia: 1894. Roy 8vo. xiv,359 pp. 90 plates. Orig cloth.
(Wheldon & Wesley) **£45 [≈ $72]**

Davies, Charles
- A Treatise on Shades and Shadows, and Linear Perspective. Hartford: A.S. Barnes & Co; New York: Wiley & Putnam, 1838. 2nd edn. Tall 8vo. 159 pp. 21 fldg plates. Occas foxing. Half leather, scuffed, front jnt weak.
(Schoyer) **$100 [≈ £62]**

Davies, D.C.
- A Treatise on Metalliferous Minerals and Mining. London: 1888. 4th edn. 8vo. Fldg plates, text ills. Orig cloth, hd of spine snagged. *(Ambra)* **£35 [≈ $56]**

Davies, E.W.L.
- Algiers in 1857. Its Accessibility, Climate and Resources described with especial reference to English Invalids. London: 1858. 1st edn. 8vo. 4 litho plates. Orig orange cloth.
(Robertshaw) **£60 [≈ $96]**

Davies, L.J. & D.O.
- South Wales Coals. Their Analyses, Chemistry and Geology. London: 1921. 8vo. 119, xliii advt pp. Sm tear in title reprd. Orig cloth. *(Henly)* **£15 [≈ $24]**

Davies, Nathaniel Edward Yorke
- Health and Condition in the Active and the Sedentary. London: Sampson Low, 1894. 250 pp. Orig blue cloth.
(C.R. Johnson) **£28 [≈ $44]**

Davies, T.
- The Preparation and Mounting of Microscopic Objects. London: [1873]. 2nd edn, enlgd. Sm 8vo. viii,214 pp. Text figs. Orig cloth, trifle faded.
(Wheldon & Wesley) **£20 [≈ $32]**

Davis, C.L.
- Laboratory Manual of Neuroanatomy. Baltimore: Wm. Wood, 1933. Part 2 Stereographic plates. Sm 4to. 30 plates on heavy stock. Orig bndg. Boxed, backstrip detached. *(Xerxes)* **$60 [≈ £37]**

Davis, Joseph Barnard
- Thesaurus Craniorum. Catalogue of the Skulls of the Various Races of Man, in the Collection of ... London: the subscribers, 1867. 8vo. 374 pp. 2 plates, num text ills. Orig bndg, some cvr stains.
(Xerxes) **$125 [≈ £78]**

Davis, Sir Robert H.
- Deep Diving and Submarine Operations. A Manual for Deep Sea Divers and Compressed Air Workers. Fifth Edition. London: 1951. Sq roy 8vo. Errata slip. Fldg frontis, cold fldg chart, num ills. Orig green cloth.
(Duck) **£110 [≈ $176]**

Davis, W.M.
- The Coral Reef Problem. New York: Amer Geog Soc, 1928. Roy 8vo. v,596 pp. 227 text figs. Orig cloth.
(Wheldon & Wesley) **£45 [≈ $72]**

Davis, William
- An Easy and Comprehensive Description of the Terrestrial and Celestial Globes ... London: for the author ..., [ca 1795]. 2nd edn. 12mo. viii,94 pp. Fldg plates. 1st few ff sl wormed at lower inner margs, & sl water stained. Contemp sheep, rebacked.
(Burmester) **£90 [≈ $144]**

Davy, Sir Humphry
- Elements of Agricultural Chemistry, in a Course of Lectures for the Board of Agriculture ... London: for Longman, 1813. 1st edn. 4to. viii,323,[1 blank],lxiii,4 pp. 10 plates (1 or 2 offset from text). Contemp diced half russia, jnts & crnrs rubbed.
(Pickering) **£700 [≈ £437]**

- Elements of Agricultural Chemistry, a Course of Lectures for the Board of Agriculture. Second Edition. London: Longman, 1814. 8vo. xii,479,[9] pp. 16 pp inserted ctlg. 10 plates. V sl foxing. New paper bds, untrimmed.
(Blackwell's) **£125 [≈ $200]**

Dawe, Edward A.
- Paper and its Uses. A Treatise for Printers Stationers and Others ... Second Edition. London: Crosby, Lockwood, 1919. 8vo. vii,161,1 pp. 35 paper samples. Orig cloth.
(Charles B. Wood) **$115 [≈£71]**

Dawkins, Sir William Boyd
- Early Man in Britain and his Place in the Tertiary Period. London: Macmillan, 1880. 1st edn. 8vo. xxiii,[1],537,[2 advt] pp. 168 ills. Minor foxing of a few ff. Orig cloth gilt, t.e.g.   *(Fenning)* **£48.50 [≈$78]**

Dawson, Sir J.W.
- Acadian Geology. The Geological Structure, Organic Remains and Mineral Resources of Nova Scotia, New Brunswick and Prince Edward Island. London: 1868. 2nd edn. 8vo. xxv,694 pp. Fldg map in pocket, 9 plates (some crnrs waterstained), 231 text ills. Orig cloth, spine relaid.   *(Henly)* **£28 [≈$44]**
- The Geological History of Plants. London: 1888. Cr 8vo. x,290 pp. Frontis, text figs. Orig cloth, spine trifle worn.
(Wheldon & Wesley) **£20 [≈$32]**
- Some Salient Points in the Science of the Earth. New York: 1894. Cr 8vo. xi,499,4 advt pp. 46 ills. Orig cloth.   *(Henly)* **£14 [≈$22]**
- The Story of the Earth and Man. London: 1874. 8vo. xiii,403 pp. 1 cold plate, text ills. Orig pict cloth gilt.   *(Henly)* **£14 [≈$22]**

Dawson, Sir Philip
- Electric Railways and Tramways, their Construction and Operation. A Practical Handbook ... London: "Engineering", 1897. 4to. xxvi,[errata ii],677,[1] pp. 40 advt pp. 503 ills. Orig half mor, gilt spine, t.e.g., minor marking to sides.
(Duck) **£150 [≈$240]**
- Electric Traction on Railways. London: "The Electrician", 1909. 1st edn. Super roy 8vo. xxxvi,856,advt pp. Erratum slip. Port frontis & 94 photo plates, 29 fldg plates, num ills. Orig half roan gilt, rubbed & chafed, hd of jnt split, back hinge cracked. *(Duck)* **£60 [≈$96]**

Day, F.
- British and Irish Salmonidae. London: Williams & Norgate, 1887. Roy 8vo. viii,299 pp. 12 plates (inc 9 chromoliths), num text figs. Orig cloth.   *(Egglishaw)* **£95 [≈$152]**
- The Fishes of Great Britain and Ireland. London: Williams & Norgate, 1880-84. 2 vols. Roy 8vo. 180 litho plates. Minor spotting. Lib stamp on title. Mor, crnrs & jnts sl rubbed.   *(Egglishaw)* **£230 [≈$368]**

- The Fishes of Malabar. London: 1865. 4to. xxxii,293 pp. 20 engvd plates. New half mor.
(Wheldon & Wesley) **£330 [≈$528]**

Deakin, R.
- Flora of the Colosseum of Rome. London: Groombridge, 1855. Cr 8vo. viii,237 pp. Frontis, 1 plain & 4 cold plates. Lacks free endpapers. Orig cloth, used.
(Wheldon & Wesley) **£35 [≈$56]**

Deaver, John B.
- Surgical Anatomy. A Treatise on Human Anatomy in its application to the Practice of Medicine and Surgery. Philadelphia: 1899. 3 vols. Num ills.   *(Rittenhouse)* **$125 [≈£78]**

De Bairacli-Levy, Juliette
- Medicinal Herbs: Their Use in Canine Ailments. New York: Sirius, 1948. 8vo. 111 pp. Orig bndg.   *(Xerxes)* **$45 [≈£28]**

de Beer, G.R.
- The Development of the Vertebrate Skull. Oxford: 1937. Roy 8vo. xxiv,552 pp. 143 plates. Orig buckram.
(Wheldon & Wesley) **£40 [≈$64]**

Deerr, N.
- Sugar and the Sugar Cane: An Elementary Treatise on the Agriculture of the Sugar Cane and on the Manufacture of Cane Sugar. Manchester: Norman Rodger, 1905. 8vo. x,396, xx,xii pp. Cold frontis, 2 b/w plates, 130 text ills, num tables. Orig cloth, stained.
(Berkelouw) **$150 [≈£93]**

De Freval, J.B.
- The History of the Heavens, Considered according to the Notions of the Poets and Philosophers, Compared with the Doctrines of Moses ... from the French of the Abbe Pluche ... London: for J. Wren, 1752. 3rd edn. 2 vols. 12mo. Frontis, 24 plates. Contemp calf, rebacked.
(Young's) **£58 [≈$92]**

De Kay, Charles
- Bird Gods [in Ancient Europe]. New York: A.S. Barnes, 1898. 8vo. xxiv,249 pp. Decs by William Wharton Edwards. Sl staining of guttering. Orig gilt dec red cloth.
(Blackwell's) **£45 [≈$72]**

De La Beche, Sir Henry Thomas
- Geological Notes. London: for Treuttel & Wurtz ..., 1830. 1st sep edn. 8vo. [ii],69, [1], xlii pp. 5 plates (4 fldg, 3 hand cold). Rec bds.
(Burmester) **£120 [≈$192]**

- The Geological Observer. Second Edition, Revised. London: Longman, 1853. Roy 8vo. xxviii, 740 pp. 308 ills. 1 v sm marg tear. Orig cloth, largely unopened, spine faded, crnrs v sl worn.   *(Duck)* **£85 [≃ $136]**
- Report on the Geology of Cornwall, Devon and West Somerset. London: 1839. 8vo. xxviii, 648 pp. Cold map, 12 plates (1 cold). A little foxing, sm blind stamp on title. Orig cloth, rebacked.   *(Baldwin)* **£90 [≃ $144]**

Delacour, Jean

- The Pheasants of the World. London: 1951. 1st imp. Cr 4to. 347 pp. 21 maps, 16 cold & 16 plain plates, diags. Orig cloth. Dw.
          (Wheldon & Wesley) **£150 [≃ $240]**
- The Pheasants of the World. London: (1951) 1965. Cr 4to. 347 pp. 21 maps, 16 cold & 16 plain plates, diags. Orig cloth.
          (Wheldon & Wesley) **£140 [≃ $224]**
- The Pheasants of the World. London: 1977. 2nd rvsd edn. Cr 4to. 432 pp. 33 plates (17 cold). Orig cloth.
           (Wheldon & Wesley) **£45 [≃ $72]**
- The Pheasants of the World. London: 1977. 2nd, rvsd & extended edn. 4to. 432 pp. 17 cold & 16 other ills. Orig simulated cloth cvrs.
            (Bow Windows) **£35 [≃ $56]**
- The Waterfowl of the World. London: Country Life, 1954. 1st edn. 3 vols. 4to. 33 maps, 16 cold plates by Peter Scott. Orig cloth gilt. Dws.   *(Hollett)* **£150 [≃ $240]**

Delany, M.C.

- The Historical Geography of the Wealden Iron Industry. London: Benn, 1921. 62 pp. Maps. Orig wraps, cvrs rather grubby.
            (Book House) **£25 [≃ $40]**

De Moivre, Abraham

- The Doctrine of Chances: or, a Method of Calculating the Probability of Events in Play. London: W. Pearson for the author, 1718. 1st edn. Lge 4to. [6],175 pp. 2 vignettes (1 on title), tailpieces. Contemp calf, spine & label reprd.   *(Rootenberg)* **$1,500 [≃ £937]**

De Morgan, Augustus

- Arithmetical Books from the Invention of Printing to the Present Time. Being Brief Notices of a Large Number of Works drawn up from Actual Inspection. London: Taylor & Walton, 1847. 1st edn. 8vo. xxviii,124 pp. Orig pebble cloth, later paper label.
          (Frew Mackenzie) **£90 [≃ $144]**
- A Budget of Paradoxes. London: 1872. 1st edn, but later advts. vii,511 pp. Advts dated July 1879. Contemp sgntr & date on title,

cuttings pasted to half-title. Orig cloth, spine relaid.   *(Whitehart)* **£60 [≃ $96]**
- The Differential and Integral Calculus ... also Elementary Illustrations of the Differentiated and Integral Calculus. London: 1842. 1st edn. 2 parts in one vol. Crnr of title reprd. Half leather.   *(Elgen)* **$275 [≃ £171]**
- The Differential and Integral Calculus. Containing Differentiation, Integration ... London: 1842. xx,784 pp. Later cloth, sm nick at hd of spine.   *(Whitehart)* **£35 [≃ $56]**
- The Differential and Integral Calculus ... London: Robert Baldwin, [1842]. 1st edn. 8vo. xx, 785,[2],64 pp. Pres bndg in full calf.
          (Rootenberg) **$385 [≃ £240]**
- An Essay on Probabilities, and on their Application to Life Contingencies and Insurance Offices. London: Longman ..., 1838. 1st edn. 8vo. xviii,306,40 appendix,16 advt pp. Engvd title, text diags. Orig cloth.
          (Rootenberg) **$275 [≃ £171]**
- An Explanation of the Gnomonic Projection of the Sphere ... London: Baldwin & Cradock, 1836. 1st edn. 8vo. Tables, diags. Orig blue-green cloth, ptd paper label, sl wear, label a trifle chipped.   *(Ximenes)* **$250 [≃ £156]**
- Trigonometry and Double Algebra. London: 1849. xi,167 pp. Frontis. Orig cloth, faded & marked, spine label rubbed, inner hinge cracked but bndg firm.
            (Whitehart) **£25 [≃ $40]**

Dempsey, G. Drysdale

- Railways. Papers on the Mechanical and Engineering Operations and Structures combined in the Making of a Railway ... London: John Weale, 1846. 4to. [ii],135,[1] pp. 27 plates. Spotting & browning. Contemp half mor, gilt spine, t.e.g., rubbed, wear to jnts.   *(Duck)* **£85 [≃ $136]**

Dendy, Walter Cooper

- The Philosophy of Mystery. London: Longman ..., 1841. 1st edn. 8vo. xii,443,16 ctlg dated April 1841 pp. Orig cloth, rear jnt & spine extremities sl worn.
            (Schoyer) **£90 [≃ £56]**

Denes, J. & Keedwell, A.D.

- Latin Squares and their Applications. Budapest: 1974. 547 pp. Sev text figs.
            (Whitehart) **£38 [≃ $60]**

Dennis, Frederic S.

- Selected Surgical Papers (1876-1914). New York: 1934. 1st edn. 2 vols. 8vo. Port, text ills. Orig cloth backed bds, faded.
            (Elgen) **$100 [≃ £62]**

Denny, Henry
- Monographia Anoplurorum Britanniae; or, an Essay on the British Species of Parasitic Insects belonging to the order Anoplura of Leach ... London: 1842. 1st edn. 8vo. 26 cold plates. Rec calf backed mrbld bds. Relevant pamphlet (1937) by G.B. Thompson inserted.
(Fenning) £125 [≈ $200]
- Monographia Pselaphidarum et Scydmaenidarum Britanniae: an Essay on the British Species of the Genera Pselaphus of Herbst and Scydmaenus of Latreille. Norwich: S. Wilkin, 1825. 8vo. vii,74 pp. 14 hand cold plates. Cloth, rebacked.
(Egglishaw) £65 [≈ $104]

Deraniyagala, P.E.P.
- A Coloured Atlas of some Vertebrates from Ceylon. Colombo: 1952-55. 3 vols. Oblong 4to. Port, 83 cold & 18 plain plates. Wraps.
(Wheldon & Wesley) £85 [≈ $136]
- The Tetrapod Reptiles of Ceylon, Vol 1, Testudinates and Crocodilians. Colombo: 1939. All published. Roy 8vo. xxxii,412 pp. 24 plates, 137 text figs, 62 tables. Cloth.
(Wheldon & Wesley) £50 [≈ $80]

Derham, W.
- Physico-Theology: or, a Demonstration of the Being and Attributes of God, from his Works of Creation. London: 1768. 8vo. xv,xvi, 444,[10] pp. 1 plate. Mod bds.
(Wheldon & Wesley) £40 [≈ $64]

Descartes, Rene
- Correspondence of Descartes and Constantyn Huygens 1635-1647. Edited from MSS now in the Bibliotheque Nationale by L. Roth. Oxford: 1926. 4to. lxxv,351 pp. 6 plates. Lib stamp on title verso. Lib cloth.
(Whitehart) £65 [≈ $104]

Description ...
- A Description of above Three Hundred Animals ... see Boreman, T.
- Description of the Patent Locomotive Steam Engine of Messrs. Robert Stephenson and Co. ... see Marshall, W.P.

Dewey, John
- Psychology. New York: Harper, 1887. 1st edn. 8vo. xii,427 pp. Orig mauve cloth, lightly shelfworn. *(Gach)* $150 [≈ £93]

Dexter, T.E.
- Animal and Vegetable Substances used in the Arts and Manufactures ... London: Groombridge, 1857. Sm 8vo. viii,184,[1 advt]

pp. Orig cloth gilt, front jnt cracking.
(Hollett) £35 [≈ $56]

Dickie, George
- A Flora of Ulster and Botanist's Guide to the North or Ireland. Belfast: C. Aitchison, 1864. 1st edn. Lge 12mo. xix,178 pp. Orig cloth.
(Fenning) £45 [≈ $72]

Dickinson, H.W.
- James Watt. Craftsman and Engineer. Cambridge: 1935. 207 pp. Ills. Foxing. Orig bndg, spine faded, cvrs rubbed.
(Book House) £25 [≈ $40]
- A Short History of the Steam Engine. Cambridge: UP, 1938. 1st edn. Roy 8vo. xvi, [ii], 255 pp. Frontis, 10 plates (1 fldg), text ills. Orig cloth, spine faded.
(Duck) £30 [≈ $48]

Dickinson, J.
- The Flora of Liverpool. London: 1851. 8vo. 166 pp. Orig cloth.
(Wheldon & Wesley) £35 [≈ $56]

Dickinson, R.E.
- Electric Trains. London: 1927. 1st edn. 8vo. xii,292 pp. Frontis, nearly 140 ills, inc 6 fldg plates. Some foxing & underlining &c. Orig cloth. *(Bow Windows)* £25 [≈ $40]

Dickinson, Robert Latou & Beam, Lura
- A Thousand Marriages. A Medical Study of Sex Adjustment. Foreword by Havelock Ellis. Baltimore: Williams & Wilkins, 1932. Roy 8vo. 482 pp. Orig bndg. Dickinson's sgnd inscrptn. *(Xerxes)* $60 [≈ £37]

Dickson, Adam
- The Husbandry of the Ancients. Edinburgh: 1788. 2 vols. 527 pp. Half calf.
(Moorhead) £195 [≈ $312]

Dickson, R.W.
- The Farmer's Companion; being a Complete System of Modern Husbandry ... Second Edition. London: Sherwood, Neely & Jones, 1813. 2 vols. 8vo. [vi],947,[i advt] pp. 103 plates (1 sl torn). Some foxing & browning. Sm hole in 1 leaf. Contemp half calf, sl worn.
(Blackwell's) £175 [≈ $280]
- Practical Agriculture; or, A Complete System of Modern Husbandry. London: for Richard Phillips, 1805. 1st edn. 2 vols. 4to. 87 tissued plates (inc 27 hand cold). Contemp calf, some wear to crnrs & hinges cracked or broken but held on cords. *(Claude Cox)* £250 [≈ $400]
- Practical Agriculture; or, A Complete System

of Modern Husbandry. London: for Richard Phillips ..., 1805. 1st edn. 2 vols. Thick 4to. 87 plates (inc 27 hand cold). Some offsetting. Contemp calf, rubbed, crnrs bumped, jnts started.   *(Elgen)* **$350 [≈ £218]**

Dickson, W.K.L. & Antonia
- The Life & Inventions of Thomas Alva Edison. London: Chatto & Windus, 1894. 4to. xvi, 362 pp. 200 ills. Orig cloth, t.e.g., dull & soiled, spine ends & crnrs sl rubbed & worn, crnrs sl bumped, hinges cracked but sound.   *(Duck)* **£65 [≈ $104]**

Dietrich, M.
- The Schulz Steam Turbine for Land and Marine Purposes. London: Owen, 1907. 73 pp. Ills. Orig bndg. *(Book House)* **£20 [≈ $32]**

Digby, Sir Kenelm
- Two Treatises: In the One of which, the Nature of Bodies, in the other, the Nature of Mans Soule, is looked into ... London: 1645. 1st English edn. 8vo. [22],429; [8],143,[26] pp. Sep title to each part. Lacks port. Contemp calf, rebacked. Wing D.1448.
                    (Rootenberg) **$550 [≈ £343]**
- Two Treatises: In the One of which, the Nature of Bodies; in the other the Nature of Mans Soule, is looked into ... London: Williams, 1658. 4to. Text diags. Panelled calf. Wing D.1450.
                    (Rostenberg & Stern) **$750 [≈ £468]**

Dircks, Henry
- Perpetuum Mobile; or, Search for Self-Motive Power, during the 17th, 18th and 19th Centuries. London: Spon, 1861. 1st edn. xli, 558,7 ctlg pp. Frontis, plates, ills. Orig cloth.   *(Wreden)* **$65 [≈ £40]**

The Diseases of Bath ...
- The Diseases of Bath. A Satire, Unadorn'd with a Frontispiece ... London: for J. Roberts ..., 1737. 1st edn. Folio. 20 pp. New mrbld bds.   *(Pickering)* **$450 [≈ £281]**

Dissertation ...
- A Dissertation on the Nature of Soils and the Properties of Manure: to which is added, the Method of Making a Universal Compost ... London: Sherwood, Gilbert & Piper, 1833. 8vo. xi, 141,45 pp. V occas foxing. MS note on endpaper. Contemp half calf, sl worn.
                    (Blackwell's) **£55 [≈ $88]**

Distant, W.L.
- Rhopalocera Malayana: a Description of the Butterflies of the Malay Peninsula. London:

1882-86. 2 vols. Roy 4to. xvi,481,[1],[4] pp. 46 cold plates, 129 text w'cuts. Green mor, spine faded.
                    (Wheldon & Wesley) **£980 [≈ $1,568]**

Disturnell, John
- Influence of Climate in North and South America ... New York: Van Nostrand, 1867. 1st edn. 334,2 advt pp. Cold frontis, fldg cold map. Lacks agricultural map. Orig cloth, a.e.g., spine ends worn, hinges cracked.
                    (Wreden) **$49.50 [≈ £31]**

Ditmars, R.L.
- The Reptiles of North America. New York: (1907) 1936. 4to. xvi,476 pp. 8 cold plates, over 400 photo ills. Orig cloth, inner jnt weak.   *(Wheldon & Wesley)* **£25 [≈ $40]**
- Reptiles of the World. New York: (1933) 1959. Rvsd edn. 8vo. xx,321 pp. Frontis, 90 plates. Orig cloth, sl stained.
                    (Wheldon & Wesley) **£25 [≈ $40]**

Dix, Thomas
- A Treatise on Land-Surveying ... Fifth Edition, Revised, Corrected, and Improved. London: 1829. Tall post 8vo. xii,216 pp. Fldg hand cold plan, frontis, ills. Contemp polished tree calf, sl rubbed, jnts cracking.
                    (Duck) **£40 [≈ $64]**

Dixon, F.
- The Geology of Sussex. Second Edition, revised by T.R. Jones. Brighton: 1878. Roy 4to. xxiv,469 pp. Frontis, fldg cold map (sl foxed), 63 plates (numbered 1-40, X-XXVIII, 18a, 18b, 32*, 38*), inc 2 hand cold. 1st plate & ends sl foxed. Orig cloth, worn.
                    (Henly) **£178 [≈ $284]**

Dixon, H.N.
- The Student's Handbook of British Mosses. London: 1924. 3rd edn. 8vo. xlvii,582 pp. 63 plates. Orig cloth, trifle used.
                    (Wheldon & Wesley) **£18 [≈ $28]**

Dixon, William Hepworth
- John Howard, and the Prison-World of Europe. From Original and Authentic Documents. By Hepworth Dixon ... New York: Robert Carter & Bros, 1849. 1st Amer edn, later iss. 12mo. [ii],401,[1],6,[2] pp. Orig embossed cloth, jnts worn.
                    (Gach) **$50 [≈ £31]**

Dobell, C.
- Antony van Leeuwenhoek and his "Little Animals", being some Account of the Father of Protozoology and Bacteriology and his

multifarious Discoveries in these Disciplines. London: 1932. 4to. vii,435 pp. 32 plates. Orig cloth, trifle soiled.
(Wheldon & Wesley) **£50 [≈ $80]**

Dobell, Horace

- Demonstrations in Diseases of the Chest, and their Physical Diagnosis. London: Churchill, 1858. 1st edn. 8vo. xv,115,32 ctlg pp. 10 chromolitho plates. Sl browning. Orig cloth, spine ends worn.   *(Elgen)* **$85 [≈£53]**

Dobson, Edward

- Brick and Tile Making. London: Crosby Lockwood, 1877. 6th edn. Sm 8vo. 276,32 advt pp. Ills. Lacks part of title. Orig bndg.
(Book House) **£16 [≈$25]**

Dobson, G.E.

- Monograph of the Asiatic Chiroptera and Catalogue of the Species of Bats in the Indian Museum, Calcutta. London: 1876. 8vo. viii, 228 pp. A few lib stamps. Orig cloth, trifle used.   *(Wheldon & Wesley)* **£60 [≈$96]**

The Doctrine of Fluxions ...

- See Emerson, William.

Dodd, A.H.

- The Industrial Revolution in North Wales. Univ of Wales Press: 1951. 2nd edn. 439 pp. Maps. Orig bndg. Dw.
(Book House) **£30 [≈$48]**

Dodd, George

- The Curiosities of Industry and the Applied Sciences. London: Routledge, 1852. 1st edn. 8vo. [4] pp, 16 sep paginated parts each with 24 pp except the last with 20. Contemp cloth.
(Rootenberg) **$750 [≈£468]**

Dodd, James Solas

- An Essay towards a Natural History of the Herring. London: for T. Vincent, 1752. 1st edn. 8vo. [8],178, pp, 12 contents ff, inc half-title, plus final 'Proposal for a Natural History of Esculent Fish' (called 'plate' in ESTC). Disbound. *(Hannas)* **£170 [≈$272]**

Dodd, Ralph

- Reports, with Plans, Sections, &c. of the Proposed Dry Tunnel, or Passage, from Gravesend, in Kent, to Tilbury in Essex ... London: for J. Taylor ..., 1798. Half-title. 1 p advts. Map, 2 plates. Edges sl browned. Orig wraps, uncut, split along spine. Cloth box.   *(Jarndyce)* **£420 [≈$672]**

Dodoens, R.

- A niewe Herball or historie of plants ... London: 1578. 1st edn of Lyte's translation. Sm folio. [xxiv],779,[24],[1] pp. W'cut title, port of Dodoens, over 800 w'cuts. A few minor reprs. Mor by Zaehnsdorf (ca 1910), a.e.g. *(Wheldon & Wesley)* **£2,000 [≈$3,200]**

Dolaeus, Johann.

- Systema Medicinale, a Compleat System of Physick, Theorical and Practical ... annexed, A Prefatory Discourse ... by William Salmon. London: 1686. 1st edn. 8vo. [30],516,360 pp. Lacks port. Sl worm. Title sl browned. Rec cloth. Wing D.1830A.
(Hemlock) **$475 [≈£296]**

Dolbear, A.E.

- The Telephone ... Boston: Lee & Shepard; New York: Charles T. Dillingham, 1877. 1st edn. 12mo. i-vi,7-128 pp. 18 ills. Orig terracotta grained cloth, black-stamped ills on sides of Dolbear's telephone, cvrs sl marked & soiled.   *(Duck)* **£350 [≈$560]**

The Domestic Habits of Birds ...

- See Rennie, J.

Donders, F.C.

- On the Anomalies of Accommodation and Refraction of the Eye ... Translated from the Author's Manuscript by William Daniel Moore. London: New Sydenham Soc, 1864. 1st edn in English. 8vo. 635 pp. Ills. Lib perf stamp on title. Orig cloth, shelf wear, front jnt starting.   *(Elgen)* **$250 [≈£156]**

Donisthorpe, H. St.J. K.

- British Ants, their Life-History and Classification. Plymouth: 1915. 1st edn. 8vo. [xvi],379 pp. 17 plates, num text figs. Orig cloth.   *(Bickersteth)* **£36 [≈$57]**
- British Ants, their Life-History and Classification. London: Routledge, 1927. 2nd edn, rvsd & enlgd. 8vo. xv,436 pp. 18 plates, 93 text diags. Orig cloth, extremities sl worn.
(Egglishaw) **£22 [≈$35]**

Donn, J.

- Hortus Cantabrigiensis, or a Catalogue of Plants, Indigenous and Exotic. Cambridge: 1809. 5th edn. 8vo. 266 pp. Calf.
(Wheldon & Wesley) **£30 [≈$48]**
- Hortus Cantabrigiensis; or an accented Catalogue of Plants, Indigenous and Exotic, cultivated in the Cambridge Botanic Garden. Edited by F. Pursch. London: 1819. 8vo. iv, 355 pp. Calf, trifle worn.

(Wheldon & Wesley) **£30 [≈ $48]**

Donnan, F.G. & Haas, Arthur (eds.)
- A Commentary on the Scientific Writings of J. Willard Gibbs. New Haven: 1936. 1st edn. 2 vols. Ills. Orig cloth. *(Elgen)* **$195 [≈ £121]**

Donovan, E.
- The Natural History of British Birds ... London: 1799. 5 vols. 8vo. 124 hand cold plates. Sl foxing. Tree calf, jnts sl rubbed.
(Wheldon & Wesley) **£1,350 [≈ $2,160]**
- The Natural History of British Fishes. London: 1802-08. 5 vols in 3. 8vo. 120 hand cold plates, all v clean without stamps. Lib blind stamps on titles. Lacks indexes to vols 2 & 4 & title to vol 5. 1 title & 1 plate reprd with tape. Half mor gilt, sl worn.
(Wheldon & Wesley) **£900 [≈ $1,440]**
- The Natural History of British Insects. London: 1793-1813. 16 vols in 8. Vol 1 'new edition'. 576 hand cold engvd plates. A few sm lib blind stamps. Green half mor gilt, a.e.g., some jnts trifle rubbed, 4 beginning to crack.
(Wheldon & Wesley) **£1,700 [≈ $2,720]**
- The Natural History of British Quadrupeds. London: 1820. 3 vols in 2. 8vo. 72 hand cold plates. Insignificant blind stamps on titles & plates. 2 title-pages transposed. Contemp half mor gilt, a.e.g.
(Wheldon & Wesley) **£1,000 [≈ $1,600]**
- The Natural History of the Nests and Eggs of British Birds. London: 1826. Oblong roy 8vo. 8 ff ptd text. 17 cold plates. 19 (of 21) mtd plate descriptions. Title & last leaf mtd, some sl soiling. Mod half mor.
(Wheldon & Wesley) **£700 [≈ $1,120]**

Donovan, Michael
- Domestic Economy. Vol.I. Brewing. Distilling. Wine-making. Baking, &c. London: Lardner's Cabinet Cyclopaedia, 1830. Vol 1 only. 1st edn. Sm 8vo. Orig cloth, ptd paper label.   *(Fenning)* **£35 [≈ $56]**
- A Treatise on Chemistry. London: Lardner's Cabinet Cyclopaedia, 1832. 1st edn. Sm 8vo. Orig cloth, ptd paper label.
(Fenning) **£45 [≈ $72]**
- A Treatise on Chemistry. London: Lardner's Cabinet Cyclopaedia, 1832. 1st edn. Sm 8vo. Rec paper bds.   *(Fenning)* **£35 [≈ $56]**

Dony, J.G.
- Flora of Bedfordshire. Luton: 1953. 8vo. 532 pp. Map, 25 plates. Orig cloth.
(Wheldon & Wesley) **£20 [≈ $32]**
- Flora of Hertfordshire. Hitchin: 1967. Roy

8vo. 112,[8] pp. 17 plates (1 cold), 706 maps. Orig cloth. *(Wheldon & Wesley)* **£15 [≈ $24]**

Dorr, J.V.N.
- Cyanidation and Concentration of Gold and Silver Ores. New York: McGraw-Hill, 1936. 1st edn. 8vo. ix,485 pp. Frontis, 132 ills. Orig cloth.   *(Gemmary)* **$60 [≈ £37]**

Dossie, Robert
- The Elaboratory Laid Open; or, the Secrets of Modern Chemistry and Pharmacy revealed ... Second Edition: With considerable Additions ... London: for J. Nourse, 1768. 8vo. xviii, [ii], 456,[viii] pp. Orig sheep, cvrs rubbed, spine ends worn, jnts cracked.
(Bickersteth) **£220 [≈ $352]**

Douglas, J.W. & Scott, J.
- The British Hemiptera, Vol 1 Hemiptera - Heteroptera. London: Ray Society, 1865. All published. 8vo. xii,628 pp. 21 plates. A little foxing. Orig cloth, sl used, jnts splitting.
(Wheldon & Wesley) **£30 [≈ $48]**

Douglas, James
- Myographiae Comparata Specimen: or, a Comparative Description of all the Muscles in a Man and in a Quadruped ... added, an Account of the Muscles peculiar to a Woman ... London: 1707. 1st edn. 12mo. xxxvi,216,16 pp. Orig calf, rebacked.
(Bickersteth) **£240 [≈ $384]**

Douglas, Richard
- Surgical Diseases of the Abdomen. Philadelphia: 1903. 1st edn. 8vo. 883 pp. 20 plates. Orig cloth, t.e.g., inner hinges cracked, sm tear at edge of spine.
(Elgen) **$100 [≈ £62]**

Doyle, Martin, et al.
- Rural Economy for Cottage Farmers and Gardeners: a Treasury of Information on keeping Sheep, Pigs, Poultry, the Horse, Pony, Ass, the Honey Bee ... London: Groombridge, [ca 1850s]. viii,303 pp. Text engvs. Orig dec cloth, sl worn.
(Box of Delights) **£20 [≈ $32]**

Drake, Daniel
- Natural and Statistical View, or Picture of Cincinnati in the Miami Country ... Cincinnati: 1815. 251,[4] pp. 2 fldg maps. Scattered foxing. Orig ptd bds, stained, spine chipped. Cloth box. *(Reese)* **$1,000 [≈ £625]**
- Natural and Statistical View, or Picture of Cincinnati and the Miami Country ... With an Appendix ... Cincinnati: 1815. 12mo.

251,[5] pp. Lacks plan & map. Scattered foxing, minor stains. Contemp tree calf, some wear. *(Hemlock)* **$225 [≈£140]**

Drake-Brockman, R.E.
- The Mammals of Somaliland. London: 1910. 8vo. Ills. Prelims v sl foxed. Orig bndg, sl mark on spine. *(Grayling)* **£75 [≈$120]**

Dresser, H.E.
- A Manual of Palearctic Birds. London: 1902-03. 2 vols in one. 8vo. 922 pp. 2 plates (1 cold, by Wolf). Cloth.
  (Wheldon & Wesley) **£40 [≈$64]**
- A Monograph of the Coraciidae, or Family of the Rollers. London: 1893. Imperial 4to. xx, 111 pp. 27 hand cold plates by Keulemans. Pp 65-72 supplied in facsimile. Lower blank marg of title restored. A few spots of foxing on 3 plates. Half mor.
  (Wheldon & Wesley) **£2,400 [≈$3,840]**

Druce, G.C.
- The Comital Flora of the British Isles. Arbroath: 1932. 8vo. xxxii,407 pp. Lge cold map, 2 ports. Orig cloth.
  (Wheldon & Wesley) **£20 [≈$32]**
- The Flora of Berkshire. Oxford: 1897. 8vo. cc, 644 pp. Map. Orig cloth.
  (Wheldon & Wesley) **£50 [≈$80]**
- The Flora of Buckinghamshire, with Biographical Notices ... Arbroath: 1926. Roy 8vo. cxxviii,437 pp. Map, 3 plates. Orig cloth, trifle used.
  (Wheldon & Wesley) **£45 [≈$72]**
- The Flora of Oxfordshire. London: 1886. 8vo. lii,452 pp. Orig cloth.
  (Wheldon & Wesley) **£50 [≈$80]**

Drummond, W.H.
- The Large Game and Natural History of South and South-East Africa. Edinburgh: Edmonston & Douglas, 1875. 1st edn. xvii,428 pp. Cold plates, text ills. Green half leather, raised bands.
  (Thornton's) **£175 [≈$280]**

Duchenne de Boulogne, G.B.
- Physiology of Motion, demonstrated by means of Electrical Stimulation and Clinical Observation and applied to the Study of Paralysis and Deformities ...Edited by Emanuel B. Kaplan. Philadelphia: 1949. 1st edn in English. One of 1500. Port. Half leather, minor ex-lib. *(Elgen)* **$90 [≈£56]**

Duck, John N.
- The Natural History of Portishead:

Comprising a Guide to the Locality, with an Appendix ... Ornithological, Entomological and Botanical ... Bristol: 1852. 12mo. 3 engvd views. Lacks map (never bound in). Orig cloth, sl loose. *(Ambra)* **£60 [≈$96]**

Dufour, John James
- The American Wine-Dresser's Guide, being a Treatise on the Cultivation of the Vine, and the Process of Wine-Making, adapted to the Soil and Climate of the United States. Cincinnati: S. Browne, 1826. [2],314,[2] pp. Occas foxing. Later half calf.
  (Reese) **$950 [≈£593]**

Dugmore, A. Radclyffe
- The Romance of the Newfoundland Caribou. London: 1913. Imperial 8vo. viii,191 pp. Cold frontis, fldg map, 63 plain plates, 39 text figs. Orig cloth.
  (Wheldon & Wesley) **£60 [≈$96]**

Duhamel du Monceau, M.
- The Elements of Agriculture. Translated from the Original French, and revised by Philip Miller. London: P. Vaillant, 1764. 1st English edn. 2 vols. 8vo. xx,445; [8],343 pp. 14 plates. Contemp half calf, rebacked & recrnrd, retaining orig mrbld paper bds.
  (Spelman) **£160 [≈$256]**

Du Moulin, Pierre
- The Elements of Logick ... now translated into English by Joshua Ahier. Oxford: Henry Hall, 1647. 2nd English edn. Sm 8vo. [10],155 pp. Title sl soiled. Qtr calf, mrbld bds. Wing D.2583. Sgntr of Richard Cope (1776-1856). *(Rootenberg)* **$200 [≈£125]**
- A Letter of a French Protestant to a Scotishman of the Covenant. London: Young & Badger, 1640. 4to. Mor. STC 7345.
  (Rostenberg & Stern) **$225 [≈£140]**

Dunbar, James
- The Practical Papermaker ... Second Edition. Leith: Reid & Son; London & New York: Spon, 1881. Sm 8vo. 72,8 advt pp. Orig cloth, cvrs faded. *(Charles B. Wood)* **$100 [≈£62]**

Duncan, Andrew
- Observations on the Distinguishing Symptoms of Three Different Species of Pulmonary Consumption. 1st American from 2nd London Edition. Philadelphia: 1819. Some foxing of endpapers. Top cvr detached.
  (Rittenhouse) **$105 [≈£65]**

Duncan, G.S.
- A Bibliography of Glass ... London: 1960.

4to. viii,544 pp. Orig buckram. Dw.
(Bow Windows) £80 [≃ $128]

Duncan, James
- Introduction to Entomology. Edinburgh:
Jardine's Naturalist's Library, 1840. 1st edn.
331 pp. Port, hand cold title vignette, 31 hand
cold & 5 plain plates. Orig cloth.
(Egglishaw) £52 [≃ $83]
- The Natural History of Bees. Edinburgh:
Jardine's Naturalist's Library, 1840. Cr 8vo.
x,301 pp. Pages 11-16 were not published.
Port, cold vignette, 19 cold & 11 plain plates.
Orig cloth, sl spotted.
(Wheldon & Wesley) £50 [≃ $80]
- The Natural History of Beetles. Edinburgh:
Jardine's Naturalist's Library, 1835. 1st edn.
Sm 8vo. 269 pp. Port, cold title vignette, 30
hand cold plates. Binder's cloth.
(Egglishaw) £34 [≃ $54]
- The Natural History of Beetles. Edinburgh:
Jardine's Naturalist's Library, 1835. 1st edn.
Sm 8vo. 269 pp. Port, cold title vignette, 30
hand cold plates. Occas spotting. Contemp
mor gilt, a.e.g.   *(Egglishaw)* £48 [≃ $76]
- The Natural History of British Butterflies ...
Edinburgh: Jardine's Naturalist's Library,
1840. 246 pp. Port, cold title vignette, 2 plain
& 32 hand cold plates. Occas spotting.
Contemp mor gilt, a.e.g.
(Egglishaw) £48 [≃ $76]
- The Natural History of British Moths,
Sphinxes, etc. Edinburgh: Jardine's
Naturalist's Library, 1841. 268 pp. Port,
addtnl engvd title, 30 hand cold plates.
Contemp mor gilt, a.e.g.
(Egglishaw) £48 [≃ $76]
- The Natural History of Foreign Butterflies.
London: 1837. 208 pp. 33 cold plates. Half
calf, some wear.   *(Moorhead)* £25 [≃ $40]
- The Natural History of Foreign Butterflies.
Edinburgh: Jardine's Naturalist's Library,
1837. 1st edn. 8vo. 208 pp. Port, hand cold
vignette title, 30 hand cold plates. Contemp
mor gilt, a.e.g.   *(Egglishaw)* £50 [≃ $80]

Duncan, James Matthews
- Researches in Obstetrics. Edinburgh: Adam
& Charles Black, 1868. 1st edn. 8vo. xiv, [ii],
467 pp,3 advt ff. 40 text figs. Orig maroon
cloth. Author's pres inscrptn.
(Bickersteth) £85 [≃ $136]

Duncan, P.M.
- The Transformations of Insects. London:
Cassell, 1882. 8vo. xii,531 pp. 40 plates, num
text ills. Orig dec cloth.
(Egglishaw) £28 [≃ $44]

Dundas, L.M.
- A Big Game Pocket Book for Kenya Colony.
London: 1927. 8vo. Ills. Orig bndg, hd of
spine chipped.   *(Grayling)* £30 [≃ $48]

Dunell, H.
- British Wire-Drawing and Wire-Working
Machinery. London: Constable, 1925. 4to. 188
pp. Ills. Orig bndg. *(Book House)* £75 [≃ $120]

Dunkin, Edwin
- The Midnight Sky: Familiar Notes on the
Stars and Planets. London: [1869]. 1st edn.
Lge 8vo. [x],326 pp. 32 plates, text figs. Orig
cloth gilt, slit in upper jnt.
(Bickersteth) £22 [≃ $35]

Dunn, Samuel
- Cosmography Epitomised, in Six Copper
Plate Delineations. London: for Robert
Sawyer, 1774. 6 dble folio engvd sheets,
partly in mezzotint, partly in line engraving.
465 x 750 mm.   *(Hannas)* £85 [≃ $136]

Duppa, R.
- The Classes and Orders of the Linnaean
System of Botany. Illustrated by Select
Specimens of Foreign and Indigenous Plants.
London: 1816. 3 vols. 8vo. 237 hand cold &
3 plain plates. A little minor foxing. Contemp
green calf gilt, trifle faded.
(Wheldon & Wesley) £380 [≃ $608]

Durant, Ghislani
- Hygiene of the Voice. New York: G.
Schirmer, 1870. 8vo. 126 pp. 8 text figs. 6 pp
inserted advts at front. Orig green cloth gilt.
Inscrbd 'Compliments of the Author'.
(Karmiole) $45 [≃ £28]

Dyar, H.G.
- The   Mosquitoes   of   the   Americas.
Washington: 1928. Roy 8vo. v,616 pp. 123
plates of 418 figs. Half cloth.
(Wheldon & Wesley) £50 [≃ $80]

Dykes, W.
- The Genus Iris. Cambridge: 1913. 1st edn.
Folio. 48 cold plates, 30 line drawings. Orig
cloth gilt, lower bd sl damped.
(Hollett) £550 [≃ $880]
- A Handbook of Garden Irises. London: 1924.
8vo. xi,250 pp. 24 plates. Orig cloth, trifle
marked.   *(Wheldon & Wesley)* £22 [≃ $35]
- Notes on Tulip Species. London: 1930.
Folio. 108 pp. 54 cold plates by E.K. Dykes.
Orig cloth, t.e.g., sl worn.
(Henly) £138 [≃ $220]

E., T.
- Some Considerations about the Reconcileableness of Reason and Religion ... see Boyle, Robert.

Ealand, C.A.
- Insect Life. London: A. & C. Black Colour Books, Twenty Shilling Series, 1921. 1st edn. 50 cold & 24 b/w plates. Orig dec cloth, spine rubbed.   *(Old Cathay)* **£25 [≈ $40]**
- Insect Life. London: A. & C. Black Colour Books, Twenty Shilling Series, 1921. 1st edn. 50 cold & 24 b/w plates. Pencilled marginalia. Orig dec cloth, sl grubby & faded with occas stains & wear to spine.
                                 (Old Cathay) **£15 [≈ $24]**

Earle, Pliny
- Memoirs of Pliny Earle, M.D., with Extracts from his Diary and Letters (1830-1892) ... Edited ... by F.B. Danborn. Boston: Damrell & Upham, [1898]. 1st trade edn. 8vo. [xvi],409,[3] pp. Frontis. Title leaf detaching. Orig olive cloth.   *(Gach)* **$75 [≈ £46]**

The Earthquake of Juan Fernandez ...
- See Sutcliffe, Thomas.

Eastern Hospitals and English Nurses ...
- See Taylor, Fanny M.

Eastman, R.
- The Kingfisher. London: 1969. 8vo. 160 pp. Cold & plain plates. Orig cloth.
                        (Wheldon & Wesley) **£18 [≈ $28]**

Eastman, Seth
- A Treatise on Topographical Drawing. New York: Wiley & Putnam, 1837. 1st edn. 68 pp. Num fldg plates. Orig cloth, worn. Inscrbd by the author.   *(Jenkins)* **$475 [≈ £296]**

Eaton, Elon Howard
- Birds of New York. New York: New York State Museum Memoir 12, 1910. 2 vols. 4to. 501; 719 pp. 106 cold & num plain plates. Orig cloth gilt, trifle worn, crnrs bumped.
                           (Hollett) **£130 [≈ $208]**

Eaton, John Matthews
- A Treatise on the Art of Breeding and Managing Tame, Domestic, Foreign and Fancy Pigeons ... London: for the author, 1858. 8vo. [1],xix,200,12 advt pp. 18 hand cold plates inc frontis. Contemp crimson calf, blindstamped panels, v fine.
                           (Rootenberg) **$550 [≈ £343]**

Eberle, John
- A Treatise of the Materia Medica and Therapeutics. Second Edition, with corrections. Baltimore & Philadelphia: 1824. 2 vols. 8vo. x,327,[1]; 401 pp. End of vol 1 foxed & sl stained. Contemp calf, not quite uniform, worn, hinges cracked.
                          (Hemlock) **$125 [≈ £78]**

Ebstein, Wilhelm
- Corpulence and its Treatment on Physiological Principles. Translated and adapted by Emil W. Hoeber. New York: Brentanos, 1884. 12mo. 49 pp. Orig wraps.
                           (Xerxes) **$75 [≈ £46]**

Eccles, A. Symons
- Difficult Digestion due to Displacements. New York: William Wood, 1898. 1st edn. 8vo. 138 pp. Ills. Orig bndg.
                           (Xerxes) **$55 [≈ £34]**

Echeverria, M. Gonzalez
- Reflex Paralysis: its Pathological Anatomy, and relation to the Sympathetic Nervous System. New York: Bailliere Bros, 1866. 8vo. 80,8 pp. 2 heliograph plates. Orig cloth, inner hinges weak.
                   (Charles B. Wood) **$275 [≈ £171]**

Eddington, A.S.
- Fundamental Theory. Cambridge: 1946. viii, 292 pp. Orig cloth, spine sl faded.
                          (Whitehart) **£45 [≈ $72]**
- The Internal Constitution of the Stars. Cambridge: (1926) 1930. 4to. x,407 pp. Orig bndg.   *(Whitehart)* **£25 [≈ $40]**
- The Nature of the Physical World. Cambridge: 1928. 1st edn. 8vo. xix,361 pp. 8 figs. Orig cloth, sl dull.
                          (Whitehart) **£15 [≈ $24]**
- The Philosophy of Physical Science. Cambridge: 1939. 1st edn. ix,230 pp. Sl foxing on endpapers & page edges. Dw sl foxed.   *(Whitehart)* **£15 [≈ $24]**
- Relativity Theory of Protons and Electrons. Cambridge: 1936. 1st edn. 4to. vi, 336 pp. Occas sl foxing. Orig cloth, worn on edges.
                          (Whitehart) **£60 [≈ $96]**
- Space, Time and Gravitation. An Outline of the General Relativity Theory. Cambridge: 1920. 1st edn. vii,218 pp. A few figs. Orig cloth, sl worn   *(Whitehart)* **£25 [≈ $40]**

Eddy, Mary Baker Glover
- Science and Health. Boston: Christian Scientist Publishing Company, 1875. 1st edn. 8vo. 456,[2] pp. Errata leaf. Owner's stamp

erased from 1st leaf of preface. Blue mor gilt, a.e.g.   *(M & S Rare Books)* **$950 [≈£593]**
- Science and Health. Third Edition, Revised. Lynn, Mass.: Dr. Asa G. Eddy, 1881. 1st 2-vol edn, with addtnl material. 2 vols. 8vo. xiv,270; [4],214 pp. Frontis. Orig green cloth gilt, extremities sl rubbed. Sales receipt from 1919 laid in ($100). *(Karmiole)* **$300 [≈£187]**
- Science & Health, with Key to the Scriptures. Boston: the author, 1886. 21st edn, rvsd. 590 pp. Port. Orig bndg.   *(Argosy)* **$75 [≈£46]**

Edge, A.B.B. & Laby, T.H.
- The Principles and Practice of Geophysical Prospecting, being the Report of the Geophysical Experimental Survey (Australia). 1931. 4to. viii,372 pp. 261 figs (4 fldg). Lib stamp on title verso & at end. Orig cloth.
(Henly) **£26 [≈$41]**

Edinburgh New Dispensatory ...
- Edinburgh New Dispensatory ... Being an Improvement upon the New Dispensatory of Dr. Lewis. Edinburgh: for Charles Elliot ..., 1786. 1st edn. 8vo. xxxii,720 pp. 3 plates. Rather browned, stained or spotted in places. Old calf, rubbed & worn.
(Hollett) **£75 [≈$120]**
- Edinburgh New Dispensatory ... Being an Improvement upon the New Dispensatory of Dr. Lewis. Edinburgh: for Charles Elliot ..., 1789. 2nd edn. 8vo. xxxii,656 pp. 6 plates. Half-title & final ff rather browned & stained. Old calf, rebacked, rubbed.
(Hollett) **£75 [≈$120]**
- Edinburgh New Dispensatory ... With a Full and Clear Account of the New Chemical Doctrines published by M. Lavoisier. Edinburgh: for William Creech, 1801. 8vo. xxxi, 622,[1] pp. 4 fldg plates (1 creased & damped). Some marg stains. Tree calf, sl bumped.   *(Hollett)* **£75 [≈$120]**

Edinburgh Philosophical Journal
- The Edinburgh Philosophical Journal exhibiting a View of the Progress of Discovery in Natural Philosophy ... Conducted by Dr Brewster and Professor Jameson ... Edinburgh: 1819-26. 1st series complete. 14 vols. 8vo. 131 (of 137) plates. Orig bds, sl rubbed.
(Pickering) **$3,500 [≈£2,187]**

Edlin, A.
- A Treatise on the Art of Bread-Making ... London: Vernor & Hood, 1805. 1st edn. 12mo. xxiv, 216,[2] pp. 5 fldg plates. Half calf, mrbld bds. Contemp annotations on last blank.   *(Rootenberg)* **$500 [≈£312]**

Edlin, H.L.
- Trees, Wood and Man. London: Collins, New Naturalist Series, 1956. 1st edn. Cold & plain plates. Orig cloth. Dw.
(Egglishaw) **£30 [≈$48]**

Edwards, E. Price & Williams, T.
- The Eddystone Lighthouses (New and Old): an Account of the Building and General Arrangement of the New Tower ... London: 1882. 1st edn. viii,186 pp. 5 plates, text ills. Orig olive cloth, sl rubbed & marked.
(Duck) **£85 [≈$136]**

Edwards, F.W., et al.
- British Bloodsucking Flies. London: BM, 1939. Roy 8vo. viii,156 pp. 44 cold & 1 plain plates, 64 text figs. Orig cloth.
(Wheldon & Wesley) **£30 [≈$48]**
- British Bloodsucking Flies. London: BM, 1939. Roy 8vo. viii,156 pp. 44 cold & 1 plain plates, 64 text figs. Orig cloth.
(Fenning) **£45 [≈$72]**

Edwards, Revd Z.I.
- The Ferns of the Axe and its Tributaries ... With an Account of the Flower Lobelia Urens found near Axminster, and nowhere else in Great Britain. London: 1862. 8vo. Ills. Occas sl spotting. Orig cloth, hd of spine sl rubbed, top outer hinge partly split.
(Ambra) **£35 [≈$56]**

Eisler, Robert
- Man Into Wolf. An Anthropological Interpretation of Sadism, Masochism and Lycanthropy. A Lecture delivered at a Meeting of the Royal Society of Medicine. London: Spring, [ca 1950]. 8vo. 284 pp. Orig bndg. Dw.   *(Xerxes)* **$45 [≈£28]**

The Elaboratory Laid Open ...
- See Dossie, Robert.

Ellerman, J.R.
- The Families and Genera of Living Rodents, with a List of Names ... by R.W. Hayman and G.W. Holt. London: 1940-41. 1st edn. 2 vols. Sm 4to. xxvi,689; xii,690 pp. 239 text figs. Orig cloth.   *(Henly)* **£98 [≈$156]**

Ellerman, J.R. & Morrison-Scott, T.C.S.
- Checklist of Palaearctic and Indian Mammals 1758 to 1946. London: British Museum, 1951. Cr 4to. 810 pp. Map. Orig buckram, spine faded. *(Wheldon & Wesley)* **£50 [≈$80]**

Elles, G.L. & Wood, E.M.
- British Graptolites. Edited by C. Lapworth.

London: 1901-18. 2 vols. 4to. clxxi, 539 pp. 52 plates, 359 figs. Cloth.
(Baldwin) **£130 [≈ $208]**

Elliot, Daniel Giraud
- Catalogue of the Collection of Mammals in the Field Columbian Museum. Chicago: 1907. 694 pp. 8vo. 92 text figs. Orig cloth, loose in cvrs.
(Wheldon & Wesley) **£25 [≈ $40]**
- A Check List of Mammals of the North American Continent, the West Indies and Neighbouring Seas. Chicago: Field Museum, 1905. 8vo. 761 pp. Orig wraps.
(Wheldon & Wesley) **£50 [≈ $80]**
- The Gallinaceous Game Birds of North America. Second Edition. New York: Francis. P. Harper, 1897. 8vo. 220,ii,i pp. 46 plates, colour chart inside rear cvr. Orig cloth.
(Charles B. Wood) **$135 [≈ £84]**
- The Land and Sea Mammals of Middle America and the West Indies. Chicago: 1904. 2 vols. 8vo. 933 pp. 68 plates, 309 figs. Plain wraps.   *(Wheldon & Wesley)* **£45 [≈ $72]**
- The Life and Habits of Wild Animals. Illustrated by Designs by Joseph Wolf. London: 1874. Imperial 4to. ix,72 pp. 20 plates. Orig dec cloth.
(Wheldon & Wesley) **£70 [≈ $112]**
- The Life and Habits of Wild Animals, illustrated by Designs by Joseph Wolf ... With Descriptive Letterpress by Daniel Giraud Elliot. London: Macmillan, 1874. Folio. vi,72 pp. 20 engvs. Orig cloth gilt, a.e.g., backstrip reprd.
(Egglishaw) **£48 [≈ $76]**
- A Review of the Primates. New York: 1912 [1913]. 3 vols. Imperial 8vo. 28 cold plates by Wolf & Smit, 140 plain plates. Sl adhesion damage to blank marg of 1 plate, sl offset from plates. Mod buckram, orig wraps (minor reprs) bound in, crnrs of vol 3 trifle bumped.
(Wheldon & Wesley) **£550 [≈ $880]**

Elliotson, John
- Address, delivered at the Opening of the Medical Session in the University of London, October 1, (1832). 4to. 16 pp. Sgntr & sm stamp on title. Later wraps.
(Hemlock) **£200 [≈ £125]**

Elliott, A.G.
- Gas and Petroleum Engines. Translated and Adapted from the French of Henry de Graffigny ... London: Whittaker, 1898. Post 8vo. x,140, advt pp. 52 w'engvd ills. Orig cloth, cvrs soiled.   *(Duck)* **£20 [≈ $32]**

Ellis, Asa
- The Country Dyer's Assistant. Brookfield, Ma.: for E. Merriam, [1798 (sic)]. 139,[4] pp. Sl chipping to edge of title. Orig bds, somewhat soiled, rebacked.
(Reese) **$1,650 [≈ £1,031]**
- The Country Dyer's Assistant. Brookfield, Ma.: E. Merriam & Co. for the author, [1799 (sic)]. 1st edn. Sm 8vo. 139,[4] pp. Rebound in half calf, antique style.
(Charles B. Wood) **$1,000 [≈ £625]**

Ellis, Havelock
- The World of Dreams. London: 1911. 1st edn. 288 pp. Orig cloth, t.e.g., crnrs bumped.
(Moorhead) **£15 [≈ $24]**

Ellis, J.
- An Essay towards a Natural History of the Corallines. London: 1755. 4to. xvii,[x], 103, [1] pp. Frontis (with offset from title), 39 plates. Title sl browned. Contemp calf.
(Wheldon & Wesley) **£250 [≈ $400]**

Ellis, J. & Solander, D.
- The Natural History of many Curious and Uncommon Zoophytes. London: Benjamin White, 1786. 4to. xii,208 pp. 63 engvd plates. Tissue guards. Half calf, mrbld bds.
(Egglishaw) **£340 [≈ $544]**

Ellis, John, M.D.
- Personal Experience of a Physician, with an Appeal to the Medical and Clerical Professions. Philadelphia: Hahnemann, 1892. 12mo. 134 pp. Wraps.
(Xerxes) **$65 [≈ £40]**

Elwes, H.J. & Henry, A.
- The Trees of Great Britain and Ireland. Edinburgh: priv ptd, 1906-13. 7 vols & general index. 4to. 7 cold titles, 5 cold frontis, port, 414 plates. Occas sl foxing. Buckram.
(Henly) **£700 [≈ $1,120]**

Emanuel, H.
- Diamonds and Precious Stones: their History, Value, and distinguishing Characteristics ... Second Edition ... London: 1867. 8vo. xxii,266,32 advt pp. Cold title, 4 plates, table of values. Some spotting. Orig cloth, marked & a little rubbed. *(Bow Windows)* **£60 [≈ $96]**

Emerson, Haven
- Selected Papers. Battle Creek, Mich.: W.K. Kellogg Foundation, 1949. 1st edn. Sm 4to. 507 pp. Orig bndg.   *(Xerxes)* **$55 [≈ £34]**

Emerson, William
- The Doctrine of Fluxions ... London: for J. Richardson, 1757. 2nd edn, enlgd. 8vo. 12 fldg plates. Contemp mottled calf gilt, sl crack in upper jnt.   *(Ximenes)* **$400** [≃ £250]
- The Elements of Geometry ... London: for F. Wingrave ..., 1794. New edn. 8vo. iii,216 pp. 14 fldg plates. Contemp calf, rebacked.
  (Young's) **£72** [≃ $115]
- The Elements of Optics. In Four Books ... London: for J. Nourse, 1768. 1st edn. 2 parts in one vol. 8vo. [iv],xii,244,[2]; vi,111 pp. 13 + 15 plates. Some sl waterstain. Contemp mrbld bds, uncut, new calf spine. The 2nd part comprises Emerson's 'Perspective'.
  (Burmester) **£150** [≃ $240]
- Miscellanies. Or a Miscellaneous Treatise; containing Several Mathematical Subjects. London: for J. Nourse, 1776. 1st edn. 8vo. ciii, 504 pp. 10 fldg plates. Contemp half calf gilt, outer hinges cracked but holding.
  (Karmiole) **$375** [≃ £234]
- The Principles of Mechanics ... Second Edition, corrected, and very much enlarged ... London: for J. Richardson, 1758. 4to. viii, [ii],284 pp. 43 fldg plates. Orig calf, rebacked.
  (Bickersteth) **£290** [≃ $464]
- The Projection of the Sphere, Orthographic, Stereographic and Gnomical ... London: J. Bettenham for W. Innys, 1749. 1st edn. 8vo. iv,52 pp. 12 fldg plates. Contemp mrbld bds, new calf spine.   *(Bickersteth)* **£95** [≃ $152]
- A System of Astronomy ... London: for J. Nourse, 1769. 8vo. x,ii,383 pp. Errata, advt leaf. 16 fldg plates. Old tree calf, edges worn & bumped, front jnt cracking.
  (Hollett) **£120** [≃ $192]

Encyclopaedia ...
- Encyclopaedia of the Social Sciences. Edited by Edwin R.A. Seligman and Alvin Johnson. New York: Macmillan, 1937. 5th printing. The dble vol iss, 13 vols in 8. Heavy 4to. Orig blue buckram.   *(Gach)* **$150** [≃ £93]

Englefield, Sir Henry C.
- A Description of the Principal Picturesque Beauties of the Antiquities and Geological Phenomena of the Isle of Wight ... London: 1816. Large Paper. Folio. Port (spotted), 3 maps, 46 plates. Sl dampstain to 1 map. Half calf. Baron Nortwick's b'plate.
  (Henly) **£450** [≃ $720]

Engleheart, N.B.
- A Concise Treatise on Eccentric Turning ... London: Holtzapffel, 1867. 2nd edn. 4to. iv, 152,[2] pp. 16 plates, a few text ills. Orig dec

royal blue cloth gilt, uncut, spine ends & jnts sl rubbed.   *(Duck)* **£135** [≃ $216]

Ennemoser, Joseph
- The History of Magic ... Apparitions, Dreams, Second Sight, Somnambulism, Predictions, Divination ... Selected by Mary Howitt. Translated from the German by William Howitt. London: Bohn, 1854. 1st edn in English. 12mo. Orig cloth, sl worn.
  (Gach) **$85** [≃ £53]

Episodes of Insect Life ...
- See Budgen, M.L.

Erichsen, Hugo
- Cremation of the Dead considered from an Aesthetic, Sanitary, Religious, Historical, Medico-Legal and Economical Standpoint. Detroit: Erichsen, 1887. 1st edn. 8vo. 264 pp. Ills. Orig bndg.   *(Xerxes)* **$75** [≃ £46]

Essay(s) ...
- An Essay concerning the Infinite Wisdom of God, manifested in the Contrivance and Structure of the Skin of Human Bodies. By a Lover of Physick and Surgery. London: for Joseph Marshall, 1724. Sole edn. 8vo. Title, vi, 63,[1 advt] pp. Sl dusty. Sewed as issued.
  (Bickersteth) **£580** [≃ $928]
- Essays on Agriculture and Planting founded on Experiments made in Ireland. By a Country Gentleman. Dublin: ptd by Graiseberry & Campbell, for William Jones, 1790. 1st edn. 8vo. [iv],xvi,127 pp. Half-title. rec bds.   *(Burmester)* **£180** [≃ $288]
- Essays on Husbandry ... see Harte, W.
- Essays Relating to Agriculture, and Rural Affairs ... see Anderson, James.
- Essays, by a Society of Gentlemen, at Exeter ... see Polwhele, Richard (ed.).

Euclid
- The Elements of Euclid Explain'd, in a New, but most Easie Method ... Written in French ... Third Edition. Oxford: L.L. for M. Gillyflower & W. Freeman, 1700. 8vo. iv,380 pp. Sl waterstained at beginning.
  (Bickersteth) **£150** [≃ $240]
- The Elements of Euclid ... Edited by Robert Simson. Glasgow: Robert & Andrew Foulis, 1756. 1st English edn of Simson's Euclid. Large Paper. 4to. [8],431 pp. Diags. Calf, spine edges sl worn. *(Elgen)* **$350** [≃ £218]
- The Elements of Euclid, with Dissertations ... By James Williamson ... Oxford: Clarendon Press, 1781-88. 1st edn thus. 2 vols in one. 4to. [2],iii,96, 145,68;

[2],xxiii,367 pp. Contemp green half vellum. *(Fenning)* **£125 [≈ $200]**
- The Elements of Euclid, viz, the First Six Books, together with the Eleventh and Twelfth ... By Robert Simpson. To this Seventh Edition are also annexed Elements of ... Trigonometry. Edinburgh: 1787. 8vo. vii,[2], 10-520 pp. Contemp tree calf, crnrs bumped. *(Spelman)* **£50 [≈ $80]**
- See also Simson, Robert.

Euler, Leonard
- Elements of Algebra ... Translated from the French by Rev. J. Hewlett ... London: 1828. 4th edn. xxx,593 pp. Mod bds, paper label. *(Whitehart)* **£60 [≈ $96]**

Evans, A.H.
- A Fauna of the Tweed Area. Edinburgh: 1911. 8vo. xxviii,262 pp. Dec title, cold map, 21 plates, 7 text figs. Some foxing. Orig cloth, spine faded. *(Wheldon & Wesley)* **£55 [≈ $88]**
- A Fauna of the Tweed Area. Edinburgh: Douglas, 1911. 8vo. xxviii,262 pp. Dec title, cold map (on linen), 21 plates, 7 text figs. Orig cloth, hd of spine sl nicked. *(Egglishaw)* **£46 [≈ $73]**

Evans, A.H. & Buckley, T.E.
- A Vertebrate Fauna of the Shetland Islands. Edinburgh: 1899. 8vo. xxix,248 pp. Map, 15 plates. Half-title reprd. Orig cloth. *(Egglishaw)* **£42 [≈ $67]**

Evans, Thomas W.
- History of the American Ambulance established in Paris during the Siege of 1870-71 ... London: Chiswick Press, Sampson Low ..., 1873. 1st edn. 4to. xxxviii,694 pp. Errata leaf. Frontis, 10 plates, 61 text w'cuts. Orig cloth, sl worn. *(Elgen)* **$400 [≈ £250]**

Evans, William Julian
- The Sugar-Planter's Manual, being a Treatise on the Art of obtaining Sugar from the Sugar Cane. Philadelphia: Lea & Blanchard, 1848. 1st Amer edn. 8vo. 264,[24 ctlg dated Aug 1851] pp. 2 fldg plates. Orig cloth, spine ends & ft of 1 jnt sl defective. *(Bickersteth)* **£85 [≈ $136]**

Eve, A.S. & Creasy, C.H.
- Life and Work of John Tyndall. London: 1945. 1st edn. xxxii,404 pp. Port frontis, plates. Dw. *(Elgen)* **$65 [≈ £40]**

Eve, Paul F.
- A Collection of Remarkable Cases in Surgery. Philadelphia: 1857. 1st edn. 8vo. 858 pp. Text ills. Some browning. Orig sheep, sl scuffed. *(Elgen)* **$275 [≈ £171]**

Evelyn, John
- Kalendarium Hortense: or, the Gard'ners Almanac, directing what he is to do monthly throughout the year. And what Fruits and Flowers are in Prime. London: 1673. 5th edn. Sm 8vo. 127,[8] pp. Lacks half-title. Blind stamp on title. New calf. *(Wheldon & Wesley)* **£120 [≈ $192]**
- Kalendarium Hortense: or, the Gardeners Almanac ... Seventh Edition, with many useful Additions. London: 1683. 8vo. 127,[ix] pp. Intl blank. Some v sl marg worm. Contemp calf, rebacked, crnrs worn. Wing E.3497. Contemp MS list of 13 gardening books. *(Clark)* **£200 [≈ $320]**
- Kalendarium Hortense: or, the Gard'ner's Almanac, directing what he is to do monthly throughout the year; and what Fruits and Flowers are in Prime. London: 1706. 10th edn. 8vo. [vi],x,[xiv], 170,[14] pp. Frontis, plate. Mod half calf. *(Wheldon & Wesley)* **£85 [≈ $136]**
- The Miscellaneous Writings. Now First Collected, with Occasional Notes by William Upcott. London: Colburn, 1825. 1st edn. Tall thick 4to. 875 pp. 4 plates (inc 1 mezzotint). Tan half mor, orig bds. *(Hartfield)* **$395 [≈ £246]**
- A Philosophical Discourse of Earth ... London: John Martyn, 1676. 1st edn. Sm 8vo. 182 pp. Imprimatur leaf laid down. Title margs reprd. Lacks blank M4. Contemp sheep, rebacked, new endpapers. Contemp MS notes. Wing E.3507. *(Blackwell's)* **£750 [≈ $1,200]**
- A Philosophical Discourse of Earth, relating to the Culture and Improvement of it for Vegetation. London: 1676. Sm 8vo. 182 pp. Contemp inscrptns on endpapers. Contemp roan. *(Wheldon & Wesley)* **£350 [≈ $560]**
- Sylva, or a Discourse of Forest Trees and the Propagation of Timber in His Majesties Dominions ... Pomona ... Kalendarium Hortense. London: 1670. 2nd edn, enlgd. Sm folio. [xlvii],247,[iv],67,33,[1] pp. 5 engvs. A few marg stains. Contemp calf, jnt cracked. *(Wheldon & Wesley)* **£170 [≈ $272]**
- Sylva, or, A Discourse of Forest-Trees, and the Propagation of Timber in His Majesties Dominions ... Second Edition much Inlarged and Improved. London: 1670. Folio. xlvii, 247,[v], 67,[i],33,[ii] pp. Title vignette, 4

plates. Occas marks. Contemp calf, worn.
(Finch) **£250 [≈ $400]**

- Sylva, or a Discourse of Forest-Trees ... Terra
... Pomona ... Kalendarium Hortense.
London: 1679. 3rd edn. Sm folio. [lxx],412,
38, [1] pp. Sl worming of lower marg at
beginning. New calf backed bds.
(Wheldon & Wesley) **£175 [≈ $280]**

- Silva ... Terra ... Pomona ... Acetaria ...
Kalendarium Hortense ... London: 1706. 4th
edn. Sm folio. [xxxv],384, [iii],275,[v] pp.
Port. 1 or 2 marg tears & wormholes.
Contemp panelled calf, rebacked, trifle worn.
(Wheldon & Wesley) **£250 [≈ $400]**

- Sylva: or A Discourse of Forest-Trees ...
London: 1776. 1st Hunter edn. 4to. liv,649,ix
pp. Fldg table, 40 plates. Contemp diced calf,
spine relaid. *(Henly)* **£245 [≈ $392]**

- Sylva, or a Discourse of Forest Trees. Reprint
of the Fourth Edition, 1706, with an Essay on
the Life and Works of the Author by J.
Nisbet. London: [1908]. 2 vols. Roy 8vo.
Minor foxing. Orig cloth.
(Wheldon & Wesley) **£35 [≈ $56]**

Everard, Anne
- Flowers from Nature, with the Botanical
Name, Class, and Order; and Instructions for
Copying ... London: Dickinson, 1835. Slim
folio. [6],13 pp. Hand cold frontis (sl foxed),
12 hand cold litho plates, each with leaf of
text. Orig cloth, spine faded & sm repr.
(Spelman) **£1,200 [≈ $1,920]**

Ewell, Thomas
- Plain Discourses on the Laws or Properties of
Matter ... Addressed to all American
Promoters of Useful Knowledge. New York:
Brisban & Brannan, 1806. Only edn. 8vo.
469,iii pp. 2 plates. Sheep, black label, sm
chip in hd of spine.
(Charles B. Wood) **$400 [≈ £250]**

Exner, John
- Rorschach Systems. New York & London:
Grune & Stratton, 1969. 1st edn. 8vo. 381 pp.
Orig bndg. *(Xerxes)* **$45 [≈ £28]**

Experiments ...
- Experiments and Observations on the
Malvern Waters ... see Wall, John.

- Experiments lately made by several Eminent
Physicians, on the Surprising and Terrible
Effects, of Almond-Water and Black-Cherry-
Water ... London: for J. Huggonson, 1741.
Sole edn. 8vo. Title,62 pp. Lower marg of
title soiled. Old plain wraps, soiled.
(Bickersteth) **£280 [≈ $448]**

An Explanation ...
- An Explanation of the Nature of Equation of
Time, and of the Use of the Equation Table
for adjusting Watches and Clocks to the
Motion of the Sun ... London: for F. Clay,
1731. Sole edn. Sm 4to. [iv],22 pp. 2 engvd
plates on leaf facing title. Wraps, sl worn &
soiled. *(Bickersteth)* **£550 [≈ $880]**

Fabre, J.H.
- Fabre's Book of Insects. Retold from
Alexander Teixeira de Matto's Translation ...
by Mrs Rodolph Stawell. Illustrated by E.J.
Detmold. London: Hodder & Stoughton,
(1921). 4to. 12 cold plates. Foredges foxed.
Orig pict cloth gilt. *(Stewart)* **£95 [≈ $152]**

- Fabre's Book of Insects. Retold from A.T. de
Matto's Translation, by Mrs Stawell.
Illustrated by E.J. Detmold. New York: 1936.
x, 271 pp. 12 cold plates. Orig green cloth
gilt. *(Henly)* **£50 [≈ $80]**

- Insect Life. Souvenir of a Naturalist. London:
1901. Cr 8vo. xii,320 pp. 16 plates. Orig cloth
gilt, a.e.g. *(Henly)* **£20 [≈ $32]**

Fairholt, Frederick William
- Tobacco: Its History and Associations.
London: 1859. 1st edn. 8vo. Engvd frontis,
100 w'engvd ills by the author. Orig cloth.
(Robertshaw) **£24 [≈ $38]**

Fairley, W.
- The Practice and Science of Mining
Engineering. A Manual adapted for the Use
of Mine Owners, Mining Engineers, Colliery
Managers ... Newcastle on Tyne: James
Fairley, 1896. 8vo. 366 pp. Orig bndg.
(Book House) **£20 [≈ $32]**

A Familiar Essay on Electricity ...
- See Dale, G.E.

Faraday, Michael
- Chemical Manipulation; being Instructions
to Students in Chemistry ... London: W.
Phillips ..., 1827. 1st edn. 8vo. 656 pp. W'cut
text ills. Later blue half mor, gilt spine.
(Frew Mackenzie) **£350 [≈ $560]**

- Chemical Manipulation; being Instructions
to Students in Chemistry ... London: W.
Philips ... W. Tait, Edinburgh; Hodges &
M'Arthur, Dublin, 1827. 1st edn. 8vo.
W'engvd text ills. Rec half calf, uncut.
(Pickering) **$1,200 [≈ £750]**

- A Course of Six Lectures on the Chemical
History of a Candle ... Edited by William
Crookes ... London: Griffin, Bohn, 1861. 1st
edn. Cr 8vo. viii,208,[8 advt] pp. 38 ills. Orig

red cloth, a little dull. *(Fenning)* **£85 [≈$136]**

- Faraday's Diary ... Foreword by Sir William H. Bragg. London: G. Bell & Sons, 1932-36. 8 vols (inc index). Lge 8vo. Frontises, text ills. Orig cloth. Dws, as new.
(Rootenberg) **$600 [≈£375]**

- Experimental Researches in Chemistry and Physics. Reprinted from the Philosophical Transactions ... and other Publications. London: Taylor & Francis, 1859. 1st edn. 8vo. viii, 496 pp. 3 plates (1 fldg). Orig cloth.
(Bickersteth) **£365 [≈$584]**

- Experimental Researches in Chemistry and Physics ... London: Taylor & Francis ..., 1859. 1st edn of the collection. 8vo. viii, 496 pp. Erratum slip at p 445. 3 fldg plates. Lib stamp at ft of title. Orig cloth, worn.
(Pickering) **$650 [≈£406]**

Farish, William
- A Plan of a Course on Lectures on Arts and Manufactures, more particularly such as relate to Chemistry. Cambridge: J. Smith, 1813. 8vo. 37 pp. Sev lib stamps. Interleaved copy with notes by a student. Contemp mrbld paper bds, rebacked with cloth.
(Offenbacher) **$350 [≈£218]**

Farmer's ...
- The Farmer's Guide ...see Young, Arthur.
- The Farmer's Kalendar ... see Young, Arthur.

- The Farmer's Tour ... see Young, Arthur.

Farrar, John
- An Experimental Treatise on Optics, comprehending the Leading Principles of the Science ... Cambridge, MA: 1826. 8vo. vii,350 pp. 61 plates, 183 figs. Orig bds, front cvr detached, rear hinge somewhat loose.
(Rittenhouse) **$125 [≈£78]**

Farrer, Reginald
- The English Rock-Garden. London: (1919) 1930. 2 vols. Roy 8vo. 100 plates. 1 or 2 pp carelessly opened. Orig cloth, trifle scuffed.
(Wheldon & Wesley) **£50 [≈$80]**

Farrington, O.C.
- Gems and Gem Minerals. Chicago: Mumford, 1903. 8vo. xii,229 pp. 16 cold plates, 61 text figs. Rebound in cloth, good ex-lib. *(Gemmary)* **$75 [≈£46]**

Farrow, E.P.
- Plant Life on East Anglian Heaths. Cambridge: 1925. Roy 8vo. 108 pp. 23 plates. Orig cloth. *(Wheldon & Wesley)* **£30 [≈$48]**

Fay, Sir Samuel
- A Royal Road: Being the History of the London & South Western Railway, from 1825 to the Present Time. Kingston-on-Thames: W. Drewett, 1883. [iv],138,[2] pp. 4 plates inc frontis. Orig gilt dec royal blue cloth, spine ends & crnrs v sl rubbed. *(Duck)* **£55 [≈$88]**

Featherstonehaugh, G.W.
- Geological Report of an Examination made in 1834 of the Elevated Country between the Missouri and Red Rivers ... Washington: 1835. 8vo. 97 pp. Fldg cold section (nearly 10 feet long). Sm lib stamp in text & title verso. Rebound in qtr calf, new endpapers.
(Henly) **£120 [≈$192]**

Featon, Mr. & Mrs E.H.
- The Art Album of New Zealand Flora; being a Systematic and Popular Description of the Native Flowering Plants of New Zealand and the adjacent Islands. Vol 1 [all published]. Wellington: 1889. 4to. xviii,180 pp. Cold frontis & 39 cold plates. Mor gilt, sl loose.
(Wheldon & Wesley) **£300 [≈$480]**

Fellows, Reginald B.
- History of the Canterbury and Whitstable Railway ... Canterbury: Jennings, 1930. 1st edn. 4to. 94 pp. 2 fldg plans & gradient section, 11 plates, text ills. Orig blue cloth, faded & sl soiled, spine ends sl rubbed, endpapers sl spotty. *(Duck)* **£45 [≈$72]**

The Female Instructor ...
- The Female Instructor; or, Young Woman's Companion: being a Guide to all the Accomplishments which adorn the Female Character ... added, Useful Medicinal Receipts ... Cookery ... Liverpool: [ca 1846?]. 560 pp. Frontis, 7 plates. Qtr roan, worn.
(Box of Delights) **£45 [≈$72]**

Fenning, Daniel
- The Young Man's Book of Knowledge: being a proper Supplement to the Young Man's Companion ... Fourth Edition, Revised, Corrected, and greatly Improved ... London: Crowder & Collins, 1786. 12mo. xiv,381,errata pp. Fldg tables, diags. Contemp calf, rebacked. *(Lamb)* **£45 [≈$72]**

Ferenczi, Sandor
- Thalassa. A Theory of Genitality. Translated by Henry Alden Bunker. New York: Psychoanalytic Quarterly, 1938. 8vo. 110 pp. Orig bndg. *(Xerxes)* **$45 [≈£28]**

Ferguson, James
- Analysis of a Course of Lectures on Mechanics, Hydrostatics, Hydraulics, Pneumatics, Electricity, and Astronomy. London: ptd in the year, 1771. 7th edn. 8vo. 44 pp. Rec bds.   *(Burmester)* **£60 [≈ $96]**
- Astronomy Explained upon Sir Isaac Newton's Principles ... Sixth Edition. London: Strahan ..., 1778. Tall thick 8vo. [viii], 501,[15] pp. Fldg frontis, 17 fldg plates. Contemp tree calf, leather label.
      (Hartfield) **$295 [≈ £184]**
- Astronomy Explained upon Sir Isaac Newton's Principles ... London: A. Strahan ..., 1803. 11th edn. 8vo. [viii],503,[17] pp. Fldg frontis, 17 fldg plates. 19th c blue half calf, sl rubbed.   *(Burmester)* **£85 [≈ $136]**
- An Introduction to Electricity. In Six Sections ... London: for W. Strahan ..., 1775. 2nd edn. 8vo. [iv],140 pp. 3 fldg plates. Endpapers & title sl soiled in margs. Contemp calf, minor wear at hd of spine, lacks label.
      (Burmester) **£90 [≈ $144]**
- Lectures on Select Subjects in Mechanics, Hydrostatics, Pneumatics and Optics: with the Use of the Globes, the Art of Dialling ... London: A. Millar, 1760. 1st edn. 8vo. [8], 417, [7] pp. 23 fldg plates (4 sl frayed along foredge). Old calf, rebacked.
      (Claude Cox) **£135 [≈ $216]**
- Lectures on Select Subjects in Mechanics, Hydrostatics, Pneumatics, and Optics: with the Use of the Globes, the Art of Dialing ... London: A. Millar, 1760. 1st edn. 8vo. [x], 417, [6] pp. 23 fldg plates. A few pp sl spotted. Orig calf, rebacked.
      (Bickersteth) **£280 [≈ $448]**
- The Young Gentleman and Lady's Astronomy, familiarly explained in Ten Dialogues between Neander and Eudosia ... Eighth Edition. Dublin: 1808. 8vo. [viii],248 pp. 8 fldg plates. Contemp polished sheep, jnts sl cracked but firm. *(Rankin)* **£40 [≈ $64]**

Ferguson, John
- Bibliotheca Chemica. A Bibliography of Books on Alchemy Chemistry and Pharmaceutics. London: 1954. 2 vols. Lge 8vo. xxi,487; 598 pp. Orig buckram, red labels.   *(Frew Mackenzie)* **£90 [≈ $144]**
- Bibliotheca Chemica. London: Vandenhoeck & Ruprecht, 1954. Reprint edn. 2 vols. Lge 8vo. Orig cloth, crnrs bumped. Dw vol 1.
      (Gach) **$150 [≈ £93]**

Ferriar, John
- An Essay towards a Theory of Apparitions. London: Cadell & Davies, 1813. 1st edn. 8vo.

x,[2],139,[5] pp. Qtr calf.
      (Rootenberg) **$350 [≈ £218]**

Feuchtersleben, Baron Ernst von
- Health and Suggestion: The Dietetics of the Mind. Translated and edited by Ludwig Lewisohn. New York: Huebsch, 1910. 1st edn. Sm 8vo. 168 pp. Orig bndg.
      (Xerxes) **$75 [≈ £46]**
- The Principles of Medical Psychology being the Outline of a Course of Lectures ... Revised and edited by B.G. Babington. London: Sydenham Society, 1847. Orig green cloth.   *(Robertshaw)* **£25 [≈ $40]**

Figuier, Louis
- The Ocean World. London: Chapman & Hall, 1869. 8vo. xii,615 pp. 32 plates, 396 text figs. Orig cloth, rear hinge reprd.
      (Egglishaw) **£20 [≈ $32]**
- The World before the Deluge. London: 1867. 8vo. viii,439,1 pp. 33 plates, 201 text figs. Orig cloth, foredges of cvrs sl dampstained.
      (Henly) **£16 [≈ $25]**

Finch, Thomas
- Elements of Self-Improvement comprising a Familiar View of the Intellectual Powers and Moral Characteristics of Human Nature. London: 1822. 1st edn. 12mo. Contemp calf.
      (Robertshaw) **£32 [≈ $51]**

Finn, Frank
- The Wild Beasts of the World. London: Jack, [1909]. 4to. xi,404 pp. 100 cold ills. Some sl browning of 1st & last ff. Orig dec cream cloth, dust soiled.   *(Blackwell's)* **£35 [≈ $56]**

Finsch, O.
- On a Collection of Birds from North-Eastern Abyssinia and the Bogos Country, with Notes by the Collector W. Jesse. London: Zoo. Trans., 1870. 134 pp. 4to. Map, 4 cold plates. Wraps.   *(Wheldon & Wesley)* **£50 [≈ $80]**

Firebaugh, Ellen
- The Physician's Wife and the Things that pertain to Her Life. Philadelphia: F.A. Davis, 1904. 8vo. 186 pp. Ills. Orig bndg, front cvr soiled, front hinge cracking.
      (Xerxes) **$65 [≈ £40]**

First Russian Railroad ...
- First Russian Railroad from St. Petersburg to Zarscoe-Selo and Pawlowsk, Established by Imperial Decree of 21st March, 1836 ... Translated from the Russian. St. Petersburg, 1837. London: Skipper & East, [March 1837].

44 pp. Dampstain. Orig wraps, worn.
(Duck) **£350 [≈ $560]**

Fisher, J.
- The Fulmar. London: New Naturalist, 1952.
1st edn. 8vo. xv,496 pp. 4 cold & 48 plain
plates, 70 maps & diags. Orig cloth, trifle
used, faded. *(Wheldon & Wesley)* **£60 [≈ $96]**

Fisher, J. & Lockley, R.M.
- Sea-Birds. London: Collins, New Naturalist,
1954. 1st edn. 8vo. xvi,320 pp. Num ills.
Orig cloth. *(Wheldon & Wesley)* **£50 [≈ $80]**

Fisher, Ronald Aylmer
- The Genetical Theory of Natural Selection.
Oxford: Clarendon Press, 1930. 1st edn. 8vo.
[8], xiv,272 pp. 2 cold plates, text ills. Orig
cloth, fine. W.H. Beveridge's b'plate.
(Rootenberg) **$250 [≈ £156]**

Fisher, W. Clark
- The Potentiometer and its Adjuncts. London:
[ca 1893]. x,194 pp. 88 text ills (2 fldg). Orig
cloth, spine ends frayed. *(Elgen)* **$65 [≈ £40]**

Fisher, William Logan
- Pauperism and Crime. Philadelphia: the
author, 1831. 1st edn. Sm 8vo. [120] pp.
Cloth backed bds.   *(Gach)* **$150 [≈ £93]**

Fiske, John
- Darwinism and Other Essays. London:
Macmillan, 1879. 1st edn. 8vo. viii,283, [5],
[advts dated March 1879] pp. Orig blue cloth.
Rowland G. Hazard's copy.
(Gach) **$125 [≈ £78]**

Fitter, R.S.R.
- London's Natural History. London: Collins,
New Naturalist Series, 1945. 1st edn. Cold &
plain plates, maps & diags. Name on
endpaper. Orig cloth. Dw.
(Egglishaw) **£12 [≈ $19]**

Fitton, E. & S.M.
- Conversations on Botany. London: 1818. 2nd
edn. Post 8vo. xxii,215 pp. 20 hand cold
plates. Some faint offsetting on the plates.
Orig bds, uncut, spine sl defective.
(Wheldon & Wesley) **£45 [≈ $72]**
- Conversations on Botany. London: Longman,
1834. 8th edn. Sm 8vo. xvi,284 pp. 22 hand
cold plates (one on 2 ff). Half calf.
(Egglishaw) **£75 [≈ $120]**

Fitton, W.H.
- A Geological Sketch of the Vicinity of

Hastings. London: 1833. 1st edn. 12mo. viii,
94 pp. Fldg plate illustrating 4 cold sections,
sev text figs. Orig cloth, faded & dull, spine
ends a little rubbed.
(Bow Windows) **£50 [≈ $80]**

Fitzgibbon, later Chambelon, Agnes
- Canadian Wild Flowers ... see Traill, C.P.

Fitzroy, Robert
- The Weather Book: A Manual of Practical
Meteorology. London: Longman, Green,
1863. 1st edn. 8vo. 464 pp. 16 fldg plates (1
on blue paper). Orig blue cloth, inner hinges
sl strained.   *(Moon)* **£95 [≈ $152]**

Fitzsimons, F.W.
- The Natural History of South Africa -
Mammals. London: 1919-20. 4 vols. 8vo. 222
ills. Orig cloth, spines faded, press marks on
spines.   *(Wheldon & Wesley)* **£75 [≈ $120]**

Flagg, John Foster Brewster
- Ether and Chloroform: their Employment in
Surgery,   Dentistry,   Midwifery,
Therapeutics, Etc. Philadelphia: Lindsay &
Blakiston, 1851. 1st edn. Sm 8vo. 189,2 advt
pp. Occas foxing. Three qtr leather, rubbed.
(Elgen) **$400 [≈ £250]**

Flatters, A.
- Methods in Microscopical Research -
Vegetable Histology. London: Sherratt &
Hughes, 1905. 1st edn. 4to. x,116 pp. 23 cold
plates, 29 text figs. Orig red cloth, spine & bd
edges worn.   *(Savona)* **£25 [≈ $40]**

Fleet, John
- A Discourse relative to the Subject of
Animation ... Boston: John & Thomas Fleet,
1797. 1st edn. 8vo. [3],25,[1] pp. Half- title.
Appendix. Mod cloth.
(Rootenberg) **$350 [≈ £218]**

Fleming, John
- A History of British Animals ... Edinburgh:
for Bell & Bradfute ..., 1828. 1st edn.
xxiii,565 pp, corrigenda. Orig cloth & bds,
untrimmed, paper spine label worn.
(Schoyer) **$60 [≈ £37]**

Fletcher, John
- Mirror of Nature, Part I. Presenting a Brief
Sketch of the Science of Phrenology. Boston:
Cassady, 1839. 8vo. 23 pp. Frontis. A few ink
numbers in text. Wraps. *(Xerxes)* **$50 [≈ £31]**

Fleure, H.J.
- A Natural History of Man in Britain. London: Collins, New Naturalist Series, 1951. 1st edn. 8vo. Cold & plain plates. Orig cloth. Dw.   *(Egglishaw)* **£20 [≃ $32]**

Flint, Austin
- On Variations in Pitch in Percussion and Respiratory Sounds and their Application to Physical Diagnosis. Buffalo: for private distribution, 1852. 1st edn. 8vo. 47 pp. Orig limp black cloth.   *(Hemlock)* **$300 [≃ £187]**
- A Treatise on the Principles and Practice of Medicine. Philadelphia: Henry C. Lea, 1866. 1st edn. 867,32 ctlg pp. Orig cloth, spine ends sl worn.   *(Elgen)* **$275 [≃ £171]**

Flint, Austin, Jr.
- Collected Essays and Articles on Physiology and Medicine. New York: 1903. 1st edn. 2 vols. Port. Orig armorial cloth, t.e.g. Sgnd inscrptn by Flint.   *(Elgen)* **$250 [≃ £156]**

Flintoft, J.
- Complete Collection of the British Ferns and their Allies in the English Lake District collected and arranged by J. Flintoft, Keswick, Cumberland. [Cumberland: ca 1850]. 4to. Ptd title. 37 ff of mtd specimens (1 unmtd but intact). Orig cloth.
   (Charles B. Wood) **$300 [≃ £187]**

Flora Europaea ...
- Flora Europaea. Edited by T.G. Tutin, V.H. Heywood and others. Cambridge: (1964-80). 5 vols. 4to. Orig cloth.
   (Wheldon & Wesley) **£320 [≃ $512]**

Floyd, W.
- Hints on Dog Breaking. London: 1882. 8vo. Ills. Orig bndg.   *(Grayling)* **£30 [≃ $48]**

Fomon, Samuel
- The Surgery of Injury and Plastic Repair. Baltimore: 1939. 1st edn. Lge 8vo. 1409 pp. Num ills (some cold). Orig cloth.
   (Elgen) **$150 [≃ £93]**

Foord, A.H.
- The Carboniferous Cephalopoda of Ireland. London: Pal. Soc. Monograph, 1897-1903. 4to. 234 pp. 49 plates. New cloth.
   (Baldwin) **£65 [≃ $104]**

Foot, Jesse
- The Life of John Hunter. London: T. Becket, 1794. 287 pp. Contemp tree calf, red label, spine sl rubbed. *(C.R. Johnson)* **£475 [≃ $760]**

Forbes, Arthur Litton Armitage
- Diseases of the Nose and Naso-Pharynx. London: Henry Renshaw, 1889. 1st edn. Sm 8vo. 156 pp. 39 text figs. Orig cloth, spine & foredges of cvr sl waterstained.
   (Bickersteth) **£35 [≃ $56]**

Forbes, Edward
- Echinodermata of the British Tertiaries. London: 1852. 4to. vii,36 pp. 4 plates. Half leather, rubbed.   *(Henly)* **£40 [≃ $64]**
- A History of British Starfishes, and other Animals of the Class Echinodermata. London: Van Voorst, 1841. 1st edn. Tall 8vo. xx,267, [2], 2 advt pp. Num ills, inc tail-pieces. Orig cloth.   *(Schoyer)* **$50 [≃ £31]**
- A Monograph of the British Naked-Eye Medusae, with Figures of all the Species. London: Ray Society, 1848. Folio. 104 pp. 13 cold plates. Trifle foxed & used. Orig cloth.
   (Wheldon & Wesley) **£50 [≃ $80]**
- On the Tertiary Fluvio-Marine Formation of the Isle of Wight. London: 1856. 8vo. xviii, 162 pp. Hand cold geological map, 2 hand cold sections, 7 plates, 20 w'cut ills. Floral cloth.   *(Henly)* **£68 [≃ $108]**

Forbes, Edward & Hanley, S.
- A History of British Mollusca and their Shells. London: [1848]-1853. Large Paper. 4 vols. Roy 8vo. 202 hand cold & 2 plain plates. Some (mostly sl) foxing of plates. Contemp half mor gilt, jnts a little rubbed & beginning to crack.
   (Wheldon & Wesley) **£900 [≃ $1,440]**
- A History of British Mollusca and their Shells. London: Van Voorst, 1848-1853. 4 vols. 8vo. 203 plain litho plates. Lib stamp on titles. Orig cloth, partly unopened.
   (Egglishaw) **£165 [≃ $264]**

Forbes, H.O.
- A Hand Book to the Primates. London: lloyd's Natural History, 1896-97. 2 vols. 8vo. 8 fldg maps, 41 cold plates. Orig cloth, vol 2 warped.   *(Wheldon & Wesley)* **£30 [≃ $48]**

Forbes, R.J.
- Short History of the Art of Distillation ... Leiden: 1948. Roy 8vo. [7],405 pp. 203 ills. Dw. George Sarton's sgntr.
   (Elgen) **$95 [≃ £59]**

Ford, E.B.
- Butterflies. London: Collins, New Naturalist Series No 1, 1945. 8vo. 56 cold & 24 b/w plates. Orig cloth. Sl rubbed dw.
   (Moon) **£20 [≃ $32]**

Forrest, H.E.
- The Vertebrate Fauna of North Wales. London: 1907. lxxii,537 pp. 28 plates. New cloth.   *(Henly)* **£65 [≈ $104]**
- The Vertebrate Fauna of North Wales. London: 1907. 8vo. lxxiv,537 pp. Cold map, 28 plates. Orig cloth, trifle used, sound ex-lib.
               (Wheldon & Wesley) **£30 [≈ $48]**

Forrest, James
- The Recognition of Ocular Disease. A Treatise for Opticians. Birmingham & London: J. & H. Taylor, 1911. 1st edn. 8vo. [vi],170 pp, inc 9 pp illust advts. 9 cold plates, num text figs. Orig cloth, a little spotted.
               (Bickersteth) **£35 [≈ $56]**

Forshaw, J.M.
- Parrots of the World. Melbourne: 1973. 1st edn. Imperial 4to. 584 pp. 158 cold plates by W. Cooper. Distribution maps. Review pasted on half-title. Blind-stamp on half-title. Orig cloth.
               (Wheldon & Wesley) **£330 [≈ $528]**

Forster, Johann Reinhold
- Observations made during a Voyage Round the World ... London: G. Robinson, 1778. 1st edn. 4to. Title,dedic, [i]-iii,[i]-iv, [9]-649, errata, subscribers pp. Fldg map, fldg chart. New cloth backed bds, uncut.
               (Frew Mackenzie) **£1,500 [≈ $2,400]**

Forster, T.
- Observations on the Brumal Retreat of the Swallow. London: 1813. 3rd edn, crrctd & enlgd. 8vo. xiv,46 pp. Leather backed bds.
               (Wheldon & Wesley) **£48 [≈ $76]**

Forster, W.
- A Treatise on a Section of the Strata from Newcastle upon Tyne to Cross Fell. London: 1883. 3rd edn. 8vo. lvi,208 pp. 14 plates. New buckram.   *(Baldwin)* **£130 [≈ $208]**

Forsyth, A.R.
- Lectures on the Differential Geometry of Curves and Surfaces. Cambridge: 1912. xxiii, 525 pp. Lib stamps at ft of title. Orig cloth, sl traces of label removal.
               (Whitehart) **£35 [≈ $56]**
- Theory of Functions of a Complex Variable. Cambridge: 1893. 1st edn. 4to. xxii,682 pp. 125 text figs. Traces of label removal from front inner cvr. Orig cloth, edges sl worn, inner hinge cracked but firm.
               (Whitehart) **£30 [≈ $48]**

Forsyth, W.
- Observations of the Diseases, Defects, and Injuries in all kinds of Fruit and Forest Trees, with an Account of a Particular Method of Cure. London: 1791. 8vo. [iv],71 pp. Loose in wraps.   *(Wheldon & Wesley)* **£48 [≈ $76]**
- A Treatise on the Culture and Management of Fruit Trees: in which a New Method of Pruning and Training is fully described. London: 1802. 1st edn. 4to. viii,371,[1] pp. 13 plates. Sl foxing at beginning. reverse of 1 plate soiled. New cloth.
               (Wheldon & Wesley) **£100 [≈ $160]**
- A Treatise on the Culture and Management of Fruit Trees. London: 1802. 1st edn. 4to. viii, 372 pp. 13 fldg plates. Tree calf, rebacked.   *(Henly)* **£120 [≈ $192]**
- A Treatise on the Culture and Management of Fruit Trees. London: 1803. 2nd edn. 8vo. xxvii, 523 pp. 13 fldg plates (1 reprd at fold). Tree calf, rebacked.   *(Henly)* **£85 [≈ $136]**
- A Treatise on the Culture and Management of Fruit-Trees. London: 1803. 2nd edn. 8vo. xxviii, 523 pp. 13 plates. Mod half calf.
               (Wheldon & Wesley) **£55 [≈ $88]**
- A Treatise on the Culture and Management of Fruit Trees ... Third Edition, with Additions. London: Longman & Rees, 1803. 8vo. xxx, 523 pp. 13 fldg plates. Contemp elab gilt red half mor. *(Gough)* **£90 [≈ $144]**

Foster, Sir Michael
- Lectures on the History of Physiology during the Sixteenth, Seventeenth and Eighteenth Centuries. Cambridge: 1901. 1st edn. 8vo. 310 pp. Frontis. Orig cloth.
               (Elgen) **$85 [≈ £53]**

Fourcroy, Antoine Francois de
- Elements of Natural History, and of Chemistry ... Translated into English. With Occasional Notes ... London: Robinson, 1788. 4 vols. 8vo. 10 fldg tables. 2 tears without loss. Orig tree calf, contrasting labels, 3 sm rough patches on cvrs.
               (Bickersteth) **£480 [≈ $768]**

Fourier, Joseph
- The Analytical Theory of Heat. Translated with Notes by Alexander Freeman. Cambridge: UP, 1878. 1st edn in English. 8vo. xxiii,466, [2], 32 advt pp. Text diags. Orig cloth.   *(Rootenberg)* **$400 [≈ £250]**

Fowler, James
- Dictionary of Practical Medicine. London: Churchill, 1890. 1st edn. Thick sm 4to. 942 pp. Orig bndg, hinges cracked, a few stains to

bds, backstrip tearing along hinges.
(Xerxes) **$65 [≈ £40]**

Fox, George Henry
- Photographic Atlas of the Diseases of the Skin. Philadelphia: Lippincott, 1904. 4 vols. Folio. 96 photo plates. Orig bndg, ex-lib, v worn, some backstrips detached.
(Xerxes) **$75 [≈ £46]**

Fox, Joseph
- The Natural History of the Human Teeth ... The History and Treatment of the Diseases of the Teeth ... London: Thomas Cox, 1803, 1806. 1st edn. 2 vols in one. Large Paper. 4to. vii,100; [2],170 pp. 22 plates. Contemp calf, rebacked.   *(Rootenberg)* **$2,500 [≈ £1,562]**

Fox, Tilbury & T.
- Epitome of Skin Diseases, with Formulae. Philadelphia: Lea, 1879. 2nd Amer edn. 8vo. 216 pp. Some spotting. Orig bndg, both hinges cracked.   *(Xerxes)* **$50 [≈ £31]**

Fox Talbot, H.
- See Talbot, H. Fox.

Francatelli, Charles E.
- The Cook's Guide and Housekeeper's & Butler's Assistant: A Practical Treatise on English and Foreign Cookery in all its Branches. London: Bentley, 1869. 24th thousand. 8vo. xx,524,advt pp. Over 40 ills. Some marking & foxing. Orig green cloth gilt, sl shaken.   *(Moon)* **£35 [≈ $56]**
- The Modern Cook: a Practical Guide to the Culinary Art in all its Branches ... London: Bentley, [1853]. 10th edn, rvsd & enlgd. xv, 552 pp. Port, text engvs. Orig cloth, pict spine relaid, new endpapers.
(Box of Delights) **£65 [≈ $104]**
- The Modern Cook: a Practical Guide to the Culinary Art in all its Branches ... London: 1880. 26th edn. Top marg of title trimmed to just above title. Mod (not rec) cloth.
(Box of Delights) **£30 [≈ $48]**

Francis, F.
- Fish-Culture, a Practical Guide to the Modern System of Breeding and Rearing Fish. London: Routledge, 1863. 1st edn. 8vo. xviii, 267 pp. Ills. Half mor.
(Egglishaw) **£34 [≈ $54]**

Francis, G.W.
- An Analysis of the British Ferns and their Allies. Fourth Edition. London: Simpkin Marshall, 1851. 8vo. viii,88,8 ctlg pp. Pict

title, 9 plates, text ills. Orig cloth.
(Claude Cox) **£25 [≈ $40]**

Francis, J.B.
- Lowell Hydraulic Experiments being a Selection from Experiments on Hydraulic Motors on the Flow of Water over Weirs, in Open Canals, Etc. New York: Van Nostrand, 1909. 5th rvsd edn. Lge 4to. 286 pp. 23 plates. Ex-lib but vg. Orig bndg.
(Book House) **£120 [≈ $192]**

Franklin, Benjamin
- Experiments and Observations on Electricity, made at Philadelphia in America ... London: F. Newbery, 1774. 5th & most complete edn. 4to. [ii],vi,514,[16] pp. 7 plates. Occas sl foxing. Orig calf, rebacked.
(Charles B. Wood) **$1,100 [≈ £687]**
- Memoirs of the Life and Writings ... Written by Himself to a Late Period, and continued to the Time of his Death by William Temple Franklin ... London: Colburn, 1818-17. 1st edn. 3 vols. 4to. 10 plates inc map, port & facs. Occas orig damp stain. Rec half calf.
(Young's) **£250 [≈ $400]**

Fraser, Henry & Stanton, A.T.
- Collected Papers on Beri-Beri ... London: John Bale, 1924. 4to. vii,103 pp. 8 plates. A few lib stamps. Orig cloth.
(Bickersteth) **£36 [≈ $57]**

Fraser, John
- Tuberculosis of the Bones and Joints in Children. New York: 1914. 1st edn. Roy 8vo. 352 pp. 51 plates (2 cold), ills. Orig cloth.
(Elgen) **$75 [≈ £46]**

Fraser, L.
- Zoologica Typica, or Figures of New and Rare Mammals and Birds described in the Proceedings ... of the Zoological Society of London. London: [1846-48-] 1849. One of 250. Demy folio. Cold title & 70 hand cold plates. Contemp mor gilt.
(Wheldon & Wesley) **£3,000 [≈ $4,800]**

Frazer, Sir James George
- The Golden Bough. London: Macmillan, 1890. 1st edn. 2 vols. 8vo. xii,409; 2,407 pp. Frontis. Orig dec cloth gilt, fine.
(Minster Gate) **£350 [≈ $560]**
- The Golden Bough. A Study in Comparative Religion. London: Macmillan, 1890. 1st edn. 2 vols. 8vo. xii,409,[2]; 407 pp. Contemp MS notes on endpapers & 2 pp of text. Orig green cloth gilt.   *(Frew Mackenzie)* **£195 [≈ $312]**

- The Golden Bough. London: Macmillan, 1922. 12 vols. Lge 8vo. Orig cloth gilt.
(Hollett) **£120 [≈ $192]**
- The Golden Bough a Study in Magic and Religion. London: Macmillan, 1932-36. 12 vols. 8vo. Frontis in vol 1. Orig gilt dec cloth, t.e.g., 1 or 2 sl marks.
(Bow Windows) **£285 [≈ $456]**
- The Golden Bough: A Study in Magic and Religion. London: Macmillan, 1951. 3rd edn, later printing. 12 vols. 8vo. Orig green cloth gilt.
(Gach) **$325 [≈ £203]**
- Lectures on the Early History of Kingship. London: Macmillan, 1905. 1st edn. 8vo. viii, 309, [2 advt] pp. Blind 'Presentation Copy' stamp on title. Orig green cloth, sides soiled & faded, spine ends sl rubbed.
(Frew Mackenzie) **£30 [≈ $48]**

Frazer, Persifor
- Bibliotics or the Study of Documents ... Detection of Fraud and Forgery ... Third Edition, greatly enlarged ... Philadelphia: Lippincott, 1901. 8vo. xxiv,15-266 pp. Frontis, 15 plates. Orig cloth.
(Charles B. Wood) **$155 [≈ £96]**

Frazier, Charles H.
- Surgery of the Spine and Spinal Cord. New York: 1918. 1st edn. Lge 8vo. xxii,971 pp. 6 cold plates, 178 ills. Orig cloth, rubbed, inner hinges cracked.
(Elgen) **$185 [≈ £115]**

Freeman, J.
- Life of the Rev. William Kirby, Rector of Barham. London: 1852. 8vo. xii,506 pp. Port, plate, facs, table. Buckram.
(Wheldon & Wesley) **£55 [≈ $88]**

French, James Weir (ed.)
- Modern Power Generators. London: Gresham Publishing Co, 1908. 2 vols. Folio. 11 cold fold-out sectional models, num ills. Orig dec cloth gilt.
(Hollett) **£85 [≈ $136]**
- Modern Power Generators. London: Gresham Publishing Co, (1908). 2 vols. Folio. xx,204; xiv,204 pp. 11 cold fold-out sectional models, over 700 ills. Orig dec green cloth gilt.
(Karmiole) **$500 [≈ £312]**

Frerichs, Fried. Theod.
- A Clinical Treatise on Diseases of the Liver. London: New Sydenham Society, 1860-61. 2 vols. 8vo. 3 plates, 68 text figs. Orig cloth, sm lib marks.
(Bickersteth) **£38 [≈ $60]**

Fresenius, C.R.
- Instruction in Chemical Analysis.
(Quantitative). Edited by J. Lloyd Bullock. London: 1846. 1st edn in English. 8vo. [xxii], 626 pp. Over 70 figs. Orig cloth.
(Bow Windows) **£75 [≈ $120]**

Freud, Sigmund
- Three Essays on the Theory of Sexuality. Authorized Translation by James Strachey. London: Imago Publishing Co, (1949). 1st English edn. 8vo. 134 pp. Orig blue cloth. Dw.
(Karmiole) **$50 [≈ £31]**

Friedman, Joseph S.
- History of Color Photography. Boston: Amer Photographic Pub Co, 1944. 1st edn. 8vo. x, 514 pp. 109 ills. Orig cloth.
(Charles B. Wood) **$75 [≈ £46]**

Friend, H.
- Flowers and Flower-Lore. London: [1883]. 2 vols. 8vo. xvi,704 pp. Text figs. Endpapers foxed. 1 or 2 pp carelessly opened. Orig cloth, spines faded.
(Wheldon & Wesley) **£50 [≈ $80]**

Friend, Hilderic
- The Story of the British Annelids (Oligochaeta). London: Epworth Press, 1924. 8vo. 288 pp. Cold frontis, 57 text figs. Orig dec cloth. Dw.
(Blackwell's) **£35 [≈ $56]**

Froggatt, W.W.
- Some Useful Australian Birds. Sydney: NSW Dept of Agriculture, 1921. 1st edn. Tall 8vo. 85 pp. 54 cold plates. Orig red cloth gilt, spine faded & trifle worn.
(Hollett) **£50 [≈ $80]**

Frohawk, F.W.
- Natural History of British Butterflies. London: Hutchinson, [1926]. 2 vols. Folio. 59 (of 60) cold plates, 4 b/w plates. Orig blue cloth.
(Spelman) **£40 [≈ $64]**
- Varieties of British Butterflies. London: 1938. 200 pp. 48 cold plates. Some v sl foxing. Orig cloth. Dw.
(Baldwin) **£55 [≈ $88]**
- Varieties of British Butterflies. London: "1938" [≈ 1946]. Roy 8vo. 200 pp. 48 cold plates. Orig cloth. Dated 1938 but actually the 1946 reprint, with the word "ex" at the start of the 2nd line on the frontis (as opposed to the end of the 1st line).
(Henly) **£40 [≈ $64]**

The Fruit-Gardener ...
- See Gibson, John, M.D.

Fry, John Storrs
- An Essay on the Construction of Wheel-Carriages ... Bristol, Bath & London: 1820. 1st edn. 1st iss (?), without appendix. [ii], viii,121 pp. W'engvd dedic & ills. Orig bds, paper spine, label, uncut, sl wear to extremities. Author's pres copy.
(Duck) **£325 [≈$520]**

Fryer, Douglas H. & Henry, Edwin R. (eds.)
- Handbook of Applied Psychology. New York: Rinehart, [1950]. 2 vols. 8vo. Orig bndg. *(Gach)* **$50 [≈£31]**

Fuchs, Adalbert
- Diseases of the Fundus Oculi, with Atlas. Translated by Erich Pressburger ... Philadelphia: [1949]. 1st English edn. One of 995. Lge 8vo. 337 pp. 44 cold plates. Orig cloth, a.e.g. Slipcase. *(Elgen)* **$100 [≈£62]**

Fulke, William
- A Most pleasant prospect into the Garden of Naturall Contemplation; to behold the naturall causes of all kind of Meteors ... Second Edition corrected ... London: 1634. Sm 8vo. [vi],71 pp. No blanks. Cropped close, sm reprs, some soil. Mod calf. STC 11439. *(Lamb)* **£180 [≈$288]**
- A Most pleasant prospect into the Garden of Naturall Contemplation; to behold the naturall causes of all kinde of Meteors ... London: 1640. Sm 8vo. [3],71 ff. Trimmed close. Half mor over mrbld bds. STC 11441. *(Rootenberg)* **£400 [≈£250]**

Fuller, Francis
- Medicina Gymnastica: or a Treatise concerning the Power of Exercise ... and the Great Necessity of it, in the Cure of Several Distempers. Sixth Edition. London: 1728. 8vo. [36], 271,[1],[10 advt] pp. Lower marg water stained. Stamp on title verso. Rec half calf. *(Spelman)* **£55 [≈$88]**
- Medicina Gymnastica: or, a Treatise concerning the Power of Exercise ... and the Great Necessity of it in the Cure of Several Distempers. London: 1728. 6th edn. 8vo. [xxxvii], 271,[1],[10 advt] pp. New contemp style half calf. *(Young's)* **£120 [≈$192]**

Fuller, John
- Art of Coppersmithing. A Practical Treatise on Working Sheet Copper into All Forms. New York: David Williams, 1911. 4th rvsd edn. 319 pp. Ills. Orig bndg. *(Book House)* **£20 [≈$32]**

Fulop-Miller, Rene
- Triumph Over Pain. Translated by E. & C. Paul. London: (1938). 1st edn in English. 8vo. [vi],438 pp. Frontis, over 30 ills. Orig cloth, sunned. *(Bow Windows)* **£80 [≈$128]**

Fulton, Robert
- Torpedo War, and Submarine Explosions. New York: William Elliot, 1810. 1st edn. Oblong 4to. 57,[3] pp. 5 plates. 1 or 2 ff in facs. Half mor. *(Jenkins)* **$450 [≈£281]**

Fur, Feather and Fin Series
- The Grouse. London: 1895. 8vo. A few pp sl foxed. Orig cloth. *(Grayling)* **£17 [≈$27]**
- The Hare. London: 1896. 8vo. Lower edges of some pp discold. Half leather. *(Grayling)* **£15 [≈$24]**
- The Hare. London: 1896. 8vo. Orig cloth. *(Grayling)* **£20 [≈$32]**
- The Partridge. London: 1896. 8vo. Cloth. *(Grayling)* **£15 [≈$24]**
- The Pheasant. London: 1895. 8vo. Orig cloth. *(Grayling)* **£17 [≈$27]**
- The Pheasant. London: 1895. 8vo. Sl foxing. Half leather, sl rubbed. *(Grayling)* **£25 [≈$40]**
- The Rabbit. London: 1898. 8vo. Sl foxing. Portion of contents wrinkled by damp. Orig green half leather. *(Grayling)* **£15 [≈$24]**
- Snipe and Woodcock. London:1903. 8vo. Orig cloth. *(Grayling)* **£30 [≈$48]**

Furman, H. Van F.
- A Manual of Practical Assaying. New York: John Wiley, 1907. 5th edn. 8vo. vi,463 pp. Text ills. Orig cloth. *(Gemmary)* **$45 [≈£28]**

Furneaux, W.
- Butterflies and Moths. London: 1905. 8vo. xiv, 350 pp. 12 cold plates, 241 figs. V sl foxing. Prize calf gilt. *(Baldwin)* **£20 [≈$32]**

Fussell, G.E.
- The Farmers Tools, 1500-1900. The History of British Farm Implements, Tools and Machinery before the Tractor Came. London: Andrew Melrose, 1952. 1st edn. 246 pp. Plates. Orig cloth, some silverfish nibbling on cvrs. Dw soiled at back. *(Wreden)* **$47.50 [≈£30]**
- More Old English Farming Books, from Tull to the Board of Agriculture, 1731 to 1793. London: 1950. 8vo. ix,186 pp. 8 ills. Orig cloth. *(Wheldon & Wesley)* **£20 [≈$32]**

Fyson, P.F.
- The Flora of the South Indian Hill Stations ... Madras: 1932. 2 vols. 8vo. 611 plates. Orig cloth.   *(Wheldon & Wesley)* £85 [≈ $136]

Gadow, Hans
- Through Southern Mexico, being an Account of the Travels of a Naturalist. London: Witherby, 1908. 8vo. xvi,527 pp. 4 maps, 170 ills. Orig red buckram gilt, backstrip sl faded.
              (Blackwell's) £80 [≈ $128]

Gairns, J.F.
- Locomotive Compounding and Superheating. A Practical Text-Book ... London: Charles Griffin, 1907. 1st edn. xxii,190,[2],advt pp. Frontis, 148 ills. Orig crimson cloth.   *(Duck)* £20 [≈ $32]

Galilei, Galileo
- Mathematical Discourses concerning Two New Sciences ... from the Italian, by Tho. Weston ... London: 1730. 1st edn of this trans. 4to. xii, 1-360,369-497,[1 errata],[2 advt] pp, correct. Engvd table at p 436, num w'cuts in text. Contemp calf, spine sl rubbed.
              (Pickering) $5,500 [≈ £3,437]

Gall, Franz Joseph
- On the Functions of the Cerebellum, by Drs. Gall, Vimont, and Broussais. Translated from the French by Dr. George Combe ... Edinburgh & London: 1838. 1st edn in English. 8vo. xliv,340 pp. Occas smudging. Contemp cloth, paper label, chipped. edges shelfworn.   *(Gach)* $350 [≈ £218]

Galpin, Francis William
- An Account of the Flowering Plants Ferns and Allies of Harleston ... Harleston: R.R. Cann, 1888. 157 pp. Ill. Orig cloth, front cvr sl marked.   *(Lamb)* £30 [≈ $48]

Galton, Sir Francis
- Finger Prints. London: Macmillan, 1892. 1st edn. 8vo. xvi,216 pp. 16 plates (15 called for in list of plates). Orig mauve cloth, lightly shelfworn.   *(Gach)* $485 [≈ £303]
- Fingerprint Directories. London & New York: Macmillan, 1895. 1st edn. 8vo. [viii], 127,[1] pp. Orig russet cloth, shelfworn.
              (Gach) $200 [≈ £125]
- Hereditary Genius ... New and Revised Edition. New York: Appleton, 1877. 2nd US edn, 2nd printing. Sm 8vo. Orig green cloth, hinges broken.   *(Gach)* $35 [≈ £21]
- Life and Letters ... see Pearson, Karl.

Galton, Sir Francis (ed.)
- Vacation Tourists and Notes of Travel in 1862-63. London: Macmillan, 1864. Sole edn. 8vo. viii,418 pp. Contemp half calf.
              (Bickersteth) £85 [≈ $136]

Gamble, S.G.
- A Practical Treatise on Outbreaks of Fire. Being a Systematic Study of their Causes and Means of Prevention ... London: Griffin, 1941. 3rd edn. 543 pp. 348 ills. Plates. Orig bndg, rear cvr spotted.
              (Book House) £40 [≈ $64]

Gardens ...
- The Gardens and Menagerie of the Zoological Society Delineated ... see Bennett, E.T.

Gardiner, W.
- The Flora of Forfarshire. London & Edinburgh: 1848. 12mo. xxiv,308,4 pp. 2 plates. Orig cloth.
              (Wheldon & Wesley) £48 [≈ $76]

Gardner, C.A. & Bennetts, H.W.
- The Toxic Plants of Western Australia. Perth: 1956. Roy 8vo. xxix,253 pp. 52 cold plates, 48 line drawings. Half mor.
              (Wheldon & Wesley) £35 [≈ $56]

Garner, R.
- The Natural History of the County of Stafford ... London: 1844. 8vo. xii,551 pp. Fldg geological map, frontis, 8 plates, 19 ills. Orig cloth., rather worn. Pres copy. Without the 61-page Supplement publ in 1860.
              (Wheldon & Wesley) £60 [≈ $96]
- The Natural History of the County of Stafford ... London: 1844-60. 1st edn. 8vo. vii, 551,61 pp. Hand cold geological map, 9 plates, 19 text engvs. Orig cloth, spine faded.
              (Henly) £95 [≈ $152]

Garner, R.L.
- Gorillas and Chimpanzees. London: 1896. 8vo. Ills. Sl foxing. Orig cloth, sl discold.
              (Grayling) £25 [≈ $40]
- Gorillas and Chimpanzees. London: 1896. 8vo. Ills. Orig cloth, sl marked.
              (Grayling) £50 [≈ $80]

Garnett, Thomas
- A Lecture on the Preservation of Health. London: for Cadell & Davies, 1800. 2nd edn. 12mo. [iv],vi,115 pp. Port. Title vignette. Lacks half-title. 19th c coarse linen cloth.
              (Burmester) £70 [≈ $112]

Garrison, Fielding H.

- An Introduction to the History of Medicine. Philadelphia: 1929. 4th edn, rvsd & enlgd. 8vo. 996 pp. Ports, num ills. Orig cloth.
(Elgen) **$85 [≈£53]**

Garven, H.S.D.

- Wild Flowers of North China and South Manchuria. Peking: 1937. 8vo. 117 pp. 102 plates. Orig bds.
(Wheldon & Wesley) **£35 [≈$56]**

Gasquoine, C.P.

- The Story of the Cambrian. A Biography of a Railway. Wrexham & Oswestry: 1922. 1st edn. Roy 8vo. x,157 pp. Fldg map, plates. Orig cloth, edges & spine ends sl rubbed.
(Duck) **£30 [≈$48]**

Gataker, T.

- Observations on the Internal Use of the Solanum or Nightshade. London: 1757. 8vo. 34 pp. New bds.
(Wheldon & Wesley) **£38 [≈$60]**

Gatke, Heinrich

- Heligoland as an Ornithological Observatory. Edinburgh: David Douglas, 1895. 1st edn. Imperial 8vo. xii,599,20 illust ctlg pp. Port, num ills. Orig pict green cloth gilt, sl bumped.
(Gough) **£48 [≈$76]**

Gatty, Mrs Alfred

- British Sea-Weeds. Drawn from Prof. Harvey's 'Phycologia Britannica' ... London: Bell & Daldy, 1872. 2 vols. 4to. xlix,74; 93 pp. 80 cold plates. Orig cloth gilt.
(Hollett) **£150 [≈$240]**

Geikie, Sir Archibald

- The Antiquity of Man in Europe, being the Munro Lectures, 1913. London: 1914. 8vo. xx, 328 pp. 4 maps, 21 plates, 9 text figs. Orig cloth, trifle used.
(Wheldon & Wesley) **£25 [≈$40]**
- Class-Book of Geology. London: 1886. 1st edn. Cr 8vo. xviii,516 pp. 209 text figs. Orig cloth, spine soiled. *(Henly)* **£15 [≈$24]**
- Geological Map of England and Wales reduced from the One Inch Maps of the Geological Survey. London: 1896. Elephant folio. Dble-page engvd title, index of colours, 13 dble-page maps, all linen-backed. Half mor, rubbed, new backstrip.
(Henly) **£150 [≈$240]**
- Life of Sir Roderick I. Murchison ... Based on his Journals and Letters ... London: Murray, 1875. 2 vols. 8vo. xii,[1],387; vi, [1], 375 pp.

22 plates, text ills. Orig cloth, mainly unopened, worn, inside jnts cracked.
(Fenning) **£85 [≈$136]**
- Text-Book of Geology. London: Macmillan, 1882. 1st edn. 8vo. xi,971 pp. frontis, fldg table, 435 text figs. Orig cloth, endpapers slit along inner jnts. *(Bickersteth)* **£115 [≈$184]**
- Text-Book of Geology. London: 1882. 8vo. xi, 971 pp. frontis, table, 435 text-figs. Half mor gilt, fine.
(Wheldon & Wesley) **£20 [≈$32]**
- Text-Book of Geology. London: 1923. 4th edn, rvsd & enlgd. 2 vols. Plates, ills. Embossed stamp on title. Orig cloth.
(Elgen) **$65 [≈£40]**

Geikie, James

- Earth Sculpture. London: 1909. 8vo. 10 plates, text figs. Orig cloth.
(Henly) **£22 [≈$35]**
- Fragments of Earth Lore. Sketches & Addresses, Geological and Geographical. Edinburgh: 1893. 1st edn. 8vo. [vii],428 pp. 6 cold fldg maps. Orig cloth.
(Bickersteth) **£48 [≈$76]**
- Fragments of Earth Lore. London: 1893. 8vo. 6 fldg maps. Orig cloth. *(Henly)* **£18 [≈$28]**
- Prehistoric Europe, a Geological Sketch. London: Edward Stanford, 1881. 1st edn. 8vo. xviii,592 pp, 3 advt ff. 5 cold plates & maps (3 fldg). Crnr cut from flyleaf. Orig cloth. *(Bickersteth)* **£68 [≈$108]**
- Prehistoric Europe, a Geological Sketch. London: 1881. 8vo. xviii,592 pp. 5 plates. Orig cloth, trifle used.
(Wheldon & Wesley) **£30 [≈$48]**
- Structural and Field Geology for Students of Pure Applied Science. Edinburgh: 1905. 1st edn. 8vo. x,435,[1] pp. 56 plates (4 cold), 142 figs. Some foxing & dust marks. Orig cloth, dull. *(Bow Windows)* **£24 [≈$38]**

General View ...

- General View of the Agriculture of the County of Lincoln ... see Young, Arthur.
- General View of the Agriculture of the County of Norfolk ... see Young, Arthur.
- General View of the Agriculture of the County of Suffolk ... see Young, Arthur.

Genet, Edmond Charles

- Memorial on the Upward Forces of Fluids, and their Applicability to Several Arts, Sciences, and Public Improvements ... Albany: 1825. 1st edn. 112 pp. 6 plates, fldg table. Occas foxing. Orig ptd green paper over bds, v minor soiling at crnr of front bd.
(Reese) **$2,250 [≈£1,406]**

- Memorial on the Upward Forces of Fluids, and their Applicability to Several Arts, Sciences, and Public Improvements ... Albany: Packard & Van Benthuysen, 1825. 1st edn. 8vo. 112 pp. 6 plates, fldg table. Minor foxing. Red mor.
(Rootenberg) **$1,750 [≈£1,093]**

Genth, F.A.
- Preliminary Report on the Mineralogy of Pennsylvania. Harrisburg: 2nd Geol Survey of Pennsylvania, 1874. 8vo. 238 pp. Cold map frontis. Orig cloth.
(Gemmary) **$60 [≈£37]**

The Gentleman Farmer ...
- See Kames, Henry Home, Lord.

Georgical Essays ...
- Georgical Essays: in which the Food of Plants is particularly considered and a New Compost recommended upon the Principles of Vegetation. London: 1769. 8vo. [iv],66 pp. Half mor, trifle used. Not the same as A. Hunter's 2 vol work of the same title & date.
(Wheldon & Wesley) **£30 [≈$48]**

Gerarde, John
- The Herball or Generall Historie of Plants. London: 1597. 1st edn. Folio. [xx], 1392,[72] pp. Engvd title (remargd), engvd port, 2146 w'cuts in the text. Last 3 ff of index reprd with sl loss. Calf, rebacked, sl worn at jnts & edges.
(Wheldon & Wesley) **£1,700 [≈$2,720]**
- The Herball or general Historie of Plants ... Very much enlarged by Thomas Johnson ... London: 1633. 2nd edn. Folio. Engvd title, nearly 3000 w'cuts (a few with contemp colouring). Lacks the 2 blanks as usual, a few sm reprs. Mod half calf, antique style.
(Wheldon & Wesley) **£1,350 [≈$2,160]**
- The Herball or general Historie of Plants ... Very much enlarged and amended by Thomas Johnson. London: 1636. 3rd edn. Folio. Engvd title, nearly 3000 w'cuts. Lacks the 2 blanks as usual. Marg repr to 1 leaf. 18th c calf gilt, sometime rebacked, front jnt cracking.
(Wheldon & Wesley) **£1,100 [≈$1,760]**
- The Herball or Generall Historie of Plants ... London: 1636. 2nd Johnson edn. Folio. Some 2765 w'cuts. Lacks engvd title, 4 preliminary ff, 1 index leaf & final blank. A few reprs & marks. Early 19th c diced russia, rebacked, crnrs restored.   *(Sotheran)* **£700 [≈$1,120]**

Getchell, F.A. (ed.)
- An Illustrated Encyclopaedia of the Science

and Practice of Obstetrics. Philadelphia: Gebbie, 1885. 1st edn. Folio. 276 pp. 84 litho plates. Orig three qtr leather, scuffed.
(Elgen) **$325 [≈£203]**

Ghosh, C.C.
- Insect Pests of Burma. Rangoon: 1940. Roy 8vo. ii,216,xv pp. 76 cold & 11 plain plates, 106 text figs. Orig bds, sl worn, sound ex-lib.
(Wheldon & Wesley) **£60 [≈$96]**

Giacomelli, H.
- See Adams, W.H. Davenport & Giacomelli, H.

Gibbs, Josiah Willard
- The Scientific Papers. London: 1906. 1st edn. 2 vols. xxviii,434; viii,284 pp. Port. Orig cloth, uncut. Leather backed solander case. Andrade's sgntr on endpapers.
(Elgen) **$350 [≈£218]**
- See also Donnan, F.G. & Haas, Arthur (eds.).

Gibbs-Smith, Charles H.
- The Invention of the Aeroplane (1799 - 1909). London: Faber & Faber, 1965. 1st edn. xxiv, 360 pp. Num ills. Orig cloth. Dw.
(Duck) **£30 [≈$48]**

Gibson, John, M.D.
- The Fruit-Gardener, Containing the Method of Raising Stocks ... Managing Fruit-Gardens ... Training Fruit-Trees ... London: for J. Nourse, 1768. 1st edn. 8vo. [viii],lxviii,411 pp. Contemp calf, rebacked, sl rubbed. Attributed by Henrey to John Gibson, M.D.
(Burmester) **£150 [≈$240]**

Gibson, William
- A New Treatise on the Diseases of Horses ... London: A. Millar, 1751. 1st edn. Large Paper. 4to. [10],464,[12] pp. Frontis, 31 plates. Contemp MS recipes. Contemp calf, rebacked.   *(Rootenberg)* **$650 [≈£406]**

Gifford, Isabella
- The Marine Botanist; an Introduction to the Study of British Sea-Weeds. Third Edition, greatly improved and enlarged. Brighton: 1853. 8vo. xl,357 pp, errata, [8 advt] pp. 6 cold & 6 other plates. Orig cloth.
(Spelman) **£30 [≈$48]**

Gilbert, Judson Bennett
- Disease and Destiny. A Bibliography of Medical References to the Famous ... London: Dawsons, 1962. Roy 8vo. 535 pp. Orig bndg. Dw.   *(Xerxes)* **$200 [≈£125]**

Gilbreth, Frank B.

- Motion Study. A Method for Increasing the Efficiency of the Workman ... New York: Van Nostrand, 1911. 1st edn. 8vo. xxiv,116,32 advt pp. Orig green cloth, sl soiled.
(Karmiole) **$100 [≃ £62]**

Giles, G.M.

- A Handbook of the Gnats or Mosquitoes, giving the Anatomy and Life History of the Culicidae. London: 1900. Roy 8vo. xi,374 pp. 8 plates. Orig cloth, trifle used.
(Wheldon & Wesley) **£35 [≃ $56]**
- A Handbook of the Gnats or Mosquitoes. London: 1902. 2nd edn. 8vo. xii,530 pp. 17 plates, text figs. Orig cloth.
(Wheldon & Wesley) **£35 [≃ $56]**

Gillespie, W.M.

- A Manual of the Principles & Practice of Road-Making: Comprising the Location, Construction, & Improvement of Roads, (Common, Macadam, Paved, Plank, etc.) and Rail-Roads. New York: 1854. 372 pp. Ills. Orig bndg, rebacked.
(Book House) **£40 [≃ $64]**

Gimlette, John D.

- Malay Poisons and Charm Cures. London: 1915. viii,127 pp. Top crnr of prelim clipped. Orig maroon cloth, part faded.
(Lyon) **£35 [≃ $56]**

Girard, C.

- Contributions to the Natural History of the Fresh Water Fishes of North America. I. A Monograph of the Cottoids. Washington: Smithsonian Contributions, 1851. All published. 4to. 80 pp. 3 plates. Plain wraps.
(Wheldon & Wesley) **£25 [≃ $40]**

Gissing, T.W.

- Materials for a Flora of Wakefield and its Neighbourhood. London: Van Voorst, 1867. 8vo. [iv], 59 pp. Orig limp cloth.
(Lamb) **£45 [≃ $72]**

Glanvill, Joseph

- Saducismus Triumphatus; or, Full and Plain Evidence concerning Witches and Apparitions ... London: 1681. 1st edn. 8vo. [viii],58, [xvi], 180; [xvi],310,[10], 311-328,[1 advt] pp. Frontis or addtnl engvd title to parts 1 & 2. Mor by Andrew Grieve of Edinburgh.
(Pickering) **$850 [≃ £531]**
- Saducismus Triumphatus: or, a Full and Plain Evidence concerning Witches and Apparitions ... Fourth Edition, with

Additions ... London: 1726. 8vo. [12],35, [10], 161,[4], [12],223-498, [4 advt] pp. Frontis, 2 plates. Rec calf backed bds.
(Fenning) **£225 [≃ $360]**
- The Vanity of Dogmatizing ... London: 1661. 1st edn. 8vo. [xxxii],250,[6] pp. Red & black title. Title & final leaf browned & soiled. Sm tear in 1 leaf. Contemp sheep, rebacked & recrnrd. Wing G.834.
(Pickering) **$950 [≃ £593]**

Glass, Thomas

- Twelve Commentaries on Fevers. Explaining the Method of curing these Disorders, upon the Principles of Hippocrates. Translated by N. Peters. London: for S. Birt & A. Tozer, 1752. 1st edn in English. 8vo. [ii],xvi,302 pp. Sl browned towards inner margs. Con calf
(Burmester) **£250 [≃ $400]**

Glasse, Hannah

- The Art of Cookery made Plain and Easy ... A New Edition, with Modern Improvements. Alexandria: Cottom & Stewart, 1805. 1st Amer edn. Sm 8vo. 293,[16] pp. Old sheep, red label, rubbed, lacks blank front & rear flyleaves. *(Charles B. Wood)* **$1,250 [≃ £781]**

Glenny, George

- Glenny's Manual of Practical Gardening ... London: Houlston & Wright, [ca 1850]. 8vo. 384 pp. Cold frontis. Orig green cloth, new endpapers.   *(Spelman)* **£60 [≃ $96]**

Glover, Mary Baker

- See Eddy, Mary Baker Glover.

Glover, Ralph

- Owen's Moral Physiology: or, A Brief and Plain Treatise on the Population Question. With Alterations and Additions by Ralph Glover. New York: Glover, 1846. 1st edn. 12m. 143 pp. Frontis. Orig bndg.
(Xerxes) **$75 [≃ £46]**

Gnudi, Martha Teach & Webster, Jerome Pierce

- Life and Times of Gaspare Tagliacozzi, Surgeon of Bologna 1545-1599 ... New York: Reichner, 1950. 1st edn. One of 2100. 4to. 538 pp. Ills. Orig bndg, crnrs bent. Authors' pres copy.   *(Xerxes)* **$150 [≃ £93]**

Godfrey, M.J.

- Monograph and Iconograph of the Native British Orchidaceae. Cambridge: 1933. 4to. xvi, 259 pp. Port, 67 plates (57 cold). Orig cloth. Dw.
(Wheldon & Wesley) **£200 [≃ $320]**

- Monograph and Iconograph of Native British Orchidaceae. Cambridge: 1933. 4to. xvi,259 pp. Port, 67 plates (57 cold). Orig cloth. Dw.
(Egglishaw) **£195 [≈ $312]**

Godfrey, W.E.
- The Birds of Canada. Ottawa: (1966). 4to. 428 pp. 64 cold & 4 plain plates, 71 figs, num maps. Orig cloth.
(Wheldon & Wesley) **£20 [≈ $32]**

Godman, F. du Cane
- A Monograph of the Petrels (Order Tubinares). London: 1907-10. Roy 4to. lv,381 pp. 106 hand cold plates (1-103, 5a, 98a, 102a). Occas minor foxing of text, some margs a trifle darkened. Brown mor, trifle faded, orig parts wraps bound in, sm dent on front cvr.
(Wheldon & Wesley) **£2,250 [≈ $3,600]**
- Natural History of the Azores, or Western Islands. London: 1870. 8vo. vii,358 pp. 2 maps. Cloth.
(Wheldon & Wesley) **£50 [≈ $80]**

Godman, John D.
- Anatomical Investigations ... added an Account of some Irregularities of Structure and Morbid Anatomy. Philadelphia: Carey & Lea, 1824. 1st edn. 4to. 134 pp. Errata leaf. 8 engvd plates. Orig calf, front hinge rubbed with sm cracks. *(Hemlock)* **$200 [≈ £125]**

Godwin, H.
- The History of the British Flora ... CUP: 1956. 1st edn. 4to. viii,384 pp. 118 text figs, 1 table. Orig cloth gilt. Dw.
(Hollett) **£35 [≈ $56]**

Goebel, K. von
- Outlines of Classification and Special Morphology of Plants. Translated by H.E.F. Garnsey. Oxford: 1887. Roy 8vo. xii,515 pp. Text figs. Half mor.
(Wheldon & Wesley) **£28 [≈ $44]**
- Wilhelm Hofmeister. The Work and Life of a Nineteenth Century Botanist. London: Ray Society, 1926. 8vo. xi,202 pp. Port, 2 facs letters, 3 text figs. Orig cloth, spine trifle faded. *(Wheldon & Wesley)* **£35 [≈ $56]**

The Gold-Headed Cane ...
- See MacMichael, William.

Goldsmith, Oliver
- An History of the Earth and Animated Nature. London: J. Nourse, 1774. 1st edn. 8 vols. 8vo. 101 plates. Polished calf, raised

bands, gilt labels, some wear, esp to spine ends. *(Hartfield)* **$795 [≈ £496]**
- A History of the Earth, and Animated Nature. Dublin: 1776-77. 2nd edn. 8 vols. 8vo. Num plates. Contemp calf.
(Wheldon & Wesley) **£120 [≈ $192]**
- A History of the Earth and Animated Nature ... New Edition. York: Wilson & Spence, 1804. 4 vols. 8vo. 100 engvd plates. Orig tree calf, contrasting labels, spines rubbed, lacks 1 label, 1 jnt cracked.
(Bickersteth) **£120 [≈ $192]**
- A History of the Earth and Animated Nature. London: Fullarton, 1850. 2 vols. Roy 8vo. Hand cold plates. Dublin half leather.
(Emerald Isle) **£95 [≈ $152]**
- A History of the Earth and Animated Nature ... Edinburgh: Fullarton & Co, [ca 1870]. 2 vols. Roy 8vo. Port, 2 hand cold title vignettes, 72 hand cold plates. New qtr calf.
(Egglishaw) **£150 [≈ $240]**
- A History of Earth and Animated Nature ... London: A. Fullarton & Co, n.d. 2 vols. 4to. Port, 72 hand cold plates. Contemp half mor gilt, 1 jnt cracked. *(Hollett)* **£95 [≈ $152]**
- Panorama of Nature. London: R. Edwards, 1817. 1st edn thus (?). 4to. xii,812 pp. 78 hand cold & 2 plain plates. Rebound in calf, gilt spine. *(Gough)* **£165 [≈ $264]**

Good, John Mason
- The Book of Nature. New York: 1827. 8vo. 530 pp. V foxed. Mottled calf, rubbed. Forms vol 6 of 'The Study of Medicine', 1826 edn.
(Elgen) **$75 [≈ £46]**
- The Study of Medicine, with a Physiological System of Nosology. Philadelphia: 1824. 2nd Amer edn. 5 vols. 8vo. Some marg stains, sl browning, Last 30 pp of vol 5 dampstained. Orig calf, red & black labels, scuffed.
(Elgen) **$275 [≈ £171]**
- The Study of Medicine. Boston: Wells & Lilly, 1826. 4th Amer edn. 5 vols. 8vo. Foxing. Orig calf, jnts tender, vol 5 v rubbed & scuffed. *(Elgen)* **$225 [≈ £140]**

Goodchild, W.
- Precious Stones. With a Chapter on Artificial Stones by Robert Dykes. London: 1908. 1st edn. 8vo. x,309 pp. 42 figs. Some finger & ink marks. Orig cloth, some upper crnrs a little bumped. *(Bow Windows)* **£45 [≈ $72]**

Goode, G.B. & Bean, T.H.
- Oceanic Ichthyology. A Treatise on the Deep-Sea and Pelagic Fishes of the World ... Washington: Mus of Comp Zool Memoirs, 1895. 2 vols 4to. xxxv,1*-26*, 553 pp. With

atlas of 123 plates. Half mor.
(Wheldon & Wesley) **£145 [≈ $232]**

Goodland, Roger
- A Bibliography of Sex Rites and Customs ...
London: Routledge, 1931. 4to. [ii],[vi], 752
pp. Orig cloth.   *(Gach)* **$150 [≈ £93]**
- A Bibliography of Sex Rites and Customs ...
London: Routledge, 1931. 4to. 752 pp. Orig
green buckram.   *(Gough)* **£30 [≈ $48]**

Goodsir, John
- The Anatomical Memoirs of ... Edited by
William Turner ... Edinburgh: Adam &
Charles Black, 1868. 1st edn. 2 vols. 14
plates. Lacks vol 1 frontis. Orig cloth, largely
unopened, jnts strengthened.
(Elgen) **$200 [≈ £125]**

Goodyear, W.A.
- The Coal Mines of the Western Coast of the
United States. San Francisco: 1877. 153 pp.
Cloth.   *(Reese)* **$135 [≈ £84]**

Gordon, Alexander
- An Historical and Practical Treatise upon
Elemental Locomotion, by means of Steam
Carriages on Common Roads ... London:
1832. 1st edn. viii,192 pp. 14 litho plates
(somewhat browned & spotty) inc frontis. Old
style bds, uncut.   *(Duck)* **£525 [≈ $840]**

Gordon, Benjamin Lee
- Medieval and Renaissance Medicine. New
York: Philosophical Library, 1959. 1st edn.
Thick 8vo. 843 pp. Photo ills. Orig bndg. dw.
(Xerxes) **$115 [≈ £71]**
- Medieval and Renaissance Medicine. New
York: 1959. 8vo. 843 pp. Frontis, ills. Orig
cloth.   *(Elgen)* **$95 [≈ £59]**

Gordon, George
- An Introduction to Geography, Astronomy,
and Dialling ... London: J. Senex ..., 1726.
1st edn. 8vo. [xii],iv,[iv], 188,40 pp. 11 engvs
on 10 fldg plates. 19th c half calf, rubbed,
rebacked.   *(Burmester)* **£175 [≈ $280]**

Gordon, George, 1806-1879
- The Pinetum: being a Synopsis of all the
Coniferous Plants at present known, with
Descriptions, History, and Synomymes [sic]
... London: Bohn, 1858. 1st edn. 8vo. xxii,
353 pp. Orig green cloth, crnrs bumped,
spine sl sunned.   *(Finch)* **£40 [≈ $64]**

Gordon, John
- Engravings of the Skeleton of the Human

Body. Edinburgh: for William Blackwood,
1818. 1st edn. 8vo. [2],135,[1] pp. 22 plates.
Occas foxing. Sgntr clipped from hd of title.
Minor ex-lib. Rebound in buckram.
(Elgen) **$95 [≈ £59]**

Gordon, Revd M.L., M.D.
- American [Medical] Missionary in Japan.
Boston: Houghton Mifflin, 1892. 8vo. 276
pp. Orig bndg.   *(Xerxes)* **$65 [≈ £40]**

Gordon, W.J.
- Our Home Railways. London: Warne,
[1910]. Complete set of 12 parts in orig cvrs.
Each part with 3 cold plates & cold print on
cvr. Sl worn, 1 or 2 spines damaged.
(Book House) **£35 [≈ $56]**

Gorgas, Ferdinand
- Dental Medicine: A Manual of Dental
Materia Medica and Therapeutics.
Philadelphia: Blakiston, 1891. 4th edn. Sm
4to. 524 pp. Orig bndg, hinges broken, spine
ends worn.   *(Xerxes)* **$75 [≈ £46]**

Gorgas, Marie & Hendrick, B.
- William Crawford Gorgas, His Life and
Work. New York: Doubleday, Page, 1924. 1st
edn. Sm 4to. 359 pp. Photo ills. Orig bndg,
t.e.g., spine ends worn. *(Xerxes)* **$85 [≈ £53]**

Gosney, E.S. & Popenoe, E.
- Sterilization for Human Betterment. New
York: Macmillan, (1929) 1930. 8vo. 202 pp.
Occas sl marg stain. Orig bndg.
(Xerxes) **$75 [≈ £46]**

Goss, W.F.N.
- Locomotive Sparks. New York: John Wiley;
London: Chapman & Hall, 1902. 1st edn, 1st
thousand. viii,172,16 advt pp. Num ills. Few
pencil marks. Orig cloth, spine ends & crnrs
sl worn, sides sl marked, front hinge cracked
but firm.   *(Duck)* **£35 [≈ $56]**

Gosse, P.H.
- Actinologia Britannica. A History of the
British Sea-Anemones and Corals. London:
Van Voorst, 1860. 8vo. xl,362 pp. 1 plain &
11 cold plates. Orig cloth gilt.
(Egglishaw) **£60 [≈ $96]**
- The Aquarium; an Unveiling of the Wonders
of the Deep Sea. London: Van Voorst, 1854.
1st edn. 8vo. xiv,278 pp. 6 chromolitho & 6
plain plates. Orig cloth, a.e.g., worn.
(Egglishaw) **£22 [≈ $35]**
- The Aquarium; an Unveiling of the Wonders
of the Deep Sea. London: Van Voorst, 1856.

2nd edn, rvsd & enlgd. 8vo. xvi,304,advt pp. 6 chromolitho plates, w'engvs. Orig cloth, a.e.g., spine relaid. *(Egglishaw)* **£32 [≈ $51]**
- The Birds of Jamaica. London: Van Voorst, 1847. 8vo. x,errata slip,447,[2] pp. Orig cloth. *(Egglishaw)* **£60 [≈ $96]**
- Evenings at the Microscope; or, Researches among the Minuter Organs and Forms of Animal Life. London: SPCK, [1859]. 2nd iss. 8vo. xii, 510,[2] pp. 113 text figs. Orig cloth. *(Wheldon & Wesley)* **£28 [≈ $44]**
- Evenings at the Microscope. Or, Researches among the Minuter Organs and Forms of Animal Life. London: SPCK, 1895. New edn, rvsd. 8vo. xvi,434 pp. Num text figs. Orig pict cloth. *(Savona)* **£15 [≈ $24]**
- An Introduction to Zoology. London: SPCK, (1844). 2 vols. 8vo. xxi,383; iv,436 pp. Num w'engvs. Orig blind stamped cloth. *(Egglishaw)* **£45 [≈ $72]**

Goss, P.H.
- Letters from Alabama ... chiefly relating to Natural History. London: Richard Clay, 1855. 1st edn. 12,306 pp. Orig blue cloth. *(Jenkins)* **£950 [≈ £593]**
- A Naturalist's Rambles on the Devonshire Coast. London: 1853. 1st edn. 8vo. xvi,451 pp. 28 plates (12 cold). Orig cloth, spine faded, 1 hinge worn. *(Baldwin)* **£35 [≈ $56]**
- A Naturalist's Ramblings on the Devonshire Coast. London: 1853. 8vo. xvi,451 pp. 28 plates (12 cold). Inserted advts dated Dec 1857. Cream endpapers. A little minor foxing. Orig blue cloth. *(Wheldon & Wesley)* **£50 [≈ $80]**
- A Naturalist's Ramblings on the Devonshire Coast. London: 1853. 1st edn. 8vo. xvi,451 pp. 28 plates (12 cold). Inserted advts dated Dec 1860. Orig green cloth, spine evenly faded. Freeman & Wertheimer 58g. *(Fenning)* **£75 [≈ $120]**
- The Romance of Natural History [First & Second Series]. London: Nisbet, 1861. 2 vols. 8vo. xvi,372; xii,393 pp. 21 ills. Orig pict gilt cloth, hd of 1 spine sl nicked. *(Egglishaw)* **£25 [≈ $40]**
- Tenby: A Sea-Side Holiday. London: 1856. 12mo. 400 pp. 20 cold & 4 plain plates. Orig cloth. *(Henly)* **£100 [≈ $160]**
- A Year at the Shore. London: Strahan, 1865. 1st edn. 8vo. Frontis & 35 other cold plates. Orig green cloth gilt, t.e.g., sm nick at hd of spine, front free endpaper loosening. The primary bndg. *(Sanders)* **£70 [≈ $112]**
- A Year at the Shore. London: 1865. 1st edn. 8vo. xii,330 pp. 36 colour-printed plates. Orig green cloth. *(Wheldon & Wesley)* **£60 [≈ $96]**

- A Year at the Shore. London: Daldy, (1865) 1877. 8vo. xii,330 pp. 36 cold litho plates. Orig cloth gilt, a.e.g. *(Egglishaw)* **£32 [≈ $51]**

Gosse, Philip
- The Squire of Walton Hall. The Life of Charles Waterton. London: 1940. 8vo. ix,324 pp. Ills. Trifle foxed at beginning & end. Orig cloth. *(Wheldon & Wesley)* **£20 [≈ $32]**

Gottlieb, Bernhard & Orban, B.
- Biology and Pathology of the Tooth and its Supporting Mechanism. Translated and edited by Moses Diamond. New York: Macmillan, 1938. 8vo. 195 pp. Photo ills. Orig bndg. *(Xerxes)* **$75 [≈ £46]**

Gould, J.
- A Monograph of the Odontophorinae or Partridges of America. London: 1850. Imperial folio. 32 hand cold plates. 2 sm marg tears in 1 plate reprd. Contemp mor gilt, new endpapers. *(Wheldon & Wesley)* **£7,500 [≈ $12,000]**

Gow, J.
- A Short History of Greek Mathematics. Cambridge: 1884. 1st edn. xiv,323 pp. Few text figs. Ink notes on 1 page of prelims. Orig cloth, spine relaid. *(Whitehart)* **£45 [≈ $72]**
- A Short History of Greek Mathematics. Cambridge: 1884; reprinted New York: 1923. xvi, 323 pp. Sev figs. Half cloth, bds, sl rubbed, amateurish replacement of inside front cvr. *(Whitehart)* **£35 [≈ $56]**

Graham, John Ryrie
- A Treatise on the Australian Merino. Melbourne: Clarson, Massina, 1870. 8vo. [4 advt], 101,[i imprint],[3 advt] pp. Orig brown cloth. *(Blackwell's)* **£155 [≈ $248]**

Grahame-White, C.
- The Story of the Aeroplane. Boston: Small, Maynard, (1911). 1st edn. Thick 8vo. xii,390 pp. Frontis, 35 ills, tables. Orig cloth, uncut. *(Berkelouw)* **$125 [≈ £78]**

Grandin, Egbert H. (ed.)
- Cyclopaedia of Obstetrics and Gynecology. New York: Wood, 1887. 12 vols. Orig bndg, sev cvrs waterstained. *(Elgen)* **$300 [≈ £187]**

Granger, B.
- Address to the Public relative to some Supposed Failures of the Cow-Pox, at Repton and its Neighbourhood, with Observations on the Efficacy and General Expediency of

Vaccination ... Burton-upon-Trent: Thomas Wayte, 1821. Half-title,34 pp. Disbound.
(C.R. Johnson) **£100 [≃ $160]**

Grant, Robert
- History of Physical Astronomy, from the Earliest Ages to the Middle of the Nineteenth Century ... London: Bohn, (1852). Lge 8vo. xx, 638,advt pp. Orig purple cloth, faded, extremities a bit frayed & chipped.
(Karmiole) **$50 [≃ £31]**

Granville, A.B.
- The Spas of Germany. Second Edition. London: Colburn, 1838. 8vo. lviii,[2],516 pp. 4 maps, 14 plates, 21 vignettes. Contemp half calf, rubbed but sound.
(Claude Cox) **£110 [≃ $176]**

Grattan, J.H.G. & Sinfer, Charles
- Anglo-Saxon Medicine Illustrated specially from the Semi-Pagan Text 'Lacnunga'. London: OUP for Wellcome, 1952. 1st edn. 8vo. xii, 234, [2] pp. 6 plates, 44 text ills. Sl damp staining to hd of 1st few ff. Orig cloth, flecked.
(Gach) **$37.50 [≃ £23]**

Gray, A.
- Darwiniana: Essays and Reviews pertaining to Darwinism. New York: 1876. 8vo. 396 pp. Sgntr on title. Orig brown cloth.
(Wheldon & Wesley) **£60 [≃ $96]**

Gray, Asa
- Elements of Botany. New York: G. & C. Carvill, 1836. 1st edn. Sm 8vo. xiv,428 pp. 125 text figs. Sl foxing. Orig brown cloth, somewhat soiled. Qtr mor slipcase.
(Karmiole) **$750 [≃ £468]**

Gray, G.R.
- A Fasciculus of the Birds of China. London: [1871]. 4to. 8 pp. 12 hand cold litho plates by W. Swainson. Red half red mor. Inscrbd & sgnd by Gray on half-title.
(Wheldon & Wesley) **£600 [≃ $960]**
- A Fasciculus of the Birds of China. London: [1871]. 4to. 8 pp. 12 plain litho plates by W. Swainson. Mod bds, orig paper label.
(Wheldon & Wesley) **£250 [≃ $400]**

Gray, Henry
- Anatomy, Descriptive and Surgical. Edited by T. Pickering Pick. A New American from the Eleventh English Edition ... Philadelphia: Lea Brothers & Lea, 1887. Lge thick 8vo. 1100 pp. Cold ills. Sl used. Orig sheep, gilt spine (scuffed). Boxed.
(Heritage) **$200 [≃ £125]**

Gray, J.E.
- Catalogue of the Carnivorous, Pachydermatous and Edentate Mammalia in the British Museum. London: 1869. 8vo. viii,398 pp. 47 w'cuts. Orig cloth.
(Wheldon & Wesley) **£35 [≃ $56]**
- Synopsis of the Species of Starfish in the British Museum. London: 1866. 4to. iv,18 pp. 16 plates. Some foxing. Bds.
(Wheldon & Wesley) **£18 [≃ $28]**

Gray, R.
- The Birds of the West of Scotland including the Outer Hebrides. Glasgow: 1871. 8vo. xi,520 pp. 15 plates. Some foxing. Orig cloth.
(Wheldon & Wesley) **£70 [≃ $112]**

Gray, S.F.
- A Natural Arrangement of British Plants, as pointed out by Jussieu, de Candolle, Brown, etc. London: 1821. 2 vols. 8vo. 21 plates. Plates 2 & 3 supplied in xerox. New cloth.
(Wheldon & Wesley) **£50 [≃ $80]**

Great Exhibition
- Exhibition of the Works of Industry of All Nations. Reports by the Juries ... Presentation Copy. London: 1852. 1st edn. 8vo. [vi],cxx,867 pp. 16 advt pp at end. 3 chromolitho plates. Orig red cloth gilt, a.e.g., sl rubbed.
(Pickering) **$1,200 [≃ £750]**

Green, A.H.
- Geology for Students and General Readers. London: 1876. 1st edn. 8vo. xxviii,552,[4 advt] pp. Fldg table, over 140 figs. A few marks. Orig cloth, spine ends a trifle fingered.
(Bow Windows) **£40 [≃ $64]**

Green, C.E. & Young, D.
- Encyclopaedia of Agriculture. London: William Green & Sons, 1908-09. 1st edn of vols 3 & 4. 4 vols. Lge 8vo. Plates. Orig mor backed cloth.
(Claude Cox) **£45 [≃ $72]**

Green, C.T.
- The Flora of the Liverpool District. Arbroath: 1933. xi,163 pp. Num ills. Trifle used, some ink & pencil notes. Orig cloth.
(Wheldon & Wesley) **£15 [≃ $24]**

Green, Roland
- A Treatise on the Cultivation of Ornamental Flowers ... Boston: Russell; New York: Thorburn, 1828. 1st edn. 8vo. 59,1 pp. Orig ptd bds, linen spine, uncut, sl wear to edges of cvrs.
(Charles B. Wood) **$950 [≃ £593]**

Green, T.
- The Universal Herbal: or Botanical Medical and Agricultural Dictionary ... Second Edition. London: Caxton Press, Henry Fisher, 1824. 2 vols. 4to. 790; 885,56 pp. 2 frontis, 1 engvd title, 107 plates, all hand cold. Contemp gilt panelled tree calf, 1 jnt reprd. *(Gough)* **£950 [≈ $1,520]**

Green, T. Henry
- Introduction to Pathology and Morbid Anatomy. Revised and enlarged by Murray and Martin. Philadelphia: Lea, 1898. 8th edn. Sm 4to. 582 pp. 1 cold plate, 216 engvs (6 cold). Orig bndg, loose.
(Xerxes) **$85 [≈ £53]**

The Green-House Companion ...
- See Loudon, John Claudius.

Greenaway, T.
- Farming in India, considered as a Pursuit for European Settlers of a Superior Class, with Plans for the Construction of Dams, Tanks, Weirs and Sluices. London: Smith, Elder, 1864. xvi,132 pp. 7 text ills. Orig green cloth gilt, partly unopened.
(Blackwell's) **£35 [≈ $56]**

Greene, William Thomas
- Parrots in Captivity ... London: George Bell & Sons, 1884-87. 1st edn. 3 vols. Lge 8vo. 81 litho plates, partly ptd in colour & stamped in gilt. Orig gilt dec green cloth, fine.
(Heritage) **$2,750 [≈ £1,718]**

Greener, William
- Gunnery in 1858: Being a Treatise on Rifles, Cannon, and Sporting Arms ... London: Smith, Elder, 1858. 1st edn. xvi,440,[16,16 advt] pp. 5 plates, 37 text ills. Occas thumbed or grubby. Mod qtr mor.
(Duck) **£145 [≈ $232]**

Greenewalt, C.H.
- Humming Birds. New York: 1960. 1st printing. 4to. xxi,250 pp. 70 cold photos, num ills from drawings. Orig buckram.
(Wheldon & Wesley) **£185 [≈ $296]**

Greenhill, A.G.
- The Applications of Elliptic Functions. London: 1892. xi,357 pp.
(Whitehart) **£35 [≈ $56]**

Greenwell, A. & Elsden, J.V.
- Roads: their Construction and Maintenance. London: Whittaker, 1901. 280 pp. Fldg map,

ills. Orig bndg.   *(Book House)* **£20 [≈ $32]**

Greenwood, George
- The Tree-Lifter; or, a New Method of Transplanting Forest Trees. London: Longman, 1844. 1st edn. 8vo. 112,[32 advt dated 1845] pp. Fldg tinted litho frontis. Orig green cloth.   *(Spelman)* **£80 [≈ $128]**

Greg, R.P. & Lettsom, W.G.
- Manual of Mineralogy of Great Britain and Ireland. London: 1858. 1st edn. 8vo. xvi,483, [1] pp. Orig cloth.
(Bow Windows) **£200 [≈ $320]**

Gregory, D.F.
- Examples of the Processes of the Differential and Integral Calculus ... Edited by William Walton. Cambridge: 1846. 2nd edn. 8vo. x,529 pp. 4 fldg diags. Contemp green half calf, sl rubbed.   *(Young's)* **£24 [≈ $38]**

Gregory, D.F. & Walton, W.
- A Treatise on the Application of Analysis to Solid Geometry. Cambridge: 1845. xii,276 pp. 3 fldg plates. Orig cloth, sl stained, spine sl worn on 1 side.   *(Whitehart)* **£35 [≈ $56]**

Gregory, G.
- A Dictionary of Arts and Sciences. London: Richard Phillips, 1806. 1st edn. 2 vols. 4to. [4],960; vi,[2],928 pp. 138 engvd plates, some foxed. Contemp half calf, rebacked, sides rubbed.   *(Claude Cox)* **£120 [≈ $192]**

Gregory, W.K.
- Evolution Emerging. London: [1951] 1957. 2 vols. 4to. xxxvi,736; vii,1013 pp. Ills. Orig cloth.   *(Baldwin)* **£65 [≈ $104]**

Gregory, William
- Letters to a Candid Inquirer on Animal Magnetism. London: Taylor, Walton, & Maberley, 1851. 1st edn. 12mo. [xxiv],528 pp. Orig blue cloth, recased.
(Gach) **$85 [≈ £53]**

Greif, Samuel (ed.)
- Who's Who in Dentistry. Biographical Sketches of Prominent Dentists in the United States and Canada. New York: 1916. 1st edn. 8vo. 238 pp. Orig cloth, spine sunned, lib b'plate.   *(Elgen)* **$85 [≈ £53]**

Greville, R.K.
- Scottish Cryptogamic Flora ... Intended to serve as a continuation of [Sowerby's] English Botany. Edinburgh: 1823-28. 1st edn. 6 vols.

8vo. 360 hand cold plates. Lacks half-titles. Contemp diced calf gilt.
(Henly) **£1,000 [≈ $1,600]**

Grew, Nehemiah
- The Anatomy of Vegetables Begun ... London: for Spencer Hickman ..., 1672. 1st edn. 8vo. xxxii, '198' [≈ 186],[22] pp, inc blank ff N6 & O8. 3 fldg plates. Some foxing, water stain on imprimatur leaf. 18th c polished calf, jnts cracked. Wing G.1946.
(Pickering) **$2,500 [≈ £1,562]**

Grey, H.C.
- Hardy Bulbs, including Half-Hardy Bulbs and Tuberous and Fibrous Rooted Plants. London: 1937-38. 3 vols. Roy 8vo. 176 plates (44 cold). Some v sl foxing at beginning & end of vols 1 & 2. Orig cloth, trifle faded.
(Wheldon & Wesley) **£350 [≈ $560]**
- Hardy Bulbs ... Volume I [only, of 3], Iridaceae. London: Williams & Norgate, 1937. 4to. 47 plates, many cold. Orig buckram gilt, spine trifle faded, sm nick at top.
(Hollett) **£75 [≈ $120]**

Griesinger, Wilhelm
- Mental Pathology and Therapeutics. Translated from the German by C.L. Robertson and J. Rutherford. London: New Sydenham Soc, 1867. 1st edn. 8vo. 530 pp. 3 lib stamps, 1 on title. Orig bndg, lacks backstrip.
(Xerxes) **£95 [≈ £59]**
- Mental Pathology and Therapeutics ... Translated from the German (Second Edition). By C. Lockhart Robertson and James Rutherford. London: New Sydenham Soc, 1867. 1st edn in English. 8vo. [ii],xiv,530,[2] pp. Orig cloth, sl worn.
(Gach) **$175 [≈ £109]**

Grieve, M.
- A Modern Herbal. The Medicinal, Culinary, Cosmetic and Economic Properties, Cultivation and Folk-Lore of Herbs, Grasses, Fungi, Trees & Shrubs with all their Modern Scientific Uses. Edited by Mrs C.F. Leyel. London: 1931. 2 vols. 8vo. 96 plates. Orig cloth, spines faded.
(Wheldon & Wesley) **£100 [≈ $160]**

Grieve, S.
- The Great Auk, or Garefowl (Alca impennis, Linn.), its History, Archaeology, and Remains. London: 1885. 4to. xi,141,58 pp. Cold map, 4 plates (2 cold), 6 ills. Orig cloth, trifle used, trifle stained.
(Wheldon & Wesley) **£185 [≈ $296]**

Griffin, John Joseph
- A System of Crystallography with Its Application to Mineralogy. Glasgow: Richard Griffin, 1841. 8vo. xxvii,346,143 pp. Num diags. Orig cloth, unopened.
(Gemmary) **$450 [≈ £281]**

Griffin, William
- A Treatise on Optics ... Second Edition. Cambridge: Univ Press, 1842. 8vo. Orig cloth backed bds, worn.
(Waterfield's) **£40 [≈ $64]**

Griffith, Ivor
- Lobscows. The Clean-Up of an Editorial Kitchen. Philadelphia: International Printing, 1939. 1st edn. 8vo. 548 pp. Ills. Orig bndg, sm stain on front bd. Author's sgnd inscrptn.
(Xerxes) **$45 [≈ £28]**

Griffith, J.E.
- The Flora of Anglesey and Caernarvonshire. Bangor: [1895]. 8vo. xx,288 pp. Map. Trifle foxed. Orig cloth.
(Wheldon & Wesley) **£40 [≈ $64]**

Griffith, J.W. & Henfrey, Arthur
- The Micrographic Dictionary; a Guide to the Examination and Investigation of the Structure and Nature of Microscopic Objects. London: Van Voorst, 1856. 1st edn. 8vo. xl, 696 pp. 41 plates (12 hand cold), 816 text w'cuts. Contemp half calf, sl rubbed.
(Claude Cox) **£45 [≈ $72]**

Griffith, Richard
- Geological and Mining Survey of the Connaught Coal District in Ireland. Dublin: Graisberry, 1818. Lge fldg cold map reprd. Half calf. *(Emerald Isle)* **£125 [≈ $200]**

Griffiths, John W.
- Treatise on Marine and Naval Architecture, or Theory and Practice Blended in Ship Building. New York & London: 1853. 3rd edn. 4to. 420,[ii] pp. Litho frontis, 44 plates. 6 ff tables. Later green qtr mor.
(Duck) **£250 [≈ $400]**

Grigson, G. & Buchanan, H.
- Thornton's Temple of Flora ... described by Geoffrey Grigson. With notes by H. Buchanan. London: 1951. Folio. viii,20 pp. 12 cold & 24 plain plates. Half cloth. Dw.
(Egglishaw) **£110 [≈ $176]**

Grimes, J. Stanley
- Etherology; or, the Philosophy of Mesmerism

and Phrenology: including a New Philosophy of Sleep and of Consciousness with a Review of the Pretensions of Neurology and Phreno-Magnetism. New York: Saxon & Miles, 1845. 1st edn. 8vo. 350 pp. Frontis. Cvrs poor.
(Xerxes) **$100 [≈£62]**

Grimwade, R.
- An Anthography of the Eucalypts. Sydney: 1920. 4to. 79 plates from photos. Orig cloth backed bds. *(Wheldon & Wesley)* **£45 [≈$72]**

Grindon, L.H.
- The Manchester Flora ... London: 1859. Cr 8vo. ix,575 pp. Last 2 ff sl soiled. Orig cloth. *(Wheldon & Wesley)* **£30 [≈$48]**
- The Manchester Flora. A Descriptive List of the Plants growing Wild within Eighteen Miles of Manchester. London: William White, 1859. 1st edn. 8vo. viii,[2],575 pp. Dec title. Num w'cuts in text. Orig cloth gilt, new endpapers. *(Spelman)* **£40 [≈$64]**

Grinnel, Joseph & Miller, Alden H.
- The Distribution of the Birds of California. Berkeley: Copper Ornithological Club, 1944. 4to. 608 pp. Cold frontis, num text ills. Orig brown cloth, soiled. *(Karmiole)* **$50 [≈£31]**

Grinstein, Alexander
- The Index of Psychoanalytic Writings. New York: IUP, 1956-64-72. Only edn. 14 vols. 8vo. Orig cloth, a few spine ends worn.
(Gach) **$285 [≈£178]**

Grodzinski, P.
- Diamond and Gem Stone Industrial Production. London: N.A.G. Press, 1942. 8vo. 256 pp. 183 text figs, 32 tables. Orig cloth, spine chipped. *(Gemmary)* **$35 [≈£21]**

Gronovius, L.T.
- Catalogue of Fish collected and described by L.T. Gronow, now in the British Museum. Edited by J.E. Gray. London: 1854. 12mo. vii, 196 pp. Wraps.
(Wheldon & Wesley) **£20 [≈$32]**

Grose, Donald
- The Flora of Wiltshire. Devizes: 1957. 8vo. 824 pp. 11 plates, num maps in text. Orig cloth. *(Wheldon & Wesley)* **£40 [≈$64]**

Grosvenor, Benjamin
- Health. An Essay on its Nature, Value, Uncertainty, Preservation, and Best Improvement. London: for E. Matthews, 1716. 1st edn. 12mo. [viii],xi,[i], 242,[2] pp.

Half-title. Contemp panelled calf, rebacked in paler calf. *(Burmester)* **£60 [≈$96]**

The Grouse in Health and Disease ...
- The Grouse in Health and Disease. London: Smith, Elder, 1911. 1st edn. 2 vols. 4to. xxiv, 512; 150 pp. 41 maps, 59 plates (mostly cold), 31 text ills. Orig cloth. *(Gough)* **£95 [≈$152]**
- The Grouse in Health and Disease, being the Final Report of the Committee of Inquiry on Grouse Disease. London: 1911. 2 vols. 4to. 41 maps, 59 plates (many cold), text figs. Buckram, trifle used.
(Wheldon & Wesley) **£85 [≈$136]**
- The Grouse in Health and Disease. Being the Popular Edition of the Report of the Committee of Inquiry on Grouse Disease. Edited by A.S. Leslie and A.E. Shipley. London: Smith, Elder, 1912. Roy 8vo. xx,472 pp. 21 plates (12 cold), num text figs. Orig cloth, t.e.g. *(Egglishaw)* **£24 [≈$38]**

Grubb, N.H.
- Cherries. London: 1949. Roy 8vo. viii,186 pp. 12 cold plates, 40 photos. Orig cloth. *(Wheldon & Wesley)* **£20 [≈$32]**

Guerini, Vincenzo
- A History of Dentistry from the most Ancient Times until the End of the Eighteenth Century. Philadelphia: 1909. 1st edn. 8vo. 355 pp. 20 plates, 104 ills. Orig cloth, cvrs sl spotted. *(Elgen)* **$115 [≈£71]**
- The Life and Works of Giuseppangelo Fonzi. Philadelphia: 1925. 1st edn. 8vo. v,136 pp. Port. Orig cloth. *(Elgen)* **$50 [≈£31]**

Guerra, Francisco
- American Medical Bibliography, 1639-1783. A Chronological Catalogue and Critical and Bibliographical Study ... New York: Lathrop C. Harper, 1962. 885 pp. 187 plates. Orig bndg. Shipping carton.
(Karmiole) **$125 [≈£78]**

Guillemin, A.
- The Application of Physical Forces. Translated from the French ... By J. Norman Lockyer. London: 1877. 1st edn. Lge 8vo. [xl], 741,[1] pp. 4 cold, 21 engvd plates, over 460 figs. A few thumb prints. Contemp prize calf, a little rubbed, spine ends chipped.
(Bow Windows) **£70 [≈$112]**

Gull, William Withey
- A Collection of the Published Writings ... Edited by Theodore D. Acland. London: New Sydenham Soc, 1894-96. 2 vols. 8vo.

Port, 23 plates (mostly cold). Orig cloth.
(Elgen) **$125 [≈ £78]**

Gunther, A.
- A Catalogue of the Fishes in the British Museum. London: 1859-70. 8 vols. 8vo. Lib stamp on titles. Orig cloth.
(Egglishaw) **£160 [≈ $256]**
- The Reptiles of British India. London: Ray Society, 1864. [One of 750]. Folio. xxviii, 452 pp. 26 litho plates. A few lib stamps. Practically free from the usual foxing. Orig bds, crnr of front cvr defective.
(Wheldon & Wesley) **£150 [≈ $240]**

Gunther, Jack Disbrow & Charles O.
- The Identification of Firearms from Ammunition Fired Therein, with an Analysis of Legal Authorities. New York & London: 1935. 1st edn. Roy 8vo. xxviii,342 pp. Photo frontis, 148 ills. Orig cloth, spine faded, sl dull, sl bumped.   *(Duck)* **£30 [≈ $48]**

Gunther, R.T.
- Early British Botanists and their Gardens, based on unpublished Writings of Goodyear, Tradescant and others. Oxford: privately ptd, 1922. Roy 8vo. viii,417 pp. 9 plates, 21 other ills. Orig white buckram, a little soiled.
(Wheldon & Wesley) **£85 [≈ $136]**
- Early Science in Cambridge. Oxford: for the author, 1937. 8vo. xii,513 pp. Num plates & text figs. Some inoffensive lib stamps. New cloth.   *(Wheldon & Wesley)* **£75 [≈ $120]**
- Oxford Gardens based upon Daubeny's Popular Guide to the Physick Garden of Oxford ... Oxford: Parker, 1912. Cr 8vo. xv,288 pp. Planting charts, 33 ills. Orig gilt dec blue cloth.   *(Blackwell's)* **£50 [≈ $80]**

Gurney, John Henry
- The Gannet, a Bird with a History. London: 1913. 8vo. lii,567 pp. 137 ills. Orig bds, trifle worn.   *(Wheldon & Wesley)* **£50 [≈ $80]**
- The Gannet. Witherby, 1913. 1st edn. 8vo. li,567 pp. 1 cold & 136 b/w plates, maps & ills. Orig cloth, spine reprd.
(Gough) **£55 [≈ $88]**
- Rambles of a Naturalist in Egypt and Other Countries ... London: with ... Ornithological Notes. Jarrold, [ca 1880]. 8vo. 307,advt pp. Orig pict gilt maroon cloth, spine sl sunned, edges frayed, sm tear at ft of spine.
(Terramedia) **$100 [≈ £62]**

Gurney, John Henry, et al.
- The House Sparrow. London: William Wesley, (1885). 12mo. vii,70,[16 advt] pp.

Frontis. Orig green cloth gilt.
(Blackwell's) **£60 [≈ $96]**

Guthrie, George J.
- Lectures on the Operative Surgery of the Eye ... London: 1827. 2nd edn. 8vo. xxvii,554 pp. 7 plates. Lib stamp on title. Half mor, dec spine, lib ticket on ft of spine.
(Rittenhouse) **$600 [≈ £375]**

Guthrie-Smith, H.
- Mutton Birds and other Birds. Christchurch: 1914. 8vo. viii,206 pp. 78 plates. Orig cloth.
(Wheldon & Wesley) **£30 [≈ $48]**

Haab, O.
- An Atlas of Ophthalmoscopy, with an Introduction to the Use of the Ophthalmoscope. New York: William Wood, 1895. Sm 8vo. 55 pp. 64 cold ills. Orig bndg.
(Xerxes) **$80 [≈ £50]**

Haanel, E.
- On the Location and Examination of Magnetic Ore Deposits by Magnetometric Measurements. Ottawa: 1904. 8vo. ix,132 pp. 8 fldg plates, 6 cold & 5 plain plates. Orig cloth, a little creased.   *(Henly)* **£18 [≈ $28]**

Hachisuka, M.
- A Handbook of the Birds of Iceland. London: 1927. Roy 8vo. v,128 pp. Map, 7 plates, fldg diag. Orig cloth.
(Wheldon & Wesley) **£45 [≈ $72]**
- Variations among Birds (chiefly Game Birds). Tokyo: 1928. 8vo. x,96 pp. 24 plates (4 cold). Wraps, last leaf & back wrapper trifle defective. *(Wheldon & Wesley)* **£25 [≈ $40]**

Haeckel, Ernst
- The Evolution of Man ... Translated from the Fifth (enlarged) edition by Joseph McCabe. London: 1905. 2 vols. 8vo. 28 plates (many cold), num figs. Endpapers a little spotted. Orig cloth.   *(Bow Windows)* **£80 [≈ $128]**

Hahnemann, Samuel
- Organon Medicine. Translated from the Fifth Edition with Appendix by R.E. Dudgeon. Chicago: Medical Advance, 1895. 8vo. 178 pp. New wraps.   *(Xerxes)* **$90 [≈ £56]**

Hales, Stephen
- An Account of some Experiments and Observations on Tar-Water ... Second Edition. To which is added, a Letter from Mr. Reid ... London: for R. Manby & H.S. Cox, 1747. 8vo. [ii], 74 pp. Diags. Disbound.
(Burmester) **£60 [≈ $96]**

Hall, Sir A.D.
- The Book of the Tulip. London: 1929. 8vo. 224 pp. 12 cold & 12 plain plates. Orig cloth. *(Wheldon & Wesley)* **£25** [≈ **$40**]
- The Genus Tulipa. London: RHS, 1940. Sm 4to. viii,171 pp. 40 cold & 23 other ills. Orig cloth. *(Wheldon & Wesley)* **£75** [≈ **$120**]
- The Genus Tulipa. With Forty Illustrations in Colour by H.C. Osterstock. London: RHS, 1940. 1st edn. Lge 8vo. viii,171 pp. 40 cold plates, 23 other ills. Orig cloth, sl faded. *(Bow Windows)* **£60** [≈ **$96**]

Hall, H.B.
- The Sportsman and his Dog; Hints on Sporting. London: 1850. 8vo. Engvd frontis. Orig cloth. *(Grayling)* **£50** [≈ **$80**]

Hall, Harrison
- Hall's Distiller ... adapted to the Use of Farmers, as well as Distillers. Philadelphia: John Bioren, 1813. 1st edn. 8vo. x,244 pp. Fldg frontis, 1 plate. browned. Contemp tree calf, re-hinged. *(Charles B. Wood)* **$650** [≈ **£406**]

Hall, J.L.
- One Hundred Years of American Psychiatry. New York: Columbia UP for the Amer Psychiatric Assoc, [1944]. 1st printing. 4to. [ii],[xxvi], 629,[3] pp. Orig 2-toned cloth, spine ends shelfworn. *(Gach)* **$50** [≈ **£31**]

Hall, Marshall
- The Principles of Diagnosis. Second Edition, entirely rewritten. New York: Appleton, 1835. 1st Amer edn. 8vo. 463,4 ctlg pp. Occas sl foxing. Orig cloth, faded, spotted, worn. *(Elgen)* **$125** [≈ **£78**]
- Principles of the Theory and Practice of Medicine. First American Edition, revised by Jacob Bigelow and Oliver Wendell Holmes. Boston: 1839. 724 pp. Ills. New qtr leather. *(Argosy)* **$85** [≈ **£53**]

Hallimond, A.F.
- Manual of the Polarizing Microscope. York: Cooke, Troughton & Simms, 1953. 2nd edn. 8vo. 204 pp. 92 text figs. Orig black cloth. *(Savona)* **£25** [≈ **$40**]

Halsted, William Stewart
- Surgical Papers. Baltimore: 1924 (≈ 1952). 1st edn, 2nd printing. 2 vols. 8vo. Frontis ports, plates, num ills. 1st few pp of vol 1 dampstained. Orig cloth. *(Elgen)* **$225** [≈ **£140**]

Hamel, Gustav & Turner, Charles C.
- Flying. Some Practical Experiences. London: Longmans, Green, 1914. Roy 8vo. xii, 341 pp. 72 photo plates. Orig gilt dec blue cloth, uncut, sl dull, faded & marked, front hinge sl slack. *(Duck)* **£50** [≈ **$80**]

Hamerton, Philip Gilbert
- Chapters on Animals. London: 1874. 8vo. 20 etched plates. Orig cloth. *(Wheldon & Wesley)* **£15** [≈ **$24**]

Hamilton, Frank H.
- A Practical Treatise on Fractures and Dislocations. Philadelphia: Blanchard & lea, 1863. 2nd edn, rvsd. 8vo. xxiii,751,32 ctlg pp. 285 w'cuts. Orig cloth, v worn. *(Elgen)* **$85** [≈ **£53**]

Hamilton, Hugh
- Philosophical Essays ... Clouds ... Aurora Borealis ... Comets ... Mechanicks ... London: John Nourse, 1767. 1st London edn. 12mo. [iv],177,[1] pp. Half-title to each part. Fldg plate. Orig soft wraps, uncut, vellum spine cracked. *(Rootenberg)* **£225** [≈ **£140**]
- Philosophical Essays ... Clouds ... Aurora Borealis ... Comets ... Mechanicks ... London: for J. Nourse, 1772. 3rd edn, enlgd. Cr 8vo. 3 part titles. Final blank. Frontis. Early stain to upper marg. Mod calf spine, contemp mrbld bds, sl worn. *(Stewart)* **£325** [≈ **$520**]

Hamilton, R.
- The Natural History of British Fishes. Vol II. Edinburgh: Jardine's Naturalist's Library, 1843. 1st edn. Sm 8vo. 424 pp. Vignette title, port, 34 hand cold plates. Orig cloth. *(Egglishaw)* **£40** [≈ **$64**]
- The Natural History of the Amphibious Carnivora including the Walrus and Seals ... Edinburgh: Jardine's Naturalist's Library, 1843. Sm 8vo. Uncold port & vignette title, 31 plates (all but 1 cold). Contemp half roan, worn, jnts torn. *(Bow Windows)* **£40** [≈ **$64**]
- The Natural History of the Ordinary Cetacea or Whales ... Edinburgh: Jardine's Naturalist's Library, 1837. 1st edn. Sm 8vo. Port, cold title vignette, 2 plain & 28 hand cold plates. Contemp mor, gilt spine, a.e.g., jnts sl rubbed. *(Egglishaw)* **£60** [≈ **$96**]

Hamilton, Sir William
- Observations on Mount Vesuvius, Mount Etna and Other Volcanoes. London: 1783. 2nd edn. 8vo. Fldg map (reprd), 5 plates. Mod half calf. Anthony Michaelis's b'plate.

(Henly) **£175 [≈ $280]**

Hamilton, Sir William Rowan
- The Mathematical Papers. Edited for the Royal Irish Academy by A.W. Conway ... Cambridge: 1931-67. 1st edn. 3 vols. 4to. Frontis ports. Orig cloth. Dws.
(Elgen) **$525 [≈ £328]**

Hammond, Robert
- The Electric Light in Our Homes. London: Warne, [1884]. 6th thousand. 8vo. 188 pp. 64 figs. 1 mtd photo (only, of 2). Dust marked. Orig dec cloth, marked. *(Moon)* **£48 [≈ $76]**

Hammond, William, M.D.
- Sexual Impotence in the Male. New York: Bermingham, 1883. 1st edn. 8vo. 274 pp. Orig bndg. *(Xerxes)* **$75 [≈ £46]**

Hanbury, F.J. & Marshall, E.S.
- Flora of Kent ... London: 1899. 8vo. lxxxiv, 444 pp. 2 fldg maps (1 cold, both with sm tears). Orig cloth, trifle used.
(Wheldon & Wesley) **£45 [≈ $72]**

Hancock, H.
- Foundations of the Theory of Algebraic Numbers. New York: 1931-32. 2 vols. xxvii, 602; xxvi,654 pp. Sm lib stamps. Orig cloth, vol 2 sl marked. *(Whitehart)* **£35 [≈ $56]**

Handley, James
- Colloquia Chirurgica: or, the Whole Art of Surgery Epitomiz'd and Made Easie, according to Modern Practice. London: Bates & Bettesworth, 1705. 1st edn. 8vo. [16],192,[4] pp. Pastedowns detached. Mottled calf. *(Rootenberg)* **$350 [≈ £218]**

Handley, James E.
- Scottish Farming in the Eighteenth Century. London: Faber, 1953. 314 pp. Orig cloth. Dw. *(Wreden)* **$39.50 [≈ £25]**

Hanger, George
- To All Sportsmen, particularly Farmers and Gamekeepers, &c. &c. London: 1814. 8vo. Half mor. *(Grayling)* **£100 [≈ $160]**

Hanish, Revd Dr Otoman Zaradusht
- Mazdaznan Health and Breath Culture. Chicago: Mazdaznan Press, 1914. 1st edn. Roy 8vo. 217 pp. Ills. Limp leather, worn. Sgnd pres copy. *(Xerxes)* **$90 [≈ £56]**

Hanley, S.
- The Conchological Miscellany, illustrative of

Pandora, Amphidesma, Ostrea, Melo, the Melanidae, Ampullaria and Cyclostoma. London: 1854-58. 4to. [22] pp. 36 (of 40) hand cold plates. Sm blind stamps on title & plates. Qtr mor, rebacked, orig wraps bound in. *(Wheldon & Wesley)* **£350 [≈ $560]**

Hansen, J.E. (ed.)
- A Manual of Porcelain Enameling. Cleveland: 1937. 513 pp. Ills. Ex-lib but vg. Orig bndg. *(Book House)* **£15 [≈ $24]**

Hardwich, T. Frederick
- A Manual of Photographic Chemistry, including the Practice of the Collodion Process. London: John Churchill, 1855. 1st edn. Sm 8vo. xvi,384 pp. Occas sl text soiling. Orig green cloth, sl rubbed.
(Burmester) **£240 [≈ $384]**

Hardy, G.A.
- The Complete Ironmonger ... London: 'The Ironmonger', [1895]. 1st coll edn. 8vo. viii, 258,34 illust advt pp. 24 ills. Orig cloth.
(Fenning) **£24.50 [≈ $40]**

Hardy, G.H.
- Divergent Series. Oxford: 1949. xvi,396 pp. Orig cloth. *(Whitehart)* **£25 [≈ $40]**

Hardy, G.H., et al.
- Inequalities. Cambridge: 1934. xii,314 pp. Orig cloth, spine faded.
(Whitehart) **£25 [≈ $40]**

Harlan, Richard
- Fauna Americana: being a Description of the Mammiferous Animals inhabiting North America. Philadelphia: 1825. 318 pp, errata leaf. Old calf, sl worn, front hinge rubbed.
(Reese) **$400 [≈ £250]**

Harle, Jonathan
- An Historical Essay on the State of Physick in the Old and New Testament ... Added, a Discourse concerning the Duty of Consulting a Physician in Sickness. London: for Richard Ford, 1729. 1st edn. 8vo. viii, 179,[1] pp. Sl waterstained at beginning. Disbound.
(Burmester) **£60 [≈ $96]**

Harris, E.C.
- New Zealand Berries - New Zealand Flowers - New Zealand Ferns. Nelson, N.Z.: [1894]. 3 vols in one. 4to. 3 litho titles, 36 plates. Minor foxing. Binder's cloth.
(Wheldon & Wesley) **£85 [≈ $136]**

Harris, Joseph
- A Treatise of Optics: containing Elements of the Science. London: sold by B. White, 1775. 1st edn. 4to. [vi],282 pp. 233 diags on 23 fldg plates. Some sl browning. Clean tears in 3 ff without loss (2 with old tape reprs). Half calf, uncut. *(Burmester)* **£350 [≈ $560]**

Harris, T.M.
- The Yorkshire Jurassic Flora. London: British Museum, 1961-79. 5 vols. 4to. 29 plates, 316 text figs. Orig cloth.
(Wheldon & Wesley) **£112 [≈ $179]**

Harris, T.W.
- A Treatise on some of the Insects injurious to Vegetation. New Edition by C.L. Flint. Boston, Mass: 1862. 8vo. xi,640 pp. 8 hand cold plates, 278 text figs. Orig cloth, somewhat worn.
(Wheldon & Wesley) **£65 [≈ $104]**

Harris, Sir William Snow
- On the Nature of Thunderstorms; and on the Means of Protecting Buildings and Shipping against the Destructive Effects of Lightning. London: John W. Parker, 1843. 8vo. xvi,226 pp. Frontis, 52 text figs. Orig cloth, lacks spine tips, b'plate removed from endpaper.
(Moon) **£100 [≈ $160]**

Harrison, J.M.
- The Birds of Kent. London: 1953. 2 vols. 4to. 80 plates (37 cold). Orig blue buckram.
(Henly) **£150 [≈ $240]**

Harte, Walter
- Essays on Husbandry. London: [ptd for W. Frederick in Bath], 1764. 1st edn. 2 parts in one vol. 8vo. xviii, errata leaf, [2],213,232 pp. 5 engvd plates, 25 w'cuts in text. Minor spotting to title. Early 19th c half calf, gilt spine. *(Spelman)* **£100 [≈ $160]**
- Essays on Husbandry ... Second Edition. London: for W. Frederick ..., 1770. 8vo. Contemp calf. *(Falkner)* **£140 [≈ $224]**

Harting, J.E.
- The Birds of Middlesex. London: 1866. Cr 8vo. xvi,284 pp. Tinted frontis by J. Wolf. Orig gilt dec cloth.
(Wheldon & Wesley) **£30 [≈ $48]**
- British Animals Extinct within Historic Times with Some Account of British Wild White Cattle. London: 1880. 8vo. x,258 pp. 36 ills. Gilt dec cloth. *(Baldwin)* **£35 [≈ $56]**
- A Handbook of British Birds. London: 1872. 8vo. xxiv,198 pp. Orig cloth, trifle soiled &

faded.   *(Wheldon & Wesley)* **£15 [≈ $24]**
- A Handbook of British Birds. London: Nimmo, 1901. 3rd edn, rvsd. Thick 8vo. xxxi, 520 pp. 35 hand cold plates. Orig buckram gilt, trifle rubbed, spine a little darkened.   *(Hollett)* **£60 [≈ $96]**
- Hints on Shore Shooting. With a Chapter on Skinning and Preserving Birds. London: 1871. 8vo. Frontis. Orig cloth.
(Grayling) **£35 [≈ $56]**
- Rambles in Search of Shells, Land and Freshwater. London: 1875. Cr 8vo. viii,110,2 pp. 10 cold plates. Index & advts sl foxed. Orig dec cloth.   *(Henly)* **£25 [≈ $40]**
- Rambles in Search of Shells, Land and Freshwater. London: 1875. Cr 8vo. viii,110,2 pp. 10 cold plates. Orig dec cloth, spine sl faded.   *(Henly)* **£35 [≈ $56]**

Hartley, David
- Hartley's Theory of the Human Mind, on the Principle of the Association of Ideas; with Essays relating to the Subject of it. By Joseph Priestley. London: for J. Johnson, 1775. 1st edn. 8vo. lxii,372 pp. Calf, front bd detached.
(Gach) **$325 [≈ £203]**
- A View of the Present Evidence ... against Mrs. Stephens's Medicines, as a Solvent for the Stone ... London: for S. Harding, 1739. 1st edn. 8vo. vi,[2],204,[4] pp. Pp 195,198, 199,200 mispaginated. Rec bds.
(Rootenberg) **$400 [≈ £250]**

Hartlib, Samuel
- The Reformed Commonwealth of Bees ... with The Reformed Virginian Silk-Worm ... London: 1655. Sm 4to. iv,62,[2], [iv],40 pp. W'cuts. Title soiled & mtd. Sl soiling & staining. Sm hole in last 2 ff, with sl loss of text. New calf, antique style.
(Wheldon & Wesley) **£750 [≈ $1,200]**
- The Reformed Virginian Silk-Worm, Or, a Rare and New Discovery ... for the feeding of Silk-worms in the Woods, on the Mulberry-Tree-leaves in Virginia ... London: 1655. [4], 40 pp. Calf.   *(Reese)* **$2,750 [≈ £1,718]**
- Samuel Hartlib His Legacy of Husbandry. London: J.M. for R. Wodenothe, 1655. 3rd edn. xvi, 303 pp. Contemp calf, untrimmed, sl wear to front hinge. Half mor slipcase. Wing H.991.   *(Reese)* **$1,250 [≈ £781]**

Harvey, W.H.
- Phycologia Britannica: or a History of British Sea-Weeds ... London: Reeve & Benham, 1846-51. 1st edn. 4 vols. Roy 8vo. 360 cold plates (chromolithos finished by hand). Contemp green half calf, gilt spines, red

labels.   *(Egglishaw)* **£680 [≃ $1,088]**

Harvey, William
- De Motu Locali Animalium, 1627. Edited, translated and introduced by Gweneth Whitteridge. Cambridge: 1959. 1st edn in English. One of 1000. 8vo. 163 pp. Frontis. Orig cloth. Dw.   *(Elgen)* **$100 [≃ £62]**
- The Works translated from the Latin with a Life of the Author by Robert Willis, M.D. London: Sydenham Society, 1847. 1st complete edn in English. 8vo. xcvi,624 pp. Orig cloth, spine relaid.
          (Bickersteth) **£75 [≃ $120]**

Harvie-Brown, J.A.
- The Capercaillie in Scotland. Edinburgh: 1888. 8vo. xv,155 pp. Map, 2 plates. Orig cloth, refixed. J.E. Harting's copy with annotations.
          (Wheldon & Wesley) **£50 [≃ $80]**

Harvie-Brown, J.A. & Buckley, T.E.
- A Fauna of the Moray Basin. Edinburgh: 1895. 2 vols. 8vo. Map, 23 plates. Emblematically gilt green mor, pigskin endpapers tooled with pictorial ex libris.
          (Wheldon & Wesley) **£85 [≃ $136]**
- A Fauna of the Moray Basin. Edinburgh: Douglas, 1895. 2 vols. 8vo. Map (mtd on linen), 23 plates. Orig cloth.
          (Egglishaw) **£70 [≃ $112]**
- A Vertebrate Fauna of Sutherland, Caithness and West Cromarty. Edinburgh: Douglas, 1887. 8vo. xi,344 pp. Fldg map, dec title, 2 panoramas, 6 plates (2 cold), 4 text ills. Orig cloth, spine faded, sl wear at hd.
          (Egglishaw) **£55 [≃ $88]**

Harwood, Sir John James
- History and Description of the Thirlmere Water Scheme. Manchester: Manchester Corporation Waterworks Committee, 1895. 8vo. 277 pp. Fldg maps, ills. Orig bndg, cvrs marked.   *(Book House)* **£75 [≃ $120]**

Hassall, A.H.
- A History of the British Freshwater Algae ... London: Longman, (1845) 1857. 2 vols. 8vo. viii,462; 24 pp. 103 cold plates (some foxed). Orig cloth.   *(Egglishaw)* **£90 [≃ $144]**

Hatchett, Charles
- A Third Series of Experiments on an Artificial Substance which possesses the Principal Characteristic Properties of Tannin; with some Remarks on Coal. London: W. Bulmer & Co, from the

Philosophical Transactions, 1806. 38 pp. Disbound.   *(Jarndyce)* **£40 [≃ $64]**

Haudicquer de Blancourt, Francois or Jean
- The Art of Glass. Shewing how to make all Sorts of Glass ... now first Translated into English. With an Appendix ... London: for Dan. Brown ..., 1699. 1st English edn. 8vo. Half-title. 9 plates. A few marg tears, v sl water stains. Contemp calf, rebacked. Wing H.1150.   *(Hannas)* **£680 [≃ $1,088]**

Haughton, S.
- Manual of Geology. London: 1876. 4th edn. Cr 8vo. xvi,415 pp. Diags, tables & plates, some fldg, text figs. Orig cloth, hd of spine chipped.   *(Henly)* **£22 [≃ $35]**

Haverschmidt, F.
- Birds of Surinam. London: 1968. Roy 8vo. xxix,445 pp. Map, 30 plain & 40 cold plates by P. Barruel, num text figs. Orig cloth, trifle used.   *(Wheldon & Wesley)* **£190 [≃ $304]**
- Birds of Surinam. London: (1968) 1971. 4to. 30 plain & 40 cold plates by P. Barruel, text figs. Orig green cloth.   *(Henly)* **£190 [≃ $304]**

Hawkhead, J.C.
- Handbook of Technical Instruction for Wireless Telegraphists. London: Wireless Press, 1917. 2nd edn. 8vo. 310 pp. Ills. Orig bndg, rubbed, shaken.
          (Book House) **£20 [≃ $32]**

Hay, R. & Synge, P.M.
- The Dictionary of Garden Plants. London: 1970. One of 265, signed by both authors. 4to. xvi,373 pp. 2048 cold photo plates on 256 pp. Mor gilt, a.e.g., by Zaehnsdorf. Solander case.   *(Henly)* **£210 [≃ $336]**

Hayes, Richard
- Interest at One View, calculated to a Farthing ... Eighteenth Edition, corrected. London: Johnson & Robinson, 1789. 12mo. 384 pp. Orig sheep, hinges broken but sides held on cords.   *(Claude Cox)* **£25 [≃ $40]**

Hayes, Thomas
- A Serious Address, on the Dangerous Consequences of Neglecting Common Coughs and Colds ... Hooping Cough and Asthma. Walpole, NH: for W. Fessenden & G.W. Nichols, 1808. 12mo. 136 pp. browning. Half leather, bds v worn.
          (Elgen) **$100 [≃ £62]**

Haynes, Thomas
- A Treatise on the Improved Culture of the Strawberry, Raspberry, and Gooseberry ... Second Edition. London: Crosby, 1814. 8vo. vii, 101,[3 advt] pp. Orig bds, paper labels, sometime rebacked, sm split in upper jnt.
(Blackwell's) **£160 [≈ $256]**

Hayward, I.M. & Druce, G.C.
- The Adventive Flora of Tweedside. Arbroath: 1919. 8vo. xxxii,296 pp. 80 ills. Endpapers foxed. Orig cloth.
(Wheldon & Wesley) **£45 [≈ $72]**

Hayward, Joseph
- The Science of Horticulture ... London: for Longman ..., 1824. 2nd edn. 8vo. xxvi,275 pp. 13 fldg plates. Occas spotting. Orig bds, rebacked.   *(Young's)* **£90 [≈ $144]**

Hazlitt, William Carew
- Gleanings in Old Garden Literature. London: (1887) 1892. Post 8vo. vii,263 pp. Orig cloth, trifle used. *(Wheldon & Wesley)* **£25 [≈ $40]**

Head, Sir Francis Bond
- Stokers and Pokers: or, The London and North-Western Railway, the Electric Telegraph, and the Railway Clearing-House. London: Murray, 1849. 1st edn. 8vo. A few w'engvs in text. Contemp blue half calf, gilt spine, sl worn.   *(Sanders)* **£75 [≈ $120]**
- Stokers and Pokers: or, The London and North-Western Railway, The Electric Telegraph, and the Railway Clearing-House ... New Edition. London: 1861. Sm 8vo. 224 pp. Sm name stamp on title. Pastedowns stained. Orig green cloth.
(Bow Windows) **£48 [≈ $76]**

Head, Henry
- Aphasia and Kindred Disorders of Speech. Cambridge: UP, 1926. 1st edn. 2 vols. Roy 8vo. Orig cloth, spine ends sl worn. Vol 2 with worn dw.   *(Elgen)* **$500 [≈ £312]**

Hearne, R.P.
- Airships in Peace and War. Being the Second Edition of 'Aerial Warfare', with Seven New Chapters ... London: John Lane; The Bodley Head, 1910. xlviii,324,advt pp. Plates, ills. Orig pict dec blue cloth, edges & crnrs sl rubbed, minor marking. *(Duck)* **£75 [≈ $120]**

Heath, F.G.
- Autumnal Leaves [New Forest]. London: 1885. 3rd edn. 8vo. 352 pp. 12 cold plates. Flyleaf defective. Orig cloth.

(Wheldon & Wesley) **£18 [≈ $28]**
- Autumnal Leaves [New Forest]. London: 1899. 4th edn. 8vo. xvi,352 pp. 12 cold plates. Orig cloth, faded.
(Wheldon & Wesley) **£18 [≈ $28]**
- The Fern Portfolio. London: SPCK, 1885. Folio. 15 cold plates. 3 sm lib stamps in text. Orig pict cloth gilt, spine ends sl worn, lib label on endpaper. *(Egglishaw)* **£25 [≈ $40]**
- The Fern World. London: Sampson Low, 1878. 4th edn. 8vo. xi,459,advt pp. Frontis, 12 cold plates. Orig dec cloth, a.e.g., jnts rubbed.   *(Egglishaw)* **£12 [≈ $19]**

Heath, Robert
- Astronomia Accurata; or, The Royal Astronomer and Navigator. Containing New Improvements in Astronomy, Chronology, and Navigation ... London: for the author, 1760. 1st edn. 4to. [24],412 pp. Calf, worn, jnts weak.   *(Elgen)* **$425 [≈ £265]**

Hebert, Luke
- The Engineer's and Mechanics Encyclopaedia ... London: Thomas Kelly, 1842. 2 vols. 8vo. [iv],796; 928 pp. Frontis, num w'engvd ills. Old cloth.
(Charles B. Wood) **$200 [≈ £125]**
- The Engineer's and Mechanics Encyclopaedia ... London: Thomas Kelly, 1847-35. 2 vols. 8vo. iv,796; 928 pp. 2 frontis, ca 2000 ills. Contemp half calf.
(Fenning) **£125 [≈ $200]**

Hedrick, U.P.
- The Cherries of New York. Albany: State of New York Dept of Agriculture, 1915. 1st edn. Lge 4to. xii,371 pp. 56 litho plates. Orig green cloth gilt. *(Terramedia)* **$100 [≈ £62]**

Heer, Professor Oswald
- The Primaeval World of Switzerland. Edited by James Haywood. London: 1876. Sole English edn. 2 vols. 8vo. Fldg cold geol map, 8 tinted litho plates, 11 plates of fossils, text ills. Orig cloth gilt, hd of vol 1 spine chipped, sm lib numbers on cvrs.
(Bickersteth) **£48 [≈ $76]**

Heilmann, G.
- The Origin of Birds. London: 1927. 8vo. vii, 208 pp. 2 cold plates, 140 figs. Orig cloth.
(Wheldon & Wesley) **£45 [≈ $72]**

Heirs of Hippocrates ...
- Heirs of Hippocrates. The Development of Medicine in a Catalogue of Historic Books in the Health Sciences Library, University of

Iowa. Iowa City: 1980. 1st edn. 4to. 474 pp. Cold frontis, ills. Orig cloth.
(Elgen) **$60 [≃ £37]**

Helleck, Reuben
- Education of the Central Nervous System, a Study of Foundations, especially of Sensory and Motor Training. New York: Macmillan, 1904. 8vo. 258 pp. Orig bndg, vertical crease in backstrip.
(Xerxes) **$60 [≃ £37]**

Hellins, John
- Select Parts of Saunderson's Elements of Algebra ... London: for C. Dilly ..., 1792. 5th edn. 8vo. iv,417 pp. 2 fldg plates. Contemp tree calf.
(Young's) **£45 [≃ $72]**

Hellot, J.P., et al.
- The Art of Dying Wool, Silk, and Cotton. Translated from the French. London: R. Baldwin, 1789. 1st English edn. 8vo. xvi,508, [iv] pp. Orig half calf, rebacked.
(Charles B. Wood) **$500 [≃ £312]**

Hellyer, S. Stevens
- The Plumber & Sanitary Houses. A Practical Treatise on the Principles of Internal Plumbing Work ... London: 1877. 1st edn. viii, 144 pp. Pict title as frontis, 7 plates, 37 text ills. Orig dec cloth, hinges cracked but sound, endpapers sl marked.
(Duck) **£110 [≃ $176]**

Helmholtz, Hermann Von
- The Description of an Ophthalmoscope. Translated by Thomas Hall Shastid. Chicago: 1916. 1st edn in English. One of 500. Lge 8vo. 33 pp. Frontis, ills. Orig cloth, worn.
(Elgen) **$225 [≃ £140]**

Helsham, Richard
- A Course of Lectures in Natural Philosophy. Published by Bryan Robinson. The Fifth Edition. London: J. Nourse, 1777. 8vo. x, 404,[2 advt] pp. 11 fldg plates. Contemp sprinkled sheep, rubbed.
(Blackwell's) **£75 [≃ $120]**

Hemingway, Taylor
- Sex Control. Curious Customs of Medieval Times. Harriman, Tenn.: the author, 1953. 8vo. 46 pp. 4 ills of chastity belts. Orig bndg.
(Xerxes) **$45 [≃ £28]**

Hemmeter, John C.
- Diseases of the Intestines. Philadelphia: 1901. 1st edn. 2 vols. Thick 8vo. Plates (some cold), ills. Orig cloth, hd of 1 spine frayed.
(Elgen) **$95 [≃ £59]**

Hennedy, R.
- The Clydesdale Flora. Glasgow: 1874. 3rd edn, rvsd. Post 8vo. xxiv,248 pp. Calf.
(Wheldon & Wesley) **£28 [≃ $44]**

Henrey, B. & Bean, W.J.
- Trees and Shrubs throughout the Year. London: 1944. 4to. vi,58 pp. 85 plates (8 cold). Orig green cloth. Dw.
(Henly) **£18 [≃ $28]**

Henrey, Blanche
- British Botanical and Horticultural Literature before 1800 ... Oxford: 1975. 3 vols. 4to. Ills. Orig cloth.
(Spelman) **£160 [≃ $256]**

Henry, G.M.
- Coloured Plates of the Birds of Ceylon, with a Short Description of each Bird by W.E. Wait. Colombo: 1927-35. Parts 1-4 (all publ) in one vol. Roy 4to. 48 cold plates. Cloth.
(Wheldon & Wesley) **£260 [≃ $416]**

Henslow, J.S.
- A Dictionary of Botanical Terms. New Edition. London: [1858]. 8vo. [vi],218 pp. Ills. Orig cloth, hd of spine torn.
(Lamb) **£15 [≃ $24]**
- Report on the Diseases of Wheat. London: 1841. Offprint from the Royal Agricultural Society Journal. 8vo. 28 pp. Mod qtr mor. Orig wraps bound in.
(Lamb) **£20 [≃ $32]**

Herbart, Johann Friedrich
- A Text-Book in Psychology ... New York: Appleton, 1891. 1st edn in English. 12mo. [iv], [lxvi],200,[10] pp. Orig dec green cloth, front hinge cracked.
(Gach) **$125 [≃ £78]**

Herbert, Sidney & Nightingale, Florence
- General Report of the Commission appointed for Improving the Sanitary Conditions of Barracks and Hospitals. Presented to Both Houses of Parliament ... London: Eyre & Spottiswoode, 1861. 1st edn. Folio. [ii],338, [2] pp. 100 'cuts in text. Orig wraps, spine taped.
(Gach) **$385 [≃ £240]**
- Regulations Affecting the Sanitary Condition of the Army, the Organization of Military Hospitals, and the Treatment of the Sick and Wounded ... London: 1858. 1st edn. Folio. [iv],[lxxxvi],627,[3] pp. 17 charts & plans, 5 fldg plates. Orig wraps, cvrs taped.
(Gach) **$350 [≃ £218]**

Hermippus Redivivus ...
- See Cohausen, J.H.

Heroical Epistle from Death ...
- Heroical Epistle from Death to Benjamin Moseley M.D. on Vaccination [in verse]. With a Postscript on some collateral Subjects. London: J.J. Stockdale, 1810. 4to. 39 pp. Disbound.   *(Jarndyce)* **£280 [≈ $448]**

Herrick, C.T. (ed.)
- Consolidated Library of Modern Cooking and Household Recipes ... New York: R.J. Bodmer, 1904. 5 vols. 8vo. Num plates. Orig cloth, rubbed.   *(Berkelouw)* **$75 [≈ £46]**

Herrick, F.H.
- Natural History of the American Lobster. Washington: Bureau of Fisheries Bulletin, 1911. Roy 8vo. 260 pp. 20 plates (3 cold). Margs of a few plates damaged by adhesion, 1 plate torn without loss. Orig cloth.
        (Wheldon & Wesley) **£38 [≈ $60]**

Herring, Richard
- Paper and Paper Making, Ancient and Modern. Third Edition. London: Longman ..., 1863. 8vo. xix,134,[ii] pp. Old lib b'plate on title verso. New linen.
        (Charles B. Wood) **$90 [≈ £56]**

Herrlinger, Robert
- History of Medical Illustration from Antiquity to 1600. Medicina Rara: [ca 1970]. 1st English edn. 4to. 178 pp. 32 plates (many cold), num text ills. Leather backed linen. Slipcase.   *(Elgen)* **$125 [≈ £78]**

Herrod-Hempsall, W.
- Bee-Keeping New and Old Described with Pen and Camera. Vol 1 [only]. London: 1930. Roy 8vo. vi,772 pp. 2 ports, 708 ills. Orig cloth, rather loose. A 2nd vol was published in 1937. *(Wheldon & Wesley)* **£70 [≈ $112]**

Herschel, Sir John Frederick William
- Essays from the Edinburgh and Quarterly Reviews, with Addresses and Other Pieces ... London: Longman ..., 1857. 1st edn. 8vo. iv, 750 pp. Contemp diced crimson calf, gilt spine, Harrow prize label.
        (Pickering) **£500 [≈ £312]**
- Outlines of Astronomy. London: Longman, Brown ..., 1849. 1st edn. 8vo. xiv,661,[1],32 advt pp. Frontis (sm tear), 6 plates. Orig cloth, rebacked. *(Rootenberg)* **$200 [≈ £125]**
- Preliminary Discourse to the Study of Natural Philosophy ... New Edition. London: for Longman & John Taylor, 1830. 8vo. Contemp red calf, gilt spine.
        (Waterfield's) **£45 [≈ $72]**

- A Treatise on Astronomy. London: Longman, Cabinet Cyclopaedia, 1833. 1st edn. 8vo. viii, 422 pp, spare title-page. 3 plates, text ills. Orig cloth, worn, spine reprd.
        (Spelman) **£35 [≈ $56]**
- A Treatise on Astronomy. London: Cabinet Cyclopaedia, 1833. 1st edn. Sm 8vo. 16 advt, viii,422,[2] pp. Addtnl engvd title, 3 plates, num text ills. Rec paper bds, uncut.
        (Fenning) **£65 [≈ $104]**
- A Treatise on Astronomy. A New Edition. London: Lardner's Cabinet Cyclopaedia, 1834. Sm 8vo. Orig cloth, ptd paper label.
        (Fenning) **£35 [≈ $56]**

Herschel, Sir John Frederick William (ed.)
- A Manual of Scientific Enquiry; prepared for the use of Her Majesty's Navy; and adapted for Travellers in general. London: 1849. 1st edn, 2nd iss. 8vo. Dble-page map, plate, fldg map, 16 ills. V sl browning. Orig blue cloth with gilt anchor. Freeman (Darwin) 325.
        (Fenning) **£250 [≈ $400]**
- A Manual of Scientific Enquiry prepared for the use of Officers in Her Majesty's Navy. London: 1859. 3rd edn. 8vo. xviii,429 pp. Map, plate, text figs. Cloth. With chapter on Geology by Darwin. Freeman 328.
        (Wheldon & Wesley) **£50 [≈ $80]**

Hertwig, O.
- Text-Book of the Embryology of Man and Animals. Translated by E. Mark. London: 1912. 5th edn. 8vo. xvi,670 pp. 2 plates, 339 figs. Orig cloth.
        (Wheldon & Wesley) **£30 [≈ $48]**

Hetley, Mrs C.
- The Native Flowers of New Zealand illustrated in Colours ... from Drawings Coloured to Nature. London: [1887-] 1888. Imperial 4to. 36 chromolithos of flowers, 3 plain plates of dissections. 2 plates. Misbound. Mod cloth. Plesch b'plate.
        (Wheldon & Wesley) **£350 [≈ $560]**

Hewitson, W.C.
- Coloured Illustrations of the Eggs of British Birds. London: 1846. 2 vols. 8vo. 138 cold plates (numbered 1-131, 12★, 25★, 37★, 67★, 78★, 101★, 120★). 1 or 2 ff sl frayed. Orig cloth, rebacked.
        (Wheldon & Wesley) **£75 [≈ $120]**
- Coloured Illustrations of the Eggs of British Birds. London: 1846. 2nd edn. 2 vols. 8vo. 138 cold plates (numbered 1-131, 12★, 25★, 37★, 67★, 78★, 101★, 120★). Contemp half calf.   *(Wheldon & Wesley)* **£95 [≈ $152]**

Hewitt, W.
- An Essay on the Encroachments of the German Ocean along the Norfolk Coast. London: 1844. 8vo. 108 pp. 2 plates. Orig cloth. Author's inscrptn.
(Henly) **£65** [≃ **$104**]

Heylyn, Peter
- [Greek title: Microcosmos]. A Little Description of the Great World. The Fourth Edition, Revised. Oxford: 1629. Sm 4to. 10 ff, 807 pp,3,[4] ff, the last blank. Contemp panelled calf, spine gilt in compartments, red label. STC 13279. *(Hemlock)* **$200** [≃ **£125**]

Hibberd, Shirley
- The Book of the Aquarium and Water Cabinet ... London: Groombridge, 1856. 1st coll edn. Sm 8vo. [4],148,[32 advt] pp. 2 plates, 38 ills. Orig green cloth gilt, a.e.g., just a little dull. *(Fenning)* **£35** [≃ **$56**]
- The Book of the Aquarium and Water Cabinet. London: 1856. Post 8vo. Text figs. Orig cloth, a.e.g. *(Henly)* **£18** [≃ **$28**]
- The Fern Garden. How to make, keep, and enjoy it; or, Fern Culture Made Easy. London: Groombridge & Sons ..., 1870. 2nd edn. 8vo. vi, 148,14 advt pp. 5 cold plates, num text ills. Orig gilt dec cloth, a little rubbed. *(Young's)* **£30** [≃ **$48**]
- Field Flowers; A Handy Book for the Rambling Botanist ... London: Groombridge, [ca 1875]. 1st edn. 8vo. iv,156,16 advt pp. Orig dec green cloth gilt. *(Gough)* **£28** [≃ **$44**]
- New and Rare Beautiful Leaved Plants ... London: Bell & Daldy, 1870. Roy 8vo. viii,144 pp. 54 cold plates, 54 text ills. Orig red half mor gilt, t.e.g., crnrs bumped. *(Berkelouw)* **£350** [≃ **£218**]
- New and Rare Beautiful-Leaved Plants ... London: Bell & Daldy, 1870. 4to. viii,144 pp. 54 colour-ptd plates. Orig cloth gilt, spine ends reprd, sl loss of gilt to hd. *(Spelman)* **£95** [≃ **$152**]
- New and Rare Beautiful Leaved Plants. London: 1870. Roy 8vo. viii,144,8 pp. 54 cold plates. 3 plates sl foxed. Orig cloth gilt, spine faded, internal jnts starting to crack. *(Henly)* **£75** [≃ **$120**]
- New and Rare Beautiful-Leaved Plants ... London: Bell & Daldy, 1874. Sm 4to. viii,144 pp. 53 cold plates. Rec qtr mor. *(Terramedia)* **$300** [≃ **£187**]
- The Rose Book: a Practical Treatise on the Culture of the Rose ... London: Groombridge, 1864. 1st edn (?). xii,288,ctlg pp. Frontis, w'engvs. Orig magenta cloth,

dulled, spine faded & worn at ends, shaken, central stitching weak.
(Box of Delights) **£45** [≃ **$72**]

Hibbert, Samuel
- Sketches of the Philosophy of Apparitions; or, an Attempt to trace such Illusions to their Physical Causes. Edinburgh & London: 1825. 2nd edn, enlgd. Sm 8vo. xii,475,[3] pp. Fldg table. Mod cloth backed bds.
(Gach) **$175** [≃ **£109**]

Hickey, Baron Harden
- Euthanasia. The Aesthetics of Suicide. New York: The Truth Seeker, 1894. 1st edn. 24mo. 167 pp. Orig dec wraps, cvr & backstrip chipped. *(Xerxes)* **$85** [≃ **£53**]

Hickock, Laurens P.
- Empirical Psychology ... Schenectady, New York: Van Debogert, 1854. 1st edn. 12mo. [v]-400, [2] pp. Some foxing & marg staining. Orig brown cloth. *(Gach)* **$85** [≃ **£53**]
- Rational Cosmology ... New York: Appleton, 1858. 1st edn. 8vo. 397 pp. Light foxing. Orig ptd brown cloth. *(Gach)* **$85** [≃ **£53**]

Higgins, Charles M.
- Horrors of Vaccination Exposed and Illustrated. Petition to the President to abolish Compulsory Vaccination in the Army and Navy. Brooklyn: Higgins, 1920. 8vo. 212 pp. Photo ills. Wraps, lacks backstrip, front cvr detached, lacks lower cvr, shaken.
(Xerxes) **$75** [≃ **£46**]

Higgins, William Mullinger
- The Philosophy of Sound, and History of Music ... London: Orr, 1838. 1st edn. 8vo. Cold frontis, engvd title, text ills. Orig cloth.
(Falkner) **£40** [≃ **$64**]

Hill, C.W.
- Electric Crane Construction. London: Griffin, 1911. Lge 8vo. 313 pp. Fldg plates, 366 ills. Orig bndg.
(Book House) **£35** [≃ **$56**]

Hill, "Sir" John
- The British Herbal ... London: 1756. Folio. [iv],533,[3] pp. Frontis, 2 vignettes, 75 engvd plates of figs. Some sl water staining, last leaf shows signs of use. Mod half calf.
(Wheldon & Wesley) **£480** [≃ **$768**]
- The Family Herbal ... Bungay: C. Brightly & T. Kinnersley, [1810?]. 8vo. viii,xl,376 pp, 15 MS index ff, 17 blanks. 54 hand cold plates. Contemp tree calf, endpapers

reinforced. *(Rootenberg)* **$500 [≈ £312]**
- The Family Herbal ... Bungay: Brightly & Kinnersley, 1812. 8vo. xl,316 pp. 54 hand cold plates. 1 plate torn & reprd, a few plates sl soiled. Mod half mor gilt, orig endpapers retained. *(Hollett)* **£100 [≈ $160]**
- See also Barnes, Thomas.

Hillary, William
- Observations on the Changes of the Air, and the Concomitant Epidemical Diseases of the Island of Barbados ... added, A Treatise on the Putrid Bilious Fever ... Notes by Benjamin Rush. Philadelphia: 1811. 1st Amer edn. 8vo. [4],260,[4] pp. Contemp calf, hinges rubbed & partly cracked. *(Hemlock)* **£300 [≈ £187]**

Hillebrand, W.
- Flora of the Hawaiian Islands, a Description of their Phanerogams and Vascular Cryptogams. Heidelberg: 1888. 8vo. xcvi,673 pp. Frontis, 4 maps. Orig cloth.
 (Wheldon & Wesley) **£120 [≈ $192]**

Hillis, John D.
- Leprosy in British Guiana. An Account of West Indian Leprosy ... London: J. & A. Churchill, 1881. Sole edn. Roy 8vo. xi,264 pp. 22 cold litho plates. Orig cloth, unopened, lacks rear free endpaper.
 (Bickersteth) **£110 [≈ $176]**

Himes, Norman
- Medical History of Contraception ... New York: Gamut, 1963. 8vo. 521 pp. Ills. Tiny chip on 2 blank end pp. Orig bndg.
 (Xerxes) **$75 [≈ £46]**

Hind, J.
- The Elements of Plane and Spherical Trigonometry ... Cambridge: 1828. 2nd edn. xii,352 pp. Pencil notes. Half roan, sl rubbed.
 (Whitehart) **£25 [≈ $40]**

Hinton, John W.
- Organ Construction. London: Weekes & Co, 1902. 2nd edn. 4to. 200 pp. Port, 17 plates. Sev sgntrs on endpaper. Orig cloth gilt, spine ends worn, top of lower hinge split, crnrs bumped. *(Hollett)* **£55 [≈ $88]**
- Organ Construction. Third Edition. revised and Enlarged. London: Weekes, 1910. Last & best edn. 4to. [8],190,[4 advt] pp. Fldg plan, 37 plates & ills. Orig cloth.
 (Fenning) **£35 [≈ $56]**

Hinton, M.A.C.
- Monograph of the Voles and Lemmings

(Microtinae) Living and Extinct. Vol 1 [all published]. London: BM, 1926. 8vo. xvi,488 pp. 15 plates, 110 text figs. Lib stamp on title. Cloth. *(Egglishaw)* **£30 [≈ $48]**

Hints for the Table ...
- See Timbs, John.

Hipkins, W.E.
- The Wire Rope and its Applications. Birmingham: J.E. Wright Ltd., Universe Works, 1896. Sm 4to. 86 ills, inc 40 cold fldg. Orig cloth gilt, cloth on upper bd rather creased & spotted. *(Hollett)* **£130 [≈ $208]**

Hirsch, August
- Handbook of Geographical and Historical Pathology. London: New Sydenham Society, 1883. 3 vols. Orig dec cloth, spine ends bumped. *(John Smith)* **£75 [≈ $120]**

Hirschfeld, Magnus
- The Sexual History of the World War. New York: Falstaff Press, [1937]. 1st edn in English. 8vo. [376] pp. Orig blue cloth.
 (Gach) **$37.50 [≈ £23]**

Hirst, Barton Cooke
- Atlas of Operative Gynaecology. Philadelphia: [1919]. 1st edn. 4to. 292 pp. 164 plates, fldg diag, text ills. Orig cloth.
 (Elgen) **$75 [≈ £46]**

Historical ...
- Historical Account of the Substances which have been used to describe Events ... see Koops, Matthias.

History ...
- The History and Description of Fossil Fuel ... see Holland, John.
- The History of Inland Navigations ... see Brindley, James.

Hitchcock, Edward
- Report on the Geology, Mineralogy, Botany & Zoology of Massachusetts. Amherst: Adams, 1833. 8vo. xii,692 pp. Num w'cuts. Leather, rubbed. Without the atlas of plates.
 (Gemmary) **$60 [≈ £37]**

Hitt, Thomas
- A Treatise on Fruit-Trees. Third Edition. London: for Robinson & Roberts, 1768. 8vo. viii, 394,[v] pp. 7 fldg plates. Contemp polished calf, red mor label.
 (Bickersteth) **£125 [≈ $200]**

Hoare, Clement
- A Practical Treatise on the Cultivation of the Grape Vine on Open Walls. Second Edition. London: Longman, 1837. 8vo. vii,[5],210 pp. Orig cloth, backstrip relaid.
(Spelman) **£55 [≈ $88]**

Hobbs, William Herbert
- Characteristics of Existing Glaciers. New York: Macmillan, 1911. 8vo. xxiv,[1],301,advt pp. 34 photo plates, 140 text ills. Embossed stamp on title & preface. Orig gilt titled dark green cloth.
(Parmer) **$100 [≈ £62]**
- The Discoveries of Antarctica within the American Sector, as revealed by Maps and Documents. Philadelphia: Amer Philosoph Soc, 1939. 4to. [ii],71 pp. 32 plates (8 fldg), 10 text figs. Half cloth & mrbld paper.
(Parmer) **$125 [≈ £78]**

Hobson, C.
- Charles Waterton: His Home, Habits, and Handiwork. London: 1867. 2nd edn. 8vo. xxxii, 375 pp. 16 ills. Buckram, trifle loose.
(Wheldon & Wesley) **£30 [≈ $48]**

Hobson, E.W.
- The Theory of Functions of a Real Variable and the Theory of Fourier's Series. Cambridge: 1921-26. 2nd rvsd edn. 2 vols. xv, 671; x,780 pp. Orig cloth, dull.
(Whitehart) **£25 [≈ $40]**

Hodges, Nathaniel
- Loimologia: or, an Historical Account of the Plague in London in 1665 ... Essay on ... Pestilential Diseases ... by John Quincy. The Third Edition, with large Additions. London: 1721. 8vo. vi,iii-vi,224,76 pp. Fldg table. Contemp calf gilt. Sir Gilbert Blane's copy.
(Rankin) **£200 [≈ $320]**
- Loimologia: or, an Historical Account of the Plague in London in 1665 ... Essay on ... Pestilential Diseases ... by John Quincy ... Third Edition, with large Additions. London: Bell & Osborn, 1721. 8vo. v,224 pp. Fldg table. Old bds, rebacked.
(Bickersteth) **£110 [≈ $176]**

Hodgkinson, Eaton
- Experimental Researches on the Strength and other Properties of Cast Iron ... London: John Weale, 1860-61. 2nd edn. Roy 8vo. vii, 225-384,ctlg pp. 5 plates. Occas spotting. Orig olive cloth, partly unopened, lower crnr of cvrs sl bumped.
(Duck) **£150 [≈ $240]**

Hodgson, J.E.
- The Dredging of Gold Placers. London: Pitman, 1911. 8vo. ix,65 pp. 17 photo ills in text. Orig cloth.
(Gemmary) **$37.50 [≈ £23]**
- The History of Aeronautics in Great Britain. From the Earliest Times to the latter half of the Nineteenth Century. London: OUP, 1924. One of 1000. 4to. x,436 pp. Cold frontis, 150 plates (13 cold). Orig cloth gilt, t.e.g. Dw.
(Berkelouw) **£400 [≈ £250]**

Hoff, Ebbe C. & Fulton, J.F.
- Bibliography of Aviation Medicine. Springfield, Ill.: 1942; Washington: 1944. 4to. Orig bndg. Inc supplement.
(Argosy) **$150 [≈ £93]**

Hoffy, A.
- Hoffy's North American Pomologist ... Book 1 [all published]. Edited by William D. Brinckle. Philadelphia: A. Hoffy, 1860. 4to. [ii], vi pp. Mtd litho port, 36 hand cold litho plates, each with dust sheet & leaf of letterpress. Orig gilt dec blue cloth.
(Charles B. Wood) **$2,500 [≈ £1,562]**

Hofmann, Carl
- A Practical Treatise on the Manufacture of Paper in all its Branches. Philadelphia: Henry Carey Baird, 1873. 1st edn. 4to. iv, 398, [xxiv] pp. 5 lge fldg plates (bit frayed & outer margs sl spotted), 129 text ills. Orig cloth.
(Charles B. Wood) **$600 [≈ £375]**

Hofmeister, Wilhelm
- On the Germination, Development, and Fructification of the Higher Cryptogamia, and on the Fructification of the Coniferae. London: Ray Society, 1862. 1st edn in English. 8vo. xvii,506 pp. 65 plates. Orig cloth, faded.
(Rootenberg) **£400 [≈ £250]**

Hogg, Jabez
- The Microscope ... New Edition. London: Routledge, 1854. 6th edn. 8vo. xx,762 pp. Cold ills by Tuffen West, 500 engvs. Orig dec cloth, reprd, spine relaid.
(Savona) **£25 [≈ $40]**
- The Microscope: its History, Construction, and Application ... Third Edition. London: Routledge, 1858. Rvsd & crrctd edn, with new preface & some new ills. 8vo. xiv,[2],607,12 advt pp. Frontis, num ills. Orig red cloth gilt.
(Fenning) **£35 [≈ $56]**
- The Microscope ... New Edition. London: Routledge, 1883. New edn. 8vo. xx,762 pp. 8 cold plates by Tuffen West, 355 text figs. Orig cloth, sl worn, inner rear hinge cracked.

(Savona) **£30 [≈ $48]**
- The Microscope ... London: Routledge, 1886. 11th edn. 8vo. xx,764 pp. 8 cold plates, ca 500 engvs. Prize calf gilt.
(Egglishaw) **£40 [≈ $64]**
- The Microscope ... London: 1898. 15th edn. 8vo. xxiv,704 pp. "Upwards of Nine Hundred Engraved and Coloured Illustrations". Some foxing to endpapers, prelims & edges. Orig cloth gilt, sl worn & rubbed, spine faded. *(Dillons)* **£45 [≈ $72]**

Hogg, Robert & Johnson, George W.
- A Selection of the Eatable Funguses of Great Britain. London: Journal of Horticulture, n.d. 8vo. vii pp. 24 hand cold litho plates, each with tissue & page of text. Old half mor gilt. *(Hollett)* **£150 [≈ $240]**

Hogg, Thomas
- A Concise and Practical Treatise on the Growth and Culture of the Carnation ... and other Flowers. London: 1822. 2nd edn. Post 8vo. xxvii,304 pp. 6 hand cold plates. Sl foxing. Orig bds, uncut.
(Wheldon & Wesley) **£75 [≈ $120]**
- A Concise and Practical Treatise on the Growth and Culture of the Carnation ... and other Flowers. Fifth Edition, with additions. London: Whitaker & Treacher, [1832]. 8vo. xxviii, 275,[i],[4 advt] pp. 6 hand cold plates. Orig green cloth, jnts reprd, label rubbed.
(Spelman) **£80 [≈ $128]**

Holbrook, M.L.
- How to Strengthen the Memory; or, Natural and Scientific Methods of Never Forgetting. New York: Holbrook, 1886. 1st edn. 8vo. 152 pp. Orig bndg. *(Xerxes)* **$65 [≈ £40]**

Holden, A.E.
- Plant Life in the Scottish Highlands. Ecology and Adaptation to their Insect Visitors. London: 1952. 8vo. xv,319 pp. 64 photo plates. Orig cloth.
(Wheldon & Wesley) **£25 [≈ $40]**

Hole, S. Reynolds
- Our Gardens. London: Haddon Hall Library, 1901. 4th edn. 8vo. x,304 pp. Cold frontis, 10 plain plates & head- & tail-pieces by Arthur Rackham. Orig cloth gilt.
(Henly) **£35 [≈ $56]**

Hole, W.
- The Distribution of Gas. London: John Allan, "Gas World", 1909. 2nd edn. 839,40 advt pp. Ills. Mod cloth bds.

(Book House) **£40 [≈ $64]**

Holland, Henry
- Chapters on Mental Physiology. London: Longman ..., 1852. 1st edn. 8vo. xii,301, [3], ctlg pp. Half-title glued to pastedown, sev sgntrs sprung. Orig brown cloth.
(Gach) **$85 [≈ £53]**
- Medical Notes and Reflections. London: 1839. 1st edn. 8vo. 628 pp. Orig cloth.
(Robertshaw) **£35 [≈ $56]**

Holland, John
- The History and Description of Fossil Fuel, the Collieries, and Coal Trade of Great Britain. London & Sheffield: 1841. 2nd edn. xvi, 486,[2 advt] pp. Text ills. Sl marked. Orig cloth, uncut, sl marked, spine faded as always, its ends sl chipped.
(Duck) **£80 [≈ $128]**
- A Treatise on the Progressive Improvement and Present State of the Manufactures in Metal. London: Lardner's Cabinet Cyclopaedia, 1831-33-34. 1st edn. 3 vols. Sm 8vo. Orig cloth, ptd paper labels. Vols 1-2 Iron & Steel; vol 3 Lead, Copper, Brass, Gold, Silver. *(Fenning)* **£48.50 [≈ $78]**

Hollick, Frederick
- Diseases of Woman: Their Causes and Cure. New York: Strong, (1847) 1853. 16mo. 417 pp. Orig bndg, gilt lettered spine, front hinge cracked. *(Xerxes)* **$75 [≈ £46]**
- The Marriage Guide, or Natural History of Generation; A Private Instructor for Married Persons and those about to Marry. New York: Strong, (1850) 1853. 16mo. 432 pp. Cold plates, num engvs. Orig cloth, gilt lettered spine. *(Xerxes)* **$75 [≈ £46]**
- Matron's Manual of Midwifery, and the Diseases of Women during Pregnancy and in Childbed. New York: Strong, (1848) 1856. 16mo. 468 pp. Tinted frontis, 60 engvs. Orig cloth, gilt lettered spine, cvr faded.
(Xerxes) **$75 [≈ £46]**

Hollingworth, D.V.
- Coke Oven Gas for Public Purposes. Evidence ... to Area Gas Supply Committee of the Board of Trade. London: 1929. 40 pp. Fldg tables. Orig bndg.
(Book House) **£15 [≈ $24]**

Hollom, P.A.D.
- The Popular Handbook of British Birds. London: 1952. 8vo. xxiii,424 pp. 132 cold & 20 other plates. Orig cloth. Dw.
(Henly) **£22 [≈ $35]**

Holly, H.W.
- The Art of Saw-Filing, Scientifically treated and explained on Philosophical Principles ... New York: John Wiley, 1864. 1st edn (?). 12mo. 56 pp. 44 text w'engvs. Occas sl foxing. Orig brown cloth gilt.   *(Gough)* **£45 [≈ $72]**

Home, Henry, Lord Kames
- See Kames, Henry Home, Lord.

Hooke, Robert
- Micrographia: or some Physiological Descriptions of Minute Bodies made by Magnifying Glasses ... Medicina Rara: [ca 1970]. One of 2300. 38 plates. Half leather.
(Elgen) **$125 [≈ £78]**

Hooker, Sir Joseph Dalton
- Handbook of the Flora of New Zealand: a Systematic Description of the Native Plants. London: 1864-67. 1st iss. 8vo. 15,lxviii,798 pp. Orig cloth, sl worn. *(Henly)* **£98 [≈ $156]**
- Life and Letters ... see Huxley, Leonard.

Hooker, Sir William Jackson
- The British Flora ... Third Edition with Additions and Corrections. London: Longman ..., 1835. 8vo. xi,499 pp. Contemp half calf, jnt cracked.   *(Lamb)* **£15 [≈ $24]**
- The British Flora ... by W.J. Hooker and G.A.W. Arnott. London: Longman, 1855. 7th edn. 8vo. 618 pp. 12 hand cold plates. New qtr calf.   *(Egglishaw)* **£34 [≈ $54]**
- Companion to the Botanical Magazine ... London: Edward Couchman, 1835 [-36]. 1st edn. 2 vols. Lge 8vo. 3 ports, 32 plates (mostly cold). Orig black cloth.
(W. Thomas Taylor) **$2,500 [≈ £1,562]**
- Niger Flora, or an Enumeration of the Plants of Western Tropical Africa ... London: 1849. 8vo. xvi,587 pp. Map, 2 plates of views, 50 plates of plants on 43 ff. Frontis waterstained, plate margs foxed. Mod half calf.
(Wheldon & Wesley) **£185 [≈ $296]**

Hooker, Sir William Jackson & Taylor, T.
- Muscologia Britannica: containing the Mosses of Great Britain and Ireland. London: 1827. 2nd edn. 8vo. xxxvii,272 pp. 37 plain plates on 36 ff. New cloth.
(Wheldon & Wesley) **£75 [≈ $120]**
- Muscologia Britannica: containing the Mosses of Great Britain and Ireland. London: 1827. 2nd edn. 8vo. xxxvii,[ii],272 pp. 37 hand cold plates on 36 ff. A little minor foxing. Name on title. Orig bds, trifle worn, inner jnts taped.
(Wheldon & Wesley) **£100 [≈ $160]**

- See also Wilson, William.

Hooper, Robert, et al.
- Lexicon Medicum; or Medical Dictionary ... Fourth American, from the Last London Edition ... With additions from American Authors ... by Samuel Akerly. New York: Harper, 1834. 4th edn, later printing. 2 vols in one. 8vo. [472]; [424] pp. Contemp calf, rubbed.   *(Gach)* **$75 [≈ £46]**

Hopgood, George
- On the Management of Hair and Scalp. Isle of Wight: W. Gabell, 1856. 12mo. [4],iv, [2], 8-78 pp. Orig cloth, worn.
(Spelman) **£35 [≈ $56]**

Hopton, W.
- Conversation on Mines, &c., between a Father and Son. Manchester: Heywood, 1883. 7th edn. 8vo. 285 pp. Ills. Sm tear in 1 plan. Orig bndg.   *(Book House)* **£16 [≈ $25]**

Horblit, H.D.
- One Hundred Books Famous in Science, based on an Exhibition held at the Grolier Club. New York: 1964. One of 1000. 4to. 449 pp. Ills. Orig cloth.
(Wheldon & Wesley) **£250 [≈ $400]**

Horne, George
- A Fair, Candid and Impartial State of the Case between Sir Isaac Newton and Mr. Hutchinson ...Oxford: at the Theatre for S. Parker ..., 1753. 1st edn. 8vo. [iv],76 pp. Half-title. Old wraps, uncut.
(Burmester) **£350 [≈ $560]**

Horner, F., et al.
- Sheet-Metal Work. A Practical Treatise dealing with Every Phase of the Sheet-Metal Industry ... With Special Chapters on Plastics. London: Caxton, n.d. 3 vols. 4to. 210; 299; 247 pp. Ills. Orig bndgs. Torn dws.
(Book House) **£20 [≈ $32]**

Horner, Joseph G.
- Lockwood's Dictionary of Terms used in the Practice of Mechanical Engineering ... London: Crosby Lockwood, 1902. 3rd edn, rvsd, with appendix. 8vo. 452 pp. Orig bndg.
(Book House) **£20 [≈ $32]**

Horner, W.E.
- Lessons in Practical Anatomy, for the Use of Dissectors. Philadelphia: E. Parker, 1823. 1st edn. 8vo. xxxi,505 pp, errata leaf. Usual foxing. Tree calf, scuffed, front starting.
(Elgen) **$300 [≈ £187]**

Hornibrook, M.
- Dwarf and Slow Growing Conifers. London: 1923. 1st edn. Cr 8vo. x,196 pp. Frontis & 24 plates. V sl foxing. Orig canvas backed bds.
(Henly) **£42 [≃ $67]**

Horrax, Gilbert
- Neurosurgery. An Historical Sketch. Springfield, Ill.: Charles Thomas, 1952. 1st edn. 8vo. 135 pp. Owner's sgntr. Orig bndg.
(Xerxes) **$45 [≃ £28]**

Horsley, Samuel
- Elementary Treatises on the Fundamental Principles of Practical Mathematics for the Use of Students, by Samuel Lord Bishop of Rochester. Oxford: Clarendon Press, 1801. 8vo. 11 fldg plates. Contemp qtr vellum, lettered by hand. John Cator's b'plate.
(Waterfield's) **£75 [≃ $120]**
- Elementary Treatises on the Fundamental Principles of Practical Mathematics. Oxford: 1801. 1st edn. Large Paper. 398 pp. 11 fldg plates. Orig calf & bds, worn, rebacked. Author's pres copy. *(Elgen)* **$150 [≃ £93]**

Hortus Sanitatis ...
- Hortus Sanitatis. An Early English Version, a recent Bibliographical Discovery by N. Hudson. London: 1954. One of 350. 4to. i-xvi, 164,xvii-xxxi pp. Num ills. Orig cloth.
(Wheldon & Wesley) **£70 [≃ $112]**

Hortus Siccus Gramineus ...
- See Salisbury, W.

Hotchkiss, Robert Sherman
- Fertility in Men. Foreword by N. Eastman, MD. Philadelphia: Lippincott, 1944. 1st edn. 8vo. 216 pp. Cold frontis, photo ills. Sgnd by former owner. Orig bndg.
(Xerxes) **$55 [≃ £34]**

Houghton, W.
- British Freshwater Fishes. London: 1895. 2nd edn. Roy 8vo. xxviii,231 pp. 24 plates, num text engvs. A few pp at the end with sm marg tears. Qtr leather.
(Wheldon & Wesley) **£30 [≃ $48]**
- Country Walks of a Naturalist with his Children. London: Groombridge, 1869. 1st edn. Sm 8vo. 8 cold plates, num w'engvs. Orig pict gilt cloth, hd of spine frayed.
(Egglishaw) **£25 [≃ $40]**
- Sea-Side Walks of a Naturalist with his Children. New Edition. London: Groombridge, 1889. Sm 8vo. vi,154 pp. 6 chromolithos. Orig dec cloth gilt, a.e.g., trifle

rubbed. *(Hollett)* **£30 [≃ $48]**

House, Homer D.
- Wildflowers of New York. New York: Univ of the State of New York Memoir 15, 1923. 2 vols. Lge 4to. 363 pp. 264 cold plates. Orig cloth gilt. *(Hollett)* **£140 [≃ $224]**

The Housekeeper's Receipt Book ...
- The Housekeeper's Receipt Book, or, the Repository of Domestic Knowledge; containing a Complete System of Housekeeping ... London: the editor, 1813. 1st edn. 8vo. iv,376 pp. 2 plates (stained, 1 reprd). Some soiling & marking. Contemp calf, worn. *(Young's)* **£110 [≃ $176]**

Howard, H.E.
- An Introduction to the Study of Bird Behaviour. Cambridge: 1929. 4to. xii,136 pp. 11 plates by G.E. Lodge, 2 plans. Sl foxing. Orig cloth. *(Wheldon & Wesley)* **£40 [≃ $64]**
- The Nature of a Bird's World. London: 1935. 8vo. vi,102 pp. Orig cloth.
(Wheldon & Wesley) **£20 [≃ $32]**
- Territory in Bird Life. London: 1920. 1st edn. 8vo. xiii,308 pp. 11 plates by Lodge & Gronvold, 2 plans. Orig cloth.
(Wheldon & Wesley) **£30 [≃ $48]**

Howard, L.O., et al.
- The Mosquitoes of North and Central America and the West Indies. Washington: 1912-17. 4 vols. Roy 8vo. 164 plates. Orig cloth, good ex-lib.
(Wheldon & Wesley) **£100 [≃ $160]**

Howard, Thomas
- On the Loss of Teeth; and on the Best Means of Restoring Them. Fifth Edition. London: ptd by T. Brettell, 1853. 8vo. Final advt leaf. Orig cloth, a.e.g. *(Sanders)* **£30 [≃ $48]**
- On the Loss of the Teeth; and on the Best Means of Restoring Them. London: Simpkin & Marshall, 1857. Sm 8vo. 61 pp, advt leaf. Litho frontis with overlay. Orig cloth, a.e.g.
(Bickersteth) **£145 [≃ $232]**

Howell, Arthur H.
- Florida Bird Life. Color Plates from Original Paintings by Francis L. Jaques. New York: Coward-McCann, 1932. 1st edn. 4to. xxiv, 579 pp. 58 cold plates. Green buckram.
(Terramedia) **$80 [≃ £50]**

Howitt, R.C.L. & B.M.
- A Flora of Nottinghamshire. Newark: privately printed, 1963. 8vo. 252 pp. Orig

cloth.   *(Wheldon & Wesley)* **£25 [≃ $40]**

Hoyer, Niels
- Man Into Woman. An Authentic Record of a Change of Sex. Translated from the German by H. Stenning ... London: Jarrolds, 1933. 1st edn. 8vo. 287 pp. 25 ills. Orig bndg, cvr worn, spine wrinkled.   *(Xerxes)* **$50 [≃ £31]**

Hoyland, John
- A Historical Survey of the Customs, Habits and Present State of the Gypsies ... York: for the author ..., 1816. 1st edn. 8vo. viii, 265,1 pp. Orig bds, uncut, spine & label sl worn.
      (Young's) **£75 [≃ $120]**

Hoyne, Temple
- Venereal and Urinary Diseases. Chicago: Halsey, 1894. 2nd edn. 8vo. 133 pp. Paper browning & becoming brittle. Orig bndg.
      (Xerxes) **$65 [≃ £40]**

Huber, Francis
- Observations on the Natural History of Bees. A New Edition, with a Memoir of the Author ... London: Tegg, 1841. 12mo. xxiv,352 pp. Addtnl engvd frontis, 5 fldg plates. Lib blind stamp on title. Orig cloth, hd of spine chipped, lib number on cvr.
      (Bickersteth) **£45 [≃ $72]**

Huber, Jean Pierre
- The Natural History of Ants ... Translated from the French with Additional Notes, by J.R. Johnson. London: for Longman ..., 1820. 1st edn in English. 12mo. Frontis, 1 hand cold plate. Contemp black half calf gilt, spine gilt, trifle rubbed.   *(Ximenes)* **$250 [≃ £156]**

Hudson, C.T. & Gosse, P.H.
- The Rotifera or Wheel-Animalcules. London: Longmans, Green, 1886. 2 vols. 4to. 4 plain & 30 dble page cold plates. Lib stamps on titles & endpapers. Half calf, crnrs sl worn.
      (Egglishaw) **£135 [≃ $216]**

Hudson, G.V.
- The Butterflies and Moths of New Zealand. Wellington: 1928. 4to. xi,386 pp. 62 plates (53 cold). Orig half mor, sm crack in front jnt.
      (Wheldon & Wesley) **£180 [≃ $288]**
- New Zealand Beetles and their Larvae. Wellington: 1934. 8vo. 236 pp. 17 cold plates. Orig cloth.
      (Wheldon & Wesley) **£50 [≃ $80]**

Hudson, W.H.
- The Naturalist in La Plata. London: 1892.

2nd edn. One of 750. 8vo. x,388 pp. 27 ills. Some foxing. Orig cloth, spotted.
      (Wheldon & Wesley) **£30 [≃ $48]**

Hueppe, Ferdinand
- The Methods of Bacteriological Investigation. Translated by Hermann M. Biggs. New York: Appleton, 1890. 8vo. 218,12 ctlg pp. Text ills. Orig cloth.
      (Elgen) **$115 [≃ £71]**

Huff, Gershom
- Electro-Physiology ... or, Electricity as a Curative Agent supported by Theory and Fact. New York: Appleton, 1853. 1st edn. 8vo. 385 pp. Orig bndg, spine ends worn, back bd & spine chipped.
      (Xerxes) **$75 [≃ £46]**

Huggins, Sir William
- The Royal Society; or, Science in the State and in the Schools. London: (1906). 1st edn. 4to. xv,131 pp. 24 plates, 6 ills. Orig gilt dec cloth.   *(Elgen)* **$90 [≃ £56]**

Hughes, S.
- Gas Works - Their Construction and Arrangement, and the Manufacture and Distribution of Coal Gas. Revised by H. O'Connor. London: Crosby, 1904. 416 pp. Ills. Orig bndg.   *(Book House)* **£30 [≃ $48]**

Huish, R.
- The Cottager's Manual for the Management of his Bees, for every Month in the Year. London: 1822. 2nd edn. 8vo. 99 pp. Sl staining. Bds, spine trifle defective.
      (Wheldon & Wesley) **£30 [≃ $48]**
- A Treatise on the Nature, Economy, and Practical Management of Bees ... London: 1815. 1st edn. 8vo. xxiii,414 pp. 6 plates. Title & plates trifle foxed. Mod half calf.
      (Wheldon & Wesley) **£100 [≃ $160]**
- A Treatise on the Nature, Economy, and Practical Management of Bees ... London: 1817. 2nd edn. 8vo. xxxix,400 pp. 6 plates. Calf.   *(Wheldon & Wesley)* **£60 [≃ $96]**

Hull, Edward
- The Coal-Fields of Great Britain: Their History, Structure, and Resources ... Fifth Edition, Revised ... London: Hugh Rees, 1905. xxii,472 pp. 2 litho plates, inc frontis, 13 b/w map plates (all but 3 fldg, 1 in pocket), text figs. Orig cloth, uncut.
      (Duck) **£65 [≃ $104]**
- The Physical Geology and Geography of Ireland. London: Edward Stanford, 1878. 1st

edn. 8vo. Half-title, 3 pp advts. 2 cold maps, 26 w'engvs in text. Title sl foxed. Orig cloth, gilt spine.　　　*(Trebizond)* **$75 [≃ £46]**

Hulme, Frederick Edward
- Familiar Wild Flowers. London: Cassell, [1890s]. Series 1-8, complete. 8 vols. 320 cold plates, w'engvd decs. Orig dec green cloth, lettered in gilt, extremities rubbed.
(Claude Cox) **£45 [≃ $72]**
- Familiar Wild Flowers. London: [1906]. 8 vols. 8vo. 320 cold plates. Orig cloth.
(Sklaroff) **£50 [≃ $80]**
- Familiar Wild Flowers. London: Cassell, [1906]. Series 1-8. 1st edn. 8 vols. 320 cold plates. 2 vols lack rear endpapers. Orig pict dark green cloth, lettered in gilt, worn.
(Sklaroff) **£50 [≃ $80]**
- Familiar Wild Flowers. London: Cassell & Co, n.d. 7 vols. 8vo. 280 cold plates. Orig cloth.　　*(Wheldon & Wesley)* **£45 [≃ $72]**
- A Series of Sketches from Nature of Plant Form. London: Day, 1868. 4to. 54 pp. Litho title. 100 cold plates. Lib blind stamp on 1st 4 pp of text. Mod cloth, lib number on spine.
(Egglishaw) **£65 [≃ $104]**
- Worked Examination Questions in Plane Geometrical Drawing, for the Use of Candidates for the Royal Military Academy, Woolwich ... London: Longmans, Green, 1882. 1st edn. 4to. vi,34 pp. 61 fldg plates. Orig cloth gilt.　　*(Fenning)* **£18.50 [≃ $30]**

Humber, William
- A Comprehensive Treatise on the Water Supply of Cities and Towns ... London: Crosby Lockwood, 1876. Imperial 4to. xiv,378,[2 advt] pp. Litho frontis, 50 plates, 1 plate in text, 250 text ills. Sl marked. Orig half mor gilt, sl worn. *(Duck)* **£475 [≃ $760]**

Humber, William (ed.)
- A Record of the Progress of Modern Engineering. 1864 ... London: Lockwood & Co, 1865. Folio. viii,52 pp. Mtd photo frontis port of Robert Stephenson, 36 dble page plates. Contemp half mor.
(Fenning) **£125 [≃ $200]**

Humboldt, Alexander von
- Aspects of Nature, in Different Lands and Different Climates; with Scientific Elucidations. Translated by Mrs. Sabine. London: Longman ..., 1849. 1st English version. 2 vols. 8vo. Orig salmon cloth, faded, stains on endpapers.　　*(Clark)* **£40 [≃ $64]**
- Cosmos: a Sketch of a Physical Description of the Universe. Translated by E.C. Otte and

W.S. Dallas. London: Bohn's Scientific Library, 1849-58. 5 vols. Cr 8vo. Port. Binder's cloth.
(Wheldon & Wesley) **£60 [≃ $96]**

Hume, A.O.
- Nests and Eggs of Indian Birds. Second Edition by E.W. Oates. London: 1889-90. 3 vols. Roy 8vo. 12 ports. Orig cloth, trifle used.　　*(Wheldon & Wesley)* **£85 [≃ $136]**

Hume, A.O. & Marshall, C.H.T.
- The Game Birds of India, Burmah and Ceylon. Calcutta: 1879-81. 3 vols. Roy 8vo. 144 cold plates. Sl marg dampstaining of last few plates. Sl foxing. Orig green cloth gilt.
(Wheldon & Wesley) **£250 [≃ $400]**

Hume, Edgar Erskine
- Medical Work of the Knights Hospitallers of Saint John of Jerusalem. Baltimore: 1940. 1st edn. Tall 8vo. 371 pp. Frontis, num ills. Orig cloth, spotted.　　*(Elgen)* **$100 [≃ £62]**

Humphreys, Henry Noel
- The Butterfly Vivarium; or, Insect Home: being an Account of a New Method of Observing the Curious Metamorphoses of some of the most Beautiful of Our Native Insects ... London: William Lay, 1858. 8vo. x,288,[4 advt] pp. 8 hand cold plates. Orig dec cloth, sl, used.　　*(Hollett)* **£60 [≃ $96]**
- The Butterfly Vivarium, or Insect Home. London: William Lay, 1858. Sm sq 8vo. xii,288 pp. 8 hand cold engvs. Orig cloth gilt, a.e.g., worn.　　*(Egglishaw)* **£55 [≃ $88]**
- The Genera of British Moths. Popularly Described ... London: Paul Jerrard & Son, [1860]. 2 vols in one. Sm 4to. 206 pp. 62 hand cold plates. Orig dec cloth gilt, sl worn & faded, rebacked to style.
(Hollett) **£160 [≃ $256]**
- Ocean and River Gardens: a History of the Marine and Freshwater Aquaria. London: Sampson Low, 1857. 2 parts in one vol. 8vo. viii,117; vii,108 pp. 20 hand cold plates. Some ff frayed. Orig cloth gilt, a.e.g., rubbed.
(Egglishaw) **£85 [≃ $136]**

Humphreys, Henry Noel & Westwood, J.O.
- British Butterflies and their Transformations. London: Thomas Sanderson, 1857. New edn, rvsd. 4to. Hand cold title (sl fingered, sm marg tear, sm blind lib stamp), 42 hand cold plates. Few top margs sl dusty. Binder's cloth.
(Hollett) **£225 [≃ $360]**

Humphry, George Murray
- Old Age. The Results of Information received respecting nearly 900 persons who had attained the age of eighty years ... Cambridge: Macmillan & Bowes, 1889, Only edn. 8vo. xii,218,1 pp. 2 mtd photos, 2 ff of lithos. Orig cloth, uncut & unopened.
(Charles B. Wood) **$300 [≃ £187]**

Hun, Henry
- An Atlas of the Differential Diagnosis of the Diseases of the Nervous System (With a Physiological Introduction). Troy, New York: Southworth, 1922. 3rd edn, rvsd & enlgd. 4to. 299 pp. Orig bndg, rear hinge just beginning to crack.   *(Xerxes)* **$100 [≃ £62]**

Hunt, Robert
- British Mining. A Treatise on the History, Discovery, Practical Development and Future Prospects of Metalliferous Mines in the United Kingdom ... London: 1867. 2nd edn, rvsd. Thick roy 8vo. Num ills. Orig cloth, inner hinges tender, sm split in jnt.
(Ambra) **£260 [≃ £416]**
- The Poetry of Science, or Studies of the Physical Phenomena of Nature. London: 1849. 2nd edn. xxvi,478,16 ctlg pp. Orig cloth, spine ends worn. *(Elgen)* **$50 [≃ £31]**
- A Popular Treatise on the Art of Photography ... Glasgow: Richard Griffin, 1841. 1st edn. 8vo. viii,96 pp. Litho frontis, 29 text ills. Orig cloth, rebacked.
(Charles B. Wood) **$2,000 [≃ £1,250]**
- Researches on Light ... Chemical and Molecular Changes produced by the Influence of the Solar Rays. London: 1844. 1st edn. 8vo. vii,[1],303,[1],[32 advt dated Sept 1844] pp. Errata slip. Cold frontis. 3 lib stamps. Orig cloth, spine ends sl chipped.
(Rootenberg) **$500 [≃ £312]**
- Researches on Light in its Chemical Relations; embracing a Consideration of all the Photographic Processes. Second Edition. London: Longman ..., 1854. 2nd edn, rvsd & enlgd. 8vo. xx,396 pp. Fldg hand cold plate, num text ills. Rec cloth.
(Charles B. Wood) **$300 [≃ £187]**
- Researches on Light in its Chemical Relations; embracing a Consideration of all the Photographic Processes. London: Longman ..., 1854. 2nd edn, rvsd & enlgd. 8vo. xx,396 pp. Fldg cold plate. rec qtr calf.
(Fenning) **£165 [≃ $264]**

Hunter, Dard
- Papermaking in Pioneer America. Philadelphia: Univ of Pennsylvania Press,

1952. 1st edn. 8vo. xiv,178 pp. 22 photo plates. Orig cloth over bds.
(Karmiole) **$125 [≃ £78]**

Hunter, John
- A Treatise on the Venereal Disease. With an Introduction and Commentary; by Joseph Adams, M.D. The Second Edition, with Additions ... London: 1818. 8vo. xx,568 pp. 6 plates. Orig bds, uncut, partly unopened, rebacked, orig label preserved.
(Bickersteth) **£120 [≃ $192]**

Hunter, Richard & Macalpine, Ida
- Three Hundred Years of Psychiatry 1535-1860. Hartsdale, New York: Carlisle Publ, [1982]. Reprint edn. Crrctd reprint of the Oxford 1970 edn. 8vo. [xxviii],1107,[1] pp. Orig buckram. Dw. *(Gach)* **$95 [≃ £59]**

Huntt, Henry
- A Visit to the Red Sulphur Spring of Virginia, during the Summer of 1837: with Observations on the Water. Boston: Dutton & Wentworth, 1839. Roy 8vo. 40 pp. Frontis. Some marg waterstains. Lacks wraps, folded lengthwise at some time.
(Xerxes) **$125 [≃ £78]**

Hurlbut, C.S.
- Minerals and Man. New York: Random House, 1968. Roy 8vo. 304 pp. 217 ills (160 cold). Orig cloth. Dw.
(Gemmary) **$35 [≃ £21]**

Hurst, C.
- Valves and Valve-Gearing: A Practical Text-Book. London: Griffin, 1914. 309 pp. Fldg plates, ills. Orig bndg.
(Book House) **£25 [≃ $40]**

Hurst, C.C.
- The Mechanism of Creative Evolution. Cambridge: 1932. Roy 8vo. xxi,365 pp. Frontis, 199 text figs. Orig cloth.
(Wheldon & Wesley) **£20 [≃ $32]**

Hurst, George H.
- Silk Dyeing, Printing and Finishing. London & New York: George Bell & Sons, 1892. 1st (& only) edn. 8vo. viii,226 pp. 66 mtd fabric samples. Orig cloth.
(Charles B. Wood) **$185 [≃ £115]**

Hutchinson, J.
- British Flowering Plants, Evolution and Classification. London: 1948. 8vo. viii,374 pp. 22 cold plates, 174 other ills. Orig cloth.
(Wheldon & Wesley) **£20 [≃ $32]**

Hutchinson, Sir Jonathan
- The Pedigree of Disease; Being Six Lectures of Temperament, Idiosyncrasy and Diathesis. London: Churchill, 1884. 1st edn. 8vo. 142 pp. Orig cloth, sl shaken ex-lib.
(Elgen) **$100 [≃ £62]**
- The Pedigree of Disease ... New York: Wood, 1885. 1st Amer edn. 8vo. 113 pp. Orig cloth, sl soiled. Roswell Park's copy.
(Elgen) **$100 [≃ £62]**

Hutchinson, W.N.
- Dog Breaking. The Most Expeditious, Certain and Easy Method ... London: 1850. 2nd edn. 8vo. Ills. Orig cloth, sl rubbed & marked.
(Grayling) **£35 [≃ $56]**

Hutchison, J.
- A Botanist in Southern Africa. London: Gawthorn, 1946. 1st edn. Imperial 8vo. xii, 686 pp. 2 fldg maps, fldg chart, 532 ills. Orig green cloth gilt.
(Gough) **£50 [≃ $80]**

Hutton, Charles
- A Mathematical and Philosophical Dictionary ... London: J. Davis, 1796-95. 1st edn. 2 vols. 4to. [iii]-viii,650; [ii],756 pp. 37 plates. Lacks half-titles. Minor browning & offsetting of plates. MS on vol 1 title. Contemp calf, stoutly rebacked, crnrs sl worn.
(Clark) **£225 [≃ $360]**
- A Mathematical and Philosophical Dictionary ... London: J. Johnson, 1795. 1st edn. 2 vols. 4to. viii,650; [2],756 pp. 27 plates. Contemp calf, hinges weak.
(Charles B. Wood) **$585 [≃ £365]**
- Mathematical Tables ... London: Rivington ..., 1822. 6th edn. viii,548 pp. Orig gilt dec leather, worn, bds crudely refixed.
(Elgen) **$85 [≃ £53]**
- Tracts on Mathematical and Philosophical Subjects ... London: Rivington ..., 1812. 3 vols. 8vo. xii,486; [4],384; [4],384 pp. Frontis, 10 plates (9 fldg), num text ills. 19th c cloth, leather labels.
(Karmiole) **$200 [≃ £125]**

Hutton, F.W.
- Manual of the New Zealand Mollusca. London: 1880. 8vo. xvi,iv,224 pp. Wrappers.
(Baldwin) **£33 [≃ $52]**

Hutton, F.W. & Drummond, James
- The Animals of New Zealand. An Account of the Colony's Air-Breathing Vertebrates. Christchurch: 1904. 1st edn. 8vo. 381 pp. Frontis, 147 text ills. Orig cloth. A.F.R. Wollaston's copy. *(Bickersteth)* **£38 [≃ $60]**

- The Animals of New Zealand. An Account of the Dominion's Air-Breathing Vertebrates. Auckland: 1923. 4th edn. 8vo. 433 pp. 154 ills. Orig cloth, spine faded.
(Wheldon & Wesley) **£30 [≃ $48]**

Hutton, James
- Abstract of a Dissertation read in the Royal Society of Edinburgh, upon the Seventh of March, and Fourth of April, M,DCC,LXXXV, concerning the System of the Earth ... [Edinburgh: priv ptd, 1785]. 1st edn. 8vo. 16 ff,30 pp,final blank leaf. Sl used. Rec calf. *(Pickering)* **$14,500 [≃ £9,062]**

Hutton, W.S.
- Steam Boiler Construction. A Practical Hand-Book ... London: Crosby Lockwood, 1903. 4th edn. Lge 8vo. 616 pp. 540 ills. Orig bndg, cvr marked. *(Book House)* **£30 [≃ $48]**

Huxley, Aldous
- Doors of Perception. New York: Harper, 1954. 1st edn. 8vo. 79 pp. Orig bndg. Dw.
(Xerxes) **$40 [≃ £25]**

Huxley, J.
- Evolution, the Modern Synthesis. London: 1942. 1st edn. 8vo. 645 pp. Orig cloth. Sl soiled dw. *(Wheldon & Wesley)* **£30 [≃ $48]**

Huxley, J.S. & de Beer, G.R.
- The Elements of Experimental Embryology. Cambridge: 1934. 8vo. xiii,514 pp. 221 text figs. Orig cloth, spine faded.
(Wheldon & Wesley) **£25 [≃ $40]**

Huxley, Leonard
- Life and Letters of Sir Joseph Dalton Hooker. London: Murray, 1918. 1st edn. 8vo. x,546; vi,569,advt pp. Fldg map, 2 frontis ports, ills. Occas pencil underlining. Orig blue cloth gilt, hinges shaken. 2 ALS by Sir W.T. Dyer included.
(Terramedia) **$150 [≃ £93]**

Huxley, Thomas Henry
- Collected Essays. London: (1893-94). 9 vols. Cr 8vo. Orig red cloth.
(Wheldon & Wesley) **£75 [≃ $120]**
- Diary of the Voyage of H.M.S. Rattlesnake. Edited from the Unpublished MS. by Julian Huxley. London: 1935. 1st edn. 8vo. 372 pp. Map, cold frontis, 12 plates. A few marks. Orig cloth. Prize label. Dw torn & partly frayed away.
(Bow Windows) **£70 [≃ $112]**
- Diary on the Voyage of H.M.S. Rattlesnake.

Edited from the Unpublished MS. by Julian Huxley. London: 1935. Roy 8vo. 372 pp. Map, 13 plates. Cloth.
(Wheldon & Wesley) **£45 [≃ $72]**

- Evidence as to Man's Place in Nature. London: 1863. 1st edn. 8vo. [iv],159 pp. 32 text figs. Orig cloth.
(Wheldon & Wesley) **£150 [≃ $240]**

- Evidence as to Man's Place in Nature. London: Williams & Norgate, 1863. 1st edn. 8vo. [iv],159,8 ctlg pp. Num text ills. Ink verses on half-title. Orig cloth, spine ends sl worn, sm defective spot in cloth on upper jnt.
(Bickersteth) **£110 [≃ $176]**

- Hume. New York: Harper, 1879. 1st Amer edn. 12mo. [ii],vi,202,[8] pp. Orig ptd brown cloth, cvrs faded.
(Gach) **$25 [≃ £15]**

- An Introduction to the Classification of Animals. London: Churchill, 1869. 1st edn thus. 8vo. [vii],147,[16 ctlg dated Oct 1872] pp. Text figs. Sm name on half-title. Orig cloth, sm slits in jnts, ft of spine sl worn.
(Bickersteth) **£48 [≃ $76]**

- Lay Sermons, Essays, and Reviews. London: Macmillan, 1870. 1st edn. 8vo. xi,378,[43 ctlg] pp. Orig cloth, ft of spine v sl worn.
(Bickersteth) **£28 [≃ $44]**

- Lay Sermons, Addresses and Reviews. London: Macmillan, 1870. 1st edn. 8vo. xi, 378, ctlg pp. Orig cloth, discold, slit in cloth on lower jnt. *(Bickersteth)* **£25 [≃ $40]**

- Life and Letters ... by his son L. Huxley. London: 1900. 2 vols. 8vo. 11 plates, 1 ill. Some sl marg stains. Orig cloth, trifle used, label removed from front cvrs.
(Wheldon & Wesley) **£45 [≃ $72]**

- Life and Letters ... by his son L. Huxley. London: 1903. 2nd edn. 2 vols. Cr 8vo. 3 ports. Cloth, trifle used.
(Wheldon & Wesley) **£35 [≃ $56]**

- A Manual of the Anatomy of Invertebrated Animals. London: Churchill ..., 1877. 1st edn. 8vo in 12s. 596 pp. 16 advt pp dated Feb 1884. 158 figs. Orig brown cloth.
(Bow Windows) **£85 [≃ $136]**

- The Oceanic Hydrozoa ... observed during the Voyage of H.M.S. 'Rattlesnake'. London: Ray Society, 1859. 4to. x,143 pp. 12 plates. Severely trimmed without loss of text or plates. Half mor, 1 jnt weak.
(Wheldon & Wesley) **£60 [≃ $96]**

- The Oceanic Hydrozoa; A Description of the Calicophoridae and Physophoridae observed during the Voyage of H.M.S. "Rattlesnake" 1846-50. London: Ray Society, 1859. Folio. xi, 143 pp. 12 plates. Lib stamps. Half calf.
(Baldwin) **£80 [≃ $128]**

- On the Origin of Species: or, the Causes of

the Phenomena of Organic Nature. A Course of Six Lectures to Working Men. New York: 1863. 8vo. 150 pp. New cloth. 1st publ in London in 1862 under a different title.
(Wheldon & Wesley) **£40 [≃ $64]**

- Science and Culture and Other Essays. London: Macmillan, 1881. 1st edn. 8vo. ix, 349, 2 advt pp. Name on title. Orig cloth, sl soiled, sm defective spot at ft of spine.
(Bickersteth) **£28 [≃ $44]**

Hyams, Edward

- The English Garden. London: 1964. 4to. 288 pp. 16 cold plates, 200 other ills. Orig cloth.
(Wheldon & Wesley) **£30 [≃ $48]**

Hyams, Edward & Jackson, A.A. (eds.)

- The Orchard and Fruit Garden. London: Longmans, Green, 1961. 4to. xv,208 pp. 120 cold photos by Eric H. West. Orig green buckram gilt, t.e.g. Dw sl soiled. Slipcase.
(Blackwell's) **£30 [≃ $48]**

Hyams, Edward & Smith, E.

- English Cottage Gardens. London: 1970. 4to. 250 pp. Num photos (some cold). Orig cloth. dw. *(Wheldon & Wesley)* **£20 [≃ $32]**

Hyatt, Thaddeus

- Teeth and their Care. Brooklyn: King, 1906. 1st edn. 43 pp. Photo ills. Orig bndg.
(Xerxes) **$55 [≃ £34]**

The Ibis

- The Ibis. Third Series Vol 2. London: 1872. 8vo. iii-xiii,491 pp. 15 plates (4 of skeletons, 11 hand cold by Keulemans). Lacks title-page. Half mor, trifle rubbed.
(Wheldon & Wesley) **£65 [≃ $104]**

- The Ibis. Sixth Series Vol 2. London: 1890. 8vo. xxii,491 pp. 14 hand cold plates (12 by Keulemans). Half mor.
(Wheldon & Wesley) **£60 [≃ $96]**

- The Ibis. New Series Vol 3. London: 1867. 8vo. iii-x,490 pp. 10 hand cold plates (8 by Wolf). Lacks title-page. Half mor, rubbed.
(Wheldon & Wesley) **£65 [≃ $104]**

Ingram, C.

- The Birds of the Riviera. London: 1926. 8vo. xvi,156 pp. 6 plates, text figs. Orig cloth, good ex-lib. *(Wheldon & Wesley)* **£25 [≃ $40]**

- Birds of the Riviera. London: Witherby, 1926. 1st edn. 8vo. xvi,155 pp. 6 photo plates, text figs. Sl spotting. Orig cloth gilt, faded. *(Gough)* **£25 [≃ $40]**

Innes, C.H.
- The Centrifugal Pump, Turbines, & Water Motors. Manchester: Tech Publ, 1904. 4th enlgd edn. 8vo. 340 pp. Ills. Orig bndg.
(Book House) **£20 [≈$32]**

Inquiry ...
- An Inquiry into the Nature, Cause, and Cure of the Present Epidemic Fever ... see Barker, John.

Inwards, Richard (ed.)
- Weather Lore. A Collection of Proverbs, Sayings and Rules concerning the Weather. London: Elliot Stock, 1893. xii,190,1 advt pp. Fldg frontis. Orig pict cloth, stain on backstrip. *(Wreden)* **$65 [≈£40]**

Iredale, T.
- Birds of New Guinea. Vol 1. Melbourne: 1956. 4to. 230 pp. 15 cold plates. Half mor. Comprises Cassowaries to Swifts, including Parrots. Vol 2, concluding the work, dealt with the Passeres.
(Wheldon & Wesley) **£120 [≈$192]**

Jackson, G. Gibbard
- All About Our British Railways. London: Jack, [1922]. xii,288,[4 advt] pp. Frontis, 62 photo plates, text ills. Orig cold pict cloth, uncut. Dw worn. *(Duck)* **£25 [≈$40]**
- The Railways of Great Britain ... London: Boy's Own Paper, [1923]. 279 pp. Cold frontis, 32 photo plates. Orig blue cloth. Dw reprd. *(Duck)* **£20 [≈$32]**

Jackson, J.
- Ambidexterity or Two-Handedness and Two-Brainedness ... London: 1905. 1st edn. 8vo. xii,258 pp. 24 ills. Some dust marks. Endpapers browned. Orig cloth.
(Bow Windows) **£48 [≈$76]**

Jackson, J.B.S.
- A Descriptive Catalogue of the Anatomical Museum of the Boston Society for Medical Improvement. Boston: Ticknor, 1847. 8vo. 352 pp. 10 plates. Orig cloth, crnrs & spine ends worn. *(Elgen)* **$125 [≈£78]**

Jackson, R.T.
- Phylogeny of the Echini, with a Revision of Palaeozoic Species. London: 1912. 4to. 490 pp. 76 plates. Title & half-title soiled. Half buckram. *(Baldwin)* **£90 [≈$144]**

Jackson, Samuel
- The Principles of Medicine ... Philadelphia: Carey & Lea, 1832. 1st edn. 8vo. 630 pp, errata leaf, 24 ctlg pp. Foxed. Sheep, rubbed, new label. *(Elgen)* **$100 [≈£62]**

Jacob, William
- An Historical Inquiry into the Production and Consumption of the Precious Metals. London: Murray, 1831. 2 vols. 8vo. xvi,380; xi,415 pp. Qtr leather, rebacked.
(Gemmary) **$275 [≈£171]**

Jacobi, Abraham
- The Intestinal Diseases of Infancy and Childhood. Detroit: George S. Davis, 1887. 1st edn. 8vo. xv,[1],301,[3] pp. Stamp on title. Orig cloth. *(Rootenberg)* **$350 [≈£218]**

Jaeger, Gustav
- Health Culture. Translated and edited by Lewis Tomali. London: Adams Bros, 1907. New rvsd edn. 12mo. 201 pp. Frontis port, ills. Orig bndg. *(Xerxes)* **$55 [≈£34]**

Jahr, G.H.G.
- Manual of Homeopathic Medicine. Translated from the German. Allentown, PA: Academical Book Store, Wesselhoeft, 1838. 8vo. 598 pp. V browned, spotted. Three qtr leather, hinges cracked, needs re-covering.
(Xerxes) **$200 [≈£125]**

James, Colonel Sir Henry
- Ordnance Survey: Abstracts of the Principal Lines of Spirit Levelling in England and Wales ... London: 1861. 1st edn. 2 vols. 4to. 2 maps frontis, 24 dble-page engvd plates. Some foxing & soiling but plates clean. Orig dec cloth, worn & soiled.
(Dillons) **£175 [≈$280]**

James, John
- History of the Worsted Manufacture in England, from the Earliest Times ... Bradford: Longmans ... Stanfield, 1857. 1st edn. Large Paper. xvi,640,40 pp. 6 plates. Orig cloth, recased, new endpapers.
(Box of Delights) **£75 [≈$120]**

James, Robert
- A Dissertation of Fevers, and Inflammatory Distempers. The Eighth Edition ... added ... A Vindication of the Fever Powder, and a Short Treatise on the Disorders of Children. London: for Francis Newbery, Junior, 1778. 8vo. [viii],160 pp,final blank. Rebacked.
(Bickersteth) **£180 [≈$288]**
- A Treatise on the Gout and Rheumatism wherein a Method is laid down of relieving in an eminent Degree those excruciating

Distempers. London: T. Osborne, 1745. 1st edn. 8vo. [4],92 pp. Disbound.
(Hemlock) **$275 [≈ £171]**

James, William
- The Principles of Psychology. New York: Henry Holt, 1890. 1st edn. 2 vols. 8vo. xii, 689; vi,704,8 advt pp. Num text figs. Previous owners' sgntrs. Orig cloth, edges sl worn.   *(Rootenberg)* **$650 [≈ £406]**
- The Principles of Psychology. New York: henry Holt, [1890]. 1st edn. 3rd printing, without Psy-chology in the advt opposite the title & with 'notice, and' on line 19 p 307 vol 1. 2 vols. 8vo. Orig olive cloth, sl shelf wear to extremities.   *(Gach)* **$650 [≈ £406]**

Jamieson, Alexander
- A Manual of Map-Making and Mechanical Geography. London, Edinburgh & Dublin: A. Fullarton, 1846. 1st edn. 8vo. xvi,97 pp. Fldg frontis, 63 text ills. Orig polished half calf, scuffed. *(Charles B. Wood)* **$125 [≈ £78]**

Jamison, Alcinous
- Intestinal Irrigation or Why, How, and When to Flush the Colon. New York: the author, 1903. 1st edn. 8vo. 185 pp. Ills. Orig bndg, hinges cracked.   *(Xerxes)* **$70 [≈ £43]**

Jane's Aircraft
- Jane's All The World's Aircraft. London: 1941. Folio. Orig cloth, cvrs discold.
(Duck) **£60 [≈ $96]**
- Jane's All The World's Aircraft. London: 1965-66. Folio. Orig cloth. Dw torn.
(Duck) **£35 [≈ $56]**

Jansen, Murk
- Feebleness of Growth and Congenital Dwarfism ... London: 1921. 1st edn. Roy 8vo. 82 pp. Fldg plate, fldg table, ills. Orig cloth, spine relaid.   *(Elgen)* **$85 [≈ £53]**

Japp, Alexander Hay
- Industrial Curiosities. Glances here and there in the World of Labour. Sixth Edition. London: Fisher Unwin, 1891. xi,372 pp. Frontis, plates, ills. Orig cloth, spine faded.   *(Wreden)* **$45 [≈ £28]**

Jardine, Sir William
- Birds of Great Britain and Ireland. Edinburgh: Jardine's Naturalist's Library, [ca 1840]. 4 vols. cr 8vo. 4 ports, 4 title vignettes, 129 hand cold plates. Last plate in vol 3 stained, with sl damage to text, few other sl marg stains. New cloth.

(Wheldon & Wesley) **£175 [≈ $280]**
- Birds of Great Britain and Ireland. Part 1. Birds of Prey. Edinburgh: Jardine's Naturalist's Library, 1838. 1st edn. Cr 8vo. 315 pp. Port, title vignette, 34 hand cold plates. New half calf. *(Egglishaw)* **£60 [≈ $96]**
- Entomology. Vol 4. Bees. Edinburgh: 1840. 8vo. 30 plates, some hand cold. Orig cloth.
(Moorhead) **£35 [≈ $56]**
- Ichthyology. Edinburgh: Jardine's Naturalists Library, 1835-43. 6 vols. 207 plates, mostly hand cold. Occas lib stamp, avoiding plate surfaces. Some plates bound out of sequence, as usual. Rebound in half leather, 1 spine misnumbered.
(Moon) **£300 [≈ $480]**
- Lions, Tigers &c. &c. Edinburgh: The Naturalist's Library, Mammalia Vol 3, 1843. 8vo. Port, engvd title with cold vignette for vol 2, 35 cold plates (complete, although the title calls for 38 cold plates). Orig cloth, spine ends worn.   *(Bickersteth)* **£36 [≈ $57]**
- Mammalia. Vol 3. Part I. Ruminantia [Deer]. Edinburgh: 1835. 8vo. 33 hand cold plates. Orig bndg.   *(Moorhead)* **£35 [≈ $56]**
- Monkeys. Edinburgh: The Naturalist's Library, Mammalia Vol 1, 1833. 8vo. Port, engvd title with cold vignette, 30 cold plates. Mod cloth.   *(Bickersteth)* **£36 [≈ $57]**
- The Natural History of Fishes of the Perch Family. Edinburgh: Jardine's Naturalist's Library, 1835. 1st edn. Cr 8vo. 177,advt pp. Hand cold vignette title, port, 32 hand cold & 2 plain plates. Orig cloth.
(Egglishaw) **£42 [≈ $67]**
- The Natural History of Gallinaceous Birds ... Edinburgh: Jardine's Naturalist's Library, 1843. Cr 8vo. Port, cold title vignette, 30 hand cold plates. Half leather.
(Egglishaw) **£40 [≈ $64]**
- The Natural History of Gallinaceous Birds ... Edinburgh: Jardine's Naturalist's Library, 1836. Cr 8vo. Port, cold title vignette, 30 hand cold plates. Contemp mor, gilt spine, a.e.g.   *(Egglishaw)* **£60 [≈ $96]**
- The Natural History of Humming Birds. Edinburgh: Jardine's Naturalist's Library, 1833. 2 vols. Sm 8vo. 2 ports, 2 cold title vignettes, 62 (of 64) hand cold plates. Calf, jnts splitting.   *(Egglishaw)* **£105 [≈ $168]**
- The Natural History of Monkeys. Edinburgh: Jardine's Naturalist's Library, 1833. 1st edn. Sm 8vo. Port, cold title vignette, 30 hand cold plates. Calf, front jnt split.   *(Egglishaw)* **£40 [≈ $64]**
- The Natural History of Monkeys. Edinburgh: Jardine's Naturalist's Library, 1833. 1st edn. Sm 8vo. Port, cold title

vignette, 30 hand cold plates. Contemp mor, gilt spine, a.e.g., jnts sl rubbed.
(Egglishaw) **£52 [≈ $83]**

- The Natural History of the Felinae. Edinburgh: Jardine's Naturalist's Library, 1834. 1st edn. Sm 8vo. Port, title vignette, 36 hand cold plates. Orig cloth, spine relaid.
(Egglishaw) **£30 [≈ $48]**

- The Natural History of the Felinae. Edinburgh: Jardine's Naturalist's Library, 1834. 1st edn. Sm 8vo. Port, cold title vignette, 36 hand cold plates. Contemp mor, gilt spine, a.e.g., edges sl rubbed.
(Egglishaw) **£48 [≈ $76]**

- The Natural History of the Nectariniadae, or Sun-Birds. Edinburgh: Jardine's Naturalist's Library, 1843. 1st edn. Cr 8vo. Port, hand cold title vignette, 30 hand cold plates. Lacks rear free endpaper. Orig cloth.
(Egglishaw) **£85 [≈ $136]**

- The Natural History of the Pachydermes ... Edinburgh: Jardine's Naturalist's Library, 1837. Sm 8vo. 248 pp. Port, 26 hand cold & 4 plain plates. Contemp mor, gilt spine, a.e.g., jnts sl rubbed. *(Egglishaw)* **£48 [≈ $76]**

- The Natural History of the Ruminating Animals. Edinburgh: Jardine's Naturalist's Library, 1836-39. 2 vols. Cr 8vo. 2 ports, 2 title vignettes (1 cold), 62 hand cold & 2 plain plates. New cloth. *(Egglishaw)* **£105 [≈ $168]**

- Ornithology. Vol 4. Parrots. Edinburgh: 1836. 8vo. 30 hand cold plates (some sl foxed). Piece torn from frontis. Orig bndg, worn. *(Moorhead)* **£50 [≈ $80]**

Jay, B.A.
- Conifers in Britain. An Illustrated Guide to Identification. London: 1952. 8vo. 47 pp. 136 plates. Orig cloth. *(Henly)* **£18 [≈ $28]**

Jay, J.C.
- A Catalogue of Recent Shells, with Descriptions of New or Rare Species. New York: 1836. 2nd edn. 8vo. 82,[6] pp. 4 cold plates. Blind stamp on title. Orig cloth, lib b'plate. *(Wheldon & Wesley)* **£60 [≈ $96]**

Jeaffreson, John Cordy
- The Life of Robert Stephenson, F.R.S. ... London: Longman ..., 1864. 1st edn. 2 vols. xvi,363; xii,335 pp. 7 plates (inc 2 port frontis), 16 w'engvd text ills. Lacks half-titles. Occas minor soiling. Contemp half calf, elab gilt spines, somewhat worn.
(Duck) **£150 [≈ $240]**

Jeancon, J.A.
- Diseases of the Sexual Organs. Vol.1:

Gynecology. Cincinnati: 1887. 1st edn, 2nd printing. Folio. 112 pp. 56 plates (40 cold). Three qtr leather, rubbed.
(Elgen) **$150 [≈ £93]**

- Pathological Anatomy, Pathology and Physical Diagnosis. A Series of Clinical Reports ... Cincinnati: 1885. 1st edn, 2nd printing. Folio. 100 pp. 100 plates (94 cold). Three qtr leather, a.e.g.
(Elgen) **$225 [≈ £140]**

Jeans, J.H.
- Problems of Cosmogony and Stellar Dynamics. Cambridge: 1919. 1st edn. viii,293 pp. 5 plates. Plate versos & edges foxed. B'plate. Orig cloth, sl wrinkled & worn. *(Whitehart)* **£40 [≈ $64]**

Jeffery, Alfred
- Notes on the Marine Glue. London: J. Teulon, 1843. 1st edn. 8vo. 16 pp. 4 plates inc frontis. Disbound. *(Spelman)* **£60 [≈ $96]**

Jeffreys, George Washington
- A Series of Essays on Agriculture and Rural Affairs ... By 'Agricola,' A North Carolina Farmer. Raleigh: 1819. Title,223 pp, index leaf. Foxed. Sgntr starting. Contemp calf, outer hinges cracked but sound.
(Reese) **$1,000 [≈ £625]**

Jeffreys, John G.
- British Conchology ... Volume V. Marine Shells ... London: 1869. 1st edn. 8vo. [2], 258, [1],[2 advt] pp. 111 plates (1 cold). Orig cloth. *(Fenning)* **£35 [≈ $56]**

Jeffries, David
- A Treatise on Diamonds and Pearls. London: for R. Lea, 1800. 3rd edn. 8vo. xvi,116,30 pp. 30 pp of figs & tables. Foxed. Qtr leather.
(Gemmary) **$225 [≈ £140]**

Jekyll, Gertrude
- Colour in the Flower Garden. London: Country Life, 1908. 1st edn. 8vo. xiv,148,[10 advt] pp. 111 plates inc num fldg plans. Orig cloth gilt. *(Fenning)* **£32.50 [≈ $52]**

- Colour in the Flower Garden. London: Country Life, George Newnes, 1908. 1st edn. 8vo. Frontis, 104 other plates, 6 fldg plans, 10 plans in text. Orig blue cloth gilt, sl rubbed. *(Sanders)* **£65 [≈ $104]**

- Home and Garden. London: Longmans, Green, 1900. 8vo. 301 pp. 55 ills from photos. Buckram gilt, spine sunned.
(Moon) **£25 [≈ $40]**

- Home and Garden. London: 1901. New edn.

8vo. xv,301 pp. 46 plates. Orig cloth, faded.
(Wheldon & Wesley) **£40 [≈ $64]**
- Lilies for English Gardens. London: 1901.
1st edn. 8vo. xii,72 pp. 72 plates, text figs.
Orig cloth.   *(Henly)* **£50 [≈ $80]**
- Lilies for English Gardens. London: 1903.
2nd edn. 8vo. xii,72 pp. 61 plates, text figs.
Some foxing. Orig buckram, sl soiled.
(Wheldon & Wesley) **£30 [≈ $48]**
- Old West Surrey. London: Longman, Green,
1904. 1st edn. 8vo. xx,320 pp. Num photo
ills. Orig green buckram gilt.
(Gough) **£50 [≈ $80]**
- Wall and Water Gardens. London: [1903].
2nd edn. 8vo. xiv,177 pp. Num ills. Some
mostly light foxing. Orig cloth, sl used.
(Wheldon & Wesley) **£35 [≈ $56]**
- Wall and Water Gardens, with Chapters on
the Rock-Garden and the Heath-Garden.
London: 1913. 5th edn. 8vo. xvi,214 pp. Cold
frontis, num plates. Orig cloth, trifle marked.
(Wheldon & Wesley) **£35 [≈ $56]**

Jekyll, Gertrude & Hussey, C.
- Garden Ornament. London: 1927. 2nd edn,
rvsd. Folio. 448 pp. Over 700 ills. Half- title
creased. Orig cloth, crnrs trifle used.
(Wheldon & Wesley) **£100 [≈ $160]**

Jekyll, Gertrude & Mawley, E.
- Roses for English Gardens. London: 1902.
8vo. xvi,166 pp. Num ills. Trifle foxed. Orig
cloth, faded.
(Wheldon & Wesley) **£40 [≈ $64]**

Jekyll, Gertrude & Weaver, Lawrence
- Gardens for Small Country Houses. London:
1920. 4th edn. 4to. 1,262 pp. Cold frontis,
429 ills. Orig cloth gilt. *(Henly)* **£85 [≈ $136]**
- Gardens for Small Country Houses. London:
[1927]. 6th edn. 4to. 1,262 pp. Cold frontis,
429 ills. Lacks front free endpaper. Orig
cloth.   *(Wheldon & Wesley)* **£55 [≈ $88]**

Jellett, J.H.
- An Elementary Treatise on the Calculus of
Variations. Dublin: 1850. xx,377 pp. Fldg
plate. Ink name on title. Three qtr roan,
rebacked.   *(Whitehart)* **£40 [≈ $64]**

Jelliffe, Smith Ely & Brink, L.
- Psychoanalysis and the Drama. New York &
Washington: Nervous & Mental Disease
Publishing, 1922. Roy 8vo. 162 pp. Orig
wraps, lacks front cvr & backstrip, crnr
bumped.   *(Xerxes)* **$75 [≈ £46]**

Jenkins, C. Francis
- Radiomovies, Radiovision, Television.
Washington: privately ptd, 1929. 1st edn. 143
pp. Orig cloth.   *(Jenkins)* **$125 [≈ £78]**
- Vision by Radio Photographs, Radio
Photograms. Washington: privately ptd,
1925. 1st edn. 8vo. 140 pp. Ills. Orig cloth.
(Jenkins) **$125 [≈ £78]**
- Vision by Radio; Radio Photographs; radio
Photograms. Washington: 1925. 1st edn. 140
pp. Num ills. Orig cloth.
(Elgen) **$135 [≈ £84]**

Jenner, Edward
- An Inquiry into the Causes and Effects of the
Variolae Vaccinae ... Known by the Name of
the Cow Pox ... London: 1798. 1st edn. 4to.
[ii],iv,75,[1 errata] pp. 4 hand cold plates, orig
tissue guards. Occas foxing. Orig or contemp
mrbld bds, uncut, rebacked. Boxed.
(Pickering) **$15,000 [≈ £9,375]**

Jennings, C.
- The Eggs of British Birds ... Bath: Binns &
Goodwin, [1853]. Sm 8vo. xxx,[ii],200 pp.
Cold title, 7 cold plates. Lib blind stamp on
title. Somewhat soiled, ink annotations on
some plates. Leather backed bds, trifle worn.
(Wheldon & Wesley) **£18 [≈ $28]**

Jennings, H.S.
- Behavior of the Lower Organisms. New
York: Columbia Univ Press; London:
Macmillan, 1906. 1st edn. 8vo. [xvi],366,[2]
pp. Orig black cloth. *(Gach)* **$175 [≈ £109]**

Jephson, Henry
- The Sanitary Evolution of London. New
York: 1907. 440 pp. Fldg map. Orig cloth,
t.e.g., uncut, cvrs sl soiled, front jnt started.
(Elgen) **$50 [≈ £31]**

Jepson, W.L.
- The Trees of California. Berkeley, CA: 1923.
2nd edn. 8vo. 240 pp. 124 text figs.
Endpapers trifle browned. Orig cloth.
(Wheldon & Wesley) **£20 [≈ $32]**

Jesse, Edward
- Gleanings in Natural History; with Local
Recollections ... Maxims and Hints for an
Angler. London: Murray, 1832-34. 1st & 2nd
series. 2 vols. 8vo. Fldg facs, title vignette,
num text ills. Half roan gilt, rather rubbed.
(Hollett) **£38 [≈ $60]**

Jevons, William Stanley
- The Coal Question an Inquiry concerning the

Progress of the Nation, and the probable Exhaustion of our Coal-Mines ... Edited by A.W. Flux ... Third Edition, revised. London & New York: Macmillan, 1906. 8vo. l,[ii],467 pp. 2 plates. Orig brown cloth.
(Pickering) **$100 [≈ £62]**
- Money and the Mechanism of Exchange. London: Henry S. King, 1875. 1st edn. 8vo. xviii,349,[1 blank],48 advt pp. Sm piece cut from crnr of title. Orig cloth, spine ends a bit rubbed. *(Pickering)* **$600 [≈ £375]**
- Money and the Mechanism of Exchange... New York: Appleton, 1875. 1st US edn. 8vo. xviii, 349 pp. Orig red cloth, stamped in black, gilt spine, v slight wear to hinges & spine extremities. *(Pickering)* **$500 [≈ £312]**
- Money and the Mechanism of Exchange ... Second Edition. London: Henry S. King, International Scientific Series, 1876. 2nd edn. 8vo. xviii,349,[1 blank],48 ctlg pp. Occas foxing. Orig red cloth, spine & crnrs darkened & rubbed. *(Pickering)* **$250 [≈ £156]**
- The Principles of Science: A Treatise on Logic and Scientific Method. London: Macmillan, 1874. 1st edn. 2 vols. 8vo. xiv, 463; vii,[1],480 pp. Frontis, text diags. Lib stamps on titles, labels on pastedowns. Orig cloth, faded. *(Rootenberg)* **$550 [≈ £343]**
- The Principles of Science: A Treatise on Logic and Scientific Method. New York: Macmillan, 1874. 1st Amer edn. 8vo. xvi, [iii]-[viii], 480 pp. Orig blind-embossed pebbled ochre cloth, quite shelfworn.
(Gach) **$65 [≈ £40]**
- The State in Relation to Labour. London: Macmillan, 1882. 1st edn. 8vo. vii,[1 blank], 166, [2 advt] pp. Orig cloth, spine & crnrs sl rubbed. *(Pickering)* **$550 [≈ £343]**

Johns, Revd Charles Alexander
- The Flowers of the Field. London: SPCK, 1911. 33rd edn, rvsd by G.S. Boulger. 612 pp. Port, 64 cold plates, num text ills. Orig pict dark blue cloth gilt, spine ends v sl rubbed, name on half-title.
(Sklaroff) **£28 [≈ $44]**

Johnson, C.G.
- Migration and Dispersal of Insects by Flight. London: 1969. 8vo. 700 pp. 219 ills. Orig cloth. *(Wheldon & Wesley)* **£35 [≈ $56]**

Johnson, Charles & Sowerby, J.E.
- The Ferns of Great Britain. London: Sowerby, 1855. 1st edn. Roy 8vo. 88 pp. 49 hand cold plates. Title reprd. New cloth.
(Egglishaw) **£32 [≈ $51]**

Johnson, Charles Pierpoint
- British Wild Flowers ... see Sowerby, John Edward & Johnson, Charles Pierpoint.

Johnson, Cuthbert
- On Guano as a Fertilizer. London: James Ridgway ..., 1843. 1st edn. 8vo. 44,4 advt pp. Disbound. *(Young's)* **£55 [≈ $88]**
- On Rendering Manures more Portable and Applicable by the Drill. London: J. Ridgway ..., 1841. 1st edn. 8vo. 44 pp. Disbound.
(Young's) **£50 [≈ $80]**

Johnson, Francis R.
- Astronomical Thought in Renaissance England. A Study of the English Scientific Writings from 1500 to 1645. Baltimore: The Johns Hopkins Press, 1937. 1st edn. 8vo. xv, 357 pp. Orig cloth. Dw.
(Frew Mackenzie) **£60 [≈ $96]**

Johnson, Fred
- Anatomy of Hallucinations. Chicago: Nelson Hall, 1978. 1st edn. 8vo. 239 pp. Orig bndg. Dw. *(Xerxes)* **$50 [≈ £31]**

Johnson, G.W.
- The Gardeners' Dictionary. London: 1877. 2nd edn. 8vo. 920 pp. Orig cloth, trifle used. *(Wheldon & Wesley)* **£18 [≈ $28]**

Johnson, J.C.F.
- Getting Gold: A Practical Treatise for Prospectors, Miners, and Students. London: Charles Griffin, 1898. 2nd edn. 8vo. xii,204, 52 advt pp. 50 ills. Orig cloth.
(Gemmary) **$45 [≈ £28]**

Johnson, James
- An Essay on Morbid Sensibility of the Stomach and Bowels, as the proximate Cause, or characteristic Condition of Indigestion, Nervous Irritability, Mental Despondency, Hypochondriasis ... London: 1827. 8vo. iv,128 pp. Contemp half calf gilt, lib labels. *(Blackwell's)* **£50 [≈ $80]**
- The Influence of Tropical Climates on European Constitutions ... From the Third London Edition greatly enlarged. New York: Duycinck ..., 1826. 416pp. Usual browning & foxing. Marg stain last 30 pp. Orig mottled calf, scuffed. *(Elgen)* **$95 [≈ £59]**
- A Treatise on Derangements of the Liver, Internal Organs, and Nervous System. From the Third London edition revised and improved. Philadelphia: Carey & Lea, 1826. 12mo. xii, 223 pp. Sl browning. Orig calf, scuffed. *(Elgen)* **$85 [≈ £53]**

Johnson, T.B.
- The Gamekeeper's Directory, containing Instructions for the Preservation of Game, the Destruction of Vermin and the Prevention of Poaching ... London: 1851. 2nd edn. 8vo. Engvd frontis of wildfowling on the moon. Orig qtr calf gilt.   *(Grayling)* **£55 [≈ $88]**
- The Shooter's Companion or Directions for Breeding &c. of Setters ... Description of Winged Game ... The Fowling Piece ... Methods of making Percussion Powder ... London: 1819. 1st edn. 8vo. Engvd plates. Some ink marks to 3 pp. New cloth.
        (Grayling) **£35 [≈ $56]**
- The Shooter's Companion ... Second Edition, Improved and Enlarged. ... London: 1823. 8vo. Engvd plates. Contemp calf, showing some wear.   *(Grayling)* **£65 [≈ $104]**

Johnston, Alexander Keith
- The Physical Atlas, A Series of Maps & Illustrations exhibiting the Geographical Distribution of Natural Phenomena ... Edinburgh: Blackwood, 1849. Folio. 30 charts, all cold or partly cold. Some foxing on title. Contemp half roan, a.e.g., spine ends worn.   *(Bow Windows)* **£295 [≈ $472]**

Johnston, George
- A History of British Sponges and Lithophytes. Edinburgh: Lizars, 1842. 8vo. xii, 264 pp. 25 tinted plates, text figs. Orig cloth.   *(Egglishaw)* **£34 [≈ $54]**
- A History of British Zoophytes. London: 1847. 2nd edn. 2 vols. 8vo. 73 plates (numbered 1-34, 34*, 35-63, 66-74), correct. Some offsetting of the plates. Sl foxing at ends. Orig cloth.   *(Henly)* **£55 [≈ $88]**
- A History of the British Zoophytes. London: 1847. 2nd edn. 2 vols. 8vo. 73 plates (numbered 1-34, 34*, 35-63, 66-74), correct. Some offsetting of the plates. Sl foxing, the last few plates affected. Half mor, trifle rubbed.   *(Wheldon & Wesley)* **£60 [≈ $96]**

Johnston, Sir Harry H.
- British Mammals. London: Hutchinson, Woburn Library, 1903. 1st edn. Sm 4to. xvi, 405 pp. 16 cold plates, num b/w ills. Orig green cloth.   *(Gough)* **£20 [≈ $32]**

Johnston, James F.W.
- The Chemistry of Common Life. Edinburgh & London: Blackwood, 1855. 1st edn. 2 vols. 8vo. Num text figs. Name stamp on titles. Orig maroon cloth, faded, lower cvr of vol 1 waterstained.   *(Bickersteth)* **£60 [≈ $96]**

Johnston, John, M.D.
- Musa Medica: a Sheaf of Song and Verse. London: 1897. 1st edn. 8vo. 151 pp. Port. Orig cloth gilt.   *(Fenning)* **£35 [≈ $56]**

Johnston, Thomas
- A General View of the Agriculture of the County of Selkirk, with Observations ... London: Bulmer, 1794. 1st edn. 4to. 50 pp, final blank. Half-title. Disbound.
        (Claude Cox) **£30 [≈ $48]**
- A General View of the Agriculture of the County of Tweedale ... drawn up for the consideration of the Board of Agriculture and Internal Improvement. London: Bulmer, 1794. 4to. 42 pp. Sl foxing. Sewed.
        (Blackwell's) **£50 [≈ $80]**

Johnstone, John
- An Account of the Most Approved Method of Draining Land According to the System practised by Mr Joseph Elkington ... Edinburgh: 1797. Sm 4to. xv,183 pp. 16 plans (misnumbered?, 2 fldg). Sl spotting & browning. 1 plate v browned. Mod half calf.   *(Dillons)* **£200 [≈ $320]**
- An Account of the Mode of Draining Land. London: 1814. 4th edn. 8vo. xvi,211 pp. 19 fldg plans. Foxed at beginning. Calf.
        (Wheldon & Wesley) **£30 [≈ $48]**

Johnstone, W.G. & Croall, A.
- The Nature-Printed Sea-Weeds ... Nature-Printed by Henry Bradbury. London: 1859-60. 4 vols. Roy 8vo. 4 cold title vignettes, 1 plain & 221 cold plates (numbered 1-207 with 15 bis plates & 2 unnumbered). Some foxing. Contemp red half mor gilt, sl rubbed.
        (Wheldon & Wesley) **£300 [≈ $480]**

Joly, C.J.
- A Manual of Quaternions. London: 1905. xxvii, 320 pp. Orig cloth, 1 crnr sl knocked.   *(Whitehart)* **£18 [≈ $28]**

Jones, B.E. (ed.)
- Cassell's Reinforced Concrete. Practice and Theory of Modern Construction in Concrete-Steel. London: Waverley, [ca 1912]. 4to. 398 pp. Ills. Orig bndg. *(Book House)* **£20 [≈ $32]**

Jones, C. Bryner (ed.)
- Livestock of the Farm. London: The Gresham Publ Co, 1916. 6 vols. Roy 8vo. Each vol ca 260 pp. Num photo plates. Orig cloth.   *(Egglishaw)* **£55 [≈ $88]**
- Livestock of the Farm. London: The

Gresham Publ Co, 1919. 6 vols. Each ca 250 pp. 235 half-tone ills. Orig cloth, sl faded & grubby.   *(Claude Cox)* **£28 [≃ $44]**

Jones, F.D. (ed.)
- Ingenious Mechanisms for Designers and Inventors. London: Machinery Publ Co, 1944. 2 vols. Sm 4to. 536; 538 pp. Ills. Orig bndg.   *(Book House)* **£20 [≃ $32]**

Jones, J.P. & Kingston, J.F.
- Flora Devoniensis. London: 1829. 2 parts in one vol. 8vo. New cloth.
            (Wheldon & Wesley) **£60 [≃ $96]**

Jones, T.R.
- General Outlines of the Organisation of the Animal Kingdom. London: 1871. 4th edn. 8vo. xliii,886 pp. 571 text figs. Qtr calf, rebacked.   *(Henly)* **£25 [≃ $40]**

Jones, Thomas Wharton
- The Principles and Practice of Ophthalmic Medicine and Surgery. 3rd American from 2nd London Edition. Philadelphia: 1863. 8vo. xxiii, 455 pp. Orig cloth, jnt torn, ft of spine frayed.   *(Rittenhouse)* **$150 [≃ £93]**

Jones, William, of Nayland
- An Essay on the First Principles of Natural Philosophy ... Oxford: Clarendon Printing House ..., 1762. 1st edn. 4to. vi, 281 pp. 3 plates. Occas spotting. Contemp black half mor, crnrs worn, spine a bit rubbed.
            (Burmester) **£250 [≃ $400]**

Jongh, L.J. de
- The Three Kinds of Cod Liver Oil; comparatively considered with reference to their Chemical and Therapeutic Properties. Philadelphia: Lea & Blanchard, 1849. 1st edn in English. 211,12 ctlg pp. Some foxing. Orig cloth, v worn, lacks backstrip.
            (Elgen) **$75 [≃ £46]**

Jordan, D.S.
- A Guide to the Study of Fishes. New York: 1905. 2 vols. 4to. 2 cold frontis, 934 text figs. Orig buckram, sound ex-lib.
            (Wheldon & Wesley) **£140 [≃ $224]**

Jordan, William Leighton
- The Winds and their Story of the World. London: Hardwicke & Bogue, 1877. 1st edn. xx, 92 pp. Frontis, diags in text. Orig olive cloth, uncut.   *(Duck)* **£20 [≃ $32]**

Jorgensen, Alfred
- Micro-Organisms and Fermentation. Translated by A.K. Miller and A.E. Lennholm. Third Edition. Completely Revised. London: Macmillan, 1900. 8vo. [16],318 pp. 83 ills. Lacks a blank flyleaf. Orig cloth, lib label on spine.
            (Fenning) **£21.50 [≃ $35]**

Joule, James Prescott
- The Scientific Papers. London: The Physical Society of London, Taylor & Francis, 1884-87. 1st edn. 2 vols. 8vo. xxix,657; xii, 391 pp. Port frontis, num fldg plates, tables & figs, text ills. Orig cloth.
            (Rootenberg) **$500 [≃ £312]**

The Journal of a Naturalist ...
- See Knapp, John L.

Joy, Norman H.
- A Practical Handbook of British Beetles. London: 1932. 2 vols. Roy 8vo. Map, 169 plates. Orig cloth, vol 1 trifle loose, sound ex-lib.   *(Wheldon & Wesley)* **£40 [≃ $64]**
- A Practical Handbook of British Beetles. London: Witherby, 1932. 1st edn. 2 vols. xxvii, 622; 194 pp, with 170 pp of ills. A little fingering. Mod half calf gilt.
            (Hollett) **£125 [≃ $200]**

Judson, Adoniram Brown
- The Influence of Growth on Congenital and Acquired Deformities. New York: 1905. 1st edn. 276 pp. 134 ills. Orig cloth.
            (Elgen) **$50 [≃ £31]**

Jukes, Joseph Beete
- The Student's Manual of Geology. Edinburgh: 1857. 1st edn. 8vo. [2],xiii,607 pp, errata leaf. Ills. Orig cloth. Author's pres inscrptn.   *(Fenning)* **£55 [≃ $88]**

Jukes-Brown, A.J.
- The Student's Handbook of Physical Geology. London: Bohn Library, 1884. Sm 8vo. [xvi],514 pp. 2 plates, figs. Lib label. Orig cloth, spine dull & trifle fingered at ends.   *(Bow Windows)* **£35 [≃ $56]**

Juler, Henry Edward
- A Handbook of Ophthalmic Science and Practice. 1st American from 1st London Edition. Philadelphia: 1884. Upper hinge sl weak.   *(Rittenhouse)* **$75 [≃ £46]**

Kames, Henry Home, Lord
- The Gentleman Farmer: being an Attempt to

improve Agriculture, by subjecting it to the Test of Rational Principles. Edinburgh: Bell & Bradfute, 1802. 8vo. xxxi,438,[i advt] pp. Port frontis & 3 plates (foxed). Contemp half calf, sprinkled edges.
(Blackwell's) **£90 [≈ $144]**

Kane, H.H.
- Opium-Smoking in America and China. New York: Putnam, (1881) 1882. 12mo. 156 pp. 2 w'cut ills. Limp bds. *(Xerxes)* **$100 [≈ £62]**

Kane, W.F. de Vismes
- European Butterflies. London: Macmillan, 1885. 1st edn. 8vo. [2],xxxii,184 pp. 15 plates. Orig cloth. *(Fenning)* **£21.50 [≈ $35]**

Kanner, Leo
- Folklore of the Teeth. New York: Macmillan, 1928. 1st edn. 8vo. Orig cloth.
(Gach) **$45 [≈ £28]**
- Folklore of the Teeth. New York: (1928) 1934. 8vo. 316 pp. Ills. Orig bndg, spine ends & crnrs worn. *(Xerxes)* **$75 [≈ £46]**

Karpinski, L.C.
- Bibliography of Mathematical Works Printed in America through 1850. Ann Arbor: 1940. 1st edn. 4to. xxvi,697 pp. Num ills. Dw.
(Elgen) **$350 [≈ £218]**

Kater, Henry & Lardner, Dionysius
- A Treatise on Mechanics. London: Lardner's Cabinet Cyclopaedia, 1830. 1st edn. Sm 8vo. Orig cloth, ptd paper label.
(Fenning) **£32.50 [≈ $52]**

Kaup, J.J.
- Catalogue of Apodal Fish in the Collection of the British Museum. London: 1856. 8vo. viii, 63 pp. 19 plates. Lib stamp on title. Orig cloth, unopened. *(Egglishaw)* **£36 [≈ $57]**

Kayser, E.
- Textbook of Comparative Geology. London: 1893. 8vo. Num plates & ills. Orig cloth.
(Henly) **£15 [≈ $24]**

Kearton, R.
- British Bird's Nests; how, where and when to find and identify them. London: [1907]. New enlgd edn. 8vo. xii,520 pp. 15 cold & 6 plain plates, num text ills. Orig cloth gilt.
(Wheldon & Wesley) **£20 [≈ $32]**

Kearton, R. & C.
- Kearton's Nature Pictures ... London: Cassell, 1910. 2 vols. 4to. 96 plates. Orig

cloth gilt. *(Egglishaw)* **£32 [≈ $51]**

Keen, William Williams (ed.)
- Surgery: Its Principles and Practice By Various Authors. Philadelphia: 1914. 6 vols. 8vo. *(Rittenhouse)* **$175 [≈ £109]**

Keill, James
- The Anatomy of the Human Body Abridged. London: 1731. 8th edn. 12mo. Contemp panelled calf, rebacked.
(David White) **£60 [≈ $96]**

Keill, John
- An Introduction to Natural Philosophy ... Translated from the last Edition of the Latin. The Fifth Edition. London: for Andrew Millar ..., 1758. 5th English edn. 8vo. xii, 06 pp. W'cut text ills. Minor foxing & soiling. Rec half calf. *(Pickering)* **$350 [≈ £218]**
- An Introduction to the True Astronomy: or Astronomical Lectures, Read in the Astronomical School of the University of Oxford. London: Lintot ..., 1721. 1st edn in English. 8vo. [viii],xvi,396,[12] pp. 2 lunar maps, 26 fldg plates. Occas browning. Contemp calf, jnts reprd.
(Pickering) **$550 [≈ £343]**
- An Introduction to the True Astronomy: or, Astronomical Lectures, Read in the Astronomical School of the University of Oxford. London: Lintot, 1721. 8vo. [viii], xvi, 396,[xii] pp. 2 maps of the moon, 26 fldg plates. Period speckled calf gilt, jnts cracked but firm. *(Rankin)* **£85 [≈ $136]**
- An Introduction to the True Astronomy: or, Astronomical Lectures ... Third Edition ... London: Henry Lintot, 1739. [6],xiv,[4], 396, [10] pp, advt leaf. 29 fldg plates (plate 18 not present but not called for?). Lacks 2 moon-maps. Orig calf, worn, sp ends chipped.
(Elgen) **$200 [≈ £125]**
- An Introduction to the True Astronomy: or, Astronomical Lectures ... Sixth Edition, corrected. London: J. Buckland, 1769. 8vo. [6], xiv,[4], 396,[10] pp, advt leaf. 28 fldg plates (inc 2 not called for). Contemp calf, sl crack in upper jnt. *(Spelman)* **£95 [≈ $152]**

Keith, Arthur
- Menders of the Maimed. Philadelphia: (1952). Facs of the 1919 edn. One of 1500. 335 pp. Ports, ills. Orig cloth. Dw.
(Elgen) **$75 [≈ £46]**

Keith, P.
- A Botanical Lexicon, or Expositor of Terms, Facts and Doctrines of the Vegetable

Physiology brought down to the Present Time. London: 1837. 8vo. 416 pp. Orig cloth, spine faded.
(Wheldon & Wesley) £25 [≈ $40]

Keith, T.
- An Introduction to the Theory and Practice of Plane and Spherical Trigonometry and the Stereographic Projection of the Sphere ... London: 1810. xxviii,420 pp. 5 fldg plates, text diags. Occas sl foxing. Contemp leather, spine relaid. *(Whitehart)* £30 [≈ $48]

Kells, C. Edmund
- Dentist's Own Book. A Faithful Account of the Experiences gained during 46 Years of Dental Practice. St. Louis: Mosby, 1925. 1st edn. Sm 4to. 510 pp. 116 photo ills. Orig bndg. *(Xerxes)* $75 [≈ £46]

Kelly, Howard A.
- A Cyclopedia of American Medical Biography ... Philadelphia: 1912. 2 vols.
(Rittenhouse) $150 [≈ £93]
- Medical Gynecology. New York: 1909.
(Rittenhouse) $60 [≈ £37]

Kemp, J.F.
- The Ore Deposits of the United States and Canada. New York: The Scientific Publ Co, 1906. 3rd edn. 8vo. xxiv,481 pp. 162 ills. Orig cloth. *(Gemmary)* $35 [≈ £21]

Kendall, P.F. & Wroot, H.E.
- Geology of Yorkshire. An Illustration of the Evolution of Northern England. Privately ptd: 1924. 8vo. xxii,995 pp. Num ills. Orig cloth. *(Henly)* £88 [≈ $140]
- Geology of Yorkshire. Privately ptd: 1924. 8vo. 995 pp. Num ills. Orig cloth, trifle loose.
(Wheldon & Wesley) £50 [≈ $80]

Kennedy, Alexander W.M. Clark
- The Birds of Berkshire and Buckinghamshire ... Eton: Ingalton & Drake ..., 1868. 1st edn. xiv,232 pp. 4 tinted photos. Orig blue cloth gilt. *(Hollett)* £150 [≈ $240]

Kennedy, P.G., et al.
- The Birds of Ireland ... Edinburgh: Oliver, 1954. 437 pp. Orig bndg. Dw.
(Emerald Isle) £26 [≈ $41]

Kennedy, Rankin
- The Book of the Motor Car. London: Caxton, 1913. 1st edn. 3 vols. Cold paper model with hinged sections overlapping on each vol front pastedown, 23 full page plates, num text ills.

Orig dark maroon cloth, pict blocked in black, crnrs sl worn. *(Sklaroff)* £60 [≈ $96]
- The Modern Machine Shop. Its Tools, Practice and Design. London: Caxton, 1907. 4 vols. 4to. Ca 800 pp. Ills. Orig bndg.
(Book House) £22 [≈ $35]

Kent, A.H.
- J. Veitch & Sons Manual of the Coniferae. A New and Enlarged Edition by A.H. Kent. London: 1900. 8vo. 562 pp. Num plates & text figs. Orig cloth gilt. *(Henly)* £60 [≈ $96]

Kent, W.S.
- A Manual of the Infusoria, including a Description of all known Flagellate, Ciliate and Tentaculiferous Protozoa, British and Foreign ... London: 1880-82. 2 vols & an atlas. Roy 8vo. 1 cold & 52 plain plates. Orig cloth, vol 1 & atlas refixed in their cases.
(Henly) £160 [≈ $256]

Kentish, T.
- A Treatise on a Box of Instruments and the Slide-Rule, with the Theory of Trigonometry and Logarithms. London: Relfe & Fletcher, 1839. 8vo. iv,[2],100 pp. Fldg frontis. Orig roan backed ptd bds. *(Spelman)* £16 [≈ $25]

Kerner, A.
- Flowers and their Unbidden Guests, with a Prefatory Letter by Charles Darwin, the Translation revised and edited by W. Ogle. London: 1878. 8vo. xvi,167 pp. 3 plates. Orig cloth. Freeman 1318.
(Wheldon & Wesley) £50 [≈ $80]

Kerr, G.L.
- Practical Coal-Mining: A Manual for Managers, Colliery Engineers, and others. London: Griffin, 1922. 8th edn. 8vo. 778 pp. 755 ills. Orig bndg, sl shaken.
(Book House) £25 [≈ $40]

Kerr, John
- The Fundamentals of School Health. London: Allen & Unwin, 1926. Only edn. Roy 8vo. 3 plates. Minor marg foxing. Orig cloth. *(Fenning)* £24.50 [≈ $40]

Kerr, Robert
- General View of the Agriculture of the County of Berwick ... Drawn up for the Consideration of the Board of Agriculture and Internal Improvement ... London: 1809. 8vo. xxxii, 504,73 pp. Fldg map, 3 plates (foxed). Sm piece torn from 1 leaf. Contemp half calf, sl worn. *(Blackwell's)* £85 [≈ $136]

Keulemans, T. & Coldewey, C.J.
- Feathers to Brush, the Victorian Bird Artist John Gerrard Keulemans, 1842-1912. Deventer, 1982. One of 500, signed by both authors. 4to. xviii,94 pp. 24 cold plates, 4 ports, text figs. Orig half mor.
(Wheldon & Wesley) **£90 [≈ $144]**

Kew, H.W.
- The Dispersal of Shells. London: International Scientific Series, 1893. xiv, 291 pp. 6 figs. Orig cloth. *(Baldwin)* **£30 [≈ $48]**

Keynes, Sir Geoffrey
- A Bibliography of Dr. Robert Hooke. Oxford: Clarendon Press, 1960. 1st edn. 4to. xxiv, 115 pp. 12 plates, text ills. Orig cloth. Dw.
(Duck) **£65 [≈ $104]**
- John Ray, a Bibliography. London: 1951. One of 650. Roy 8vo. xvi,164 pp. 4 plates, 16 facs title-pages. Orig cloth.
(Duck) **£75 [≈ $120]**

Keynes, John Maynard
- Essays in Persuasion. London: Macmillan, 1931. 1st edn. 8vo. [4],376 pp. Orig cloth. Dw v sl worn.
(M & S Rare Books) **$175 [≈ £109]**
- The General Theory of Employment Interest and Money. London: Macmillan, 1936. 1st edn. 8vo. xii,403 pp. Orig green cloth, spine sl rubbed. *(Frew Mackenzie)* **£120 [≈ $192]**
- The General Theory of Employment Interest and Money ... London: Macmillan, 1936. 1st edn. 8vo. [1 advt],xii,403 pp. Orig blue cloth.
(Pickering) **$550 [≈ £343]**
- The General Theory of Employment Interest and Money ... New York: Harcourt, Brace, n.d., preface dated 1935, [1936?]. 1st Amer edn. 8vo. xii,403 pp. Orig black cloth.
(Pickering) **$300 [≈ £187]**
- The General Theory of Employment Interest and Money. London: 1936. 1st edn. 8vo. 12,403 pp. Orig cloth.
(M & S Rare Books) **$250 [≈ £156]**
- How to Pay for the War. A Radical Plan for the Chancellor of the Exchequer. London: Macmillan, 1940. 1st edn. 8vo. vi,88 pp. Orig paper cvrd bds.
(Frew Mackenzie) **£20 [≈ $32]**
- A Revision of the Treaty. Being a Sequel to The Economic Consequences of the Peace. London: Macmillan, 1922. 1st edn. 8vo. viii, 223,6 advt pp. Orig blue cloth, sl spotted, outer marg of cvrs faded.
(Frew Mackenzie) **£60 [≈ $96]**
- A Tract on Monetary Reform. London: Macmillan, 1923. 1st edn. 8vo. Orig cloth,

bds sl spotted.   *(Fenning)* **£65 [≈ $104]**

Keynes, John Neville
- The Scope and Method of Political Economy. London: Macmillan, 1891. 1st edn. 8vo. xiv, 359 pp. Orig maroon cloth, spine v rubbed.
(Burmester) **£120 [≈ $192]**
- Studies and Exercises in Formal Logic ... London: Macmillan, 1884. 1st edn. 8vo. xii, 414, [2 advt] pp. Orig maroon cloth, spine trifle faded.   *(Burmester)* **£150 [≈ $240]**

Keys, John
- The Antient Bee-Master's Farewell; or, Full and Plain Directions for the Management of Bees. London: 1796. 1st edn. 8vo. xvi,273 pp. 2 plates. Trifle foxed. Calf, reprd.
(Wheldon & Wesley) **£175 [≈ $280]**
- The Practical Bee-Master. London: 1780. 8vo. xii,390,[2] pp. Lacks the plate. New half calf.   *(Wheldon & Wesley)* **£100 [≈ $160]**
- A Treatise on the Breeding and Management of Bees. London: 1814. New edn. Sm 8vo. xvi, 272 pp. 2 plates. Orig bds, uncut, reprd.
(Wheldon & Wesley) **£75 [≈ $120]**
- A Treatise on the Breeding and Management of Bees ... A New Edition. London: 1814. Appears to be the only edn. 12mo. xvi,272 pp. 2 plates. Orig tree calf, gilt spine, mor label.   *(Bickersteth)* **£148 [≈ $236]**

Keys, Thomas
- History of Surgical Anesthesia. New York: Schuman's, 1945. 1st edn. 8vo. 191 pp. Photo ills. Owner stamped & sgnd. Orig bndg.
(Xerxes) **$125 [≈ £78]**

Kidd, John
- On the Adaptation of External Nature to the Physical Condition of Man ... Exercise of his Intellectual Faculties. London: William Pickering, 1833. 1st edn. 8vo. [8 inserted advt], xvi,375,[1] pp. Contemp blue cloth, rubbed.   *(Gach)* **$135 [≈ £84]**

Kidd, Walter
- The Direction of Hair in Animals and Man. London: 1903. 8vo. 154 pp. 33 figs.
(Rittenhouse) **$125 [≈ £78]**

Killington, F.J.
- A Monograph of the British Neuroptera. London: Ray Society, 1936-37. Orig printing. 2 vols. 8vo. 30 plates (8 cold). Sound ex-lib. Orig cloth, spines faded & soiled.
(Wheldon & Wesley) **£45 [≈ $72]**
- A Monograph of the British Neuroptera. London: Ray Society, 1936-37. 2 vols. 8vo. 30

plates (8 cold). Orig cloth.
(Wheldon & Wesley) £60 [≈ $96]

Kinahan, G. Henry
- Manual of the Geology of Ireland. London: Kegan Paul, 1878. Map, ills. Orig bndg.
(Emerald Isle) £35 [≈ $56]

King, E.S.J.
- Surgery of the Heart. Baltimore: 1941. 1st Amer edn. Tall 8vo. 728 pp. Ills. Orig cloth.
(Hemlock) $150 [≈ £93]

King, John S.
- Dawn of the Awakened Mind. New York: McCann, 1920. 1st edn. 8vo. 451 pp. Orig bndg.
(Xerxes) $65 [≈ £40]

King, W.
- The Permian Fossils of England. London: Pal. Soc. Monograph 5, 1850. 4to. xxxvii,258 pp. 28 plates (sl waterstained). Half disbound.
(Baldwin) £28 [≈ $44]

Kingsley, Charles
- Glaucus; or the Wonders of the Shore. Cambridge: Macmillan, 1855. 1st edn. 1st iss. Post 8vo. 165,[16 advt dated May 1855] pp. Frontis. Orig cloth gilt, spine a trifle faded.
(Hollett) £75 [≈ $120]
- Glaucus; or the Wonders of the Shore. Cambridge: 1855. 1st edn. Post 8vo. [viii], 165 pp. Frontis. Orig dec cloth, fine.
(Wheldon & Wesley) £33 [≈ $52]

Kirby, Mary & Elizabeth
- The Sea and its Wonders. London: Nelson, 1871. 1st edn. 8vo. 304 pp. Text figs. Orig pict cloth gilt, a.e.g. *(Egglishaw)* £22 [≈ $35]
- The Sea and its Wonders. London: Nelson, 1875. 8vo. 304 pp. Text figs. Orig pict cloth gilt, a.e.g. *(Egglishaw)* £14 [≈ $22]

Kirby, William
- Monographia Apium Angliae: or an Attempt to divide into their Natural Genera and Families, such Species of the Linnean Genus Apis as have been discovered in England. Ipswich: 1802. 2 vols. 8vo. 18 plates (4 cold). Calf. *(Wheldon & Wesley)* £175 [≈ $280]
- On the History, Habits and Instincts of Animals. London: William Pickering, Bridgewater Treatise VII, 1835. 2 vols. 8vo. cv,[i],406; viii,542 pp. 16 plates. Contemp polished calf. *(Lamb)* £30 [≈ $48]
- On the Power, Wisdom, and Goodness of God, as manifested in the Creation of Animals, and in their History, Habits and

Instincts ... New Edition, edited by Thomas Rymer Jones. London: Bohn, 1852. 2 vols. 8vo. Orig red cloth.
(Waterfield's) £35 [≈ $56]

Kirby, William & Spence, William
- An Introduction to Entomology ... London: 1815-26. 1st edn. 4 vols. 8vo. 2 ports, 30 plates (6 cold), fldg table. Appendix, 31 pp, 1816 bound in at end of vol 1. Sl foxing & offsetting, upper outer crnr of plates in vol 3 waterstained. Half calf, reprd.
(Wheldon & Wesley) £120 [≈ $192]
- An Introduction to Entomology ... London: 1816-26. Vol 1 2nd edn, rest 1st edns. 4 vols. 8vo. 2 ports, 30 plates (6 cold), fldg table. Some limited foxing esp port in vol 4. Half calf. *(Wheldon & Wesley)* £100 [≈ $160]
- An Introduction to Entomology ... Third Edition. London: Longman, 1818. 2 vols. 8vo. 5 cold plates. Vol 1 title marked. Contemp half calf, rubbed, vol 2 jnts broken.
(Waterfield's) £40 [≈ $64]
- An Introduction to Entomology ... Fifth Edition. London: Longman ..., 1828. 4 vols. 8vo. 2 ports, 30 plates (6 hand cold). Contemp half calf, red & green labels.
(Bickersteth) £70 [≈ $112]

Kirby, William Egmont
- Butterflies and Moths of the United Kingdom. London: [1912]. 8vo. lii,468 pp. 70 cold plates. Orig cloth. *(Henly)* £20 [≈ $32]
- Butterflies and Moths of the United Kingdom. London: Routledge, [1912]. 8vo. lii, 468 pp. 79 [sic] cold plates. Orig pict dec cloth. *(Egglishaw)* £25 [≈ $40]

Kirby, William Forsell
- The Butterflies and Moths of Europe. London: Cassell, 1903. 4to. lxxii,432 pp. 1 plain & 54 cold plates, text ills. Orig pict dec cloth, a.e.g. *(Egglishaw)* £140 [≈ $224]
- The Butterflies and Moths of Europe. London: Cassell, 1907. 4to. lxxii,432 pp. 1 plain & 54 cold plates, text ills. Binder's cloth.
(Egglishaw) £90 [≈ $144]
- A Hand-Book to the Order Lepidoptera. London: Lloyd's Natural History, 1896-97. 5 vols. 8vo. 158 cold plates. Sl spotting to prelims & tissues. Lib labels on front endpapers, heavy stamps on rear endpapers. Orig cloth gilt. *(Hollett)* £55 [≈ $88]
- List of Hymenoptera, with Descriptions and Figures of the typical Specimens in the British Museum. Vol. 1. Tenthredinidae and Siricidae [all published]. London: 1882. 8vo. 16 cold plates. Cloth.

(Egglishaw) **£18 [≃ $28]**

Kirk, Edward C. (ed.)
- The American Text-Book of Operative Dentistry. Philadelphia: Lea, 1897. 1st edn. Lge 8vo. 702,16 ctlg pp. 751 text engvs. Orig sheep, sl rubbed.   *(Elgen)* **$95 [≃ £59]**

Kirk, T.
- The Forest Flora of New Zealand. Wellington: 1889. Sm folio. xv,345 pp. 159 plates. Inscrptns on title. Orig cloth, trifle used.   *(Wheldon & Wesley)* **£150 [≃ $240]**
- The Students' Flora of New Zealand and the Outlying Islands. Wellington: [1899]. All published. Roy 8vo. vi,408 pp. Orig cloth, sl used.   *(Wheldon & Wesley)* **£45 [≃ $72]**

Kirkaldy, David
- Result of an Experimental Inquiry into the Comparative Tensile Strength and other Properties of Various Kinds of Wrought-Iron and Steel. Glasgow: Bell & Bain, 1862. 1st edn. [ii],i-viii, 9-100,105-212 pp, correct. 16 plates. Sl marked. Orig cloth, discold.   *(Duck)* **£150 [≃ $240]**

Kirkland, J.
- The Modern Baker, Confectioner and Caterer. London: Gresham, 1924. Rvsd edn. 4 vols. 4to. 1043 pp. Ills. Orig bndg, sl faded.   *(Book House)* **£20 [≃ $32]**

Kirkman, F.B. & Hutchinson, H.G.
- British Sporting Birds. London: 1936. 4to. xii,428 pp. 31 cold & 12 plain plates. Orig pict blue cloth.   *(Henly)* **£25 [≃ $40]**

Kirwan, Richard
- The Manures most advantageously applicable to the Various Sorts of Soils ... London: for Vernor & Hood, 1796. 4th edn. 8vo. Title sl soiled. New wraps.   *(Stewart)* **£85 [≃ $136]**

Kitchiner, William
- The Art of Invigorating Life by Food, Clothes, Air, Exercise, Wine, Sleep, etc. and Peptic Precepts ... London: Hurst, Robinson, 1822. 3rd edn, enlgd. 12mo. 298 pp. Minor spotting. Orig bds, spine v worn & chipped.   *(Xerxes)* **$175 [≃ £109]**
- The Cook's Oracle: containing Receipts for Plain Cookery on the most Economical Plan for Private Families ... Boston: 1822. viii, [9]-380 pp. Occas foxing & dampstaining. Contemp calf, outer front hinge cracked, sm chip at hd of spine, bds sl warped.   *(Reese)* **$400 [≃ £250]**

Kleen, Emil
- Handbook of Massage. Translated from the Swedish by E.M. Hartwell, MD. Philadelphia: Blakiston, 1892. 8vo. 316 pp. Lacks prelim blank. Orig bndg.   *(Xerxes)* **$50 [≃ £31]**

Kline, M.
- Mathematical Thought from Ancient to Modern Times. New York: 1972. xvii,1238 pp. A few figs. Orig cloth.   *(Whitehart)* **£35 [≃ $56]**

Knapp, F.H.
- The Botanical Chart of British Flowering Plants and Ferns ... Bath, London & Dublin: 1846. 1st edn. 8vo. [xii],90,[4 advt] pp. Some foxing. Orig cloth, a little faded & marked.   *(Bow Windows)* **£48 [≃ $76]**

Knapp, John Leonard
- The Journal of a Naturalist. London: Murray, 1829. 2nd edn. 8vo. xvi,413 [sic] pp. 8 engvd plates (1 cold) [sic]. 19th c half lilac calf, spine faded. The Syston Park copy.   *(Young's)* **£35 [≃ $56]**
- The Journal of a Naturalist. London: 1829. 2nd edn. 8vo. xvi,423 [sic] pp. 7 plates (1 cold) [sic]. Contemp annotations on title & half-title. Orig cloth, lower jnt torn.   *(Wheldon & Wesley)* **£30 [≃ $48]**
- The Journal of a Naturalist. London: 1829. 2nd edn. 8vo. 7 engvd plates (1 cold). Rec qtr calf, leather label.   *(Ambra)* **£32 [≃ $51]**
- The Journal of a Naturalist. London: 1830. 3rd edn. 8vo. xvi,440 pp. 7 plates (1 cold). Plates sl foxed. Half mor, trifle rubbed.   *(Wheldon & Wesley)* **£30 [≃ $48]**

Knight, C.W.R.
- Aristocrats of the Air. London: 1925. Cr 4to. xii,166 pp. Cold frontis, 53 ills. Orig cloth.   *(Wheldon & Wesley)* **£30 [≃ $48]**
- Aristocrats of the Air. London: 1946. 2nd rvsd edn. Roy 8vo. 150 pp. Cold frontis, 53 ills. Orig cloth.   *(Wheldon & Wesley)* **£20 [≃ $32]**

Knight, Charles
- Cyclopaedia of the Industry of all Nations. New York: Putnam; London: Charles Knight, 1851. 1st edn. Thick 8vo. xxiv,1810 pp. 37 w'engvd ills. Orig cloth, 1 or 2 cracks in spine.   *(Charles B. Wood)* **$150 [≃ £93]**

Knight, Edward H.
- Knight's American Mechanical Dictionary. Boston: Houghton, Mifflin, 1884. 3 vols. 4to.

2831 pp. 7395 w'engvd ills. Old cloth, leather labels, mrbld edges.
(Charles B. Wood) **$250 [≈£156]**

Knott, Cargill Gilston (ed.)
- Napier Tercentenary Memorial Volume. Edinburgh: 1915. 4to. ix,441 pp. 15 plates, port, other ills. A few light pencil notes. Orig cloth, dust stained & sl marked.
(Whitehart) **£45 [≈$72]**
- See also under Tait, Peter Guthrie.

Knowlson, John C.
- The Yorkshire Cattle-Doctor and Farrier ... Written in Plain Language ... Tenth Thousand. London: Longman, 1848. 8vo. xiv,272 pp. Half-title, engvd frontis. Fldg table 'The Farmer's Ready Reckoner' at end. Foxed & soiled. Orig brown cloth gilt, worn.
(Blackwell's) **£30 [≈$48]**

Knox, Alexander
- The Ballynahinch Mineral Waters ... together with Notices of the Scenery, Antiquities and Natural History. Belfast: Lamont Bros, 1846. 47,[vi] pp. Orig yellow ptd wraps, tipped into bds.
(Emerald Isle) **£75 [≈$120]**
- The Irish Watering Places. Their Climate, Scenery and Accomodations ... Analyses of their Principal Springs ... and the Various Forms of Disease to which they are adapted ... Dublin: Curry, 1845. 336 pp. Contemp travelling bndg, with flap & tie.
(Emerald Isle) **£50 [≈$80]**

Knox, Arthur E.
- Ornithological Rambles in Sussex ... London: Van Voorst, 1849. 1st edn. Cr 8vo. vi, 250,ctlg pp. 4 plates. Orig cloth gilt.
(Egglishaw) **£25 [≈$40]**
- Ornithological Rambles in Sussex. London: 1850. 2nd edn. 8vo. x,254 pp. 4 plates. Occas sl foxing. Orig cloth, trifle used.
(Wheldon & Wesley) **£20 [≈$32]**
- Ornithological Rambles in Sussex ... Third Edition. London: 1855. 8vo. xii,260 pp, advt leaf. 4 tinted litho plates. Orig gilt illust cloth.
(Bow Windows) **£40 [≈$64]**

Knox, Robert
- Engravings of the Cardiac Nerves .... copied from the Tabulae Neurologicae of Antonio Scarpa. By Edward Mitchell, Engraver ... Third Edition. Edinburgh: 1832. 4to. 23 plates & 7 supplementary plates. Sl spotting. Orig cloth, spine ends worn.
(Bickersteth) **£145 [≈$232]**

Koh, L. Winfield
- Practical Treatise on Diseases of the Digestive System. Philadelphia: F.A. Davis, 1930. Roy 8vo. 1125 pp. 542 engvs, inc 7 cold plates. Orig bndg. Sgnd pres copy.
(Xerxes) **$60 [≈£37]**

Kohler, Wolfgang
- The Mentality of Apes. Translated from the Second Revised Edition ... London: Kegan Paul, Intl. Lib, 1925. 1st English edn. 8vo. Orig cloth.
(Fenning) **£35 [≈$56]**

Kolliker, Rudolph Albert von
- Manual of Human Histology. (Translated and Edited by George Busk and Thomas Huxley). London: Sydenham Society, 1853-54. 1st edn in English. 2 vols. 8vo. Num text w'cuts. Contemp cloth.
(Rootenberg) **$250 [≈£156]**

Koops, Matthias
- Historical Account of the Substances which have been used to describe Events ... from the Earliest Date, to the Invention of Paper ... London: T. Burton, 1800. 1st edn. Sm folio. 91 pp. Old lib perf in blank marg of 2nd leaf. Rec old style qtr calf.
(Charles B. Wood) **$1,500 [≈£937]**
- Historical Account of the Substances which have been used to describe Events ... from the Earliest Date to the Invention of Paper ... London: Jaques, 1801. 2nd edn. 8vo. Frontis. Lib stamps & blind stamps on frontis & title. Qtr mor, scuffed, lib b'plate.
(Rostenberg & Stern) **$1,250 [≈£781]**
- Historical Account of the Substances which have been used to describe Events ... from the Earliest Date to the Invention of Paper ... London: 1801. 2nd edn. Variant with pp 1-258 on de-inked waste paper. 8vo. [ii],273 pp. Frontis. Rec calf backed bds.
(Charles B. Wood) **$1,100 [≈£687]**

Kovacs, Richard
- Electrotherapy and Light Therapy. Philadelphia: Lea & Febiger, 1938. 3rd edn. Thick roy 8vo. 744 pp. Cold plate, 307 ills. Orig bndg.
(Xerxes) **$75 [≈£46]**

Krafft-Ebbing, R. von
- Psychopathia Sexualis, with special reference to Contrary Sexual Instinct: a Medico-Legal Study. Authorized Translation of the 7th Enlarged and Revised German Edition ... Philadelphia: Davis; London: Rebman, 1894. xiv,436 pp. Orig cloth.
(Box of Delights) **£30 [≈$48]**

Kramer, W.
- The Aural Surgery of the Present Day. Translated by Henry Power. With Corrections and Additions by the Author. London: New Sydenham Society, 1863. vii,154 pp. 9 w'cuts. Sm lib stamps on title. Orig cloth. *(Elgen)* **$65 [≈ £40]**

Kunz, George Frederick
- The Curious Lore of Precious Stones. Philadelphia: 1913. 1st edn. Wide 8vo. xiv, 406 pp. Frontis, 5 cold & num other plates, text ills. Orig gilt dec cloth, t.e.g. *(Elgen)* **$135 [≈ £84]**
- Ivory and the Elephant in Art, in Archaeology, and in Science. New York: 1916. Roy 8vo. xxvi,527 pp. Frontis, 2 maps, 125 plates, 41 text figs. Orig dec buckram, uncut, inner jnts reprd. *(Wheldon & Wesley)* **£150 [≈ $240]**

Kurr, J.G.
- The Mineral Kingdom. Edinburgh: Edmonston & Douglas, 1859. Sm folio. 70,48 pp. 23 hand cold & 1 other plates. Half leather, rebacked. *(Gemmary)* **$1,250 [≈ £781]**

Kyan, John Howard
- On the Elements of Light, and their Identity with those of Matter, Radiant and Fixed. London: Longman ..., 1838. 8vo. xiv, 130 pp. Hand cold frontis, 3 plates (1 hand cold). Occas marg spotting, the 2 uncold plates foxed. Orig cloth, recased, new endpapers. *(Spelman)* **£160 [≈ $256]**
- On the Elements of Light, and their Identity with those of Matter, Radiant and Fixed. London: Longman ..., 1838. 1st edn. 8vo. xiv,130 pp. Hand cold frontis, 3 plates (1 hand cold). Orig cloth, dec in blind. *(Rootenberg)* **$250 [≈ £156]**

The Ladies New Dispensatory ...
- The Ladies New Dispensatory and Family Physician. London: for Johnson & Payne, 1769. 154 pp. New cloth, leather spine. *(Box of Delights)* **£120 [≈ $192]**

Laennec, R.T.H.
- A Treatise on Medical Auscultation and on Diseases of the Lungs and Heart ... Edited by Theophilus Herbert ... Notes ... of F.H. Ramadge ... London: 1846. 8vo. xxxi,[1],862 pp. Half-title. Port, 6 plates. Orig cloth, spine reprd preserving title, crnrs bumped. *(Hemlock)* **$400 [≈ £250]**

Laishley, Richard
- A Popular History of British Birds' Eggs. London: Lovell Reeve, 1858. 1st edn. Sm 8vo. 20 cold lithos. Orig cloth gilt, new endpapers. *(Hollett)* **£40 [≈ $64]**

Lakes, A.
- Prospecting for Gold and Silver in America. Scranton, PA: The Colliery Engineer Co, 1896. 2nd edn. Cr 8vo. 287 pp. 76 ills, sev fldg. Orig cloth, sl worn. *(Gemmary)* **$45 [≈ £28]**

Lamarck, J.B.
- Zoological Philosophy. An Exposition with regard to the Natural History of Animals. Translated by H. Elliot. London: Macmillan, 1914. 8vo. xcvii,410 pp. Orig cloth. *(Wheldon & Wesley)* **£70 [≈ $112]**

Lamb, M.C.
- Leather Dressing. Including Dyeing, Staining & Finishing. Third Edition. London: 1925. Roy 8vo. xii,472,30 advt pp. 4 card plates containing 20 actual mtd samples, other ills. Orig royal blue cloth, v sl marked. *(Duck)* **£110 [≈ $176]**

Lamb, Patrick
- Royal Cookery; or, the Complete Court Cook ... London: for Abel Roper ..., 1710. 1st edn. 8vo. [xvi],127,[16] pp. 35 fldg plates. Repr to crnr of 1st 3 ff. Stain on last leaf. Contemp elab gilt red mor, spine relaid, a bit rubbed. *(W. Thomas Taylor)* **$1,750 [≈ £1,093]**

Lambert, A.B.
- A Description of the Genus Pinus. London: 1832. 1st 8vo edn. 2 vols. Port, 74 hand cold plates (12 fldg). Some foxing to text. Green half mor, hd of 1 spine chipped. *(Henly)* **£2,300 [≈ $3,680]**

Lamond, Henry
- The Sea-Trout. A Study in Natural History. London: Sherratt & Hughes, 1916. 1st edn. 4to. xi,219 pp. 9 cold plates, 64 ills. Orig green cloth gilt. *(Hollett)* **£50 [≈ $80]**
- The Sea-Trout. A Study in Natural History. London: Sherratt & Hughes, 1916. 4to. xi,219 pp. 9 cold plates, 62 ills. Orig cloth. *(Egglishaw)* **£40 [≈ $64]**

La Motte, Guillaume
- See Mauquest de La Motte, Guillaume.

Lancereaux, E.
- A Treatise on Syphilis, Historical and

Practical. London: New Sydenham Society, 1868-69. 2 vols. 8vo. Orig cloth, vol 1 spine relaid, new endpapers.
(Bickersteth) **£28 [≈ $44]**

Lanchester, F.W.
- The Flying-Machine from an Engineering Standpoint ... London: Constable, 1916. 1st edn thus. viii,136,[2 advt] pp. Fldg plate, 55 text ills. Orig cloth, soiled, spine ends rubbed.
(Duck) **£30 [≈ $48]**
- The Flying-Machine. Two Papers: The Aerofoil, and the Screw Propeller. London: Inst of Automobile Engineers, Reprint from Proc Vol IX, 1915. 186 pp. Errata slip at p 90. 3 plates (2 fldg), text diags. Orig cloth. Author's 'Compliments' stamp.
(Duck) **£37.50 [≈ $60]**

Lanczos, C.
- Space through the Ages. The Evolution of Geometrical Ideas ... London: 1970. x,320 pp. 133 figs. Orig cloth, partly faded.
(Whitehart) **£25 [≈ $40]**

Landman, J.H.
- Human Sterilization. The History of the Human Sterilization Movement. New York: Macmillan, 1932. 1st edn. 8vo. 341 pp. Orig bndg.
(Xerxes) **$60 [≈ £37]**

Landsborough, D.
- A Popular History of British Seaweeds. London: 1849. 1st edn. 8vo. xx,368,4 pp. 20 hand cold plates of seaweeds & 2 anatomical plates. Contemp half calf, sl rubbed.
(Henly) **£40 [≈ $64]**
- A Popular History of British Seaweeds. London: 1857. 3rd edn. 12mo. [xvi],400 pp. 22 plates (20 hand cold). Orig cloth.
(Wheldon & Wesley) **£30 [≈ $48]**
- A Popular History of British Zoophytes, or Corallines. London: 1852. Sm 8vo. xii,404 pp. 20 cold plates. New cloth.
(Wheldon & Wesley) **£18 [≈ $28]**

Landseer, Thomas
- Characteristic Sketches of Animals, principally from the Zoological Gardens ... with Descriptive and Illustrative Notices by John Henry Barrow ... London: 1832. 1st edn. Folio. viii,[68],[ix]-x pp. Engvd & ptd titles, 32 plates, 32 vignettes. Half mor, rubbed.
(Bickersteth) **£130 [≈ $208]**

Landsteiner, Karl
- The Specificity of Serological Reactions. Springfield: (1936). 1st edn in English. 178

pp. Orig bndg.
(Elgen) **$115 [≈ £71]**

Lang, H.C.
- The Butterflies of Europe Described and Figured. London: Reeve, 1884. 2 vols. Sm 4to. 82 chromolitho plates. Mod backed cloth gilt.
(Hollett) **£285 [≈ $456]**

Lang, W.D.
- A Handbook of British Mosquitoes. London: BM (Nat. Hist.), 1920. Roy 8vo. vi,125 pp. 5 cold plates, text figs. Wraps.
(Egglishaw) **£12 [≈ $19]**

Langley, Batty
- A Sure Method of Improving Estates, by Plantations of Oak, Ash, Beech, and other Timber-Trees, Coppice-Woods, etc. London: 1728. 8vo. [viii],xx,274 pp. 1 plate. Mod half calf.
(Wheldon & Wesley) **£245 [≈ $392]**

Langley, Samuel Pierpont
- Experiments in Aerodynamics. Washington: Smithsonian Contributions to Knowledge 801, 1891. 1st edn. Imperial 4to. [iv],115 pp. 10 plates, text diags. Orig dark green cloth, back cvr sl discold. The Honeyman copy.
(Duck) **£185 [≈ $296]**
- Experiments in Aerodynamics. Washington: Smithsonian, 1902. 2nd edn. Folio. 115 pp. Plates & charts. Orig cloth.
(Jenkins) **$200 [≈ £125]**
- The Internal Work of the Wind. Washington: Smithsonian Contributions to Knowledge 884, 1893. Imperial 4to. iv,24 pp. 5 plates, 3 text figs. Orig cloth, v sl marked & worn.
(Duck) **£55 [≈ $88]**

Langley, Samuel Pierpont & Manly, Charles M.
- Langley Memoir on Mechanical Flight ... Washington: Smithsonian Contributions to Knowledge, Vol 27, No.3, 1911. Roy 4to. x, [ii],320 pp. 104 plates, text ills. Minor surface injury to p 200. Orig olive cloth, sl marked & scratched, hinges cracked.
(Duck) **£225 [≈ $360]**
- Langley Memoir on Mechanical Flight. Washington: Smithsonian, 1911. 1st edn. Folio. 320 pp. Num plates. Harvard duplicate. Later cloth.
(Jenkins) **$200 [≈ £125]**

Lankester, Mrs P.
- Wild Flowers Worth Notice ... London: Robert Hardwicke, 1861. Sm 8vo. 8 hand cold plates by J.E. Sowerby. Orig green cloth gilt, extremities sl worn, sm piece chipped

from hd of spine.   *(Hollett)* **£35 [≈ $56]**
- Wild Flowers Worth Notice. London: 1872. Cr 8vo. 8 hand cold plates. Orig dec cloth.
        (Henly) **£25 [≈ $40]**

Laplace, Pierre Simon
- Mecanique Celeste. By the Marquis de la Place ... Translated, with a Commentary, by Nathaniel Bowditch ... Boston: Hilliard ..., 1829-39. 1st edn of the 1st complete English trans. One of 250. 4 vols. 4to. 3 ports. Minor foxing. Later half mor, crnrs sl worn.
        (W. Thomas Taylor) **$5,000 [≈ £3,125]**

La Quintinye, J. de
- The Compleat Gard'ner ... abridged ... with very considerable improvements by George London and Henry Wise. London: 1699. 8vo. xxxv, "309" [≈ 325],[7] pp. With the 8 extra ff between D1 & D2. Frontis, 10 plates. Browned. Contemp calf, rebacked.
        (Wheldon & Wesley) **£150 [≈ $240]**

Lardner, Dionysius
- The Electric Telegraph Popularised. London: Walton & Maberley, 1860. 8vo. Irregular pagination. Contemp prize calf gilt, jnts & edges rubbed, name erased from prize label.   *(Sanders)* **£18 [≈ $28]**
- An Elementary Treatise on the Differential and Integral Calculus. London: 1825. xxxii, [2],520 pp. Contemp calf, rebacked.
        (Whitehart) **£40 [≈ $64]**
- Hand-Book of Natural Philosophy. Mechanics. London: 1858. Sm 8vo. xvi,403,[1] pp. Over 350 figs. Orig cloth, hd of spine & 1 jnt a little worn.
        (Bow Windows) **£30 [≈ $48]**
- Popular Geology, containing Earthquakes and Volcanoes. London: 1856. Cr 8vo. 211 figs. Sm stamp on pastedown. Orig cloth, crnr missing from endpaper. *(Henly)* **£28 [≈ $44]**
- Railway Economy: a Treatise on the New Art of Transport ... London: 1850. 1st edn. Cr 8vo. xxiii,528,[1] pp. Orig cloth, v sl wear to spine.   *(Fenning)* **£75 [≈ $120]**
- The Steam Engine Familiarly Explained and Illustrated ... Fifth Edition, considerably enlarged. London: John Taylor, 1836. 8vo. xii, 379,[1],[12 advt] pp. Tipped-in directions to binder. Frontis, engvd title, 12 plates. Some foxing. Orig cloth, spine relaid.
        (Spelman) **£40 [≈ $64]**
- A Treatise on Arithmetic, Theoretical & Practical. London: Lardner's Cabinet Cyclopaedia, 1834. 1st edn. Sm 8vo. Orig cloth, ptd paper labels.
        (Fenning) **£12.50 [≈ $20]**

- A Treatise on Hydrostatics and Pneumatics ... London: for Longman, Rees ... & John Taylor, 1831. 1st edn. 8vo. viii,353 pp. Pict engvd title, w'engvd text ills. Contemp green polished calf, gilt, the Somerhill copy with b'plate.   *(Pickering)* **$150 [≈ £93]**

Lardner, Dionysius (ed.)
- The Museum of Science and Art. London: 1859. 12 vols in 3. Contemp calf.
        (Baldwin) **£75 [≈ $120]**

Lashley, K.S.
- Brain Mechanisms and Intelligence: A Quantitative Study of Injuries to the Brain. Chicago: (1930). 1st edn, 2nd printing. 186 pp. 11 plates, text ills. Orig bndg.
        (Elgen) **$95 [≈ £59]**

Latham, P.M.
- The Collected Works. With a Memoir by Sir Thomas Watson. Edited by Robert Martin. London: New Sydenham Society, 1876-78. 2 vols. 8vo. Orig cloth, vol 1 spine relaid.
        (Bickersteth) **£35 [≈ $56]**

The Lathe and Its Uses ...
- The Lathe and Its Uses; or, Instruction in the Art of Turning Wood and Metal. London: "The English Mechanic", [1850s]. 284,advt pp. Engvs. Some foxing. New cloth bds.
        (Book House) **£50 [≈ $80]**

La Touche, J.D.D.
- A Handbook of the Birds of Eastern China ... London: 1925-34. 2 vols. 8vo. 2 maps, 26 plates. Cloth, sl used, orig wraps bound in. Author's inscrptn.
        (Wheldon & Wesley) **£225 [≈ $360]**

Laufer, Berthold
- The Domestication of the Cormorant in China and Japan. Chicago: 1931. (62) pp. 3 plates. Leather backed blue cloth.
        (Lyon) **£45 [≈ $72]**

Laurence or Lawrence, John
- The Clergy-Man's Recreation: Shewing the Pleasure and Profit of the Art of Gardening. London: Bernard Lintott, 1714. 1st edn. 8vo. [12],83,[1 advt] pp. Frontis. W'cut text diag. Light offset on title. Mod calf.
        (Rootenberg) **$750 [≈ £468]**
- The Clergy-Man's Recreation: Shewing the Pleasure and Profit of the Art of Gardening. By John Lawrence [sic] ... London: for Bernard Lintott, 1714. 1st edn. 8vo. [16],83, [1] pp. Frontis. Mod calf antique.
        (W. Thomas Taylor) **$600 [≈ £375]**

- The Fruit-Garden Kalendar ... London: Lintot, 1718. 1st edn. 8vo. Half-title. 3 pp ctlg. Fldg frontis. Text ill. Mod bds.
(Stewart) £100 [≈ $160]
- A New System of Agriculture. Being a Complete Body of Husbandry and Gardening. London: 1726. Folio. xxiv,456 pp. Frontis, 2 plates. Contemp calf, rebacked.
(Wheldon & Wesley) £180 [≈ $288]

Laurie, Joseph
- Homoeopathic Domestic Medicine ... Fifth American Edition, enlarged and improved, by A. Gerald Hull ... New York: W. Radde, 1849. 8vo. 568 pp. Minimal foxing. Half calf, hinges reprd. *(Hemlock)* $100 [≈ £62]

Lavater, Johann Kaspar
- Aphorisms on Man: translated from the Original Manuscript ... London: for J. Johnson, 1788. 1st edn in English. Vol 1, all published. Sm 8vo. vi,224 pp, errata leaf. Frontis engvd by William Blake after Fuseli. Orig tree calf, mor label.
(Bickersteth) £380 [≈ $608]
- Essays on Physiognomy ... Illustrated by Engravings ... Translated from the French by Henry Hunter. London: for John Stockdale, 1810. 1st printing. 3 vols in 5. Folio. 174 plates, num text engvs. Margs somewhat foxed. Early leather backed bds, a.e.g.
(Gach) $750 [≈ £468]

Laveran, Alphonse
- Paludism. Translated by J.W. Martin. London: New Sydenham Society, 1893. 1st edn in English. 197 pp. 6 plates, text figs. Orig cloth. *(Elgen)* $135 [≈ £84]

Lavoisier, Antoine
- Elements of Chemistry, in a New Systematic Order. Translated from the French by Robert Kerr. Edinburgh: for William Creech, 1790. 1st edn in English. 8vo. l,511 pp. Half- title. 2 fldg tables, 13 plates. Contemp calf, rebacked. *(Pickering)* $1500 [≈ £937]
- Elements of Chemistry. Translated from the French by Robert Kerr. Edinburgh: William Creech, 1792. 2nd edn. 8vo. 592 pp. 2 fldg tables, 13 fldg plates. Contemp calf, gilt spine. *(Gemmary)* $650 [≈ £406]

Law, H. & Clark, D.K.
- The Construction of Roads and Streets. In Two Parts ... London: Crosby Lockwood, 1887. 3rd edn. 345 pp. Ills. Orig bndg, shaken. *(Book House)* £18 [≈ $28]
- The Construction of Roads and Streets.

Revised, with Additional Chapters, by A.J. Wallis-Tayler. London: Crosby Lockwood, 1914. 8th edn. 520 pp. Ills. Orig bndg, cvrs marked. *(Book House)* £20 [≈ $32]

LaWall, Charles
- Four Thousand Years of Pharmacy. Philadelphia: Lippincott, 1927. 3rd edn. Thick roy 8vo. 665 pp. Photo ills. Orig bndg, front hinge cracked. *(Xerxes)* $75 [≈ £46]

Lawrence, John
- The New Farmer's Calendar; or Monthly Remembrancer, for all Kinds of Country Business ... by a Farmer and Breeder. Britons! Honour the Plough. London: C. Whittingham, 1800. 8vo. [viii],616 pp. Fldg frontis. Contemp half calf.
(Lamb) £75 [≈ $120]
- See also Laurence, John.

Lawrence, John, "William Henry Scott'
- The Sportsman's Calendar, or Monthly Remembrancer of Field Diversions. By the Author of British Field Sports. London: Sherwood, Neely & Jones, 1818. Sm 8vo. [iv], 172,[4 advt] pp. 10 w'cut vignettes by Scott & Bewick. Mod crushed mor gilt.
(Blackwell's) £210 [≈ $336]

Lawrence, Robert Means
- Primitive Psycho-Therapy and Quackery. Boston: Houghton Mifflin, 1910. 8vo. 276 pp. Owner's b'plate & a few notes on rear pp. Orig bndg. *(Xerxes)* $100 [≈ £62]

Lawrence, William
- Lectures on Physiology, Zoology, and the Natural History of Man, delivered at the Royal College of Surgeons. London: 1822. 8vo. xi, [i],500 pp. 12 plates. Orig green mor pict gilt (view of King's College Chapel, Cambridge), a.e.g., rubbed, sl worn at crnrs.
(Bickersteth) £80 [≈ $128]
- Lectures on Physiology, Zoology, and the Natural History of Man, delivered at the Royal College of Surgeons. London: 1823. 3rd edn. 8vo. xix,496 pp. 13 plates. A little foxing & soiling, sm tear in 1 plate without loss. Contemp half calf, sl rubbed.
(Wheldon & Wesley) £60 [≈ $96]

Lawson, A.C. & Reid, H.F.
- The California Earthquake of April 18th, 1906. Report of the ... Investigation Committee. Washington: 1908-10. 2 vols in 3 (only, without the atlas vol). 147 plates. Orig wraps, loose, vol 1 lacks upper wrapper.
(Henly) £30 [≈ $48]

Lawson, W.
- A New Orchard and Garden ... with the Country Housewife's Garden ... Preface by E.S. Rohde. London: 1927. One of 650. Sm 4to. xxvi,116 pp. W'cuts. Half parchment.
(Wheldon & Wesley) **£45** [≈ $72]

Layton, T.B.
- Catalogue of the Onodi Collection in the Museum of the Royal College of Surgeons of England. London: 1934. xxiv,131 pp. 47 plates (54 figs). Orig bndg, ex-lib.
(Elgen) **$100** [≈ £62]

Leadbetter, Charles
- Astronomy; or, the True System of the Planets Demonstrated ... London: for J. Wilcox ..., 1727. 1st edn. 8vo. [vi],viii, [iv],120 pp. Engv on p 18v, 9 plates (all but 2 fldg). Contemp panelled calf, spine v sl worn.   *(Pickering)* **$1,500** [≈ £937]

Le Blanc & Armengaud
- The Engineer and Machinist's Drawing Book, a Complete Course of Instruction ... Glasgow: Blackie, 1855. Folio. viii,116 pp. Engvd title, frontis, 70 plates (3 tinted), 246 text figs. Occas foxing, sl marg waterstain on 6 plates. Contemp half calf.
(Spelman) **£140** [≈ $224]

Le Clerc, Sebastian
- Practical Geometry: or, a New and Easy Method of treating that Art. The Third Edition. London: T. Bowles, 1727. Sm 8vo. [2],195,[i],[6] pp. 82 plates. Rebound in full calf, front end paper stained.
(Spelman) **£220** [≈ $352]
- Practical Geometry: or, a New and Easy Method of treating that Art. The Fifth Edition. London: for John Bowles, 1768. Fcap 8vo. 195,[5] pp. 80 plates within the pagination. Later calf.
(Spelman) **£80** [≈ $128]

Lee, A.B.
- The Microtomist's Vade-Mecum, a Handbook of the Methods of Animal and Plant Microscopic Technique. Edited by J.B. Gatenby and T.S. Painter. London: 1937. 10th edn. 8vo. xi,784 pp. 11 ills. Orig cloth.
(Wheldon & Wesley) **£35** [≈ $56]
- The Microtomist's Vade-Mecum, a Handbook of the Methods of Animal and Plant Microscopic Technique. London: 1950. 11th edn, rvsd. 8vo. xiv,753 pp. 8 ills. Orig cloth. *(Wheldon & Wesley)* **£50** [≈ $80]

Lee, J.R.
- The Flora of the Clyde Area. London: 1933. Cr 8vo. xvi,391 pp. Orig cloth.
(Henly) **£18** [≈ $28]

Lee, James
- An Introduction to Botany ... Extracted from the Works of Dr. Linnaeus ... London: 1760. 1st edn. 8vo. xvi,320 pp. 12 plates. Contemp calf, a little worn.
(Wheldon & Wesley) **£160** [≈ $256]
- An Introduction to Botany ... Extracted from the Works of Dr. Linnaeus ... Third Edition, corrected, with large additions. London: Rivington ..., 1776. 8vo. xxiv,432 pp. 12 plates. Contemp calf, mor label, hinges cracking, lacks free endpapers.
(Claude Cox) **£55** [≈ $88]

Lee, John Edward
- Note-Book of an Amateur Geologist. London: Longman, Green, 1881. 1st edn. 8vo. v,90 pp. Woodburytype frontis, 209 litho plates. Orig green cloth gilt, sm split in upper hinge.
(Hollett) **£35** [≈ $56]

Lee, Mrs R.
- Trees, Plants, and Flowers: their Beauties, Uses, and Influences. The Illustrations Drawn and Coloured by James Andrews. London: Grant & Griffith, 1854. 8vo. viii,464 pp. 8 cold plates. Orig green cloth gilt, a.e.g., lower cvr v sl discold.
(Bickersteth) **£110** [≈ $176]

Lee, Robert
- Lectures on the Theory and Practice of Midwifery. delivered in the Theatre of St. George's Hospital. Philadelphia: Barrington & Haswell, 1844. 1st Amer edn. 540 pp. 67 w'cuts. Orig stiff wraps, worn.
(Elgen) **$150** [≈ £93]

Lees, F.A.
- The Flora of West Yorkshire. London: 1888. 8vo. xii, 843 pp. Cold map. Inscrptn on reverse of half-title. Orig cloth.
(Wheldon & Wesley) **£45** [≈ $72]

Le Fanu, W.R.
- A Bio-Bibliography of Edward Jenner 1749-1823. London: 1951.
(Rittenhouse) **$100** [≈ £62]

Leffel, James, & Co. (publishers)
- The Construction of Mill Dams ... Springfield, Ohio: James Leffel & Co, 1874. 1st edn. 312,313-336 advt pp. Ills. Orig cloth,

spine faded, spine ends & crnrs sl rubbed, front hinge cracked but firm.
(Duck) **£150 [≈ $240]**

Lefroy, H.M.
- Indian Insect Pests. Calcutta: 1906. Roy 8vo. viii,318 pp. 365 text figs. Sl foxed at beginning. Orig dec cloth.
(Wheldon & Wesley) **£40 [≈ $64]**

Legge, W.V.
- A History of the Birds of Ceylon. London: 1878-80. 3 vols in one. Roy 4to. Cold map, 1 plain plate, 34 hand cold plates by Keulemans. New half mor.
(Wheldon & Wesley) **£1,400 [≈ $2,240]**

Leggett, William F.
- Ancient and Medieval Dyes. Brooklyn: 1944. 1st edn. 12mo. vi,95 pp. Endpapers with remains of dw.   *(Elgen)* **$55 [≈ £34]**

Le Grand, Anthony
- An Entire Body of Philosophy, According to the Principles of the Famous Renate Des Cartes ... London: 1694. Only edn in English. Folio. [xxx],403,[iii], 263,[i] pp. Frontis (creased, remtd), 100 plates. Contemp calf, rebacked, crnrs v sl worn. Wing L.950.
(Clark) **£600 [≈ $960]**

Leibnitz, Gottfried Wilhelm von & Clarke, Samuel
- A Collection of Papers which passed between the late Learned Mr Leibnitz and Dr Clarke in the Year 1715 and 1716 ... London: Knapton, 1717. 1st edn. 8vo. xiii,[iii], 416, 46 pp,advt leaf. Marg worm in 2 sections. Contemp calf, rebacked & reprd.
(Bickersteth) **£385 [≈ $616]**

Leidy, J.
- A Flora and Fauna within Living Animals. Washington: Smithsonian Contributions, [1853]. 4to. 67 pp. 10 plates (1 cold). Plates rather foxed & with waterstain in lower margs. New cloth.
(Wheldon & Wesley) **£35 [≈ $56]**

Leigh, Charles
- The Natural History of Lancashire, Cheshire and the Peak, in Derbyshire. Oxford: 1700. Folio. Port, cold map, 24 plates. Some signs of use, outer marg of title trimmed. Calf, rebacked. *(Wheldon & Wesley)* **£180 [≈ $288]**

Leitze, Ernst
- Modern Heliographic Processes: a Manual of Instruction in the Art of Reproducing

Drawings, Engravings, Manuscripts, etc. by the Action of Light ... New York: Van Nostrand, 1888. Lge 8vo. viii,143,[iv] pp. 10 mtd specimen heliographs, 32 text ills. Orig cloth.   *(Charles B. Wood)* **$350 [≈ £218]**

Le Mesurier, A.
- Game, Shore and Water Birds of India. London: 1904. 4th (last) edn. Roy 8vo. xvi, 323 pp. 180 figs. Sm stamp on title, sl foxing. Orig cloth. *(Wheldon & Wesley)* **£25 [≈ $40]**
- Game, Shore and Water Birds of India. London: 1904. 4th edn. 8vo. xvi,324,8 advt pp. 5 fldg tables, 180 ills. Sm worm hole in top marg. Ends a little browned. Rebound in cloth.   *(Henly)* **£50 [≈ $80]**

Leonardo Da Vinci
- The Literary Works. New York: 1939. 2nd edn, enlgd & rvsd. 2 vols. Lge 4to. 135 plates, num text ills. Orig pict gilt cloth, t.e.g. Dws.
(Elgen) **$250 [≈ £156]**

Leoning, G.C.
- Military Aeroplanes. Simplified - Enlarged ... Boston: 1918. 1st edn. Roy 8vo. viii,202 pp. Frontis, num plates, text ills. Orig cloth gilt, hd of spine sl worn.
(Berkelouw) **$200 [≈ £125]**
- Monoplanes and Biplanes. Their Design, Construction and Operation ... New York: 1911. 1st edn. 8vo. xiv,332,2 pp. Frontis, num plates & text ills. Orig illust cloth.
(Berkelouw) **$200 [≈ £125]**

Leslie, A.S. & Shipley, A.E. (eds.)
- See The Grouse in Health and Disease.

Le Soeuf, A.S. & Burrell, H.
- The Wild Animals of Australasia, embracing the Mammals of New Guinea and the nearer Pacific Islands. London: 1926. 8vo. 388 pp. 105 ills. Orig cloth. Dw.
(Wheldon & Wesley) **£50 [≈ $80]**

Letchworth, William Pryor
- The Insane in Foreign Countries. New York & London: Putnam's, 1889. 2nd edn. Heavy 8vo. xvi,400,7,[3] pp. 21 plates. Tipped-in ptd note. Orig green cloth, recased.
(Gach) **$275 [≈ £171]**

Letter(s) ...
- A Letter from a Physician in London to his Friend in the Country ... see Brown, Richard.

- Letters concerning the Present State of the French Nation ... see Young, Arthur.

Lettsom, John Coakley
- The Naturalist's and Traveller's Companion, containing Instructions for Collecting and Preserving Objects of Natural History ... Second Edition corrected and enlarged. London: 1774. 8vo. xvi,89,[9] pp. Hand cold title & frontis. Rec half calf by Middleton.
(Burmester) **£120 [≈ $192]**

Levens, Peter
- A right profitable Booke for all Diseases called, The path-way to health ... London: John Beale for Robert Bird, 1632. 4to. [2], 114,[6] ff. Ornamental title-border. Black Letter. Tears in blanks. Contemp calf, sl warped.   *(Rootenberg)* **$1,200 [≈ £750]**

Levick, George Murray
- Antarctic Penguins. New York: McBride, Nast, 1914. 1st Amer edn. 8vo. x,140 pp. Num photo ills. Orig cloth, spine ends worn, cvr spotting, some foxing to foredge.
(Parmer) **$65 [≈ £40]**

Levy, S.I.
- The Rare Earths: Their Occurrence, Chemistry, and Technology. London: Edward Arnold, 1915. 8vo. xiv,345 pp. 8 ills. Orig cloth.   *(Gemmary)* **$60 [≈ £37]**

Levy, S.J.
- Broken Bridges. A Collection of Short Stories and Plays of Dental Life. Brooklyn: the author, 1924. 1st edn. 8vo. Orig bndg, worn & backstrip edges tearing.
(Xerxes) **$75 [≈ £46]**
- Story of the Allied Dental Council, being the Story of Dentistry in the New York Metropolitan Area. New York: Dental History, 1944. 8vo. 363 pp. Orig bndg.
(Xerxes) **$75 [≈ £46]**

Lewer, S.H. (ed.)
- Canaries, Hybrids and British Birds in Cage and Aviary ... see Robson, J., et al.

Lewes, G.H.
- Sea-Side Studies at Ilfracombe, Tenby, The Scilly Isles, and Jersey. Edinburgh: Blackwood, 1858. 1st edn. 8vo. ix,414,advt pp. 7 plates. Orig cloth gilt, used.
(Egglishaw) **£24 [≈ $38]**
- Studies in Animal Life. London: Smith, Elder, 1862. 1st edn. 8vo. 196,16 ctlg pp. Hand cold frontis, text ills. Orig dec green cloth, crnrs bumped, spine ends sl worn.
(Savona) **£18 [≈ $28]**

Lewis, G.R.
- The Stannaries - A Study of the English Tin Miner. Harvard: 1924. 8vo. 299 pp. Orig bndg.   *(Book House)* **£35 [≈ $56]**

Lewis, Gilbert Newton
- Valence and the Structure of Atoms and Molecules. New York: Chemical Catalog Co, 1923. 1st edn. Orig dark blue cloth gilt, spine lettering rubbed, spine ends sl worn, minor stains & sl wrinkling of cloth on upper bd. Arthur Chapman's copy with pencil notes.
(Sklaroff) **£65 [≈ $104]**

Lewis, William
- A Course of Practical Chemistry ... London: Nourse, 1746. 1st edn. 8vo. [xx],432, [30] pp, advt leaf. 9 plates. Title marg browned. 1 leaf torn, no loss. Orig calf, spine ends sl worn, upper jnt cracked.
(Bickersteth) **£300 [≈ $480]**
- An Experimental History of the Materia Medica, or of the Natural and Artificial Substances made use of in Medicine ... London: H. Baldwin, for the author ..., 1761. 1st edn. 4to. xxii,advt & privilege ff,591pp, [31],errata leaf. V occas sl foxing. Leather, wear to jnts.   *(Elgen)* **$350 [≈ £218]**
- The New Dispensatory: containing I. The Elements of Pharmacy. II. The Materia Medica ... III. The Preparations and Compositions of the new London and Edinburgh Pharmacopoeias ... London: Nourse, 1770. 3rd edn. 8vo. viii, [iv],692 pp. Lacks endpapers. Orig calf.
(Bickersteth) **£85 [≈ $136]**

Leybourn, William
- Dialing ... shewing how to make all such Dials, and to adorn them with all Useful Furniture ... London: 1682. 1st edn. Folio. [20], 76,89-187, [12],189-192,12, 181-226, 273-330 pp. Frontis port, 23 plates, num text ills. Dampstains lower margs. Old calf, rebacked.   *(Rootenberg)* **$1,800 [≈ £1,125]**

Liautard, A.
- Animal Castration. New York: William Jenkins, 1884. 1st edn. 12m. 148 pp. 44 plates. Orig bndg.   *(Xerxes)* **$95 [≈ £59]**

Libby, Willard F.
- Radiocarbon Dating. Chicago: (1952). 1st edn. viii,124 pp. Plates, map. Orig bndg, sl rubbed.   *(Elgen)* **$135 [≈ £84]**

Liebig, Justus von
- Animal Chemistry, or Organic Chemistry in

its Application to Physiology and Pathology. London: Taylor & Walton, 1842. 1st English edn. 8vo. xxiv,354,[2 advt] pp. Orig cloth, sl faded. *(Rootenberg)* **$300 [≃£187]**

- Chemistry in its Application to Agriculture and Physiology ... Edited from the Manuscript of the Author by Lyon Playfair ... Third Edition. London: for Taylor & Walton, 1843. 8vo. Orig green cloth.
(Waterfield's) **£40 [≃$64]**

Life ...
- The Life of Sir Kenelm Digby ... see Longueville, T.

Lilford, Lord
- Coloured Figures of the Birds of the British Islands. London: R.H. Porter, 1885-97. 1st edn. 7 vols. Port, 421 cold plates (64 hand cold) by Thorburn (268) & Keulemans (125). Sl foxing at outer edges of some plates. Contemp gilt dec red half mor.
(Gough) **£2,250 [≃$3,600]**
- Coloured Figures of the Birds of the British Islands. London: 1885-97. 1st edn. 7 vols. Roy 8vo. Port, 421 cold plates by Thorburn & Keulemans. Somewhat foxed, mainly on the text. Half calf gilt, a.e.g. ALS by Lilford inserted.
(Wheldon & Wesley) **£1,850 [≃$2,960]**
- Lord Lilford on Birds. being a Collection of Informal and Unpublished Writings. Edited by A. Trevor-Battye. London: 1903. Cr 4to. xvii,312 pp. 13 plates by Thorburn. Orig cloth, trifle used.
(Wheldon & Wesley) **£20 [≃$32]**

Lilley, A.E.V. & Midgley, W.
- A Book of Studies in Plant Form with some Suggestions for their Application to Design. London: 1895. 8vo. 147 pp. Text figs. Orig cloth, trifle worn.
(Wheldon & Wesley) **£16 [≃$25]**

Lima, E. da Cruz
- Mammals of Amazonia. Vol 1, General Introduction and Primates [all published]. Rio de Janeiro: 1945. One of 975. Cr folio. 274 pp. 42 cold plates. Orig wraps.
(Wheldon & Wesley) **£90 [≃$144]**

Lincoln, E.H.
- Wild Flowers of New England Photographed from Nature. Pittsfield, Mass.: 1911-14. 2nd edn. 16 parts in 8 vols. Atlas 4to. 400 platinum prints, mtd on hand-made paper. Half mor.
(Wheldon & Wesley) **£3,000 [≃$4,800]**

Lind, James
- An Essay on Diseases Incidental to Europeans in Hot Climates ... Third Edition, Enlarged and Improved. London: 1777. 8vo. xvi, 379 pp,4 ff. Lib stamp on title verso. Repr to 1 leaf without loss. Contemp calf, hinges rubbed, paper label on spine.
(Hemlock) **$325 [≃£203]**

Lindemann, F.A.
- The Physical Significance of the Quantum Theory. Oxford: 1932. vii,148 pp. Traces of label removal on endpaper. Orig cloth, water stain on outer edges, 2 jnts beginning to crack. *(Whitehart)* **£38 [≃$60]**

Lindgren, W.
- Mineral Deposits. New York: McGraw-Hill, 1913. 1st edn. 8vo. xv,883 pp. 257 ills. Orig cloth, rubbed. *(Gemmary)* **$35 [≃£21]**
- Mineral Deposits. New York: McGraw-Hill, 1928. 3rd edn. 8vo. xx,1049 pp. 317 ills. Orig cloth. *(Gemmary)* **$40 [≃£25]**
- Mineral Deposits. New York: McGraw-Hill, 1933. 4th edn. 8vo. xvii,930 pp. 333 ills. Orig cloth. *(Gemmary)* **$45 [≃£28]**
- The Tertiary Gravels of the Sierra Nevada of California. Washington, DC: U.S.G.S. Professional Paper 73, 1911. 4to. 226 pp. Pocket map, 36 plates, 16 text figs. Orig wraps. *(Gemmary)* **$80 [≃£50]**

Lindley, George
- A Guide to the Orchard and Kitchen Garden ... Edited by John Lindley. London: for Longman, 1831. 1st edn. 8vo. xxxii,601,[1] pp. Orig cloth backed bds, spine & label sl worn. *(Burmester)* **£45 [≃$72]**

Lindley, John
- Flora Medica; A Botanical Account of all the more Important Plants used in Medicine, in different Parts of the World. London: Longmans ..., 1838. 1st edn. 8vo. xiii,656 pp. Orig cloth gilt, spine faded, recased.
(Hollett) **£38 [≃$60]**
- Flora Medica; a Botanical Account of all the more Important Plants used in Medicine, in different Parts of the World. London: 1838. Sole edn. 8vo. [xvi],655,[1] pp. Orig cloth, gilt lettered spine faded, spine ends sl worn. *(Bickersteth)* **£55 [≃$88]**
- Ladies' Botany, or a Familiar Introduction to the Study of the Natural System of Botany. London: Ridgway, [ca 1835]. 2nd edn. 8vo. xvi, 302 pp. 25 hand cold plates. Occas sl foxing. Cloth, rebacked retaining orig backstrip. *(Egglishaw)* **£50 [≃$80]**

- Ladies Botany: or, a Familiar Introduction to the Study of the Natural System of Botany. Second Edition. London: James Ridgway, [ca 1835]. 2 vols. 8vo. xvi,299; viii,280 pp. 50 hand cold plates. Old green half calf, spine extremities rubbed.
  (Karmiole) **$350 [≈ £218]**
- Ladies Botany or a Familiar Introduction to the Study of the Natural System of Botany. London: 1841. 2nd edn. 8vo. 50 hand cold text figs. Orig cloth. *(Henly)* **£75 [≈ $120]**
- Ladies' Botany or a Familiar Introduction to the Study of the Natural System of Botany. London: [1848]. 5th edn. The original edn, not Bohn's remainder. 2 vols. 8vo. 50 hand cold plates. Minor foxing. Orig cloth, trifle used. *(Wheldon & Wesley)* **£100 [≈ $160]**
- Medical and Oeconomical Botany. Second Edition. London: Bradbury & Evans, 1856. 8vo. 274,16 advt pp. 363 w'engvs in text. Orig green cloth, sl worn. *(Gough)* **£25 [≈ $40]**
- Medical and Oeconomical Botany. London: 1856. 2nd edn. 8vo. 274 pp. 363 figs. Orig cloth, reprd.
  (Wheldon & Wesley) **£45 [≈ $72]**
- The Theory of Horticulture. London: 1840. 8vo. xvi,387 pp. 37 text figs. Sm blind stamp on endpapers & title. Half calf, rubbed.
  (Wheldon & Wesley) **£38 [≈ $60]**
- The Theory and Practice of Horticulture. London: 1855. 8vo. xvi,606 pp. Orig cloth, sl faded, sm tear in jnt. *(Gough)* **£35 [≈ $56]**

Lindley, John & Hutton, William

- The Fossil Flora of Great Britain ... London: 1831-37. 3 vols. 8vo. li,223; xxviii, 208; 207 pp. 230 plates. Calf gilt, spines relaid.
  (Baldwin) **£175 [≈ $280]**

Lindley, John & Moore, T.

- The Treasury of Botany, a Popular Dictionary of the Vegetable Kingdom, with Glossary of Botanical terms. London: 1870. New edn. 2 vols. Post 8vo. 20 plates. Calf gilt.
  (Wheldon & Wesley) **£35 [≈ $56]**
- The Treasury of Botany; a Popular Dictionary of the Vegetable Kingdom, Glossary of Botanical Terms. London: 1884. 4th edn. 2 vols. Cr 8vo. 20 plates, num text figs. Orig cloth, spines relaid.
  (Henly) **£38 [≈ $60]**

Lindsay, W.L.

- Mind in the Lower Animals in Health and Disease. London: 1879. 2 vols. 8vo. Orig cloth, sl used.
  (Wheldon & Wesley) **£45 [≈ $72]**
- A Popular History of British Lichens.

London: Reeve, 1856. 1st edn. Sm 8vo. xxxii, 352 pp. 22 hand cold litho plates. Orig dec cloth. *(Egglishaw)* **£32 [≈ $51]**

Linnaeus, Carl

- A Dissertation on the Sexes of Plants. Translated from the Latin. By James Edward Smith. London: for the author, & sold by George Nicol, 1786. 1st edn in English. 8vo. xv,62 pp, advt leaf. Lacks half-title. Disbound. *(Bickersteth)* **£110 [≈ $176]**
- A General System of Nature ... By W. Turton. London: 1806. 6 vols. 8vo. Port of Linnaeus & 10 plates. Lacks half-titles. Half calf. The complete zoology & botany, but without vol 7 (minerals).
  (Wheldon & Wesley) **£375 [≈ $600]**
- A Generic and Specific Description of British Plants. Translated from the Genera et Species Plantarum, with Notes and Observations, by J. Jenkinson. Kendal: 1775. 8vo. xxviii,[4],258,[9] pp. 5 plates. Title & 1st plate browned. New half calf.
  (Wheldon & Wesley) **£120 [≈ $192]**
- Lachesis Lapponica, or a Tour in Lapland, now first published from the Original Manuscript Journal ... by J.E. Smith. London: 1811. 2 vols. 8vo. 55 text ills. Contemp calf gilt, reprd.
  (Wheldon & Wesley) **£350 [≈ $560]**
- A System of Vegetables .. Translated from the ... Systema Vegetabilium ... and from the Supplementum Plantarum ... By a Botanical Society at Lichfield. Lichfield: 1783. 2 vols. 8vo. Half-titles. 11 plates. Sl soiled, sm ink spot on 1 title. Mod qtr calf.
  (Wheldon & Wesley) **£250 [≈ $400]**

Linnell, E.H.

- The Eye as an Aid in General Diagnosis. Philadelphia: Edwards & Docker, 1898. 1st edn. 8vo. 248 pp. 4 cold plates. Text ills. Orig bndg. *(Xerxes)* **$65 [≈ £40]**

Linton, E.F.

- Flora of Bournemouth, including the Island of Purbeck. London: [1900]. Cr 8vo. ix,290 pp. Map. Orig cloth, back cvr trifle stained.
  (Wheldon & Wesley) **£20 [≈ $32]**

Lipson, E.

- The History of the Woollen and Worsted Industries. London: Black, 1921. 8vo. 273 pp. Fldg frontis, ills. Some pencil underlining. Orig bndg.
  (Book House) **£30 [≈ $48]**

Lisle, Edward
- Observations in Husbandry. London: J. Hughs, 1757. 4to. Port frontis. Errata leaf. Contemp calf, jnts cracked, crnrs worn, new label. *(Stewart)* **£175 [≈ $280]**
- Observations in Husbandry ... Second Edition ... London: J. Hughs ..., 1757. 2 vols. 8vo. xxii,[23]-398,[2]; 406,[2] pp. Fldg port frontis. Contemp sprinkled calf, gilt spines, mor labels, red sprinkled edges, fine. Sir James Colquhoun of Luss b'plate.
  (Finch) **£450 [≈ $720]**

Lisney, A.
- Bibliography of British Lepidoptera 1608 to 1799. London: Chiswick Press, privately ptd for the author, 1960. One of 500. Roy 8vo. 320 pp. Frontis, 39 ills. Buckram.
  (Wheldon & Wesley) **£60 [≈ $96]**

Livestock of the Farm ...
- See Jones, C. Bryner (ed.).

Lloyd, H. Alan
- Some Outstanding Clocks over Seven Hundred Years, 1250-1950. London: Leonard Hill, 1958. 1st edn. [xx],160 pp. 173 plates, num text figs. Orig russet cloth gilt. Dw. *(Sklaroff)* **£38 [≈ $60]**

Lobb, Theophilus
- A Practical Treatise of Painful Distempers, with some effectual Methods of curing them, exemplified in a great Variety of suitable Histories. London: for James Buckland, 1739. 1st edn. 8vo. xxx,[ii], 320, [xiv] pp. Orig calf. *(Bickersteth)* **£165 [≈ $264]**

Lock, Charles G. Warnford (ed.)
- Spon's Encyclopaedia of the Industrial Arts, Manufactures and Commercial Products. London: Spon, 1882. 5 vols. 4to. Num ills. Orig cloth. *(Charles B. Wood)* **£250 [≈ $156]**

Locke, John
- A Collection of Several Pieces never before printed, or not extant in his Works ... London: J. Bettenham for R. Francklin, 1720. 1st edn, 1st iss. 8vo. [xxxv], xxiv, [ii], 362,[xviii] pp, 1 page of errata, 3 advt pp. Plate at p 187. Orig calf, sl worn.
  (Bickersteth) **£450 [≈ $720]**

Lockley, R.M.
- Shearwaters. London: (1942). 8vo. xi,238 pp. Ills. Orig cloth.
  (Wheldon & Wesley) **£18 [≈ $28]**

Lockyer, J. Norman
- Meteoric Hypothesis. A Statement of a Spectroscopic Inquiry into the Origin of Cosmical Systems. London: Macmillan, 1890. 1st edn. xvi,560,59 advt pp. 7 plates, 101 text w'cuts. Orig green cloth, inner hinges starting. *(Karmiole)* **$75 [≈ £46]**
- Studies in Spectrum Analysis. London: 1878. 1st edn. Sm 8vo. xii,258 pp. Tinted frontis, 7 plates (1 cold, 5 tinted), 51 text ills. Prize tree calf gilt. *(Elgen)* **$100 [≈ £62]**
- Studies in Spectrum Analysis. London: Intl Scientific Series, 1886. 4th edn. xii,258 pp. 8 plates, 51 text ills. Orig cloth, spine sl worn at hd. *(Whitehart)* **£15 [≈ $24]**

Lodge, R.B.
- Bird-Hunting through Wild Europe. London: Robert Culley, (1908). 8vo. 333,[2 advt] pp. 124 photo ills. Orig gilt dec cloth, t.e.g. *(Blackwell's)* **£40 [≈ $64]**

Loeb, Jacques
- Comparative Physiology of the Brain and Comparative Psychology. New York: Putnam's, 1900. 1st edn. 8vo. [iv],[xii],309,[3] pp. Orig cloth. *(Gach)* **$85 [≈ £53]**

Loftus, William
- The Brewer. A Familiar Treatise on the Art of Brewing ... London: Loftus, 1857. New edn. 8vo. iv,192 pp. Orig cloth, sl spotted.
  (Young's) **£60 [≈ $96]**

Lomax, Montagu
- The Experiences of an Asylum Doctor with Suggestions for Asylum and Lunacy Law Reform. London: Allen & Unwin, [1921]. 1st edn. 8vo. 255, [1] pp. Orig black cloth, sl musty. *(Gach)* **$35 [≈ £21]**

Longfield, C.
- The Dragonflies of the British Isles. London: Warne, Wayside and Woodland Series, 1949. 2nd edn. Cr 8vo. 256 pp. 16 cold plates, 42 other plates, text figs. Orig cloth.
  (Egglishaw) **£24 [≈ $38]**

Longhurst, Henry
- Adventure in Oil. The Story of British Petroleum. With a Foreword by The Rt. Hon. Sir Winston Churchill. London: Sidgwick & Jackson, 1959. 1st edn. 286 pp. 49 plates inc frontis, map endpapers. Orig green cloth. Dw sl worn. *(Duck)* **£15 [≈ $24]**

Longridge, C.C.
- Gold & Tin Dredging & Mechanical

Excavators. London: The Mining Journal, [ca 1912]. 3rd edn. Roy 8vo. xvi,425 pp. Num plates & diags, some fldg. Orig cloth.
(Gemmary) **$100 [≈ £62]**

Longstaff, G.B.
- Butterfly Hunting in Many Lands. Notes of a Field Naturalist. London: 1912. 8vo. xx,729 pp. 16 plates (7 cold). Sl foxing. Orig cloth, trifle used. *(Wheldon & Wesley)* **£50 [≈ $80]**

Longueville, T.
- The Life of Sir Kenelm Digby. By One of His Descendants. London: Longmans, Green, 1896. 1st edn. 8vo. xiii,310,24 ctlg pp. Frontis, 6 plates. Orig cloth.
(Elgen) **$95 [≈ £59]**

Lorenz, Konrad
- Evolution and Modification of Behavior. Chicago & London: Univ of Chicago Press, [1965]. 1st edn, 1st printing. Thin 8vo. [vi], [122] pp. Orig green bds. Author's pres inscrptn. *(Gach)* **$85 [≈ £53]**

Loudon, Jane
- Instructions in Gardening for Ladies. London: 1840. 1st edn. 12mo. xii,406 pp. Engvd title, 16 text ills. Minor foxing. Orig cloth, rebacked preserving most of orig spine.
(Wheldon & Wesley) **£35 [≈ $56]**
- The Ladies' Companion to the Flower Garden ... Third Edition. London: 1844. Sm 8vo. viii, 346 pp. Intl advt leaf. Cold frontis, 70 ills. Orig cloth gilt, headband just a little worn. *(Fenning)* **£38.50 [≈ $62]**
- The Ladies' Companion to the Flower Garden. London: 1849. 5th edn. Post 8vo. viii, 351 pp. Cold frontis. Orig cloth, a.e.g., sl worn, jnts beginning to split.
(Wheldon & Wesley) **£30 [≈ $48]**

Loudon, John Claudius
- Arboretum et Fruticetum Britannicum; or, the Trees and Shrubs of Britain delineated and described. London: 1838. 1st edn. 8 vols. 8vo. 412 plates, 2546 text figs. Sl foxing. Orig cloth, 3 vols reprd.
(Wheldon & Wesley) **£150 [≈ $240]**
- Arboretum et Fruticetum Britannicum; or, the Trees and Shrubs of Britain, Native and Foreign, Hardy and Half-Hardy. London: 1844. 2nd edn. 8 vols. 8vo. 412 plates, correct. Sl foxing. Orig cloth.
(Wheldon & Wesley) **£150 [≈ $240]**
- An Encyclopedia of Agriculture ... London: Longman ... & sold by Carey & Lea, Philadelphia, 1826. Thick 8vo. xvi,1226,[ii]

pp. 823 w'engvd ills. Orig half mor, sl scuffed.   *(Charles B. Wood)* **$200 [≈ £125]**
- An Encyclopaedia of Agriculture ... London: Longman ..., 1831. 2nd edn. Thick 8vo. xl, 1282 pp. Over 1100 ills. Mod half levant mor gilt. *(Hollett)* **£85 [≈ $136]**
- An Encyclopaedia of Agriculture. London: Longman ..., 1844. 5th edn. Lge 8vo. 1375,32 ctlg dated Oct 1847 pp. Over 1200 ills. Orig cloth, extremities sl worn, lib b'plate, sm cloth stamp on spine. *(Schoyer)* **$100 [≈ £62]**
- An Encyclopaedia of Gardening ... New Edition considerably Improved and Enlarged. London: [1835]. 8vo. 1271 pp. 981 ills. New buckram. *(Wheldon & Wesley)* **£60 [≈ $96]**
- An Encyclopaedia of Gardening ... New Edition. Edited by Mrs Loudon. London: Longman, 1860. Thick 8vo. xl,1278,40 advt pp. Ills. Orig green cloth, sm reprs to jnts.
(Spelman) **£75 [≈ $120]**
- An Encyclopaedia of Plants ... London: Longman ..., 1829. 1st edn. Thick roy 8vo. [16 advt],xviii,1159 pp. Num ills. Orig vellum backed mrbld bds, rather worn, front jnt broken.   *(Hollett)* **£60 [≈ $96]**
- Encyclopaedia of Plants. London: 1855. New impression. 8vo. xxii,1574 pp. Over 2000 text figs. New cloth.
(Wheldon & Wesley) **£60 [≈ $96]**
- The Green-House Companion; comprising a general Course of Green-House and Conservatory Practice throughout the Year ... London: Whitaker, Treacher & Co, 1832. 3rd edn. xii,408 pp. Hand cold frontis. Orig binder's cloth, a little worn.
(Hollett) **£35 [≈ $56]**
- Observations on the Formation and Management of Useful and Ornamental Plantations ... Landscape Gardening ... Edinburgh: Constable; London: Longman ..., 1804. 1st edn. 8vo. 342,[18] pp. 10 plates. rec calf antique.
(Charles B. Wood) **$950 [≈ £593]**
- Self-Instruction for Young Gardeners, Foresters, Bailiffs, Land-Stewards, and Farmers, in Arithmetic etc. ... With a Memoir of the Author. London: 1845. 8vo. liii,240 pp. Port, text figs. Orig cloth.
(Wheldon & Wesley) **£50 [≈ $80]**
- Trees and Shrubs: an Abridgment of the Arboretum et Fruticetum Britannicum. London: Warne, 1875. 8vo. lxxii,1162 pp. Over 2000 text figs. New cloth, old calf backstrip laid down. *(Egglishaw)* **£45 [≈ $72]**

Love, John
- Geodaesia: or, the Art of Surveying and Measuring Land Made Easy ... London: for

Innys & Richardson, 1753. 6th edn. 8vo. [20], 196, [16], 4,[36],8 pp. 4 full-page ills, tables, diags. Contemp leather, sl worn, cvrs re-attached with rice-paper hinges. Cloth case.
(Schoyer) **$110 [≈ £68]**

Lovell, Archibald
- A Summary of Material Heads Which may be Enlarged and Improved into a Compleat Answer to Dr. Burnet's Theory of the Earth ... by a Pensioner of the Charter-House. London: by T.B. ..., 1696. Only edn. 4to. [iv],22 pp. Sl browned. Mod qtr calf. Wing L.3242. *(Finch)* **£175 [≈ $280]**

Lovett, William
- Elementary Anatomy and Physiology, for Schools and Private Instruction ... London: Simpkin, Marshall, 1853. 2nd edn. xxiv,159, [i] pp. 10 cold plates. Orig cloth gilt, sl bumped & faded at edges, spine ends sl defective. *(Hollett)* **£55 [≈ $88]**

Lowe, E.J.
- Beautiful Leaved Plants ... London: Groombridge, 1861. Tall 8vo. ii,144 pp. 60 cold plates. A little spotting. Orig half calf gilt, rebacked. *(Hollett)* **£85 [≈ $136]**
- Beautiful Leaved Plants ... see also under Lowe, E.J. & Howard, W.
- The Coming Drought: Or, the Cycle of the Seasons, with a Chronological History of all the Droughts and Frosts ... London: Bemrose ..., [1879]. 1st edn. 8vo. 31 pp. Ptd wraps, trifle soiled. *(Young's)* **£30 [≈ $48]**
- Ferns: British and Exotic. London: 1872. 8 vols. 479 cold plates. [With] A Natural History of New and Rare Ferns ... London: 1871. 72 cold plates. Together 9 vols. Some v limited foxing. Orig cloth.
(Wheldon & Wesley) **£250 [≈ $400]**
- A Natural History of British Grasses. London: Groombridge, 1858. Roy 8vo. [vi],245 pp. 74 cold engvs. Orig green cloth gilt, faded, sl soiled. *(Blackwell's)* **£40 [≈ $64]**
- A Natural History of British Grasses. London: 1858. 1st edn. Roy 8vo. vii,245 pp. 74 cold plates. 1st 3 ff foxed, sl foxing elsewhere. Contemp half mor.
(Wheldon & Wesley) **£35 [≈ $56]**
- A Natural History of British Grasses. London: 1858. 1st edn. Roy 8vo. 74 cold plates. 10 plates sl foxed. Orig green cloth, spine faded. *(Henly)* **£85 [≈ $136]**
- A Natural History of New and Rare Ferns, none of which are included in ... "Ferns, British and Exotic". London: Groombridge, 1862. Roy 8vo. viii,192 pp. 72 cold plates,

text w'cuts. Half mor, sl rubbed.
(Egglishaw) **£50 [≈ $80]**
- A Natural History of New and Rare Ferns: containing Species and Varieties, none of which are included in any of the Eight Volumes of "Ferns, British and Exotic". London: 1865. Roy 8vo. viii,192 pp. 72 cold plates. Orig cloth.
(Wheldon & Wesley) **£40 [≈ $64]**
- Our Native Ferns, or a History of the British Species and their Varieties. London: 1865-67. 2 vols. Roy 8vo. 79 [sic] cold plates, 909 w'cuts. Orig cloth, sl used.
(Wheldon & Wesley) **£35 [≈ $56]**
- Our Native Ferns: or a History of the British Species and their Varieties. London: 1867-69. 2 vols. Lge 8vo. 348; 492 pp. 76 [sic] tissued cold plates, num other ills. Orig pict green cloth gilt. *(Terramedia)* **$150 [≈ £93]**

Lowe, E.J. & Howard, W.
- Beautiful Leaved Plants. London: 1864. Roy 8vo. 60 cold plates finished by hand. Sl foxing affecting text & 2 plates. Orig cloth, spine faded. *(Henly)* **£85 [≈ $136]**
- Beautiful Leaved Plants ... London: Nimmo, 1891. Roy 8vo. viii,144 pp. 60 cold plates, num text ills. Orig cloth, spine ends worn.
(Berkelouw) **$300 [≈ £187]**

Lowe, F.A.
- The Heron. London: New Naturalist, 1954. 1st edn. 8vo. xiii,177 pp. Cold frontis, 28 ills. Orig cloth. *(Wheldon & Wesley)* **£60 [≈ $96]**

Lowe, John
- The Yew-Trees of Great Britain and Ireland. London: Macmillan, 1897. 1st edn. xiv,270 pp. 23 photo plates, 16 text ills. Orig green cloth gilt, uncut, endpapers browned & spotty. *(Duck)* **£40 [≈ $64]**

Lowne, B.T.
- The Anatomy and Physiology of the Blow-Fly (Musca Vomitoria, Linn.). London: 1870. 8vo. viii,121 pp. 10 plates (some cold). Lib b'plate, no lib stamps. Orig cloth, trifle used.
(Wheldon & Wesley) **£35 [≈ $56]**
- The Anatomy, Physiology, Morphology and Development of the Blow-Fly ... London: for the author by R.H. Porter, 1890-95. 2 vols. 8vo. xii,viii,778 pp. 52 plates, 108 text figs. Cloth. *(Egglishaw)* **£50 [≈ $80]**

Lozano, Pedro
- A True and Particular Relation of the Dreadful Earthquake which happened at Lima ... Description ... of ... Peru ... London:

Osborne, 1748. 1st English edn. xxiii,341,3 advt pp. 9 fldg maps & plates. Occas browning & perf lib marks. 19th c half mor, rehinged.   *(Young's)* **£190 [≈ $304]**

Lubbock, Sir John, Baron Avebury
- Monograph of the Collembola and Thysanura. London: Ray Society, 1873. 8vo. x,276 pp. 78 plates (31 hand cold). Some foxing as usual. Orig cloth, jnts sl worn.
          (Egglishaw) **£68 [≈ $108]**
- The Scenery of England and the Causes to which it is due. London: 1902. 1st edn. 8vo. xxvi,534 pp. Fldg map, frontis, 197 text ills. Orig cloth gilt.   *(Henly)* **£18 [≈ $28]**

Lubbock, Richard
- Observations on the Fauna of Norfolk. Norwich: Charles Muskett, 1845. 1st edn. 8vo. viii,156,4 advt pp. Fldg map, 2 lithos (sl foxed). Orig green cloth. *(Gough)* **£40 [≈ $64]**
- Observations on the Fauna of Norfolk, and more particularly on the District of the Broads. Norwich: 1848. 8vo. viii,156 pp. Map (rather foxed), 2 plates. Orig cloth.
          (Wheldon & Wesley) **£50 [≈ $80]**

Lucas, W.J.
- The Book of British Hawk Moths. London: Upcott Gill, 1895. Sm 8vo. x,157 pp. Frontis, 14 plates, text ills. Orig dec cloth gilt.
          (Egglishaw) **£18 [≈ $28]**
- British Dragonflies (Odonata). London: Upcott Gill, 1900. 1st edn. 27 cold plates, 57 figs. Orig dec buckram gilt.
          (Hollett) **£140 [≈ $224]**
- British Dragonflies (Odonata). London: Upcott Gill, 1900. xiv,356 pp. 27 cold plates, 57 text figs. Orig dec buckram, jnts weak.
          (Egglishaw) **£60 [≈ $96]**

Lufkin, Arthur Ward
- A History of Dentistry. Philadelphia: [1938]. Sm 8vo. 255 pp. Ills. Orig cloth.
          (Elgen) **$30 [≈ £18]**

Lunge, G.
- Coal-Tar and Ammonia. London: Gurney & Jackson, 1916. 5th enlgd edn. 3 vols. 1718 pp. Ills. Orig bndg. *(Book House)* **£45 [≈ $72]**

Lupton, A.
- A Practical Treatise on Mine Surveying. London: Longmans, Green, 1902. 8vo. vii,414 pp. 209 text figs. Orig cloth.
          (Gemmary) **$60 [≈ £37]**

Lupton, Thomas
- A Thousand Notable Things of Sundry Sorts, enlarged ... London: Wotton & Conyers, 1686. 8vo. [vi],301,[303],[21] pp. Browned, occas spotting. A few marg reprs. A few margs frayed. Contemp sheep, spine ends sometime reprd, crnrs worn. Wing L.3501.
          (Finch) **£120 [≈ $192]**

Lurie, E.
- Louis Agassiz. A Life in Science. London: 1960. 8vo. xiv,449 pp. Ills. Orig cloth. Dw.
          (Baldwin) **£33 [≈ $52]**

Lydekker, Richard
- Catalogue of the Heads and Horns of Indian Big Game bequeathed by A.O. Hume. London: British Museum, 1913. Port. Orig bndg, trifle used.
          (Wheldon & Wesley) **£20 [≈ $32]**
- The Game Animals of Africa. London: Rowland Ward, 1908. 484 pp. Ills. Orig bndg.   *(Trophy Room Books)* **$325 [≈ £203]**
- The Game Animals of Africa. London: 1926. 2nd edn. 8vo. Ills. Orig cloth, v sl marked.
          (Grayling) **£150 [≈ $240]**
- The Game Animals of Africa. London: 1926. 8vo. Ills. Some foxing & dampstaining to edges. Orig bndg, rubbed, hinges sprung.
          (Grayling) **£50 [≈ $80]**
- The Game Animals of India, Burma, Malaya and Tibet. Second Edition, revised by J.G. Dollman. London: 1924. 8vo. ills. Orig bndg, v sl bumped.   *(Grayling)* **£110 [≈ $176]**
- The Game Animals of India, Burma, Malaya and Tibet. Second Edition, revised by J.G. Dollman. London: 1924. 8vo. ills. Sl affected by damp. Orig buckram, v faded.
          (Grayling) **£55 [≈ $88]**
- A Hand-Book to the British Mammalia. London: Lloyd's Natural History, 1896. 32 cold plates. Orig cloth.
          (Wheldon & Wesley) **£15 [≈ $24]**
- A Hand-Book to the British Mammalia. London: Lloyd's Natural History, 1896. 8vo. xii,339 pp. 32 cold plates. Cloth.
          (Egglishaw) **£12 [≈ $19]**
- A Hand-Book to the Carnivora. Part I. Cats, Civets, and Mungooses. London: Lloyd's Natural History, 1896. 1st edn. Cr 8vo. viii, 312 pp. 32 cold plates. Orig cloth, a little wear to spine.   *(Fenning)* **£28.50 [≈ $46]**
- A Handbook to the Marsupialia and Monotremata. London: Allen's Naturalist's Library, 1894. 8vo. xvi,302 pp. 38 cold plates. Qtr blue cloth & mrbld bds.
          (Egglishaw) **£28 [≈ $44]**

- A Handbook to the Marsupialia and Monotremata. London: Lloyd's Natural History, 1896. Reprint of the 1894 edn in Allen's Natural History with the same plates, plus an appendix. Cr 8vo. xvi,320 pp. 38 cold plates. Orig cloth.
(Wheldon & Wesley) **£30 [≈ $48]**
- Mostly Mammals. Zoological Essays. London: 1903. 8vo. Ills. Orig bndg.
(Grayling) **£25 [≈ $40]**
- Mostly Mammals, Zoological Essays. London: 1903. 8vo. ix,383 pp. 16 plates. Orig cloth. *(Wheldon & Wesley)* **£20 [≈ $32]**
- The Sportsman's Bird Book. London: Rowland Ward, 1908. 1st edn. Lge 8vo. xviii,620 pp. Num ills. Orig red pict cloth gilt, sl sunned. *(Hollett)* **£40 [≈ $64]**
- The Sportsman's British Bird Book. London: Rowland Ward, 1908. 620 pp. Num ills. Orig bndg. *(Trophy Room Books)* **$175 [≈ £109]**

Lydekker, Richard (ed.)
- The Royal Natural History. London: 1893-96. 6 vols. Roy 8vo. 72 cold plates, 1600 engvs. Some (mostly sl) foxing. Plain buckram. *(Wheldon & Wesley)* **£80 [≈ $128]**
- Royal Natural History. London: 1893-96. 6 vols. Imperial 8vo. 72 cold plates, 1600 text figs. Sl foxing. Contemp half mor, trifle rubbed. *(Wheldon & Wesley)* **£80 [≈ $128]**

Lydekker, Richard & Blaine, G.
- Catalogue of the Ungulate Mammals in the British Museum. London: 1913-16. 5 vols. 8vo. 225 text figs. Lib stamps on vols 1-3 titles. Orig cloth. *(Egglishaw)* **£80 [≈ $128]**

Lydston, G. Frank
- The Diseases of Society (The Vice and Crime Problem). Philadelphia: Lippincott, 1904. 1st edn. 8vo. 623 pp. Orig bndg.
(Xerxes) **$85 [≈ £53]**

Lyell, Sir Charles
- Elements of Geology, or the Ancient Changes of the Earth and its Inhabitants as illustrated by Geological Monuments. London: 1865. 6th edn. 8vo. xvi,794 pp. 769 text figs. Rather used ex-lib. Orig cloth, worn.
(Wheldon & Wesley) **£30 [≈ $48]**
- The Geological Evidences of the Antiquity of Man. London: 1863. 1st edn. 8vo. x,518 pp. 2 plates, 58 w'cuts. Sm stain in crnr of plates. New buckram. *(Baldwin)* **£60 [≈ $96]**
- The Geological Evidences of the Antiquity of Man. Philadelphia: 1863. 1st Amer edn. 8vo. x,518 pp. 2 plates, 58 text figs. Orig cloth gilt, spine ends sl worn. *(Henly)* **£135 [≈ $216]**

- The Geological Evidence of the Antiquity of Man. London: 1873. 4th edn, rvsd. 8vo. xix, 572 pp. 2 plates, 56 text figs. Orig cloth, sl used. *(Wheldon & Wesley)* **£60 [≈ $96]**
- A Manual of Elementary Geology. London: 1852. 4th edn. 8vo. xxxi,512 pp. 500 w'cuts. Some pencil notes, outer upper crnr of 1st few ff sl waterstained. New cloth.
(Wheldon & Wesley) **£40 [≈ $64]**
- A Manual of Elementary Geology ... Fifth Edition, Greatly Enlarged ... London: Murray, 1855. xvi,655,[1] pp. Frontis, 750 ills. Contemp calf gilt, elab gilt spine, minor scuffing to sides. *(Duck)* **£135 [≈ $216]**
- Principles of Geology ... London: 1830-33. 1st edn. 3 vols. 8vo. 11 plates & maps (4 cold). Half-title in vols 1 & 3 only, correct. Mod half calf gilt.
(Wheldon & Wesley) **£950 [≈ $1,520]**
- Principles of Geology ... London: 1830-32-33. 1st edn. 3 vols. 8vo. 3 frontis (2 cold), 3 maps (2 fldg), 8 engvd plates. Rec bds, mostly uncut.
(Rootenberg) **$1,800 [≈ £1,125]**
- Principles of Geology ... Third [sic] Edition. London: Murray, 1834. 4 vols. Tall cr 8vo. xxx,[ii],420; [ii],453; [ii],426; [ii],393,[3 advt] pp. 14 plates, inc vol 1 frontis. Minor spotting. Contemp polished calf gilt, by A. Banks, Edinburgh, v sl worn.
(Duck) **£375 [≈ $600]**
- Principles of Geology ... London: 1867. 10th edn. 2 vols. Stamp on titles. Orig cloth.
(Baldwin) **£60 [≈ $96]**
- The Student's Elements of Geology. London: 1874. 2nd edn. 12mo. Frontis, 645 text figs. Orig cloth gilt. *(Henly)* **£25 [≈ $40]**
- The Student's Elements of Geology. London: 1874. 2nd edn. 12mo. xix,672 pp. 645 text figs. Orig cloth gilt. *(Henly)* **£15 [≈ $24]**

Lyman, R.S., et al.
- Social and Psychological Studies in Neuropsychiatry in China. Peking: Henri Vetch, 1939. 8vo. [16],382 pp. Over 30 plates, sev fldg charts. Orig green cloth, sl soiled, label removed from spine.
(Karmiole) **$85 [≈ £53]**

Lynn, Walter
- Niktalopsia: or, the Use and Abuse of Snuffers. London: A. More, 1726. 1st edn. 1st iss (?), with an incorrect spelling of the title. 8vo. [2],28 pp. 2 w'cut ills. Title & last page soiled. Num authorial (?) MS crrctns. Mrbld bds. *(Rootenberg)* **$950 [≈ £593]**

Lyon, P.
- Observations on the Barrenness of Fruit Trees, and of the means of Preventions and Cure. Edinburgh: C. Stewart, for William Blackwood ..., 1813. 1st edn. 8vo. [iv],80, [16 advt] pp. Frontis. Orig bds, ptd label.
(Burmester) **£110 [≈ $176]**

Lysachy, A.M.
- Joseph Banks in Newfoundland and Labrador, 1766. His Diary, Manuscripts and Collections. London: Faber, 1971. 1st edn. 4to. 512 pp. 12 cold plates, 9 facs, 105 monochrome ills & maps, 6 text figs. Orig cloth gilt. Dw.   *(Hollett)* **£60 [≈ $96]**

M'Adam, John Loudon
- Remarks on the Present System of Road Making; with Observations ... Bristol: J.M. Gutch, 1816. 1st edn. 8vo. 32 pp. Title sl foxed. Mod half calf.
(Charles B. Wood) **$1,500 [≈ £937]**
- Remarks on the Present System of Road Making; with Observations ... Third Edition, Carefully Revised, with Considerable Additions, and an Appendix. London: Longman ..., 1820. 8vo. Minor repr to title, occas sl foxing. Contemp half calf, gilt spine.
(Clark) **£75 [≈ $120]**

M'Alpine, Daniel
- Biological Atlas. A Guide to the Practical Study of Plants and Animals. London: 1880. 1st edn. 4to. ix,49,4 pp. 423 cold figs on 24 plates. Sm lib stamp on title. Orig dec green cloth.   *(Henly)* **£50 [≈ $80]**
- The Botanical Atlas. A Guide to the Practical Study of Plants. Edinburgh: 1883. 2 vols in one. Sm folio. Cold frontis, 52 cold plates. New cloth. *(Wheldon & Wesley)* **£45 [≈ $72]**
- The Botanical Atlas ... Edinburgh: W. & A.K. Johnston, 1883. Folio. Cold & ptd titles, 52 cold plates. Orig cloth gilt, spine ends strengthened with tape, inner jnts strengthened.   *(Bickersteth)* **£35 [≈ $56]**
- The Botanical Atlas ... Edinburgh: 1883. 2 vols. Imperial 4to. Addtnl cold titles, 52 cold plates. Sl foxing at ends. Orig dec cloth gilt.
(Henly) **£85 [≈ $136]**

Macaulay, J. (ed.)
- Modern Railway Working. A Practical Treatise by Engineering and Administrative Assistants. London: Gresham, 1912. 8 vols. 4to. Ills. Orig dec bndg.
(Book House) **£40 [≈ $64]**

MacCallum, W.G.
- William Stewart Halsted, Surgeon. Baltimore: 1930. 1st edn. 8vo. 241 pp. Port, 13 plates, fldg chart. Orig cloth.
(Elgen) **$75 [≈ £46]**

McCarrison, Sir Robert
- Nutrition and National Health being the Cantor Lectures delivered before the Royal Society of Arts, 1936. London: 1944. 8vo. 75 pp. 3 plates, text figs. Orig cloth.
(Bickersteth) **£18 [≈ $28]**

MacCulloch, John
- Remarks on the Art of Making Wine ... London: Longman ..., 1816. 1st edn. 12mo. [2], vi,261,[1] pp. Rec half calf, mrbld bds.
(Rootenberg) **£550 [≈ $343]**

McCulloch, John Ramsay
- A Dictionary, Practical, Theoretical, and Historical, of Commerce and Commercial Navigation ... Edited by Henry Vethake. Philadelphia: 1840. 2 vols. xi,[1],767; 803 pp. Foxed, 2 sgntrs starting. Orig cloth, edgeworn.   *(Reese)* **$200 [≈ £125]**
- The Principles of Political Economy ... Second Edition, corrected and greatly enlarged. London, Edinburgh & Dublin: 1830. 2nd edn. 8vo. [iii]-xii,563 pp. Lacks half-title. Lib label on pastedown. Half calf over old mrbld bds, rebacked & recrnrd.
(Pickering) **$500 [≈ £312]**

McDonald, A.
- A Complete Dictionary of Practical Gardening: comprehending all the Modern Improvements in the Art. London: 1807. 2 vols. 4to. 74 plain plates (61 by Sydenham Edwards). Vol 1 title creased. Some foxing & sl waterstaining. London: 1807.
(Wheldon & Wesley) **£200 [≈ $320]**

McDonald, D.
- A History of Platinum. London: Johnson, Matthey, 1960. Cr 4to. [10],254 pp. 71 ills. Orig cloth. Dw.   *(Gemmary)* **$75 [≈ £46]**

Macdonald, John Denis
- Sound & Colour, their Relations, Analogies & Harmonies. Gosport: 1869. 1st edn. 8vo. [6], 86 pp, imprint leaf. 2 plates, 16 diags in text (13 hand cold), 4 tables (1 hand cold). Orig gilt lettered purple cloth.
(Spelman) **£120 [≈ $192]**
- Sound & Colour, their Relations, Analogies and Harmonies. London: Longman, Green ... 1869. 1st edn. 8vo. [viii],86,[ii] pp. 2 plates,

16 text diags (15 hand cold). Orig cloth, somewhat worn.
(Charles B. Wood) **$285 [≈£178]**

Macer Floridus
- Macer Floridus. De Viribus Herbarum. A Middle English Translation. Edited by G. Frisk. Upsala: 1949. 8vo. 338 pp. Cloth.
(Wheldon & Wesley) **£35 [≈$56]**

M'Farlane, P.
- Antidote against the Unscriptural and Unscientific Tendency of Modern Geology ... London: 1871. 1st edn. Sm 8vo. xii,362,[2] pp. Some random spots, a few pencil notes. Orig cloth. *(Bow Windows)* **£35 [≈$56]**

Macgillivray, W.
- Descriptions of the Rapacious Birds of Great Britain. Edinburgh: 1836. Sm 8vo. vii, [i],482 pp. 2 plates, 21 text figs. Orig cloth, worn & reprd. *(Wheldon & Wesley)* **£35 [≈$56]**

McIlwraith, T.
- The Birds of Ontario with a Description of their Nests and Eggs. Toronto: 1894. 2nd edn, enlgd & rvsd. 8vo. 426 pp. Port, text figs. Orig cloth.
(Wheldon & Wesley) **£35 [≈$56]**

M'Intosh, Charles
- The New and Improved Practical Gardener, and Horticulturist ... London: Thomas Kelly ..., 1851. 8vo. iv,15,972 pp. Port frontis, 10 cold plates (dated 1838-39), num text ills. Contemp half calf. *(Young's)* **£60 [≈$96]**

McIver, J.R.
- Gems, Minerals, and Rocks. Johannesburg: Horters Printers, 1966. Roy 8vo. 268 pp. Num cold & b/w photos. Orig cloth. Dw.
(Gemmary) **$75 [≈£46]**

Mackenzie, Colin
- Five Thousand Receipts in all the Useful and Domestic Arts ... A New Edition. London: for Sir Richard Phillips & Co, 1829. Sq 12mo. Contemp calf, rebacked.
(Sanders) **£85 [≈$136]**
- One Thousand Processes in Manufactures and Experiments in Chemistry ... Fourth Edition. London: for Sir Richard Phillips, 1823. 8vo. iv, 29,xxxii,646 pp. Fldg cold plate, 19 engvd & 2 w'cut plates, text figs. Orig half calf, rubbed, upper jnt cracked.
(Bickersteth) **£125 [≈$200]**

Mackenzie, J.S.E.
- British Orchids: How to Tell One from Another. London: Unwin Bros, [1918]. Lge 4to. 40 pp. 10 tissued cold ills laid down on art paper. Orig cloth, buckram spine, cvr faded, inner hinge tender.
(de Beaumont) **£68 [≈$108]**

Mackenzie, James
- Essays: on Retirement from Business; on Old Age; and on Employment of the Soul after Death ... by a Physician. Fourth Edition. London: Rivington, 1812. 8vo. xii,180 pp. Contemp calf, mor label, hinges cracked, endpapers spotted. *(Claude Cox)* **£18 [≈$28]**
- The History of Health, and the Art of preserving it ... Edinburgh: William Gordon ..., 1759. 2nd edn. 8vo. xii,436 pp. Sl brown stain on title. Contemp calf, rebacked.
(Burmester) **£80 [≈$128]**

Mackenzie, Sir James
- The Future of Medicine. London: 1919. 1st edn. 238 pp. Orig bndg, spine ends sl frayed, few spots on cvr. *(Elgen)* **£35 [≈£21]**
- Principles of Diagnosis and Treatment in Heart Affections. London: 1916. 1st edn. 264 pp. Ills. Orig bndg. *(Elgen)* **£100 [≈£62]**
- Principles of Diagnosis and Treatment in Heart Affections. London: 1917. 1st edn, 3rd imp. 264 pp. Ills. Orig bndg.
(Elgen) **£50 [≈£31]**

Mackenzie, William
- A Practical Treatise on Diseases of the Eye. Boston: 1833. 1st Amer edn. Tall 8vo. Half-title,xii,719 pp. Orig cloth backed bds, slit in centre of spine, jnts starting, sl soiled.
(Elgen) **$325 [≈£203]**
- A Practical Treatise on Diseases of the Eye. London: 1840. 3rd edn. Sm rubber stamps on title. Upper hinge cracked, some soil & wear on cvr. *(Rittenhouse)* **$200 [≈£125]**

MacKinney, Loren
- Medical Illustrations in Medieval Manuscripts. Berkely: Univ of Calif Press, 1965. Lge 8vo. xviii,264 pp. 104 ills, 18 cold. Orig red cloth. Dw sl soiled.
(Karmiole) **$75 [≈£46]**

Mackintosh, L.J.
- Birds of Darjeeling and India. Calcutta: Banerjee, 1915. 1st edn. 8vo. 233,lxviii pp. Contemp blue calf gilt. *(Gough)* **£28 [≈$44]**

Maclaurin, Colin
- An Account of Sir Isaac Newton's

Philosophical Discoveries, in Four Books ... London: 1748. 1st edn. 4to. [viii],xx,[xx], 392 pp. 6 fldg plates. Sl marg water stain on a few pp, outer edge of ff in 1 sgntr sl soiled & chipped. Qtr calf antique.
(Bickersteth) £365 [≈ $584]

- A Treatise of Algebra, in Three Parts ... Appendix. London: Millar & Nourse, 1748. 1st edn. 8vo. xiv,366,[i], 65,[i] pp. 12 fldg plates, text figs. Orig calf, upper jnt cracked, sl wear at hd of spine, label rubbed.
(Bickersteth) £220 [≈ $352]

Maclean, John S.
- The Newcastle & Carlisle Railway, 1825-1862. The First Railway across Britain ... Newcastle upon Tyne: R. Robinson, 1948. Super roy 8vo. 121 pp. Port frontis, plates, fldg table, ills. Orig dec bds, sl rubbed, lower crnrs sl bumped.   *(Duck)* £40 [≈ $64]

Maclean, Magnus (ed.)
- Modern Electrical Practice. London: [ca 1903]. 6 vols. 4to. Ills. Orig bndgs, some cvrs rubbed.   *(Book House)* £25 [≈ $40]
- Modern Electrical Engineering. London: Gresham, 1919. 1st edn (?). 6 vols. Super roy 8vo. 66 plates, v num ills. Orig cloth, sl soiled.   *(Duck)* £40 [≈ $64]

Mcleod, A.
- Useful Minerals and Rare Ores. New York: John Wiley, 1917. 2nd edn. 12mo. xv,254 pp. Orig cloth.   *(Gemmary)* $35 [≈ £21]

Macleod, Henry Dunning
- The Elements of Banking ... London: Longmans, Green, 1876. 1st edn. 8vo. [ii], xiii,[1 blank],270,[32 advt dated Nov 1875] pp. Orig cloth, spine a bit rubbed & sunned, inner hinges beginning to crack.
(Pickering) $300 [≈ £187]

MacMichael, William
- The Gold-Headed Cane. Edited by William Munk, M.D.. London: Longmans, Green, 1884. 1st Munk edn. 8vo. xvi,246 pp. Orig cloth gilt.   *(Fenning)* £21.50 [≈ $35]

McNicoll, D.H.
- Dictionary of Natural History Terms with their Derivations. London: 1863. 8vo. Lib stamp on title. Orig cloth, reprd.
(Henly) £15 [≈ $24]

Macnish, Robert
- The Philosophy of Sleep. Glasgow: W.R. M'Phun, 1830. 1st edn. 12mo. xi,[i],268 pp.

Lacks half-title. Contemp half calf, gilt spine, trifle scuffed.   *(Burmester)* £80 [≈ $128]

Macoun, J. & J.M.
- Catalogue of Canadian Birds. Ottawa: 1909. Rvsd edn. Roy 8vo. 761,xviii pp. 1 or 2 sm lib stamps. New cloth.
(Wheldon & Wesley) £35 [≈ $56]

Macpherson, H.A.
- A Vertebrate Fauna of Lakeland. Edinburgh: David Douglas, 1892. 1st edn. 8vo. civ,552,19 illust ctlg pp. Fldg cold map, 2 hand cold plates by Keulemans, 6 sepia engvs, 9 w'engvs. Orig green cloth, spine faded.   *(Gough)* £60 [≈ $96]

Maddock, Alfred Beaumont
- Practical Observations on the Efficacy of Medicated Inhalations in the Treatment of Pulmonary Consumption, Asthma, Bronchitis, Chronic Cough ... and in affections of the Heart ... London: 1844. 1st edn. 4to. [8],121 pp. Cold frontis, text ills. Orig cloth, sl worn. *(Hemlock)* $400 [≈ £250]

Mager, Henri
- Water Diviners and their Methods. Translated from the Fourth Edition ... London: G. Bell, 1931. 1st (?) English edn. xii,308 pp. 8 plates, inc frontis, 57 text ills. Orig cloth, uncut, sl rubbed.
(Duck) £60 [≈ $96]

Magoun, F.A.
- The Frigate "Constitution" and other Historic Ships. Salem, MA: Marine Research Society, 1928. 1st trade edn. Folio. 157 pp. 16 plans, 30 plates, num text ills. Orig cloth. dw.
(Schoyer) $200 [≈ £125]

Maham, D.H.
- An Elementary Course of Civil Engineering. London: Fullarton & Co ..., 1859. 2nd edn. 4to. xiii,[ii],211 pp. 15 plates. Orig cloth, spine damaged. Edited by Peter Barlow.
(Young's) £75 [≈ $120]

Major, Ralph H.
- A History of Medicine. Springfield: (1954). 1st edn. 2 vols. Ports, num ills. Orig cloth. Dws.   *(Elgen)* $140 [≈ £87]

Malloch, P.D.
- Life History and Habits of the Salmon, Sea Trout, Trout and other Freshwater Fish. London: Black, 1910. Roy 8vo. xvi,264 pp. 239 ills. Orig cloth with 2 mtd photos.
(Egglishaw) £40 [≈ $64]

- Life History and Habits of the Salmon, Sea Trout, Trout and other Freshwater Fish. London: Black, 1912. 2nd edn. Roy 8vo. xix, 294 pp. 274 ills. Orig cloth with 2 mtd photos. *(Egglishaw)* **£45 [≃ $72]**

Malthus, Thomas Robert
- Definitions in Political Economy ... London: Murray, 1827. 1st edn. 8vo. viii,261, [3] pp. Lacks half-title. Title & endpapers sl foxed. Contemp half calf over mrbld bds, crnrs & hd of spine worn, lacks label, sm piece of hd of spine chipped. *(Pickering)* **$2,500 [≃ £1,562]**
- An Essay on the Principle of Population ... London: for J. Johnson & T. Bensley, 1803. 2nd edn. Lge 4to. viii,[4],610 pp. Tear in last leaf reprd. Contemp calf, rebacked. *(Heritage)* **$3,500 [≃ £2,187]**
- An Essay on the Principle of Population ... London: 1806. [With] Additions to the Fourth and Former Editions of An Essay on the Principle of Population ... London: 1817. 1st edn. 3 vols. 8vo. A few lib stamps. Old mrbld bds, rebacked in cloth, lacks most endleaves. *(M & S Rare Books)* **$600 [≃ £375]**
- An Essay on the Principle of Population ... Fourth Edition. London: for J. Johnson, by T. Bensley, 1807. 2 vols. 8vo. xvi,580; vii,[1],484,[60] pp. Contemp red half calf, spines sl faded & with a few old scratches, crnrs sl bumped. *(Finch)* **£370 [≃ $592]**
- An Essay on the Principle of Population ... Fourth Edition. London: for J. Johnson ..., 1807. 4th edn. 2 vols. 8vo. xvi,580; vii,[1], 484,[60] pp. Light waterstain on marg of endpapers. Half calf gilt, rebacked, crnrs rubbed. *(Pickering)* **$1,000 [≃ £625]**
- Principles of Political Economy considered with a View to their Practical Applications ... Boston: Wells & Lilly, 1821. 1st Amer edn. 8vo. viii,472 pp. Browned, quite heavy spotting in last qtr. Contemp half leather, spine & crnrs rubbed, hd of spine chipped. *(Pickering)* **$1,000 [≃ £625]**
- Principles of Political Economy ... Second Edition with considerable additions ... London: William Pickering, 1836. 2nd edn. 8vo. liv,[ii],446 pp. Contemp calf gilt, rebacked. *(Pickering)* **$2,000 [≃ £1,250]**

Maltz, Maxwell
- New Faces - New Futures [plastic surgery]. New York: Smith, 1936. 1st edn. 8vo. 315 pp. Photo ills. Orig bndg. *(Xerxes)* **$65 [≃ £40]**

Manby, G.W.
- An Essay on the Preservation of Shipwrecked

Persons. With a Descriptive Account of the Apparatus, and the Method of Applying it ... London: Longman ..., 1812. Only edn. 8vo. xiii,94 pp. Text w'engvs. Orig bds, uncut, rebacked. *(Young's)* **£175 [≃ $280]**

Mandeville, Bernard de
- The Fable of the Bees: or, Private Vices, Publick Benefits. London: 1723. 8vo. [viii], 428,[11] pp. Panelled calf, jnts cracked. *(Wheldon & Wesley)* **£150 [≃ $240]**

Manilius, Marcus
- The five Books ... Containing a System of the ancient Astronomy and Astrology ... Done into English Verse ... By Mr. Tho. Creech. London: 1700. 2nd iss. 8vo. [ii],68,32, 1-48, 47-134,[viii],88 pp. Frontis, 5 plates. Sl ageing. Contemp calf, rebacked. Wing M.431. *(Clark)* **£85 [≃ $136]**

Mansell-Pleydell, J.C.
- Flora of Dorsetshire. Dorchester: for private circulation, 1895. 2nd edn. 8vo. xxxviii,345,xxv pp. 2 maps. Orig cloth. *(Wheldon & Wesley)* **£50 [≃ $80]**

Mantell, Gideon Algernon
- The Fossils of the South Downs; or Illustrations of the Geology of Sussex. London: 1822. 4to. Cold map, 41 plates (5 hand cold). Sm section cut from dedic leaf. Waterstains on plates. Half calf, spine relaid. *(Henly)* **£180 [≃ $288]**
- Geological Excursions round the Isle of Wight and along the adjacent Coast of Dorsetshire. London: 1854. 3rd edn. Cr 8vo. xxii,356,2,32 advt pp. Fldg cold map, 19 plates, text figs. Orig cloth. *(Henly)* **£45 [≃ $72]**
- The Geology of the South-East of England. London: 1833. 8vo. xix,415 pp. Frontis, fldg hand cold map, 5 plates. Sm lib stamps on title, half-title & frontis verso. Orig bds, rebacked with cloth, new endpapers, orig label preserved. *(Henly)* **£85 [≃ $136]**
- The Geology of the South-East of England. London: 1833. xix,404 pp. Fldg cold map & sections, 5 plates (stained & foxed). Title foxed. Contemp half calf. *(Baldwin)* **£90 [≃ $144]**
- Medals of Creation. London: 1844. 1st edn. 2 vols 8vo. Orig gilt dec cloth. *(Baldwin)* **£55 [≃ $88]**
- Petrifactions and their Teachings. London: 1851. 8vo. xi,496 pp. 115 w'cuts. Light stain on frontis. Orig cloth. *(Baldwin)* **£45 [≃ $72]**
- Petrifactions and their Teachings or a Guide

to the Organic Remains of the British Museum. London: 1851. Cr 8vo. xi,496 pp. Frontis, 115 text figs. Sm lib stamp on title verso. Orig cloth.   *(Henly)* **£40 [≈ $64]**

- A Pictorial Atlas of Fossil Remains, consisting of Coloured Illustrations selected from Parkinson's "Organic Remains of a Former World". London: 1850. 4to. 75 hand cold plates. Worm hole in top marg. Orig cloth, rebacked preserving part of orig spine.
        (Henly) **£185 [≈ $296]**

- A Pictorial Atlas of Fossil Remains. London: 1850. 4to. 207 pp. Cold frontis, 74 uncold litho plates. New cloth gilt.
        (Egglishaw) **£150 [≈ $240]**

- Thoughts on Animalcules, or a Glimpse of the Invisible World revealed by the Microscope. London: 1846. xvi,144,3,8 pp. 12 hand cold plates. Sm lib stamp on title. Half-title & 6 plates sl foxed. Orig green cloth, spine faded & worn.
        (Henly) **£80 [≈ $128]**

- Wonders of Geology. London: 1848. 6th edn. 2 vols. Cr 8vo. Frontis in vol 1, 6 plates (3 hand cold). Orig cloth, spines sl faded.
        (Henly) **£48 [≈ $76]**

Manual of Gunnery ...

- Manual of Gunnery for Her Majesty's Fleet. By Authority of the Lords Commissioners of the Admiralty. London: HMSO, 1886. Super roy 8vo. [ii],xii,498 pp. 99 plates. Card in pocket. 2 ff dogeared. Orig cloth, rebacked, hd of spine nicked.   *(Duck)* **£135 [≈ $216]**

Maple, William

- A Method of Tanning without Bark. Dublin: A. Rhames, 1729. 1st edn. 8vo. [6],35,[5] pp. 1 hand cold plate. Front pastedown detached & partly lacking. Gilt dec mottled calf.
        (Rootenberg) **$950 [≈ £593]**

Maplet, A.

- A Greene Forest or a Naturall Historie ... Reprinted from the Edition of 1567, with an Introduction by W.H. Davies. London: Hesperides Press, 1930. One of 500. Sm 4to. ix, 184,[2] pp. Orig buckram.
        (Wheldon & Wesley) **£40 [≈ $64]**

Marcet, Jane

- Conversations on Chemistry ... (New Haven): From Sydney's Press, 1814. 8vo. 12 plates. Frontis sl soiled. Orig calf, leather spine label, sl rubbed.   *(Karmiole)* **$100 [≈ £62]**

- Conversations on Natural Philosophy ... Fifth Edition. London: Longman, Rees ..., 1827. x,[2],429 pp. 23 plates (browned). Contemp

polished calf gilt, a.e.g., crnrs sl rubbed.
        (Claude Cox) **£25 [≈ $40]**

Marcet, William

- An Experimental Inquiry into the Action of Alcohol on the Nervous System. London: J.E. Adlard, 1860. 1st edn. 8vo. 20 pp. Rec bds.
        (Rootenberg) **$150 [≈ £93]**

Marey, E.J.

- Movement. Translated by Eric Pritchard. London: Heinemann, 1895. 1st English edn. 8vo. xvi,332 pp. 200 text ills. Orig cloth, spine sl faded.
        (Charles B. Wood) **$225 [≈ £140]**

Marey, Etienne Jules

- Animal Mechanism. A Treatise on Terrestrial and Aerial Locomotion. London: Henry S. King, Intl Scientific Library, 1874. 1st edn in English. 8vo. xvi,283,[32 advt] pp. 117 w'engvs. Orig red cloth.
        (Rootenberg) **$250 [≈ £156]**

Marie, Pierre & Souza-Leite, J.D.

- Essays on Acromegaly. With Bibliography and Appendix of Cases by Other Authors. London: New Sydenham Soc, 1891. 1st edn in English. 182,6 annual report,38 ctlg pp. 2 fldg plates. Orig cloth. *(Elgen)* **$125 [≈ £78]**

Marion, F.

- Wonderful Balloon Ascents: or, the Conquest of the Skies. A History of Balloons and Balloon Voyages. New York: 1870. 1st edn. 8vo. xvi,218,38 pp. Frontis, 39 plates. Orig cloth, spine faded, lower spine worn.
        (Berkelouw) **$150 [≈ £93]**

Marker, Melvin

- Dependable Dentistry. New York: Dennison, 1930. 1st edn. 4to. 114 pp. Num photo ills. A few spots. Orig bndg.   *(Xerxes)* **$80 [≈ £50]**

Markham, Gervase

- Markhams Master-Piece: containing all Knowledge belonging to the Smith, Farrier, or Horse-Leech, touching the curing of all Diseases in Horses. London: 1662. 9th imp. Sm 4to. [xiv],591,[21] pp. Frontis (with expl leaf), fldg plate, text ills. Calf, reprd.
        (Wheldon & Wesley) **£220 [≈ $352]**

Marks, Anne

- The Cat in History, Legend and Art. London: Elliot Stock, 1909. 1st edn. 4to. 85 pp, list of subscribers. Ills. Orig pict green cloth.   *(Moon)* **£25 [≈ $40]**

Marks, G.C.
- Hydraulic Power Engineering. A Practical Manual on the Concentration and Transmission of Power by Hydraulic Machinery. London: Crosby Lockwood, 1905. 2nd enlgd edn. 8vo. 388 pp. 240 ills. Piece cut from fly. Orig bndg, sl marked.
(Book House) **£28 [≈ $44]**

Marks, George Edwin
- A Treatise on Marks' Patent Artificial Limbs with Rubber Hands and Feet. New York: A.A. Marks, 1888. 1st edn. 397 pp. Frontis, text w'cuts. Ex-lib. Orig bndg, lib number on spine. *(Elgen)* **£125 [≈ £78]**

Marmery, J. Villin
- Progress of Science. Its Origin, Course, Promoters and Results. London: 1895. 1st edn. xxxi,358,40 ctlg pp. Orig bndg.
(Elgen) **$65 [≈ £40]**

Marquand, E.D.
- Flora of Guernsey and the lesser Channel Islands. London: 1901. 8vo. viii,501 pp. 5 maps. Orig cloth, v sl stained.
(Wheldon & Wesley) **£50 [≈ $80]**

Marriott,
- The Parrot-Keeper's Guide: comprising the Natural History of Macaws, Cockatoos, Parrots ... Treatment ... Diseases ... Cure ... By Marriott, an Experienced Dealer. London: Dean & Son, [1876]. Sm 8vo. 48 pp. 16 uncold lithos. Orig green cloth gilt.
(Blackwell's) **£85 [≈ $136]**

Marryat, Thomas
- Therapeutics: or, the Art of Healing. To which is added, a Glossary ... Bristol: R. Edwards; London: Lee & Hurst, 1798. 14th edn. 12mo. 227 pp. Contemp tree calf, gilt spine, sl worn, lower jnt cracked but sound.
(Burmester) **£40 [≈ $64]**

Marsh, G.P.
- Man and Nature; or, Physical Geography as Modified by Human Nature. London: Sampson Low, 1864. 1st edn. 8vo. xv,560 pp. Contemp red half mor gilt.
(Gough) **£45 [≈ $72]**

Marsh, John, of Chichester
- The Astrarium Improved; or, Views of the Principal Fixed Stars and Constellations ... London: George & John Cary, 1833. Sm 4to. 12 pp. 12 plates, the constellation stars have been joined up & others inserted by a previous owner. Orig cloth, v faded.

(Blackwell's) **£150 [≈ $240]**

Marshall, Alfred
- Elements of Economics of Industry being the First Volume [all published] of Elements of Economics ... London: Macmillan, 1892. 1st edn. 8vo. xiv,416 pp. Pencil & ink underlinings & annotations in 1st half, some ink spotting. Orig cloth, v worn & soiled.
(Pickering) **$370 [≈ £231]**
- Fiscal Policy of International Trade ... Ordered by the House of Commons to be printed, 11 November 1908. London: HMSO, 1908. Parliamentary Paper. Folio. 29 pp. Lib stamps. A little dustsoiled. Stab sewn as issued. *(Pickering)* **$150 [≈ £93]**
- Industry and Trade. A Study of Industrial Technique and Business Organization ... London: Macmillan, 1919. 1st edn. 8vo. xxiv, 875 pp. Orig blue cloth, spine sunned, spine ends a bit rubbed. *(Pickering)* **$300 [≈ £187]**
- Money Credit and Commerce ... London: Macmillan, 1923. 1st edn. 8vo. xv,[i],369, [1],[2 advt] pp. Orig blue cloth, sm piece lacking from headband, sm nick in top of upper bds. *(Pickering)* **$200 [≈ £125]**

Marshall, Arthur
- Explosives. Their Manufacture, Properties, Tests and History. London: Churchill, 1915. 4to. xvi,624 pp. 136 ills. Orig cloth, spine faded, hinges cracked, cvrs sl soiled.
(Duck) **£35 [≈ $56]**

Marshall, C.
- A Plain and Easy Introduction to the Knowledge and Practice of Gardening, with Hints on Fish-Ponds. London: 1813. 5th edn. 8vo. iv,448 pp. New cloth.
(Wheldon & Wesley) **£28 [≈ $44]**

Marshall, C.F. Dendy
- Centenary History of the Liverpool & Manchester Railway ... London: Locomotive Publ Co, 1930. 4to. x,192 pp. Map frontis, 28 plates (20 cold), text ills. 3 neat ink underlinings. Orig dec cloth gilt, 2 minor surface injuries, lower crnrs bumped.
(Duck) **£85 [≈ $136]**

Marshall, W.P.
- Description of the Patent Locomotive Steam Engine of Messrs. Robert Stephenson and Co., Newcastle upon Tyne ... London: 1838. 67 pp. 4 fldg engvd plates (neat reprs to verso), text ills. Occas foxing. Lib blindstamps & lending card at end. Later cloth. *(Reese)* **$500 [≈ £312]**

Marshall, William
- Planting and Ornamental Gardening; a Practical Treatise. London: Dodsley, 1785. 1st edn. 8vo. xi,[5],638 pp. Half-title. Contemp calf, jnts, spine ends, & crnrs reprd.
(Spelman) **£180 [≈ $288]**
- Planting and Ornamental Gardening; a Practical Treatise. London: 1785. 8vo. xv,638 pp. Mod bds.
(Wheldon & Wesley) **£75 [≈ $120]**
- Planting and Rural Ornament: being a Second Edition, with large additions, of Planting and Ornamental Gardening, a Practical Treatise. London: G. Nichol, 1796. 2 vols. 8vo. xxxii,408,[8]; xx,454,[4] pp. Contemp half calf gilt, gilt spine & labels.
(Spelman) **£200 [≈ $320]**
- Planting and Rural Ornament, being a Second Edition with large additions of Planting and Ornamental Gardening, a Practical Treatise. London: 1796. 2 vols. 8vo. Contemp calf, trifle worn, front cvr of vol 2 rather loose.
(Wheldon & Wesley) **£75 [≈ $120]**
- The Rural Economy of Norfolk ... London: Cadell, 1787. 1st edn. 2 vols. 8vo. xix,400; [xvi],392,[4] pp. Fldg map. Sl worm in vol 1 gutters & some margs of vol 2. Section Cc vol 2 misfolded. Orig bds, untrimmed.
(Blackwell's) **£140 [≈ $224]**
- The Rural Economy of the West of England. London: for G. Nicol ..., 1796. 1st edn. 2 vols. 8vo. 2 advt ff in vol 2. Fldg map. Sl spotting of vol 1 title. Mod mottled calf, uncut.   *(Claude Cox)* **£140 [≈ $224]**

Marsham, T.
- Entomologia Britannica ... Tomus I Coleoptera [all published]. London: White, 1802. xxxi,548 pp. Lib stamp on 1 page. Disbound.   *(Egglishaw)* **£58 [≈ $92]**

Martin, Benjamin
- A Panegyrick on the Newtonian Philosophy ... London: for W. Owen ..., 1749. 1st edn. 8vo. 63,[1 advt] pp. Paper browned. Minor loss to lower blank margs of 1st few ff. A few tears reprd. Bds.   *(Pickering)* **$500 [≈ £312]**
- The Philosophical Grammar; being a View of the Present State of Experimental Physiology, or Natural Philosophy. In Four Parts ... London: 1755. 5th edn. 362,index pp. 26 fldg plates, 2 fldg tables. Occas sl marks. New half leather.   *(Rittenhouse)* **$195 [≈ £121]**
- A Plain and Familiar Introduction to the Newtonian Philosophy ... London: for W. Owen, 1754. 2nd edn. 8vo. [viii],164,[4] pp. 6 fldg plates. 19th c mrbld bds, rebacked.

(Burmester) **£225 [≈ $360]**
- The Young Gentleman and Lady's Philosophy, in a Continued Survey of the Works of Nature and Art ... London: W. Owen & the author, 1759-63. 1st edn. 2 vols. 8vo. xii,410,[iv]; [ii],412,[viii] pp. Frontis, 52 numbered plates, 1 supplementary plate. Contemp calf, sl worn. *(Clark)* **£125 [≈ $200]**

Martin, E.A.
- A Bibliography of Gilbert White. London: 1934. 8vo. vii,194 pp. Ills. Orig blue cloth.
(Henly) **£38 [≈ $60]**
- A Bibliography of Gilbert White ... of Selborne. London: 1934. 8vo. viii,194 pp. Ills. Orig cloth.
(Wheldon & Wesley) **£30 [≈ $48]**

Martin, E.K.
- Atlas of Pathological Anatomy. Bristol, & Baltimore: 1930-35. 2 vols. Over 600 plates (mostly cold). Orig bndg.
(Elgen) **$125 [≈ £78]**

Martin, Dr Franklin
- The Joy of Living. An Autobiography. New York: Doubleday, Doran, 1933. 1st edn. Ltd edn. 2 vols. 8vo. 491; 526 pp. Photo ills. Orig bndg. Author's sgnd inscrptn.
(Xerxes) **$70 [≈ £43]**

Martin, H.M.
- The Design and Construction of Steam Turbines. London: "Engineering", 1913. 4to. 372 pp. Fldg plates, ills. Mod cloth. Ex-libris C.A. Parsons.   *(Book House)* **£45 [≈ $72]**

Martin, J.H.
- Microscopic Objects Figured and Described. London: 1870. 8vo. vi,114 pp. 97 plates. Half calf, lacks label.
(Wheldon & Wesley) **£35 [≈ $56]**

Martin, P.I.
- A Geological Memoir on a Part of Western Sussex. London: 1828. 4to. xi,100 pp. Cold geological map, 3 plates of sections, table. Title sl spotted. Orig cloth, paper label.
(Baldwin) **£95 [≈ $152]**

Martindale, Adam
- The Country Survey-Book: or Land-Meters Vade-Mecum. Wherein the Principles and Practical Rules for Surveying of Land, are so plainly delivered ... Appendix ... London: 1692. 2nd edn. 12mo. 1st few ff sl wormed. Contemp calf, rebacked, some rubbing. Wing M.854A.   *(Ximenes)* **$675 [≈ £421]**

Martine, George, 1702-1741
- Essays and Observations on the Construction and Graduation of Thermometers, and on the Heating and Cooling of Bodies. Third Edition. Edinburgh: Alexander Donaldson, 1789. 12mo. vi,177 pp. 1 plate. Orig tree calf, red mor label.
(Charles B. Wood) **$200 [≈£125]**

Martyn, Thomas
- Flora Rustica: Exhibiting Figures of such Plants as are either Useful or Injurious in Husbandry. London: 1792-94. 4 vols in 1. 8vo. 144 hand cold plates by F.P. Nodder. 1 plate rather foxed, occas minor foxing elsewhere. Contemp calf gilt.
(Wheldon & Wesley) **£450 [≈$720]**
- Flora Rustica: Exhibiting Figures of such Plants as are either Useful or Injurious in Husbandry. London: 1792-94. 4 vols. 8vo. 144 hand cold plates by F.P. Nodder. A little browning & foxing. Qtr calf, uncut, rebacked.
(Henly) **£550 [≈$880]**
- Thirty-Eight Plates, with Explanations; intended to illustrate Linnaeus's System of Vegetables ... London: 1788. 1st edn. 8vo. 72, [2] pp. 38 hand cold plates. Calf, rebacked.
(Hemlock) **$250 [≈£156]**
- Thirty-Eight Plates, with Explanations; intended to illustrate Linnaeus's System of Vegetables ... London: for J. White, 1799. 8vo. Advt leaf A4. 38 plates. Sl foxing. Mod bds.
(Stewart) **£90 [≈$144]**
- Thirty-Eight Plates with Explanations; intended to illustrate Linnaeus's System of Vegetables ... London: J. White, 1799. 8vo. vi,72 pp. 38 plates. Sl foxing. 19th c half calf, gilt spine.
(Spelman) **£70 [≈$112]**

Maskell, W.M.
- An Account of the Insects Noxious to Agriculture and Plants in New Zealand; The Scale-Insects (Coccidiae). Wellington: 1887. 8vo. 116 pp. 20 cold & 3 plain plates. Sl foxing. Orig cloth.
(Wheldon & Wesley) **£45 [≈$72]**

Massee, G.
- British Fungi. London: Routledge, n.d. Roy 8vo. 551 pp. 40 cold plates. Lib stamp on front pastedown & last page. Orig gilt dec bds.
(Savona) **£28 [≈$44]**
- European Fungus Flora, Agaricaceae. London: 1902. Cr 8vo. vi,274 pp. Orig cloth, sl soiled.
(Henly) **£18 [≈$28]**

Mathew, M.A.
- The Birds of Pembrokeshire and its Islands.

London: Porter, 1894. 1st edn. 8vo. lii,131 pp. 2 fldg cold maps, mtd photo frontis, key, 2 orig photo plates. Orig cloth, rather mottled, sm piece torn from ft of spine.
(Gough) **£35 [≈$56]**

Mathews, G.B.
- Theory of Numbers. Part I. Cambridge: 1892. 1st edn. xii,323 pp. A few marg ink notes. Orig cloth, sl worn, lib labels on spine & endpaper, shaky. *(Whitehart)* **£35 [≈$56]**

Matho ...
- See Baxter, Andrew.

Matteson, Antonette
- Occult Family Physician and Botanic Guide to Health. New York: the author, 1894. 1st edn. 8vo. 317 pp. Port frontis. Orig bndg, front hinge cracked, rear hinge cracking.
(Xerxes) **$60 [≈£37]**

Matthews, L.H.
- British Mammals. London: Collins, New Naturalist Series, 1952. 1st edn. 16 cold & 48 plain plates, 92 text figs. Orig cloth. Dw.
(Egglishaw) **£25 [≈$40]**

Matthews, L.H. (ed.)
- The Whale. London: 1968. Oblong 4to. 287 pp. Num cold & plain ills. Orig cloth.
(Wheldon & Wesley) **£25 [≈$40]**

Matthiesen, P. & Palmer, R.S.
- The Shorebirds of North America ... see Stout, G.D. (ed.).

Maudsley, Henry
- Body and Mind: An Inquiry into their Connection and Mutual Influence, specially in reference to Mental Disorders: An enlarged and revised edition ... added, Psychological Essays. New York: Appleton, 1890. x,275,[8] pp. "Orig binder's cloth".
(Hemlock) **$75 [≈£46]**
- The Physiology of Mind. London: Macmillan, 1876. Orig cloth, spine ends frayed, spine & crnrs worn.
(Patterson) **£50 [≈$80]**

Maunder, Samuel
- The Treasury of Natural History ... New Edition revised and corrected, with an extra Supplement. London: Longmans, Green, (1848) 1878. Sm 8vo. xvii,809 pp. Over 900 w'cut ills. Calf, red label.
(Egglishaw) **£17 [≈$27]**

Maupertius, Pierre Louis Moreau de
- The Figure of the Earth, determined from Observations ... at the Polar Circle ... London: for T. Cox ..., 1738. 1st edn in English. 8vo. vii,[i],232 pp. Fldg map, 9 fldg plates. Title backed. Contemp calf, sl later mor label.   *(Burmester)* £250 [≈ $400]

[Mauquest de] La Motte, Guillaume
- A General Treatise on Midwifry ... Translated into English by Thomas Tomkyns, Surgeon. London: for James Waugh, 1746. 1st edn in English. 8vo. xx,536 pp. Orig calf, sl rubbed, sm defective spot on spine, lower jnt slit at ft.
          (Bickersteth) £550 [≈ $880]

Mauriceau, Francis
- The Diseases of Women with Child, and in Child-Bed ... Fourth Edition, Corrected and Augmented ... London: Andrew Bell, 1710. 8vo. xliv,373,[10] pp. 10 plates. Browned. Contemp panelled calf, hinges split but holding.   *(Hemlock)* $550 [≈ £343]

Maury, M.G.
- The Physical Geography of the Sea. London: 1870. 8vo. xvi,493 pp. 13 plates. Some spotting. Orig cloth, spine ends lightly fingered.   *(Bow Windows)* £28 [≈ $44]

Mawe, John
- Familiar Lessons on Mineralogy and Geology ... added A Practical Description of the Lapidary's Apparatus ... London: Longman ..., 1822. 4th edn. Sm 8vo. viii,111 pp. 4 pp of hand cold plates. New qtr leather.
          (Gemmary) $225 [≈ £140]
- The Voyager's Companion, or Shell Collector's Pilot ... London: the author ..., 1825. 4th edn. Sm 8vo. viii,75,[i advt] pp. 2 hand cold plates. Frontis offset. Drab mor grain paper (?) bds, title label to front cvr, later cloth spine. *(de Beaumont)* £78 [≈ $124]

Mawe, Thomas & Abercrombie, John
- Every Man his Own Gardener. Being a New, and much more Complete Gardener's Calendar, and General Director ... London: for B. Law ..., 1797. 15th edn. 8vo. [iv],702,[48] pp. Frontis. Old calf, front jnt cracked but holding.   *(Young's)* £45 [≈ $72]
- Every Man his own Gardener. Being a New, and much more Complete Gardener's Kalendar, and General Director, than any one hitherto published ... Dublin: for P. Byrne ..., 1798. 14th edn. 8vo. [iv],626,[19] pp. Old sheep, rubbed.   *(Young's)* £42 [≈ $67]

- Every Man His Own Gardener ... Seventeenth Edition. London: 1803. Lge 12mo. vii,758,[96] pp. Frontis. Rec half calf.
          (Bow Windows) £75 [≈ $120]
- Every Man His Own Gardener; Being a New, and much more Complete Gardener's Calender and General Directory than any one hitherto published ... London: Rivington ..., 1813. 20th edn. 8vo. vi,812 pp. Port frontis. Contemp sheep, red label, hd of spine worn.
          (Young's) £55 [≈ $88]

Mawson, T.H.
- The Art and Craft of Garden Making. London: 1901. 2nd edn, rvsd & enlgd. Roy 4to. xviii,252 pp. 178 ills. Orig buckram, trifle used, spine faded.
          (Wheldon & Wesley) £60 [≈ $96]

Maxim, Sir Hiram S.
- Artificial and Natural Flight. London & New York: Whittaker, 1908. 1st edn. xvi,166 pp. Num ills. Orig cloth, partly faded, front hinge cracked but firm.   *(Duck)* £80 [≈ $128]

Maxwell, Sir Herbert
- British Fresh-Water Fishes. London: Hutchinson, Woburn Library, 1904. Roy 8vo. viii, 316 pp. 12 cold plates. Orig cloth gilt.   *(Egglishaw)* £24 [≈ $38]
- Trees. A Woodland Notebook containing Observations on certain British and Exotic Trees. Glasgow: 1915. Sm 4to. xvi,236 pp. 50 plates (17 cold). Orig cloth, somewhat discold.   *(Wheldon & Wesley)* £20 [≈ $32]

Maxwell, James Clerk
- An Elementary Treatise on Electricity. Oxford: Clarendon Press, 1881. 1st edn. 8vo. xvi,208,[39 advt] pp. 6 litho plates, text ills. Orig cloth.   *(Rootenberg)* £350 [≈ £218]
- The Scientific Papers ... Edited by W.D. Niven. Cambridge: UP, 1890. 1st edn. 2 vols. Lge 4to. xxix,[3],607; vii,[1],806,[2 advt] pp. 3 ports, 14 litho plates, num text ills. Orig cloth, uncut, cvrs sl discold, sm tear vol 1 backstrip.   *(Rootenberg)* $850 [≈ £531]
- The Scientific Papers ... Paris: (1927). 2 vols. 4to. [xxii],607; [viii],806 pp. 3 ports, 14 plates, num text figs. Tiny piece torn from 2 ff. Contemp green half calf, t.e.g., by Riviere, a little faded & marked.
          (Bow Windows) £385 [≈ $616]
- A Treatise on Electricity and Magnetism. Third Edition. Oxford: Clarendon Press, 1892. 2 vols. 8vo. Errata slip, advt ff. 20 plates, num figs. Minor underlining & marginalia. Orig cloth, just a trifle rubbed.

(Bow Windows) £200 [≈ $320]

May, J.B.
- The Hawks of North America ... New York: Natl Assoc of Audubon Societies, 1935. Roy 8vo. xxxii,140 pp. 37 cold & 4 plain plates. Orig cloth.   *(Egglishaw)* £45 [≈ $72]

Maynard, Charles J.
- Eggs of North American Birds. Illustrated with Ten Hand-Colored Plates. Boston: De Wolfe, Fiske, 1890. iv,159 pp. 10 hand cold plates showing over 75 eggs, Blank lower crnr of frontis chipped. Orig brown cloth, stamped in black & gold.   *(Karmiole)* £100 [≈ £62]

McNab, Robert
- The North British Cultivator: A Treatise on Gardening, Agriculture, and Botany. Perth: J. Dewar, & T. Richardson, 1842. 8vo. Sl foxing. Orig cloth, paper label sl chipped, some wear to headcap, inkstain on lower cvr.
                            (Rankin) £45 [≈ $72]

Mead, Richard
- A Discourse Concerning the Action of the Sun and Moon on Animal Bodies; and the Influence which this may have in many Diseases ... London: ptd in the year 1708. 1st edn. 8vo. 32 pp. Sl foxed. Rec morocco backed bds.   *(Pickering)* $950 [≈ £593]
- A Discourse on the Plague. The Ninth Edition corrected and enlarged. London: for A. Millar, & J. Brindley, 1744. 8vo. [viii], xl, 164 pp. Misbound but complete. Contemp calf, lacks label.   *(Bickersteth)* £40 [≈ $64]
- The Medical Works ... Dublin: for Thomas Ewings, 1767. 1st edn thus. 8vo. xix, [1 blank], 511,[1 blank, 59 index, 1 blank] pp. 4 plates, plan. Contemp calf, rebacked, crnrs worn.   *(Bow Windows)* £240 [≈ $384]

Medical Botany ...
- Medical Botany: or, History of Plants in the Materia Medica of the London, Edinburgh and Dublin Pharmacopoeias, arranged according to the Linnaean System. London: 1821. 2 vols in one. 8vo. 2 ports, 138 hand cold plates. Minor foxing & offsetting. Mod qtr leather.
                    (Wheldon & Wesley) £850 [≈ $1,360]

Meigs, Arthur V.
- The Origin of Disease ... Philadelphia: Lippincott, 1897. 1st edn. xiv,229 pp. 61 plates. Orig cloth, lower spine edge frayed.
                            (Elgen) $55 [≈ £34]

Meigs, Charles D. (ed.)
- The History, Pathology, and Treatment of Puerperal Fever and Crural Phlebitis ... Philadelphia: Barrington & Haswell, 1842. 1st edn thus. 338,2 advt pp. Occas foxing. Minor ex-lib. Rebound in three qtr mor.
                            (Elgen) $150 [≈ £93]

Meinertzhagen, R.
- Nicoll's Birds of Egypt ... see Nicoll, Michael & Meinertzhagen, R.
- Pirates and Predators. The Piratical and Predatory Habits of Birds. London: 1949. Cr 4to. ix,230 pp. 18 cold & 26 plain plates. Orig cloth, cvrs sl tape-marked.
                    (Wheldon & Wesley) £150 [≈ $240]
- Pirates and Predators. The Piratical and Predatory Habits of Birds. London: Oliver & Boyd, 1949. 1st edn. 4to. ix,230 pp. 41 cold plates. Orig cloth gilt. Dw.
                            (Hollett) £180 [≈ $288]

Meinzer, O.E.
- Hydrology. New York: 1942. 1st edn. Cr 4to. ix,712 pp. Num text figs. Orig cloth.
                            (Henly) £15 [≈ $24]

Melland, Frank
- Elephants in Africa. London: 1938. 8vo. xv,186 pp. 12 plates. Orig cloth.
                    (Wheldon & Wesley) £38 [≈ $60]
- Elephants in Africa. London: 1938. 8vo. Ills. Orig bndg.   *(Grayling)* £25 [≈ $40]

Mendes Da Costa, Emanuel
- Elements of Conchology: or, an Introduction to the Knowledge of Shells ... London: for Benjamin White, 1776. 1st edn. 8vo. viii,[iii-]vi,318,[2 errata] pp, advt leaf. 2 fldg charts, 7 fldg plates. Contemp calf, mor label.
                            (Claude Cox) £110 [≈ $176]

The Mental Guide ...
- The Mental Guide, Being a Compend of the First Principles of Metaphysics ... Predicated on the Analysis of the Human Mind. Boston: Marsh & Capen, & Richardson & Lord, 1828. 8vo. 384 pp. Sl foxed. Contemp calf.
                            (Gach) $175 [≈ £109]

Mercer, S.D.
- Spinal Curvatures, and Treatment of Spinal Diseases by Plaster of Paris Jacket ... Omaha, Nebraska: Republican Steam Book and Job Print, 1878. 8vo. 32,[8] pp. 8 full page & 2 text engvs. Orig green cloth, sl rubbed & soiled. Author's inscrptn.
                            (Karmiole) $45 [≈ £28]

Meredith, Louisa Anne, nee Twamley
- Our Wild Flowers. By Louisa Anne Twamley. London: 1839. Post 8vo. 12 hand cold plates. Mor elab gilt, a.e.g., lacks front free endpaper.
(Wheldon & Wesley) **£100** [≈ $160]
- Some of My Bush Friends in Tasmania, Native Flowers, Berries and Insects drawn from Life. London: 1860. Imperial 4to. Cold title & 14 cold plates inc 2 page-borders. Occas sl foxing & soiling. Orig cloth gilt, a.e.g., rebacked.
(Wheldon & Wesley) **£280** [≈ $448]

Merrett, H.S.
- Land and Engineering Surveying. A Practical Treatise ... London: Spon, 1875. 2nd edn. Lge 8vo. 317 pp. 41 plates, some fldg. Mod cloth.   *(Book House)* **£22.50** [≈ $36]
- A Practical Treatise on the Science of Land and Engineering Surveying, Levelling, Estimating Quantities, &c. ... Fourth Edition, revised and corrected with an Appendix ... London: Spon, 1887. Lge 8vo. xiv, 346 pp. 42 plates. Orig gilt dec cloth.
(Spelman) **£30** [≈ $48]

Merrill, E.D.
- A Flora of Manila. Manila: 1912. 8vo. 490 pp. Binder's cloth, trifle spotted.
(Wheldon & Wesley) **£45** [≈ $72]

Merrill, G.P.
- The Non-Metallic Minerals: their Occurrence and Uses. New York: John Wiley, 1904. 1st edn. 8vo. xi,414,16 advt pp. 32 plates, 28 text figs. Orig cloth.
(Gemmary) **$37.50** [≈ £23]

A Method ...
- A Method of Tanning without Bark ... see Maple, William.

Metz, R.
- Precious Stones and Other Crystals. New York: Viking Press, 1965. 4to. 191 pp. 89 mtd cold plates. Orig cloth.
(Gemmary) **$50** [≈ £31]

Meyen, Franz J.F.
- Outlines of the Geography of Plants ... Translated by Margaret Johnston. London: Ray Society, 1846. 1st English edn. 8vo. [2],x, 422,[16] pp. Fldg table. Orig cloth gilt, t.e.g., sl wear to headbands. *(Fenning)* **£45** [≈ $72]

Meyer, A.B. & Wiglesworth, L.W.
- The Birds of Celebes and the Neighbouring Islands. Berlin: 1898. 2 vols. Roy 4to. 7 cold maps, 42 hand cold & 3 plain plates. Red half mor, edges trifle faded.
(Wheldon & Wesley) **£2,000** [≈ $3,200]

Meyer, H.L.
- Coloured Illustrations of British Birds and their Eggs. London: 1853-57. 7 vols. 8vo. 322 hand cold plates of birds, 100 hand cold plates of eggs, 8 plain plates. Some sl foxing & signs of use. Green mor gilt extra, fine.
(Wheldon & Wesley) **£975** [≈ $1,560]

Meyer de Schauensee, R.
- The Birds of Colombia and adjacent areas of South and Central America. Narberth, Pa.: 1964. 8vo. xvi,430 pp. 20 plates (11 cold), 87 text figs. Orig cloth.
(Wheldon & Wesley) **£70** [≈ $112]

Meyerhof, Otto Fritz
- Chemical Dynamics of Life Phenomena. Philadelphia & London: Lippincott, [1924]. 1st edn. 8vo. 110 pp. Orig cloth.
(Bickersteth) **£50** [≈ $80]

Meyrick, W.
- The New Family Herbal; or, Domestic Physician. Birmingham: 1790. 1st edn. 8vo. xxiv,498 pp. Frontis (margs stained), 14 plates. Contemp calf. Other material bound at end.   *(Wheldon & Wesley)* **£160** [≈ $256]

Miall, L.C.
- The Geology, Natural History and Pre-Historic Antiquities of Craven in Yorkshire, from Whitaker's History of Craven. Leeds: 1878. 4to. 42 pp. Cold geol map & section. Orig cloth, rebacked.   *(Henly)* **£38** [≈ $60]

Michaelis, Sebastien
- The Admirable Historie of the Possession and Conversion of a Penitent woman. Sedvced by a Magician that made her to become a Witch ... London: 1613. 1st edn, iss with dated title. 4to. Sl fraying. Contemp limp vellum, sl worn, lacks ties. STC 17854A.
(Finch) **£1,200** [≈ $1,920]

Michaux, F. Andrew
- The North American Sylva ... Philadelphia: 1817-19. 3 vols in 2 (7 half-vols). Roy 8vo. [iv],xii,268; ii,ii,ii,250; iv,285 pp, errata page. 128 hand cold & 28 plain plates. Foxed. Contemp half calf, worn, jnts cracking but firm.   *(Sotheran)* **£2,400** [≈ $3,840]
- The North American Sylva ... Philadelphia: 1871. 5th edn. 5 vols inc Thomas Nuttall's Supplement. Cr 8vo. Port, 277 hand cold

plates. Blind-stamped mor, a.e.g.
(Henly) **£2,000 [≈ $3,200]**

Michelet, Jules

- The Bird. With 210 Illustrations by Giacomelli. London: Nelson, 1869. 2nd English edn. Lge 8vo. 340 pp. Contemp elab gilt mor, a.e.g., spine darkened, cvrs sl spotted. Slipcase.   *(Heritage)* **$150 [≈ £93]**
- The Insect. London: Nelson, 1875. Roy 8vo. xii,368 pp. 140 ills by Giacomelli. Orig pict gilt cloth, a.e.g.   *(Egglishaw)* **£18 [≈ $28]**
- The Insect. With 140 Illustrations by Giacomelli. London: 1875. 1st English edn. Lge 8vo. 368 pp. Last few ff sl foxed. Contemp elab gilt mor, a.e.g., sl rubbed, hinges cracking. Slipcase.
(Heritage) **$150 [≈ £93]**

Michell, W.

- On Difficult Cases of Parturition; and the Use of Ergot of Rye. London: 1828. Sole edn. 8vo. xv,128 pp. Disbound.
(Bickersteth) **£38 [≈ $60]**

Middlemiss, H.S. (ed.)

- Narcotic Education. Edited Report of the Proceedings of the First World Conference on Narcotic Education Philadelphia, Pennsylvania Jul 5-9, 1926. Washington, DC: Middlemiss, 1926. 1st edn. 8vo. 403 pp. Orig bndg.   *(Xerxes)* **$60 [≈ £37]**

Milbank, Jeremiah, Jr.

- The First Century of Flight in America. Princeton: UP, (1943). 8vo. x,248 pp. Num plates. Orig cloth gilt. Dw.
(Berkelouw) **$100 [≈ £62]**

Miles, W.J.

- Modern Practical Farriery ... London: William Mackenzie, n.d. 4to. vii,536,90 pp. 40 plates (14 hand cold). Contemp half mor gilt, cloth a little faded.
(Hollett) **£120 [≈ $192]**

Mill, James, 1773-1836

- Analysis of the Phenomena of the Human Mind. London: Baldwin & Cradock, 1829. 1st edn. 2 vols. 8vo. [iv],320; [iv],312 pp. Sl foxing & edge-browning to front & rear ff. Mod cloth backed bds. *(Gach)* **$650 [≈ £406]**

Mill, John Stuart

- Autobiography. London: Longmans, Green, Reader, & Dyer, 1873. 1st edn. 8vo. [ii],vi, 313,[5] pp. B'plate, stamp & owner inscrptn on title, some early pencil marginalia. Orig

pebbled green cloth.   *(Gach)* **$75 [≈ £46]**
- On Liberty. London: Longman ..., 1865. People's Edition. 8vo. [iv],68 pp. Ptd in dble column. Orig purple cloth, uncut, upper cvr gilt lettered, brown endpapers, spine sl sunned.   *(Finch)* **£20 [≈ $32]**
- The Positive Philosophy of Auguste Comte. Boston: William V. Spencer, 1866. 1st Amer edn. 8vo. 182 pp. Orig burgundy cloth gilt.
(Karmiole) **$100 [≈ £62]**
- Principles of Political Economy. London: Parker, 1852. 3rd edn. 2 vols. 8vo. 604; 571 pp. Orig bndg.   *(Xerxes)* **$150 [≈ £93]**
- The Subjection of Women ... Philadelphia: Lippincott, 1869. 1st Amer edn. Cr 8vo. 174, [6] pp. 1st leaf (half-title) loose. Orig dark green cloth, rubbed, negligible hole in spine.
(Finch) **£150 [≈ $240]**

Millais, J.G.

- British Deer and their Horns. London: 1897. Folio. xviii,224 pp. Cold frontis, num text & full-page ills. Orig buckram, recased, sl signs of use.   *(Wheldon & Wesley)* **£175 [≈ $280]**
- British Deer and their Horns. London: Sotheran, 1897. Imperial 4to. xviii,224 pp. Cold frontis, 10 electrogravures, 185 text & full-page ills. Orig buff buckram, spine sometime relaid, soiled, crnrs reprd, new endpapers.   *(Blackwell's)* **£200 [≈ $320]**
- Game Birds and Shooting Sketches. London: 1892. Folio. Num cold & other plates. Light foxing. Orig bndg, some rubbing, upper hinge sl split.   *(Grayling)* **£200 [≈ $320]**
- The Mammals of Great Britain and Ireland. London: 1905. 3 vols. Folio. Num cold & other plates by A. Thorburn & others. Sl foxing. Orig buckram, rubbed, spine snagged, needs some restoration.
(Grayling) **£300 [≈ $480]**
- The Mammals of Great Britain and Ireland. London: 1904-06. One of 1025. 3 vols. Imperial 4to. 62 cold & 62 photogravure plates by Millais, Thorburn & others, & 125 plates from photos. Some dustsoiling vol 1. Half mor gilt, gilt dec spines, sm fault in 1 hinge.   *(Henly)* **£350 [≈ $560]**
- The Mammals of Great Britain and Ireland. London: Longmans, Green, 1904. 3 vols. Lge thick 8vo. 31 cold & 63 plain plates, 18 gravures. A few edges reprd. Mod three qtr blue levant mor gilt. *(Hollett)* **£450 [≈ $720]**
- The Natural History of British Game Birds. London: 1909. One of 550. Folio. 18 cold plates, 17 photogravures by Archibald Thorburn & J.G. Millais. Some sl browning to margs of 1 or 2 cold plates. Orig buckram, sl rubbed.   *(Grayling)* **£450 [≈ $720]**

- The Natural History of British Game Birds. London: 1909. One of 550. Folio. 18 cold plates, 17 photogravures, 2 other ills, by Archibald Thorburn & J.G. Millais. Orig buckram, trifle faded, crnrs trifle bumped.
(Wheldon & Wesley) **£480 [≃ $768]**
- The Natural History of British Surface-Feeding Ducks. London: Longmans Green, 1902. One of 600 Large Paper. 4to. 41 cold plates, 6 gravures, 25 other ills. Crease in blank crnr of frontis. Orig 2-tone cloth, t.e.g., spine ends & crnrs sl rubbed.
(Claude Cox) **£450 [≃ $72]**
- The Natural History of the British Surface-Feeding Ducks. London: Longmans, Green, 1902. One of 600. Roy 4to. Pict half-title, 6 gravures, 41 cold plates, 25 other ills. Orig cloth, t.e.g.   *(Egglishaw)* **£500 [≃ $800]**

Millar, G.H.
- A New ... Body or System of Natural History. London: [1785]. Folio. [iv],5-618, [2] pp. Engvd frontis, 85 plates of figs. A few plates mtd. Calf, rebacked.
(Wheldon & Wesley) **£200 [≃ $320]**

Miller & Fink
- Neon Signs Manufacture - Installation - Maintenance. New York: McGraw Hill, 1935. 1st edn, 4th imp. 8vo. xiii,288 pp. Frontis, 12 tables, 103 text ills. Orig green cloth gilt.   *(Moon)* **£50 [≃ $80]**

Miller, F.T.
- The World in the Air. The Story of Flying in Pictures ... New York: 1930. 2 vols. 4to. Over 1200 ills, illust endpapers. Orig cloth, spines faded & somewhat worn at ends.
(Berkelouw) **$125 [≃ £78]**

Miller, Gerrit S.
- Catalogue of the Mammals of Western Europe (Europe exclusive of Russia) in the Collection of the British Museum. London: 1912. 8vo. xv,1019 pp. 213 ills. Orig cloth gilt, sl worn & v sl stained.
(Fenning) **£45 [≃ $72]**
- The Families and Genera of Bats. London: (1907) 1967. xviii,282 pp. 14 plates. Orig bndg.   *(Wheldon & Wesley)* **£35 [≃ $56]**

Miller, Hugh
- Sketch Book of Popular Geology. London: 1859. 1st edn. 8vo. xxxvi,358 pp. Orig cloth.
(Baldwin) **£25 [≃ $40]**
- Sketchbook of Popular Geology. Edinburgh: 1859. 1st edn. 8vo. Section cut from front endpaper. Orig cloth, hinges starting to crack.

(Henly) **£22 [≃ $35]**
- The Testimony of the Rocks, or Geology in its Bearings on the Two Theologies, Natural and Revealed. Edinburgh: Constable, 1857. 1st edn. 8vo. xi,500 pp. 152 text figs. Orig cloth, spine ends worn.
(Egglishaw) **£24 [≃ $38]**
- The Testimony of the Rocks, or Geology in its Bearings on the Two Theologies, Natural and Revealed. London: 1858. 24th thousand. xii, 501 pp. Frontis, 153 text figs. Orig cloth, rebacked.   *(Henly)* **£15 [≃ $24]**

Miller, Philip
- The Abridgement of the Gardener's Dictionary ... London: for the author ..., 1763. 1st 4to edn of the abridgement. Frontis, 10 fldg plates, 1 other plate. Sl marg worming, sl dog-eared & dampstained at end, final leaf reprd. Mod calf antique.
(Heritage) **$275 [≃ £171]**
- The Gardener's Dictionary: containing the Methods of Cultivating the Kitchen, Fruit and Flower Garden, etc. London: 1731. 1st edn. Folio. xvi,[iv],[841] pp. Frontis, 4 plates (frontis & 2 plates in xerox). Occas water staining & signs of use. New cloth.
(Wheldon & Wesley) **£120 [≃ $192]**
- The Gardener's Dictionary: containing the Methods of Cultivating the Kitchen, Fruit and Flower Garden ... London: 1731. 1st edn. Folio. xvi,[iv],B1-8D2 pp. Frontis, 4 plates. Minor staining & marg worming. Sm hole in 1 leaf. Contemp calf, rebacked, crnrs reprd.
(Clark) **£250 [≃ $400]**
- The Gardener's Dictionary. London: 1733. 2nd edn. Folio. Frontis, 4 plates. Contemp calf, trifle worn, rebacked.
(Wheldon & Wesley) **£185 [≃ $296]**
- The Gardener's Dictionary ... abridged from the last Folio Edition. London: 1754. 4th edn. 3 vols. 8vo. Frontis, 3 plates. Contemp calf, lacks labels, jnts beginning to crack.
(Wheldon & Wesley) **£160 [≃ $256]**
- The Gardener's Kalendar ... Thirteenth Edition, adapted to the New Style ... prefixed, a Short Introduction to the Science of Botany ... London: for the author, 1762. 8vo. xv,47,369,[11] pp. Frontis, 5 plates. Contemp calf, sl worn, hinge cracked but firm.
(Claude Cox) **£75 [≃ $120]**
- The Gardeners Kalendar. London: 1769. 15th edn. 8vo. lxvi,382,[21] pp. Frontis, 5 plates. New half calf, antique style.
(Wheldon & Wesley) **£45 [≃ $72]**

Miller, Thomas
- Common Wayside Flowers. London:

Routledge ..., 1860. 1st edn thus. 4to. [viii],185 pp. 24 cold ills by Edmund Evans after Birket Foster. Some spotting. Orig elab gilt cloth, paper onlays, a.e.g., rubbed, gilt dulled, later endpapers, spine extremities rubbed.   *(de Beaumont)* **£98 [≃ $156]**

Millikan, Robert Andrews
- The Electron. Its Isolation and Measurement and the Determination of some of its Properties. Chicago: Aug 1917. 1st printing. Sm 8vo. xii,268 pp. Plates, ills. Orig cloth, inner hinges starting.   *(Elgen)* **$125 [≃ £78]**

Millingen, J.G.
- Curiosities of Medical Experience. Second Edition, revised and considerably augmented. London: 1839. 8vo. Some foxing at beginning & end. Orig cloth, front inner hinge broken.
                          (Robertshaw) **£24 [≃ $38]**

Mills, W.H.
- Railway Construction. London: Longman, 1900. 2nd edn. 8vo. 365 pp. Ills. Mod cloth.
                        (Book House) **£35 [≃ $56]**

Mills, Wesley
- The Nature and Development of Animal Intelligence. London: Fisher Unwin, 1898. 1st edn. 8vo. xii,[308] pp. Orig dec green cloth gilt, edges bumped. *(Gach)* **$75 [≃ £46]**

Milne, E.A.
- Kinematic Relativity. Oxford: 1948. vi,238 pp. Orig bndg.   *(Whitehart)* **£35 [≃ $56]**
- Relativity, Gravitation and World Structure. Oxford: 1935. viii,365 pp. 4 plates. Occas sl foxing. Orig cloth.   *(Whitehart)* **£40 [≃ $64]**

Minerals and Metals ...
- Minerals and Metals. London: John W. Parker, 1847. 5th edn. 12mo. xii,255 pp. 9 plates. Qtr leather. *(Gemmary)* **$100 [≃ £62]**

Minot, H.D.
- The Land-Birds and Game-Birds of New England. Second Edition, edited by W. Brewster. Boston: 1895. 8vo. xxiv,492 pp. Plate, text figs. Orig cloth, trifle used.
                     (Wheldon & Wesley) **£30 [≃ $48]**

Miscellanies ...
- Miscellanies. Or a Miscellaneous Treatise; containing Several Mathematical Subjects ... see Emerson, William.

Mises, Ludwig von
- Theory and History. An Interpretation of

Social and Economic Evolution. New Haven: Yale UP, 1957. 1st edn. 8vo. x,384 pp. Orig cloth. Sl frayed dw. *(Pickering)* **$150 [≃ £93]**

Mitchell, Sir Arthur
- About Dreaming, Laughing and Blushing. London: William Green & Sons, 1905. 1st edn. 157 pp. Orig cloth. *(Hollett)* **£30 [≃ $48]**

Mitchell, John M.
- The Herring. Its Natural History and National Importance. London: Longman, Green ..., 1864. xii,372 pp. Fldg cold frontis, 5 tinted lithos. Orig blue cloth, gilt spine, extremities sl rubbed. *(Karmiole)* **$65 [≃ £40]**

Mitchell, P.C.
- Centenary History of the Zoological Society of London. London: 1929. 8vo. xi,307 pp. 32 ports, 9 plans, chart. Orig cloth, trifle used.
                     (Wheldon & Wesley) **£38 [≃ $60]**

Mitchell, Silas Weir
- Some Recently Discovered Letters of William Harvey with Other Miscellanea ... With a Bibliography of Harvey's Works ... Philadelphia: 1912. 1st edn. Sq 8vo. [ii],59, [3] pp. 2 plates, ptd tissue guards. Orig ptd stiff wraps.   *(Gach)* **$65 [≃ £40]**

Mitzsch, C.L.
- Pterylography. Translated from the German by W.S. Dallas. Edited by P.L. Sclater. London: Ray Society, 1867. Folio. x,181 pp. 10 plates. Lib stamp on title. Orig bds, unopened.   *(Egglishaw)* **£36 [≃ $57]**

Mivart, St. George
- Dogs, Jackals, Wolves and Foxes: a Monograph of the Canidae. London: 1890. 4to. xxvi,216 pp. 45 hand cold plates by Keulemans, text figs. Orig cloth, spine reprd.
                (Wheldon & Wesley) **£750 [≃ $1,200]**
- A Monograph of the Lories, or Brush-Tongued Parrots, composing the Family Loridae. London: 1894. Roy 4to. liii,193 pp. 4 maps, 61 hand cold plates. Not of the earliest colouring. Brown cloth.
            (Wheldon & Wesley) **£1,850 [≃ $2,960]**
- On Truth: A Systematic Inquiry. London: Kegan Paul, Trench, 1889. 1st edn. 8vo. [ii], x, 580 pp. Blue buckram, lightly shelfworn.
                          (Gach) **$65 [≃ £40]**
- The Origin of Human Reason, being an Examination of Recent Hypotheses concerning it. London: Kegan Paul, 1889. 1st edn. 327 pp. Marg inkstain on 1st 53 pp. Orig bndg, spine ends sl frayed.

(Elgen) **$55 [≈ £34]**
- Types of Animal Life. London: 1893. 8vo. viii,374 pp. Num text figs. Orig cloth.
(Wheldon & Wesley) **£25 [≈ $40]**

The Modern Dictionary ...
- The Modern Dictionary of Arts and Sciences; or, Complete System of Literature ... London: for the authors, & sold by G. Kearsly, 1774. 1st edn. 4 vols. 8vo. Frontis in vol 1, 46 plates, tables in text. Occas v sl mark. Contemp calf, mor labels.
(Finch) **£220 [≈ $352]**

Moffet, Thomas
- Insectorum sive Minimorum Animalium Theatrum ... London: Thom. Cotes, 1634. Sm folio. W'cut title vignette, ca 500 w'cuts in text. Old stain to final ff. Contemp reversed sheep, sl rubbed. STC 17993.
(Stewart) **£950 [≈ $1,520]**

Moholy, Luca
- An Hundred Years of Photography, 1839-1939. Harmondsworth: Penguin Books, 1939. 1st edn. 12mo. 182,[x] pp. 35 ills. Paper browned as usual. Orig ptd wraps. Dw.
(Charles B. Wood) **$50 [≈ £31]**

Moivre, Abraham de
- See De Moivre, Abraham.

Moll, J.W.
- Phytography as a Fine Art comprising Linnean Descriptions, Micrography and Pen Portraits. Leyden: 1934. Roy 8vo. xix,534 pp. 7 plates. Orig cloth.
(Wheldon & Wesley) **£25 [≈ $40]**

Monginot, F. de
- A New Mystery in Physick, discovered, by curing of Fevers & Agues by Quinquiana or Jesuites Powder. Translated from the French by Dr. Belon, with additions. London: Will. Crook, 1681. 1st English edn. 12mo. 30ff,99,9 ctlg pp. Lib stamp on title verso. Mod three qtr calf. *(Hemlock)* **$850 [≈ £531]**

Monro, Alexander, 1697-1767
- The Works ... Published by his Son ... Life of the Author. Edinburgh: Charles Elliot, 1781. 1st edn. 4to. [2],xxiv,791,[1] pp. Frontis port, 7 fldg plates (some offsetting). Sl browning. Three qtr calf, mrbld bds, 5 raised bands, red label.   *(Elgen)* **$700 [≈ £437]**

Montagu, George
- Supplement to the Ornithological Dictionary, or Synopsis of British Birds. London: ptd by S. Woolmer, Exeter, & sold by S. Bagster ...; 1813. 1st edn. [Ca 480] pp. Errata leaf at end. 24 plates. Half calf, gilt spine.
(Claude Cox) **£55 [≈ $88]**

Monteath, Robert
- The Forester's Guide and Profitable Planter ... Second Edition, with important Additions and Improvements. Edinburgh & London: 1824. 8vo. lvi,395 pp. Frontis & 14 plates. Occas sl spotting. Contemp paper bds, ptd paper label, jnts cracked, spine ends worn.
(Finch) **£48 [≈ $76]**

Moore, C.W.
- A Practical Guide for Prospectors, Explorers, and Miners. London: Kegan Paul, 1893. 8vo. xiv,286,56 advt pp. 5 cold plates, 149 text figs. Orig cloth.   *(Gemmary)* **$55 [≈ £34]**

Moore, David
- Concise Notices of the Indigenous Grasses of Ireland, best suited for Agriculture, with Dried Specimens of Each Kind. Glasnevin [i.e. Dublin]: R. Purdue Mullingar, 1843. 1st edn. Sm folio. 8 pp, 36 ff with 48 mtd specimens. Old bds, rebacked.
(Charles B. Wood) **$800 [≈ £500]**

Moore, Henry Charles
- Omnibuses and Cabs. Their Origin and History. London: Chapman & Hall, 1902. 1st edn. xiv,282 pp. Ills. Orig pict cloth, spine faded, stain on front cvr.
(Wreden) **$75 [≈ £46]**

Moore, John, 1730-1802
- Medical Sketches: in Two Parts. Providence: Carter & Wilkinson, 1794. 1st Amer edn. [8],271 pp. Contemp mottled calf.
(Jenkins) **$200 [≈ £125]**

Moore, T.
- British Wild Flowers. London: 1867. 8vo. xxvii,424 pp. 24 hand cold plates depicting 96 species. Orig cloth gilt. *(Henly)* **£60 [≈ $96]**
- The Ferns of Great Britain and Ireland. Edited by J. Lindley. Nature-Printed by H. Bradbury. London: 1857. 2nd edn. Imperial folio. 51 cold plates. Occas marking & sl dustsoiling but plates clean. A few sm reprs. Contemp red mor gilt extra, a.e.g., some wear. *(Wheldon & Wesley)* **£1,200 [≈ $1,920]**
- Nature Printed British Ferns ... Nature Printed by Henry Bradbury. London: 1859-[60]. 2 vols. 122 cold plates. Some foxing, on plate 51 quite heavily. Sm lib stamps on front endpapers. Orig red cloth

gilt, new endpapers. *(Henly)* **£130 [≃ $208]**
- A Popular History of British Ferns. London:
[1859]. 3rd edn, rvsd. Post 8vo. xvi, 394,6 pp.
22 cold plates. Orig red cloth gilt, soiled.
(Henly) **£25 [≃ $40]**

Moquin-Tandon, C.H.B.A.
- The World of the Sea. London: Cassell,
(1869). Roy 8vo. viii,448 pp. 18 cold plates,
text figs. Frontis foxed. 1 leaf reprd. Orig
cloth, faded, hd of spine & crnrs worn.
(Egglishaw) **£25 [≃ $40]**

More, Alexander Goodman
- Life and Letters of Alexander Goodman
More, F.R.S.E., F.L.S., M.R.I.A. With
Selections from his Zoological Writings.
Edited by C.B. Moffat ... Dublin: Hodges,
Figgis, 1898. 1st edn. 8vo. xii,642 pp. Port,
map. Orig cloth, t.e.g.
(Fenning) **£68.50 [≃ $110]**

Moreau, R.E.
- The Palaearctic-African Bird Migration
Systems. New York: 1972. 8vo. xvi,384 pp.
Num maps & plates. Orig cloth.
(Wheldon & Wesley) **£60 [≃ $96]**

Morell, John Daniel
- Historical and Critical View of the
Speculative Philosophy of Europe in the
Nineteenth Century. New York: Robert
Carter & Bros, 1856. 1st Amer edn. 8vo. Orig
cloth, jnts chipped. *(Gach)* **$50 [≃ £31]**

Moreton, C.O.
- The Auricula: Its History and Character.
London: 1965. Folio. 90 pp. 17 cold plates by
R. McEwen. Wraps.
(Wheldon & Wesley) **£60 [≃ $96]**
- Old Carnations and Pinks. London: 1955. Sm
folio. xi,51 pp. 8 cold plates by R. McEwen.
Orig cloth backed bds.
(Wheldon & Wesley) **£45 [≃ $72]**

Morgagni, Giovanni Battista
- The Seats and Causes of Diseases investigated
by Anatomy: In Five Books ... Translated
from the Latin ... by Benjamin Alexander.
London: A. Millar ..., 1769. 1st edn in
English. 3 vols. 4to. Sl browned. Contemp
calf, 19th c reback, some pp unopened.
(Rootenberg) **$2,000 [≃ £1,250]**

Morgan, Augustus de
- See De Morgan, Augustus.

Morgan, Conway Lloyd
- Animal Life and Intelligence. London:
1890-91. 1st edn. 8vo. xvi,512 pp. Frontis, 40
figs. Frontis & title sl foxed. Orig cloth.
(Wheldon & Wesley) **£25 [≃ $40]**
- Habit and Instinct. London & New York:
Edward Arnold, 1896. 1st edn. 8vo. [viii],
[352] pp. Frontis. Blue cloth.
(Gach) **$85 [≃ £53]**

Morgan, Lewis H.
- The American Beaver and His Works.
Philadelphia: 1868. 330 pp. Frontis, fldg
map, ills. Orig cloth gilt, extremities frayed.
(Reese) **$175 [≃ £109]**

Morgan, William Gerry
- The American College of Physicians, its First
Quarter Century. Philadelphia: 1940. 1st
edn. 4to. 275 pp. Ills. Orig bndg, minor spine
wear. *(Xerxes)* **$45 [≃ £28]**

Morison, James
- Morisonia; or, Family Advisor of the British
College of Health ... With an Appendix ...
Small Pox ... Fourth Edition. London: for the
College of Health, 1833-34. 2 vols. Frontis
port, 3 plates. Occas foxing. Sm hole in 1 page
with sl loss. Polished calf, gilt spine.
(Elgen) **$225 [≃ £140]**

Morrice, Alexander
- A Treatise on Brewing; wherein is exhibited
the whole Process of the Art and Mystery of
Brewing the various Sorts of Malt Liquor ...
London: Sherwood, Neely & Jones, 1815.
8vo. xxiv,179,[3 advt] pp. 4 w'cuts. Sl foxing.
Orig bds, uncut, backstrip defective.
(Blackwell's) **£135 [≃ $216]**

Morris, F.O.
- A History of British Birds. London:
[1863-67]. Cabinet edn. 8 vols. 358 hand cold
plates. Orig dec red cloth, spines a little
faded. *(Henly)* **£420 [≃ $672]**
- A History of British Birds. London: George
Bell, 1870. 2nd edn. 6 vols. Lge 8vo. 365 cold
plates. Some spotting. Orig pict cloth gilt.
(Hollett) **£520 [≃ $832]**
- A History of British Birds. Leeds & London:
n.d. Cabinet Edition. 8 vols. Cr 8vo. 358
hand cold plates. Some foxing, mostly
confined to the text. Vols 1-4 half mor,
rubbed, vols 5-8 cloth, worn, sellotape &
labels removed from endpapers.
(Wheldon & Wesley) **£350 [≃ $560]**
- A History of British Birds. London: John C.
Nimmo, 1895. 4th edn. 6 vols. 8vo. 365 hand

cold plates. Orig pict green cloth gilt, fine.
(Gough) £650 [≃ $1,040]
- A History of British Birds. London: Nimmo, 1895-97. 4th edn, rvsd & enlgd. 6 vols. Roy 8vo. 394 hand cold plates. Occas sl marg spotting. Dark blue half mor gilt, t.e.g., vol 1 spine sl faded. *(Egglishaw)* £580 [≃ $928]
- A History of British Birds. London: Groombridge & Sons, 1851-57. 1st edn. 6 vols. 358 hand cold plates with litho cream background. Contemp green half mor, gilt spines, a.e.g., extremities sl rubbed.
(Claude Cox) £500 [≃ $800]
- A History of British Butterflies. London: [1852-] 1853. 1st edn. Roy 8vo. 71 cold plates of insects. 2 plain plates of apparatus (lower margs browned). Offsetting on 3 plates. Contemp half calf.
(Wheldon & Wesley) £100 [≃ $160]
- A History of British Butterflies. London: Groombridge, 1853. 8vo. 71 hand cold plates. Occas sl spotting. Late 19th c mor, rubbed, gilt on spine rather indistinct.
(Spelman) £80 [≃ $128]
- A History of British Butterflies. London: 1870. Roy 8vo. viii,159 pp. 72 cold & 2 plain plates. Orig cloth, faded.
(Wheldon & Wesley) £95 [≃ $152]
- A History of British Butterflies. London: 1872. 4th edn. Roy 8vo. 72 cold & 2 plain plates. Orig cloth, spine reprd.
(Wheldon & Wesley) £85 [≃ $136]
- A History of British Butterflies. London: Nimmo, 1895. 8th edn, rvsd & enlgd. Roy 8vo. viii,235 pp. 79 hand cold & 2 plain plates. Binder's cloth, spine faded.
(Egglishaw) £60 [≃ $96]
- A History of British Butterflies. London: 1904. 9th edn. Roy 8vo. viii,235 pp. 79 hand cold & 2 plain plates. Foxing at ends not affecting plates. Orig cloth.
(Henly) £80 [≃ $128]
- A History of British Butterflies. London: Routledge, 1908. Sm 4to. 79 hand cold plates. Occas sl spotting. Orig cloth gilt, extremities rubbed, jnts cracked. *(Hollett)* £75 [≃ $120]
- A History of British Moths. London: 1896. 5th edn. 4 vols. Imperial 8vo. 132 hand cold plates containing 1933 distinct specimens. Orig cloth gilt. *(Henly)* £145 [≃ $232]
- A Natural History of the Nests and Eggs of British Birds. London: Groombridge, 1853. 1st edn. 3 vols. 76 chromolitho plates, finished by hand. Contemp green half calf.
(Gough) £95 [≃ $152]
- A Natural History of the Nests and Eggs of British Birds. Second Edition. London: George Bell & Sons, 1875. 3 vols. Lge 8vo.

Orig cloth gilt, spines sl worn.
(Spelman) £70 [≃ $112]
- A Natural History of the Nests and Eggs of British Birds. Fourth Edition. Revised and Corrected by W.B. Tegetmeier. London: John C. Nimmo, 1896. Lge 8vo. 248 plates, chiefly cold by hand. Orig green cloth gilt, spine ends rubbed. *(Spelman)* £80 [≃ $128]

Morris, Sir Henry
- Surgical Diseases of the Kidney and Ureter ... Chicago: 1904. 2 vols. 2 cold plates, over 200 engvs. Orig bndg, sm cvr stain vol 1.
(Elgen) $100 [≃ £62]

Morris, J.
- A Catalogue of British Fossils. London: 1843. 8vo. 221 pp. A few lib stamps. Orig cloth, rubbed. *(Baldwin)* £20 [≃ $32]

Morris, R.
- The Botanist's Manual: a Catalogue of Hardy, Exotic and Indigenous Plants. London: 1824. 8vo. [iv],189 pp. Contemp half calf, upper jnt cracked.
(Wheldon & Wesley) £20 [≃ $32]

Mortimer, John
- The Whole Art of Husbandry; or, the Way of Managing and Improving of Land ... Second Edition. London: J.H. for H. Mortlock ..., 1708. 8vo. Contemp calf.
(Falkner) £110 [≃ $176]

Morton, G.H.
- The Geology of the Country around Liverpool. London: George Philip, 1891. 2nd edn. 8vo. [ix],287 pp. 20 plates. A few spots. Orig cloth gilt. *(Hollett)* £40 [≃ $64]

Morton, John Chalmers
- A Cyclopaedia of Agriculture, Practical and Scientific ... Glasgow: Blackie & Sons, 1855. 1st edn. 2 vols in 3. Lge 8vo. 52 steel engvd plates, text ills. Sl foxing. Contemp purple diced calf, gilt dec spines, extremities rubbed.
(Heritage) $450 [≃ £281]

Morton, Leslie Thomas (compiler)
- Garrison and Morton's Medical Bibliography: An Annotated Check-List of Texts Illustrating the History of Medicine. London: Gower, 1954. 2nd rvsd & enlgd edn. 8vo. [xiv],655,[3] pp. Orig cloth backed bds.
(Gach) $45 [≃ £28]

Morton, Samuel George
- Crania Aegyptiaca; or, Observations on

Egyptian Ethnography derived from Anatomy, History and the Monuments. Philadelphia: John Penington, 1844. Sm folio. 67 pp. 14 plates (sl foxed). Title edges sl frayed. Orig bds, becoming disbound, needs new backstrip. *(Xerxes)* **$200 [≈ £125]**

Morton, Thomas

- Engravings Illustrating the Surgical Anatomy of the Head and Neck, Anxilla, Bend of Elbow, and Wrist, with Descriptions. London: Taylor & Walton, 1845. 1st edn. 8vo. [2],24,8 ctlg pp. 8 hand cold fldg plates. Orig cloth, sl damaged.
(Offenbacher) **$300 [≈ £187]**

Morton, Thomas George

- The History of the Pennsylvania Hospital 1751-1895 ... Philadelphia: Times Printing House, 1895. 2nd rvsd edn. Heavy 8vo. [iv], [x], 591,[3] pp. 21 plates on 20 ff, text ills. Orig cloth backed bds, cvrs flecked.
(Gach) **$85 [≈ £53]**

Morton, William J.

- The X-Ray or, Photography of the Invisible and its Value in Surgery. New York: 1896. Ills. Upper hinge loose, some wear.
(Rittenhouse) **$175 [≈ £109]**

Moseley, H.N.

- Notes by a Naturalist. An Account of Observations made during the Voyage of H.M.S. "Challenger" round the World in the Years 1872-1876. London: 1892. New rvsd edn. 8vo. xxiv,540 pp. Fldg cold map, port, text figs. Orig cloth, trifle used.
(Wheldon & Wesley) **£40 [≈ $64]**

Motte, Andrew

- A Treatise of the Mechanical Powers, wherein the Laws of Motion, and the Properties of those Powers are explained and Demonstrated ... London: Motte, 1727. 1st edn. 8vo. [6],222,[2] pp. Errata & advt leaf. 3 plates, text w'cuts. Contemp calf, rebacked.
(Rootenberg) **$950 [≈ £593]**

Mottelay, Paul Fleury

- Bibliographical History of Electricity & Magnetism Chronologically Arranged. London: 1922. 1st edn. xx,673 pp. 14 plates. Orig bndg, t.e.g., front hinge starting.
(Elgen) **$375 [≈ £234]**

Moubray, Bonington

- A Practical Treatise on Breeding, Rearing, and Fattening all kinds of Domestic Poultry, Pheasants, Pigeons and Rabbits ... Fourth

Edition; with Additions ... London: 1822. 8vo. xii,312 pp. Hand cold frontis. Orig paper backed bds, spine somewhat defective.
(Frew Mackenzie) **£80 [≈ $128]**

- A Treatise on Domestic Poultry, Pigeons and Rabbits. London: 1815. 8vo. vii,218 pp. 1 sm tear reprd. Orig bds, rebacked. Pres copy.
(Wheldon & Wesley) **£95 [≈ $152]**

Moulen, Fred

- Orchids in Australia. Sydney: Edita S.A. Lausanne, 1958. 4to. 148 pp. Orig bndg. Dw.
(Terramedia) **$70 [≈ £43]**

Mountfort, G.

- Tigers. London: 1973. 8vo. Num cold ills. Orig bndg. Dw.   *(Grayling)* **£15 [≈ $24]**

Moxon, Elizabeth

- English Housewifery exemplified in about 450 Receipts ... with an Appendix ... Leeds: Thomas Wright, 1790. 13th edn, crrctd. viii, 6-203, supplement title,2-37,[15] pp. Fldg plate of fish, 6 plates of table settings. Lacks 1st page of introduction. Old sheep, worn.
(Box of Delights) **£80 [≈ $128]**

Moxon, Joseph

- A Tutor to Astronomy and Geographie ... London: ptd by Joseph Moxon ..., 1659. 1st edn. 4to. xiv,224,40 pp. Engvd factotum title, 12 text engvs. Somewhat stained & soiled. Contemp vellum bds, recased. Wing M.3021.
(Pickering) **$1,850 [≈ £1,156]**

- A Tutor to Astronomy and Geography ... Fourth Edition, Corrected and Enlarged. London: S. Roycroft for Joseph Moxon, 1686. 8vo. [8],272,[8] pp. Port frontis, num text engvs. Old calf, rebacked. Wing M.3025.
(Karmiole) **$375 [≈ £234]**

Mudie, Robert

- The Feathered Tribes of the British Islands. London: Whittaker, 1834. 1st edn. 2 vols. 8vo. 2 engvd titles with cold vignettes by Baxter, 19 hand cold engvd plates. Half calf, green labels.   *(Egglishaw)* **£135 [≈ $216]**

- The Feathered Tribes of the British Islands. London: 1878. 4th edn. 2 vols. 8vo. Engvd titles with hand cold vignettes, 19 plates of birds, 7 plates of eggs all cold by hand. Orig cloth, spines relaid.   *(Henly)* **£60 [≈ $96]**

- The Feathered Tribes of the British Islands. Fourth Edition by W.C.L. Martin. London: Bohn's Illustrated Library, 1888. Re-issue of the 1854 edn. 2 vols. 8vo. 2 cold vignettes, 2 cold frontis, 24 cold plates, 11 text figs, as is correct. Binder's cloth.
(Wheldon & Wesley) **£60 [≈ $96]**

Muirhead, G.

- The Birds of Berwickshire ... Edinburgh: Douglas, 1889-95. One of 100 Large Paper. 2 vols. Roy 8vo. 3 maps, 10 plates, num other ills. Orig buckram, t.e.g.
(Egglishaw) **£120 [≈ $192]**

Mullens, W.H. & Swann, H.K.

- A Bibliography of British Ornithology from the Earliest Times to the End of 1912. London: [1916-] 1917. Orig edn. 8vo. xx,691 pp. New cloth.
(Wheldon & Wesley) **£200 [≈ $320]**
- A Bibliography of British Ornithology from the Earliest Times to the End of 1912. London: [1916-] 1917, facs reprint 1986. 8vo. xx,691 pp. Orig cloth.
(Wheldon & Wesley) **£35 [≈ $56]**

Muller, F.

- Facts and Arguments for Darwin. Translated by W.S. Dallas. London: 1869. Cr 8vo. vii,144 pp. 67 figs. Orig cloth.
(Wheldon & Wesley) **£50 [≈ $80]**

Muller, H.

- The Fertilisation of Flowers, translated by D'Arcy W. Thompson, with a Preface by Charles Darwin. London: 1883. 8vo. xiii,669 pp. 186 figs. Orig cloth.
(Wheldon & Wesley) **£90 [≈ $144]**

Muller, John

- A Treatise Containing the Elementary Part of Fortification, Regular and Irregular ... For the Use of the Royal Academy of Artillery at Woolwich ... London: for C. Nourse ..., 1782. 4th edn. 8vo. xvi,240 pp. 34 engvs, many fldg. Orig bds, uncut.
(Young's) **£105 [≈ $168]**

Muller, William

- The Elements of the Science of War; containing the Modern, Established, and Approved Principles of the Theory and Practice of the Military Sciences. London: Longman ..., 1811. 3 vols. 8vo. 75 plates. Contemp calf, spines faded, sm snag in 1 jnt.
(Waterfield's) **£200 [≈ $320]**

Munk, William

- The Roll of the Royal College of Physicians of London ... Second Edition, revised and enlarged. London: the College, 1873. 3 vols. 8vo. Title vignettes. Sm blind lib stamp on titles. Orig green cloth gilt, spines relaid.
(Bickersteth) **£160 [≈ $256]**

Murchison, Sir Roderick Impey

- Life ... Based on his Journals and Letters ... see Geikie, Sir Archibald.
- Outline of the Geology of the Neighbourhood of Cheltenham. New Edition, revised by J. Buckman and H.E. Strickland. London: 1845. 110 pp. Fldg cold map, 13 plates. Orig cloth.
(Baldwin) **£70 [≈ $112]**
- Siluria. The History of the Oldest Known Rocks with Organic Remains ... London: 1854. 3rd edn. 8vo. xx,592 pp. Cold geol map, 41 plates. Frontis sl stained, a few plates lightly spotted. New qtr mor.
(Baldwin) **£90 [≈ $144]**
- Siluria. London: 1859. 3rd edn. 8vo. xx, 592 pp. Text figs. Lacks the plates & maps. Orig cloth.
(Henly) **£18 [≈ $28]**
- The Silurian Region and Adjacent Counties of England & Wales Geologically Illustrated ... London: (1839). Fldg hand cold map, on 3 sheets, mtd on linen. Minor dampstain on 1 fold. Orig 4to calf backed case. Sgnd by Murchison. The map for 'The Silurian System'.
(Fenning) **£195 [≈ $312]**

Murdoch, Patrick

- Mercator's Sailing applied to the Time Figure of the Earth with an Introduction concerning the Discovery and Determination of that Figure. London: for A. Millar, 1741. 4to. xxvii,38,[2 advt] pp. 3 fldg plans & diags. Paper cvrd bds.
(Lamb) **£125 [≈ $200]**

Murphy, R.C.

- Oceanic Birds of South America, a Study of Species of the related Coasts and Seas, including the American Quadrant of Antarctica. New York: Amer Mus Nat Hist & Macmillan, [1948]. 2 vols. 4to. 16 cold & 72 plain plates, 80 text figs. Orig buckram.
(Wheldon & Wesley) **£150 [≈ $240]**

Murphy, R.C. & Amadon, D.

- Land Birds of America. New York: 1953. 4to. Num cold ills. Orig cloth.
(Wheldon & Wesley) **£20 [≈ $32]**

Murray, A.

- The Northern Flora; or, a Description of the Wild Plants belonging to the North and East of Scotland. Part I [all published]. Edinburgh: 1836. 8vo. xvii,150,xvi pp. 2 plates. Orig ptd bds, cloth back.
(Wheldon & Wesley) **£50 [≈ $80]**

Murray, Lady Charlotte

- The British Garden: A Descriptive Catalogue of Hardy Plants, Indigenous, or Cultivated ...

London: for Thomas Wilson ..., 1808. 3rd
edn. 2 vols. 8vo. xxxi,767,[26] pp. Contemp
tree calf.   *(Young's)* **£70 [≃ $112]**

Murray, J.
- A Portrait of Geology. By a Fellow of the
Geological Society. London: 1838. 8vo. vii,
[1], 216 pp. Cold frontis & 2 plates. Without
the extra plate of Pentacrinus briareus
sometimes found. Orig cloth, trifle used.
          (Wheldon & Wesley) **£28 [≃ $44]**
- Researches in Natural History. London:
1830. 2nd edn. 8vo. x,146 pp. Orig cloth.
          (Wheldon & Wesley) **£20 [≃ $32]**

Murray, John
- Practical Remarks on Modern Paper ...
Edinburgh: Blackwood; London: Cadell,
1829. 1st edn. 8vo. [iv],119 pp. Sev inscrptns
on pastedown. Orig bds, untrimmed.
          (Charles B. Wood) **$1,500 [≃ £937]**

Murray, Sir John & Pullar, L.
- Bathymetrical Survey of the Scottish Fresh-
Water Lochs. Report on the Scientific
Results. Edinburgh: Challenger Office, 1910.
6 vols. Roy 8vo. 239 plates of cloth-backed
maps (in vols 3-6). Half pigskin gilt.
          (Egglishaw) **£945 [≃ $1,512]**

Murray, Margaret Alice
- The Witch-Cult in Western Europe. A Study
in Anthropology. OUP: 1921. 1st edn. 8vo.
303 pp. Orig cloth gilt, sl rubbed at hd & ft.
          (Minster Gate) **£36 [≃ $57]**

Musgrave, Samuel
- An Essay on the Nature and Cure of the (so
called) Worm-Fever. London: T. Payne ...,
1776. 8vo. Light stain on title. Mod bds.
          (Stewart) **£85 [≃ $136]**

Muspratt, Sheridan
- Chemistry Theoretical, Practical & Analytical
as applied and relating to the Arts and
Manufactures. Glasgow, Edinburgh, London
& New York: William Mackenzie, [ca 1860].
2 vols. 4to. [vi],836; 880 pp. 8 ports, 452 +
506 w'engvd ills. Orig mrbld bds, mor spines.
          (Charles B. Wood) **$200 [≃ £125]**

Muybridge, Edward
- The Horse in Motion ... see Stillman, J.D.B.

Myer, Jesse S.
- Life and Letters of Dr. William Beaumont ...
Introduction by Sir William Osler. St. Louis:
1912. 1st edn. 8vo. xxv,317 pp. Frontis port,

plates, ports, ills. Orig cloth.
          (Elgen) **$115 [≃ £71]**

Myrick, Herbert
- Sugar. A New and Profitable Industry in the
United States ... New York: Orange Judd,
1897. Roy 8vo. x,160 pp. Advts. Num ills.
Rebound in cloth. *(Berkelouw)* **$100 [≃ £62]**

Nagel, O.
- The Mechanical Appliances of the Chemical
and Metallurgical Industries. New York:
1908. 311,advt pp. 292 ills. Orig bndg.
          (Book House) **£25 [≃ $40]**
- Producer Gas Fired Furnaces. New York: PP,
1909. 192,advt pp. 237 ills. Orig bndg.
          (Book House) **£15 [≃ $24]**

Naismith, John
- General View of the Agriculture of the
County of Clydesdale ... Drawn up for the
Consideration of the Board of Agriculture.
Brentford: P. Norbury, 1794. 1st 4to edn. 81
pp. Half-title. Final blank fragmented. Sewed
as issued.   *(Blackwell's)* **£45 [≃ $72]**

Nall, G.H.
- The Life of the Sea Trout, especially in
Scottish Waters ... London: Seeley, Service,
1930. 8vo. 335 pp. 95 plates, 15 diags. Orig
cloth.   *(Egglishaw)* **£30 [≃ $48]**

Napier, James
- A Manual of Dyeing and Dyeing Receipts ...
London: Griffin, 1875. 3rd edn. 8vo. xxviii,
420,32 advt pp. 57 examples of fabrics laid in.
Orig cloth gilt, recased, sl dulled & worn,
front jnt cracked but sound.
          (Hollett) **£95 [≃ $152]**
- A Manual of Dyeing Receipts, for General
Use ... London & Glasgow, 1858. 2nd edn,
rvsd & enlgd. 12mo. [viii],88 pp. 56 mtd cloth
samples. Rec bds.
          (Charles B. Wood) **$150 [≃ £93]**

Nasmyth, James
- James Nasmyth Engineer. An
Autobiography. Edited by Samuel Smiles.
London: Murray, 1883. 1st edn. 8vo.
xviii,[1],456,[4 advt] pp. Port, 10 plates, 81
ills. Orig cloth gilt, recased.
          (Fenning) **£45 [≃ $72]**

Nasmyth, James & Carpenter, James
- The Moon: considered as a Planet, a World,
and a Satellite ... Second Edition. London:
1874. 2nd edn. 4to. xvi,189,[1],[2 advt] pp.
Frontis, 23 plates (inc 20 mtd woodbury-

types), num w'engvs in text. Orig dec blue cloth, spine ends & crnrs a little worn.
(Pickering) **$850 [≃ £531]**

The Natural History of Bees ...
- See Bazin, G.A.; Duncan, J.

The Naturalist ...
- The Naturalist. A Popular Monthly Magazine Illustrative of the Animal, Vegetable and Mineral Kingdoms. London: 1851-58. 8 vols. Num ills. Discreet lib stamps. Half calf.   *(Baldwin)* **£70 [≃ $112]**

Neal, Ernest
- The Badger. London: Collins, New Naturalist Series, 1948. 1st edn. 8vo. xv,158 pp. 30 photos (1 cold). Orig cloth, trifle used.
(Wheldon & Wesley) **£12 [≃ $19]**
- The Badger. London: Collins, New Naturalist Monographs No 1, 1948. 1st edn. 158 pp, 1 blank leaf. 12 maps & diags, 1 cold & 29 b/w photos. Orig green buckram, ft of upper cvr faded, minor splash marks on cvrs. Dw.   *(Sklaroff)* **£21 [≃ $33]**

Needham, J.
- A History of Embryology. Cambridge: 1934. 8vo. xviii,274 pp. 16 plates, 40 text figs. Orig cloth, sl used.
(Wheldon & Wesley) **£35 [≃ $56]**

Neele, G.P.
- Railway Reminiscences of Half a Century's Progress in Railway Working and of a Railway Superintendent's Life, principally on the LNWR. Privately Printed: 1904. Lge 8vo. 544 pp. Ills. Sl foxing of prelims & edges. Orig bndg.   *(Book House)* **£40 [≃ $64]**

Negretti & Zambra
- A Treatise on Meteorological Instruments ... London: Negretti & Zambra, 1864. 1st edn. Roy 8vo. xii,152,[8 ctlg] pp. 98 text w'cuts. Orig cloth, spine faded.
(Rootenberg) **$425 [≃ £265]**

Neilson, R.M.
- The Steam Turbine. London: Longmans, 1903. 8vo. 294 pp. Fldg plates, ills. Orig bndg.   *(Book House)* **£20 [≃ $32]**

Neligan, J. Moore
- A Practical Treatise on Diseases of the Skin. Philadelphia: Blanchard & Lea, 1852. 1st Amer edn. 133,12 ctlg pp. Sl browning. Orig cloth, spine ends & 1 edge frayed.
(Elgen) **$75 [≃ £46]**

Nelson, T.H. & Clarke, W.E.
- Birds of Yorkshire. London: A. Brown, 1907. 1st edn. Large Paper. 2 vols. 4to. xlv,374; xii,375-843 pp. 2 cold frontis & titles, 209 b/w plates & ills. Orig green cloth.
(Gough) **£115 [≃ $184]**
- The Birds of Yorkshire. London: 1907. 2 vols. 8vo. 2 cold frontis, 74 plates. Orig green cloth, trifle used & faded.
(Wheldon & Wesley) **£70 [≃ $112]**

Nesbit, A.
- A Complete Treatise on Practical Land Surveying, in Six Parts: designed chiefly for the Use of Schools. York: Thomas Wilson, for Longman, Hurst ..., 1810. 275 pp. 16 pp engvd Field Book at end. 7 plates. Sl browned & marked. Contemp half leather, rubbed.
(Moon) **£95 [≃ $152]**

Nesbit, J.C.
- On Agricultural Chemistry, and the Nature and Properties of Peruvian Guano. Third Edition. London: Longman, [1856]. 8vo. [iii], 128 pp. Orig pale green cloth gilt, spine sl faded.   *(Blackwell's)* **£45 [≃ $72]**

Nethersole-Thompson, D.
- The Greenshank. London: New Naturalist, 1951. 1st edn. 8vo. xii,244 pp. 4 cold & 14 plain plates. Orig cloth, trifle faded.
(Wheldon & Wesley) **£40 [≃ $64]**

New ...
- New Cyclopaedia of Botany ... see Brook, Richard.
- The New Dispensatory ... see Lewis, William.
- The New Farmer's Calendar ... see Lawrence, John.
- The New Handmaid to Arts, Sciences, Agriculture, etc. in Nine Books ... Husbandry ... Painting ... Dying ... Farriery ... Manchester: ptd by A. Swindells ..., [ca 1800]. 12mo. 96 pp. Outer margs trimmed just touching text on 10 ff. Disbound.
(Burmester) **£120 [≃ $192]**
- A New System of Domestic Cookery ... see Rundell, Maria Eliza.
- A New System of Natural History. Edinburgh: for Peter Hill, Edinburgh, & Thomas Cadell, London, 1791-92. 3 vols. 8vo. xviii,587; viii,568; viii,579 pp. 149 etched plates (some light offsetting). Contemp three qtr leather, rubbed.
(Schoyer) **$300 [≃ £187]**

Newbery, J.
- The Newtonian System of Philosophy ... Six Lectures read to the Lilliputian Society, by Tom Telescope ... Third Edition. London: J. Newbery, 1766. 140 pp. Frontis, 8 plates. Contemp green stained parchment backed blue bds, sl worn.
(Minster Gate) **£685 [≈ $1,096]**

Newhall, C.S.
- The Trees of North-Eastern America. New York: 1890. 8vo. xvi,250 pp. 166 figs. Orig cloth gilt. *(Henly)* **£32 [≈ $51]**

Newlands, James
- The Carpenter and Joiner's Assistant ... London: Blackie, [ca 1860]. Folio. xiii,291 pp. Addtnl engvd title, 115 plates, 828 ills. Contemp half calf. *(Fenning)* **£85 [≈ $136]**
- The Carpenter and Joiner's Assistant. London: Blackie, n.d. New edn. Sm folio. xii, 254 pp. 100 engvd plates. Half calf gilt, trifle worn. *(Hollett)* **£65 [≈ $104]**

Newlands, John A.R.
- On the Discovery of the Periodic Law, and on Relations among the Atomic Weights. London: Spon, 1884. 1st edn. 8vo. viii,39,[16 advt] pp. 2 fldg tables. Orig cloth.
(Rootenberg) **$250 [≈ £156]**
- On the Discovery of the Periodic Law and on Relations among the Atomic Weights. London: 1884. 1st edn thus. 12mo. viii,39,[16 advt] pp. 2 fldg plates. Orig cloth.
(Elgen) **$225 [≈ £140]**

Newman, Edward
- A Familiar Introduction to the History of Insects; being a new and greatly improved edition of the Grammar of Entomology. London: Van Voorst, 1841. 8vo. [xvi],288 pp, 2 advt ff. Num text engvs. Orig cloth, spine ends sl worn. *(Bickersteth)* **£35 [≈ $56]**
- A History of British Ferns. London: Van Voorst, 1840. 1st edn. 8vo. xxxvi,[ii],104 pp. Errata slip. W'engvd text figs & vignettes. Orig cloth, sl faded & spotted.
(Bickersteth) **£24 [≈ $38]**

Newman, John B.
- Fascination, or, the Philosophy of Charming, illustrating the Principles of Life in Connection Spirit and Matter. New York: Fowler & Wells, 1847. 1st edn. Sm 8vo. 176 pp. Ills. Orig bndg. *(Xerxes)* **$95 [≈ £59]**

Newstead, R.
- Guide to the Study of Tsetse-Flies.

Liverpool: 1924. Roy 8vo. xi,332 pp. 4 maps, 28 plates (3 cold). Wraps, trifle worn.
(Wheldon & Wesley) **£20 [≈ $32]**

Newton, Alfred
- A Dictionary of Birds. London: 1896. 8vo. xii, 1088 pp. Fldg map, text figs. Orig cloth. *(Henly)* **£20 [≈ $32]**
- A Dictionary of Birds. London: Adam & Charles Black, 1893-96. 1st edn. 4 vols. Tall 8vo. 1088,124 pp. Fldg cold map, num ills. Orig red cloth spines, ptd bds, rubbed, Linnean Society b'plate.
(Schoyer) **$75 [≈ £46]**
- Ootheca Wolleyana: an Illustrated Catalogue of the Collection of Birds' Eggs formed by the late John Wolley. Part 1 Accipitres. London: 1864. Roy 8vo. viii, 180, [2] pp. 10 cold & 8 plain plates. Orig bds, trifle worn.
(Wheldon & Wesley) **£75 [≈ $120]**

Newton, Sir Isaac
- The Chronology of Ancient Kingdoms Amended ... London: J. Tonson ..., 1728. 1st edn. Sm 4to. xiv,[2],376 pp. 3 fldg plates. Occas sl foxing. Contemp calf, rebacked, jnts & crnrs rubbed. *(Heritage)* **$450 [≈ £281]**
- The Correspondence ... Edited by H.W. Turnbull. Cambridge: UP for the Royal Society, 1959-60-61-67. 4 vols. 4to. Frontis, ports, ills. Orig cloth. Dws, as new.
(Rootenberg) **$225 [≈ £140]**
- The Correspondence. Edited by H.W. Turnbull. Volume I. 1661-1675. Cambridge: 1959. Frontis, 6 plates. Orig cloth.
(Whitehart) **£80 [≈ $128]**
- The Correspondence. Edited by H.W. Turnbull. Volume II. 1676-1687. Cambridge: 1960. Frontis, 6 plates. Orig cloth.
(Whitehart) **£80 [≈ $128]**
- The Correspondence. Edited by H.W. Turnbull. Volume III. 1688-1694. Cambridge: 1961. Frontis, 6 plates. Orig cloth. *(Whitehart)* **£80 [≈ $128]**
- The Mathematical Principles of Natural Philosophy ... Translated into English by Andrew Motte ... New Edition ... London: 1803. 3 vols. 8vo. Frontis port, 59 fldg plates, 2 fldg tables, w'engvd text diags. Minor foxing, plates sl browned. Contemp calf, sl worn. *(Pickering)* **$900 [≈ £562]**
- Opticks ... Fourth Edition. London: for William Innys ..., 1730. 8vo. [8], 382,[2 advt] pp. 12 plates. Sl foxed. Mod calf.
(Heritage) **$1,250 [≈ £781]**
- Opticks ... The Fourth Edition, Corrected. London: for William Innys ..., 1730. 8vo. [8], 382,[2 advt] pp. 12 fldg plates. Occas foxing,

mostly light. Contemp calf, rebacked.
(Heritage) **$1,000 [≈£625]**
- Opticks ... The Fourth Edition, Corrected ...
London: for William Innys ..., 1730. 4th edn.
8vo. viii,382,2 advt pp. 12 fldg plates.
Contemp calf, rebacked.
(Pickering) **$1,500 [≈£937]**
- A Treatise of the System of the World ...
Translated into English. London: F. Fayram,
1728. 1st edn in English of De Mundi
Systemate. 8vo. xxiv,154 pp. 2 plates, num
text w'cut diags. Contemp calf, rebacked,
crnrs reprd.
(W. Thomas Taylor) **$1.750 [≈£1,093]**

Newton, John Frank
- The Return to Nature, or, a Defence of the
Vegetable Regimen. Part the First [all
published]. London: Cadell & Davies, 1811.
1st edn. vi,160 pp. Contemp half calf,
rebacked.   *(Claude Cox)* **£85 [≈$136]**

Newton, Thomas
- An Illustration of Sir Isaac Newton's Method
of Reasoning, by Prime and Ultimate Ratios
... Leeds: ptd by Edward Baines ..., 1805. 1st
edn. 8vo. xiii,[i],60,[2] pp. Errata slip. 2 fldg
plates. Disbound.   *(Burmester)* **£35 [≈$56]**

The Newtonian System of Philosophy ...
- See Newbery, J.

Nichol, J.P.
- The Phenomena and Order of the Solar
System. Edinburgh: William Tait, 1838. [2
advt],241 pp. 21 plates. Orig cloth gilt, faded,
hd of spine reprd, ft of spine sl chipped.
(Hollett) **£55 [≈$88]**
- Thoughts on Some Important Points relating
to the System of the World. Edinburgh: 1848.
2nd edn. 8vo. 280 pp. 14 plates (some fldg),
text figs. Orig half calf.
(Bickersteth) **£38 [≈$60]**

Nichol, John
- Francis Bacon. His Life and Philosophy.
Edinburgh: 1888-89. 2 vols. 12mo. Frontis
ports. Orig dec cloth.   *(Elgen)* **£60 [≈£37]**

Nichols, J.T.
- The Fresh-Water Fishes of China. New York:
1943. 4to. xxxvi,322 pp. 10 cold plates, 143
text figs. Sm ownership stamps. Orig cloth.
(Wheldon & Wesley) **£150 [≈$240]**

Nichols, Thomas
- Observations on the Propagation and
Management of Oak Trees in general; but

more immediately applying to his Majesty's
New Forest, in Hampshire ... Southampton:
T. Baker, [1791]. 1st edn. 8vo. 42,[2] pp.
Disbound.   *(Burmester)* **£90 [≈$144]**

Nicholson, A. & Lydekker, R.
- Manual of Palaeontology. London: 1889. 3rd
edn. 2 vols. 8vo. Over 1400 text figs. Sl
foxing. Orig cloth, vol 2 front cvr damp
stained, vol 2 spine relaid.
(Henly) **£48 [≈$76]**

Nicholson, G. (ed.)
- The Illustrated Dictionary of Gardening.
London: [1884-85]. 4 vols in 8. Cr 4to. 24
cold plates, over 2300 text figs. Orig green
cloth, vol 1 rebacked, vol 3 trifle worn.
(Wheldon & Wesley) **£35 [≈$56]**
- The Illustrated Dictionary of Gardening.
London: 1884-88. 4 vols. Cr 4to. 24 cold
plates, 2377 text ills. Green half mor gilt, sl
rubbed.   *(Henly)* **£48 [≈$76]**
- The Illustrated Dictionary of Gardening.
London: 1884-88. 8 vols & the Century
Supplement 4 vols. 4to. 36 cold plates, num
text figs. Orig cloth gilt.
(Henly) **£65 [≈$104]**

Nicholson, H.A.
- A Manual of Palaeontology ... Second Edition
Revised and Enlarged. Edinburgh: 1879. 2
vols. 8vo. Erratum slip in vol 2. 722 figs. Orig
cloth, a little rubbed & dull, inner hinges a
little shaken.   *(Bow Windows)* **£55 [≈$88]**

Nicholson, William
- American Edition of the British Encyclopedia
or Dictionary of Arts and Sciences ...
Philadelphia: Mitchell, Ames & White,
1819-21. 12 vols. 8vo. "Upwards of 180
engravings". Contemp sheep, gilt spines, a
few hinges cracked.
(Charles B. Wood) **$300 [≈£187]**
- A Dictionary of Practical and Theoretical
Chemistry, with its Application to the Arts
and Manufactures ... London: for Richard
Phillips, 1808. 8vo. 12 engvd plates, 13 ptd
fldg tables. Orig tree calf, rebacked, old spine
preserved.   *(Bickersteth)* **£98 [≈$156]**
- A Dictionary of Chemistry. London: G.G.J.
& J. Robinson, 1795. 1st edn. 2 vols. 4to. viii,
1132 pp. 4 plates. Rec old style qtr calf.
(Charles B. Wood) **$750 [≈£468]**
- An Introduction to Natural Philosophy.
London: J. Johnson, 1787. 2nd edn. 2 vols.
8vo. xx,367; xi,376 pp. 25 fldg plates.
Contemp calf, spine ends worn, front jnts
cracked, vol 1 front cvr almost detached.

Cloth case. *(Schoyer)* **$300 [≈£187]**

Nicol, J.
- Guide to the Geology of Scotland. London: 1844. 8vo. 272 pp. 9 cold plates. Lacks map. Orig cloth. *(Baldwin)* **£33 [≈$52]**

Nicol, Walter
- The Gardener's Kalendar; or Monthly Directory of Operations in every Branch of Horticulture. Edinburgh: ptd by A. Neill ..., 1810. 1st edn. 8vo. xxiii,i,648 pp. Half- title. Sl spotting. Contemp qtr calf, sl rubbed. *(Young's)* **£95 [≈$152]**
- The Planter's Kalendar; or the Nurseryman's & Forester's Guide. Edited and Completed by Edward Sang. Edinburgh: 1812. 8vo. xx,595 pp, advt. 2 cold & 1 plain plates. Contemp calf, red gilt label. *(Spelman)* **£95 [≈$152]**

Nicoll, Michael & Meinertzhagen, Richard
- Nicoll's Birds of Egypt. London: Hugh Rees, 1930. 1st edn. 2 vols. Folio. xvi,348; 349-700 pp. Port, 3 maps, 37 plates (31 cold). Orig gilt dec green cloth. *(Terramedia)* **$800 [≈£500]**
- Nicoll's Birds of Egypt. London: 1930. 2 vols. 4to. 3 maps, 31 cold plates by G.E. Lodge, 6 photogravures, 88 text figs. Light foxing at ends. Orig green cloth, new endpapers. *(Henly)* **£275 [≈$440]**

Nicols, A.
- Zoological Notes on the Structure, Affinities, Habits, and Mental Faculties of Wild and Domestic Animals. London: 1883. 8vo. vii,370 pp. Ills. Orig bds, refixed. The author's copy with annotations. *(Wheldon & Wesley)* **£20 [≈$32]**

Nightingale, Florence
- Notes on Nursing: What It Is, And What It Is Not. London: Harrison & Sons, [1859]. 79 pp. Advts for 1860 publications on flyleaves. The right of Translation is reserved. Orig black pebble cloth, rebacked. *(Moon)* **£55 [≈$88]**
- Notes on Nursing ... London: Harrison, 59, Pall Mall, Bookseller to the Queen, [1860]. 1st edn. "The right of Translation is reserved" on title. 8vo. 79 pp. Strip cut from title. Orig cloth, yellow endpapers & pastedowns with advts, pastedowns sl defective. *(Bickersteth)* **£45 [≈$72]**
- Notes on Nursing: What It Is, and What It Is Not. New York: 1860. 1st Amer edn. Orig green cloth, yellow endpapers, hd of spine reprd. *(David White)* **£95 [≈$152]**
- Notes on Nursing ... New Edition, Revised

and Enlarged. London: Harrison ..., 1860. 2nd, enlgd, edn. 8vo. xvi,222,[2 advt for 'cheap' edn, colophon on verso] pp. Orig purple wavy grain cloth, jnts rubbed. Inscrbd by the author. *(Pickering)* **$1,600 [≈£1,000]**
- Notes on Nursing: What it is, and What it is not. New York: Appleton, 1860. 1st Amer edn. 8vo. Intl pp sl foxed. Orig bndg. *(Xerxes)* **$120 [≈£75]**
- Organization of Nursing ... Liverpool: A. Holden; London: Longman ..., 1865. 1st edn. 8vo. [8],102 pp. Frontis (sl foxed). Qtr mor over cloth. With the compiler's inscrptn on endpaper. *(Rootenberg)* **$550 [≈£343]**
- See also Herbert, Sidney & Nightingale, Florence.

Niktalopsia ...
- See Lynn, Walter.

Nisbet, J.
- Our Forests and Woodlands. London: Haddon Hall Library, 1900. 8vo. x,340 pp. 12 plates, head- & tail-pieces by Arthur Rackham. Orig cloth gilt, unopened. *(Henly)* **£45 [≈$72]**

Nissen, C.
- Herbals of Five Centuries. Munich: 1958. One of 100. Lge folio. 50 orig ff from early herbals mtd with ptd captions. Num ills on reverse sides. Unbound. Orig box. Includes a vol of text. *(Wheldon & Wesley)* **£1,200 [≈$1,920]**

Noble, Sir Andrew
- Artillery and Explosives ... London: Murray, December 1906. 1st edn. Super roy 8vo. xvi,548 pp. 79 plates, text figs. Orig cloth, sl soiled & rubbed, rear hinge cracked but sound. *(Duck)* **£85 [≈$136]**

Nordenskiold, E.
- The History of Biology: a Survey. New York: (1928) 1946. 8vo. xii,629,xv pp. 32 ports. Orig cloth. *(Wheldon & Wesley)* **£38 [≈$60]**

North, F.J., et al.
- Snowdonia, the National Park of North Wales. London: New Naturalist, 1949. 1st edn. 8vo. xviii,469 pp. 6 maps, 40 cold & 32 plain plates, 25 diags. Orig cloth, sl faded. Dw. *(Wheldon & Wesley)* **£28 [≈$44]**

Nott, J.F.
- Wild Animals, Photographed and Described. London: 1886. Roy 8vo. xi,568 pp. 38 plates (2 cold). Red morocco gilt.

(Wheldon & Wesley) **£30 [≈ $48]**

Nourse, T.
- Campania Felix, or, a Discourse of the Benefits and Improvements of Husbandry ... to which are added Two Essays: I, Of a Country- House. II. Of the Fuel of London. London: 1700. 8vo. [vi],354 pp. Frontis. Sl foxing. Contemp calf, rebacked.
(Wheldon & Wesley) **£200 [≈ $320]**

Nutt, Thomas
- Humanity to Honey-Bees: or, Practical Directions for the Management of Honey-Bees upon an Improved and Humane Plan ... Wisbech: 1832. 1st edn. 8vo. xxiii,240 pp. Frontis, 10 ills. Mod half calf.
(Wheldon & Wesley) **£100 [≈ $160]**
- Humanity to Honey-Bees. Wisbech: 1834. 2nd edn. 8vo. xxii,269 pp. Frontis, text ills. Amateur cloth.
(Wheldon & Wesley) **£70 [≈ $112]**
- Humanity to Honey-Bees ... Wisbech: John Leach, for the author ..., 1837. 4th edn, rvsd & enlgd. 8vo. xxx,281 pp. Frontis, fldg plate, 10 w'engvd text ills. A few ff carelessly opened. Binder's cloth.
(Burmester) **£60 [≈ $96]**

O'Brien, C. & Parkinson, C.
- Wild Flowers of the Undercliff, Isle of Wight. London: 1881. 8vo. vii,143 pp. 8 cold plates. Orig cloth. *(Wheldon & Wesley)* **£30 [≈ $48]**

Observations on Modern Gardening ...
- See Whately, Thomas.

Oetteking, Bruno
- Human Craniology. A Somato-Morphological Interpretation of the Human Cranium. New York: Chiropractic Institute, 1957. 1st edn. Sm 4to. 144 pp. Ills. Wraps. Sgnd pres copy.   *(Xerxes)* **$50 [≈ £31]**

Ogilvie-Grant, W.R.
- A Hand-Book to the Game-Birds. London: Edward Lloyd, 1896. 1st edn. 2 vols. 8vo. xvi,304; xvi,316 pp. 42 [sic] chromolitho plates. Orig maroon cloth, sl rubbed & marked, snag in 1 backstrip.
(Claude Cox) **£22 [≈ $35]**
- A Hand-Book to the Game-Birds. London: Lloyd's Natural History, 1896-97. 2 vols. Cr 8vo. 48 [sic] cold plates. Orig cloth, neatly refixed.   *(Wheldon & Wesley)* **£40 [≈ $64]**

Oldham, C.F.
- The Sun and the Serpent: A Contribution to

the History of Serpent-Worship. London: Constable, 1905. 1st edn. 8vo. 217 pp. Orig bndg.   *(Gach)* **$35 [≈ £21]**

Oliver, C.P.
- Meteors. Baltimore: Williams & Wilkins, 1925. 8vo. xvii,276 pp. 22 plates. Orig cloth. *(Gemmary)* **$50 [≈ £31]**

Oliver, W.R.B.
- New Zealand Birds. Wellington, 1955. 2nd edn, rvsd & enlgd. 4to. 661 pp. 12 cold plates, num ills. Orig cloth, crnrs sl bumped.
(Wheldon & Wesley) **£50 [≈ $80]**

Ommundsen, H. & Robinson, Ernest H.
- Rifles and Ammunition, and Rifle Shooting. London: Cassell, 1915. 1st edn. Cr 4to. xvi, [ii],335 pp. 64 plates, fldg diag plate, 37 text ills. Occas sl marks. Orig dec cloth gilt, t.e.g., sl marked & soiled.   *(Duck)* **£95 [≈ $152]**

Onania ...
- Onania; or, the Heinous Sin of Self-Pollution, and all its Frightful Consequences, in both Sexes, Consider'd ... Sixth Edition, Corrected and Enlarged. London: T. Crouch, 1722. 8vo. viii,104 pp. Title & last page soiled. Stitched as issued.
(Bickersteth) **£185 [≈ $296]**
- Onania Examined, and Detected: or, The Ignorance, Error, Impertinence, and Contradiction of a Book call'd Onania, Discovered and Exposed ... By Philo-Castitatis. London: 1723. Apparently sole edn. 12mo. [xii],129,3 advt pp. Title stained. Sewed as issued. *(Bickersteth)* **£220 [≈ $352]**

Oppian
- Oppian's Halieuticks of the Nature of the Fishes and Fishing of the Ancients in V. Books. Translated from the Greek. Oxford: 1722. 8vo. [viii],13, 232,[vii] pp. Contemp red mor gilt, trifle worn, rebacked.
(Wheldon & Wesley) **£120 [≈ $192]**

Origin ...
- The Origin and Proceedings of the Agricultural Associations in Great Britain, In which their Claims to Protection against Foreign Produce, Duty-free, are fully and ably set forth ... London: Ruffy & Evans, [1820]. 8vo. 46 pp. Some dust soiling. Sewed.
(Blackwell's) **£60 [≈ $96]**

Ormerod, Edward Latham
- British Social Wasps. London: 1868. xi,270 pp. 4 cold & 10 plain plates. Orig bndg, trifle used.   *(Wheldon & Wesley)* **£20 [≈ $32]**

- British Social Wasps: an Introduction ... London: 1868. Sole edn. 8vo. xi,270 pp. 14 plates (4 hand cold). Orig pict cloth, sl loose. *(Bickersteth)* £35 [≈ $56]

Ormerod, Eleanor A.
- A Manual of Injurious Insects with Methods of Prevention and Remedy for their Attacks on Food Crops, Forest Trees, and Fruit. To which is appended a Short Introduction to Entomology. London: 1890. xv,410,[1 advt] pp. Port, text engvs. Orig gilt dec cloth. *(Blackwell's)* £40 [≈ $64]

Osborn, Henry Fairfield
- The Origin and Evolution of Life. On the Theory of Action Reaction and Interaction of Energy. London: 1918. 8vo. xxxi,322 pp. Text figs. Cloth. *(Bickersteth)* £22 [≈ $35]

Osler, Sir William
- Bibliotheca Osleriana. A Catalogue of Books Illustrating the History of Medicine and Science. Montreal & London: McGill - Queen's UP, 1969. Thick lge 4to. 792 pp. Orig bndg. *(Xerxes)* $150 [≈ £93]
- Principles and Practice of Medicine. New York: 1896. 2nd edn. New lib buckram. *(Rittenhouse)* $100 [≈ £62]
- Principles and Practice of Medicine. New York: Appleton, 1907. 6th edn. 1143 pp. Orig bndg. *(Xerxes)* $115 [≈ £71]
- The Principles and Practice of Medicine ... Ninth Thoroughly Revised Edition. New York & London: 1921. Thick 8vo. xxiv,1168 pp. Endpapers marked. Orig cloth, a little dull. *(Bow Windows)* £45 [≈ $72]
- See also Abbott, Maude E. (ed.).

Osmond, Floris
- Microscopic Analysis of Metals. Edited by J.E. Stead. London: 1904. 1st English edn. Cr 8vo. x,178 pp. 3 fldg plates, 90 ills. Sm stamp on title. Orig cloth gilt. *(Fenning)* £45 [≈ $72]

Otter, William
- The Life and Remains of the Rev. Edward Daniel Clarke, LL.D. Professor of Mineralogy in the University of Cambridge. London: J.F. Dove, 1824. 1st edn. 4to. xii,670 pp. Port frontis. New cloth, a.e.g. *(Gemmary)* £125 [≈ $78]
- The Life and Remains of the Rev. E.D. Clarke, Professor of Mineralogy in the University of Cambridge. London: 1824. 4to. xii, 670 pp. frontis. Frontis & title foxed & browned, other pp lightly spotted. New qtr calf gilt. *(Baldwin)* £75 [≈ $120]

Overton, L.J.
- Heating and Ventilating. Manchester: Sutherland, [1920s?]. 8vo. 270,advt pp. Ills. Orig bndg. *(Book House)* £12 [≈ $19]

Owen, C.
- An Essay towards a Natural History of Serpents. London: 1742. 4to. xxiii,240,[12] pp. 7 engvd plates. Mod half calf, antique style. *(Wheldon & Wesley)* £280 [≈ $448]

Owen, David D.
- Report of a Geological Survey of Wisconsin, Iowa, and Minnesota; and incidentally of a Portion of Nebraska Territory. Philadelphia: Lippincott, 1852. 1st edn. Folio. 638,[1] pp. Fldg cold maps, 27 plates, ills. *(Jenkins)* $275 [≈ £171]

Owen, Sir Richard
- Description of the Fossil Reptilia of South Africa in the Collection of the British Museum. London: BM, 1876. xii,86 pp. 70 plates (many fldg). Binder's cloth, good ex-lib copy. *(Wheldon & Wesley)* £250 [≈ $400]
- Memoir on the Pearly Nautilus ... with Illustrations of its External Form and Internal Structure. London: 1832. 4to. 68 pp. 8 plates (7 duplicated in outline). Orig cloth, rebacked, trifle warped & soiled, sound ex-lib. *(Wheldon & Wesley)* £150 [≈ $240]
- Odontography; or a Treatise on the Comparative Anatomy of the Teeth ... London: Bailliere, 1840-45. 1st edn. 2 vols (text & atlas). Lge 8vo. xix,[3],lxxiv,655,[3]; 37, [1] pp. 168 litho plates (numbered 1-150). Perf lib stamp on titles. Sl foxing. Orig half mor. *(Rootenberg)* $750 [≈ £468]
- On the Anatomy of Vertebrates Vol 1. Fishes and Reptiles. London: 1866. Vol 1 only (of 3). 8vo. xlii,650 pp. 452 text figs. Orig cloth, trifle worn, jnts loose, hd of spine sl defective. *(Wheldon & Wesley)* £50 [≈ $80]
- On the Classification and Geographical Distribution of the Mammalia ... appendix "On the Gorilla' and "on the Extinction and Transmutation of Species". London: Parker, 1859. 103,4 advt pp. Lib stamp on title. Orig cloth gilt. *(Egglishaw)* £55 [≈ $88]
- Palaeontology or a Systematic Summary of Extinct Animals and their Geological Relations. London: 1861. 2nd edn. 8vo. xvi, 463 pp. 173 figs. Orig cloth. *(Baldwin)* £55 [≈ $88]

Owen, Richard, Professor of Geology, Nashville
- Key to the Geology of the Globe. Nashville:

1857. 8vo. 256 pp. 2 fldg geological maps (1 cold). Some marg stains & browning, a few folds with tears. Orig cloth, spine relaid.
(Baldwin) **£45 [≈ $72]**

Owen, Richard Startin
- Life of Richard Owen by his Grandson ... London: Murray, 1894. 1st edn. 2 vols. 8vo. 409; 393 pp. Port, ills. Orig bndg, 1 hinge tender.   *(Xerxes)* **$150 [≈ £93]**
- The Life of Richard Owen by his Grandson ... The Scientific Portions revised by C. Davies Sherborn ... Also an Essay ... by the Right Hon. T.H. Huxley ... London: Murray, 1894. 1st edn. 2 vols. 8vo. 2 port frontis, 5 plates, text ills. Orig cloth gilt, sl rubbed.   *(Burmester)* **£32 [≈ $51]**

Ozanam, J.
- Recreations in Mathematics and Natural Philosophy. London: 1840. xiv,826 pp. Over 400 w'cut ills. Orig cloth, dust stained & worn, hd of spine defective.
(Whitehart) **£40 [≈ $64]**

Padgett, Earl C.
- Surgical Diseases of the Mouth and Jaws. Philadelphia: 1938. 1st edn. 807 pp. 334 ills. Orig bndg.   *(Elgen)* **$150 [≈ £93]**

Page, D.
- Economic Geography. London: 1874. Cr 8vo. xv,336,16 advt pp. Cold geol map in pocket, text figs. Orig cloth gilt.
(Henly) **£15 [≈ $24]**
- Hand-Book of Geological Terms, Geology and Physical Geology. London: 1865. 2nd edn. 8vo. 506,4 advt pp. Orig cloth, new endpapers.   *(Henly)* **£18 [≈ $28]**
- The Past and Present Life of the Globe. London: 1861. 1st edn. Cr 8vo. 256,16 advt pp. Num w'cut text ills. V sl foxing at ends. Orig pict cloth gilt.   *(Henly)* **£15 [≈ $24]**

Paget, C.J. & James
- Sketch of the Natural History of Yarmouth ... London: [ptd by F. Skill, Yarmouth] Longman, Rees, 1834. 8vo. xxii,88 pp. Sm blank piece cut from title. Half mor. A few early MS notes. From the library of C.G. Doughty.   *(Lamb)* **£65 [≈ $104]**

Paget, Sir James
- Memoirs and Letters of Sir James Paget. Edited by [his son] Stephen Paget. London: 1901. 1st edn. 438,32 ctlg pp. 3 ports, 7 plates. Orig cloth, sl rubbed.
(Elgen) **$75 [≈ £46]**

Paine, Martyn
- Physiology of the Soul and Instinct, as Distinguished from Materialism. With Supplementary Demonstrations of the Divine Communication of the Narratives of Creation and the Flood. New York: Harper, 1872. 1st edn. 8vo. 707,[3],9,[3] pp. Orig cloth.
(Gach) **$85 [≈ £53]**

Pajot des Charmes, C.
- The Art of Bleaching Piece-Goods, Cottons, and Threads, of Every Description, rendered more easy ... Translated from the French [by William Nicholson] ... London: Robinson, 1799. 8vo. xvi,351 pp. 9 fldg plates. Orig pink bds, uncut, respined in paper.
(Charles B. Wood) **$350 [≈ £218]**

Palgrave, R.H. Inglis
- Dictionary of Political Economy ... London & New York: Macmillan, 1894. 1st edn. 3 vols & Appendix to vol 3 (reprint 1909). 8vo. Orig dark blue cloth, appendix in orig faded grey-blue wraps, spine of 1st 2 vols reprd.
(Pickering) **$750 [≈ £468]**

Palmer, E. & Pitman, N.
- Trees of South Africa. Cape Town: 1972-73. 2nd edn. 3 vols. 4to. 24 cold plates, num other ills. Orig cloth.
(Wheldon & Wesley) **£220 [≈ $352]**

Palmer, T.S.
- Index Generum Mammalium, A List of the Genera and Families of Mammals. Washington: North American Fauna No 23, 1904. Orig printing. 8vo. 984 pp. Qtr cloth.
(Wheldon & Wesley) **£65 [≈ $104]**

Pardo-Castello, V.
- Diseases of the Nails. Springfield, Ill.: Charles Thomas, 1941. 2nd edn. 8vo. 193 pp. Photo ills. Orig bndg.   *(Xerxes)* **$95 [≈ £59]**

Pareto, Vilfredo
- The Mind and Society ... Edited by Arthur Livingston, Translated by Andrew Bongiorno and Arthur Livingston. London: Cape, 1935. 1st edn in English. 4 vols. 8vo. Frontis port in vol 1. Orig black cloth.
(Pickering) **$600 [≈ £375]**

Paris, J.A.
- Pharmacologia; Comprehending the Art of Prescribing ... London: W. Phillips, 1822. 3rd edn (?). 2 vols. 8vo. xii,448; 464,3,[4 advt] pp. Occas staining, mostly light. Orig bds, uncut.   *(Gough)* **$95 [≈ £152]**

- Pharmacologia; corrected & extended in accordance with the London Pharmacopoeia of 1824 ... New York: Samuel Wood, 1825. "3rd Amer edn". 2 vols. Volvule mtd inside front cvr. Some foxing & browning, marg stain on 40 ff. Cloth backed bds, worn. *(Elgen)* **$175 [≈£109]**
- Pharmacologia. New York: Duyckinck ..., 1828. "3rd Amer edn". 2 vols. Volvule in facs, mtd inside front cvr. Some foxing & browning. Contemp calf, scuffed. *(Elgen)* **$125 [≈£78]**
- Philosophy in Sport made Science in Earnest: being an Attempt to Implant in the Young Mind the First Principles of Natural Philosophy ... London: Murray, 1853. 7th edn, enlgd. xxiv,528,32 ctlg pp. Ills by Cruikshank. Orig cloth, jnts reprd, new endpapers. *(Box of Delights)* **£35 [≈$56]**

Parish, Edmund

- Hallucinations and Illusions: A Study of the Faculties of Perception. London: Walter Scott, 1897. 1st edn in English. 12mo. [xvi], 390,[18] pp. 4 pp advts at front. Orig red cloth. *(Gach)* **$65 [≈£40]**

Park, J.

- The Cyanide Process of Gold Extraction. London: Charles Griffin, 1913. 4th English edn. 8vo. xiv,239 pp. 12 plates (mostly fldg), 8 w'cuts. Orig cloth. *(Gemmary)* **$65 [≈£40]**

Parker, H. & Bowen, Frank C.

- Mail and Passenger Steamships of the Nineteenth Century. The Macpherson Collection. With Iconographical and Historical Notes. London: Sampson Low, 1928. Sm folio. xxviii,324 pp. 186 plates (16 cold). Orig cloth, sl soiled. *(Karmiole)* **$300 [≈£187]**

Parker, Langston

- Modern Treatment of Syphilitic Diseases both Primary and Secondary. London: Churchill, 1845. 2nd edn. 8vo. 228 pp. Front & rear blank pp gone. Orig bndg, hinges cracked. *(Xerxes)* **$60 [≈£37]**

Parkes, Oscar & Perkins, E.

- The Detection of Disease. London: Sampson, Low, 1930. 1st edn. 8vo. 116 pp. Photo ills. Orig bndg, bds sl warped. *(Xerxes)* **$80 [≈£50]**

Parkinson, James

- Fossil Organic Remains. London: 1830. 2nd edn. 8vo. viii,350 pp. 10 plates. Orig cloth. *(Baldwin)* **£60 [≈$96]**

- Organic Remains of a Former World ... London: 1833. 2nd edn. 3 vols. 4to. 3 frontis (2 cold), 3 title vignettes, 51 hand cold plates. Lacks half-titles vols 1 & 3. Some offsetting from the plates, a pencil notes. Orig cloth, trifle worn. *(Wheldon & Wesley)* **£650 [≈$1,040]**
- Organic Remains of a Former World ... London: 1833. 2nd edn. 3 vols. 4to. 3 frontis (2 cold), 50 plates (49 hand cold). Lacks half-titles vols 1 & 3. Blind stamp on title & title-pages. Lib b'plates. Buckram. *(Baldwin)* **£650 [≈$1,040]**
- Outlines of Oryctology. Fossil Organic Remains ... with the Formation of the Earth ... London: 1822. 1st edn. 8vo. vii,[1],3436 [sic] pp. 10 engvs. Half calf. *(Hemlock)* **$400 [≈£250]**

Parkinson, John

- Paradisi in Sole Paradisus Terrestris. Faithfully reprinted from the Edition of 1629. London: 1904. Folio. Engvd title, port, num full-page w'cuts. Orig bds, uncut, trifle used, trifle loose. *(Wheldon & Wesley)* **£130 [≈$208]**
- Paradisi in Sole Paradisus Terrestris. Faithfully reprinted from the Edition of 1629. London: Methuen, 1904. Folio. Facs of engvd title, num ills. Orig cloth backed bds, uncut, minor foxing on edges. *(Elgen)* **$295 [≈£184]**
- Theatrum Botanicum: The Theater of Plants, or, an Herball of a large Extent ... London: 1640. Folio. [xx],1746 pp. Front errata leaf (mtd). Addtnl engvd title, over 2500 w'cuts in text. Sl marg staining, minor marg tears. Calf, rebacked. *(Wheldon & Wesley)* **£1,500 [≈$2,400]**

Parkinson, Richard

- The Experienced Farmer, an Entire New Work in which the Whole System of Agriculture, Husbandry, and Breeding of Cattle is explained ... London: Robinson, 1798. 1st edn. 2 vols. 8vo. xx,302; [iii],315,[3] pp. Subscribers, errata, advt leaf. Contemp calf gilt, sl worn. *(Hollett)* **£295 [≈$472]**

Parnell, R.

- The Grasses of Britain. London: [1842-] 1845. Roy 8vo. xxvii,xxi,311 pp. 142 plates. New cloth. Comprises: Grasses of Scotland 1842 & Grasses of Britain 1845. *(Wheldon & Wesley)* **£60 [≈$96]**
- The Grasses of Scotland. Edinburgh: Blackwood, 1842. Roy 8vo. xxi,152 pp. 66 plates. Prize calf gilt, rebacked. *(Egglishaw)* **£50 [≈$80]**

Parry, Caleb Hillier
- Cases of Tetanus and Rabies Contagiosa, or Canine Hydrophobia, with Remarks ... Bath: 1814. 1st edn. 8vo. vi,218 pp. Stamp on title verso. Orig bds, paper label, uncut, sm defects to front hinge & spine.
(Hemlock) **$275 [≈ £171]**

Parshall, H.F. & Hobart, H.M.
- Electric Railway Engineering. London: Constable, 1907. 1st (only?) edn. Lge imperial 8vo. xxiv,475 pp. 437 ills. Orerig cloth, gilt spine, edges sl rubbed & marked.
(Duck) **£85 [≈ $136]**

Parsons, Frank Alvah
- The Psychology of Dress. Garden City, New York: Doubleday, Page, 1921. 1st edn. 8vo. [xxiv], 358,[5] pp. Frontis, 140 pp of halftone ills. Orig green cloth. *(Gneach)* **$35 [≈ £21]**

Parsons, R.H.
- The Early Days of the Power Station Industry. Cambridge: Babcock & Wilcox & Univ Press, 1939. Lge 8vo. 217 pp. Ills. Orig bndg. *(Book House)* **£25 [≈ $40]**

Parsons, Talcott
- The Structure of Social Action ... New York & London: McGraw-Hill, 1937. 1st edn, 1st printing. 8vo. [xiv],818 pp. Orig black cloth. Dw lightly worn. *(Gach)* **$125 [≈ £78]**

Parsons, William Barclay
- Rapid Transit in Foreign Cities. New York: privately issued, 1894. Sm 4to. 66 pp. 53 pp ills. Mod cloth. *(Book House)* **£85 [≈ $136]**

Pass, Crispin de
- Hortus Floridus. Contayning a very lively and true description of the Flowers of the Springe, Summer, Autumn and Winter. Translated from the Latin by S. Savage. London: Cresset Press, 1928-29. One of 500. 2 vols. Oblong 4to. Orig half leather, vol 2 refixed. *(Wheldon & Wesley)* **£100 [≈ $160]**

Passages ...
- Passages from the Diary of a Late Physician ... see Warren, Samuel.

Pasteur, Louis
- Studies on Fermentation, The Diseases of Beer, Their Causes, and the Means of Preventing Them .... London: Macmillan, 1879. 1st edn in English. 8vo. [xvi],418. ,[2 advt] pp. 12 plates, text ills. Orig brown cloth, extremities sl rubbed, edges sl foxed.
(Heritage) **$600 [≈ £375]**

Patchell, W.H.
- Application of Electric Power to Mines and Heavy Industries. London: Constable, 1913. 8vo. 333 pp. Ills. Orig bndg.
(Book House) **£15 [≈ $24]**

Patents for Inventions
- Abridgements of Specifications relating to Carriages and Other Vehicles for Railways. A.D. 1807-1866. London: Commissioners for Patents, 1871. xxxiv,[ii],1496,15 pp. 1 leaf torn with sl loss. Orig cloth, discold, hd of spines frayed. *(Duck)* **£75 [≈ $120]**
- Abridgements of Specifications relating to Raising, Lowering, and Weighing. A.D. 1617 - 1865. London: Commissioners for Patents, 1867. xxx,[ii],1204,[errata 2],8 pp. Part of contents sl bumped at top edge. Orig cloth, discold, top edge of rear cvr frayed.
(Duck) **£75 [≈ $120]**

Patten, C.J.
- The Aquatic Birds of Great Britain and Ireland. London: Porter, 1906. 8vo. xxx,590 pp. 56 plates, 68 text figs. New cloth.
(Egglishaw) **£28 [≈ $44]**

Patterson, A.H.
- Nature in Eastern Norfolk. London: 1905. 8vo. 352 pp. 12 cold plates. Sl foxing. Orig cloth. *(Baldwin)* **£25 [≈ $40]**
- Notes of an East Coast Naturalist. London: 1904. Cr 8vo. xiii,304 pp. 12 cold plates by Frank Southgate. Orig cloth, trifle used.
(Wheldon & Wesley) **£30 [≈ $48]**
- Wild Life on a Norfolk Estuary. London: 1907. 8vo. xv,352 pp. 40 ills. Half-title rather foxed, minor foxing elsewhere. Orig cloth, trifle worn. *(Wheldon & Wesley)* **£35 [≈ $56]**

Patterson, David
- Colour-Matching on Textiles. A Manual intended for the Use of Dyers ... London: Scott, Greenwood, 1901. 1st edn. 8vo. xii, 128, 36 pp. Chromolitho frontis, 29 text ills, 14 textile samples mtd on 4 plates. Orig cloth.
(Charles B. Wood) **$150 [≈ £93]**
- The Science of Colour Mixing. A Manual intended for the Use of Dyers ... London: Scott, Greenwood, 1900. 1st edn. 8vo. xii, 128, 32 pp. 5 chromolitho plates, 41 text ills, 4 pattern plates with 11 mtd specimens. Half-title clipped. Orig cloth.
(Charles B. Wood) **$100 [≈ £62]**
- Textile Colour Mixing. A Manual intended for the Use of Dyers ... Second Revised

for the Use of Dyers ... Second Revised
Edition. London: Scott, Greenwood, 1915.
8vo. xii,128,24 pp. 5 chromolitho plates, 41
text ills, 4 pattern plates with 11 mtd
specimens. Orig cloth.
(Charles B. Wood) **$85 [≈ £53]**

Patterson, R.
- Letters on the Natural History of the Insects
mentioned in Shakspeare's Plays with
Incidental Notices of the Entomology of
Ireland. London: 1838. Post 8vo. xv,270 pp.
Half calf.   *(Wheldon & Wesley)* **£30 [≈ $48]**

Patterson, R. Hogarth
- The New Golden Age and Influence of the
Precious Metals upon the World ...
Edinburgh & London: Blackwood, 1882. 1st
edn. 2 vols. 8vo. xiv,487; vi,542 pp. Orig
turquoise cloth, gilt spines, some cockling on
vol 2.   *(Pickering)* **$500 [≈ £312]**

Pavlov, I.P.
- Conditioned Reflexes ... Translated and
Edited by G.V. Anrep. OUP: 1927. 1st edn.
Lge 8vo. [xvi],430,[2] pp. Orig black cloth. Sl
defective dw, edges & jnts of dw chipped, sm
section cut from spine. *(Gach)* **$250 [≈ £156]**
- Lectures on Conditioned Reflexes ...
Translated from the Russian by W. Horsley
Gantt ... New York: International Publishers,
[1928]. 1st edn in English. 8vo. 414,[2] pp. 4
plates. Orig cloth. Dw. *(Gach)* **$150 [≈ £93]**
- Work of the Digestive Glands. Translated by
W.H. Thompson. London: Charles Griffin,
1910. 2nd English edn. 8vo. 266 pp. Ills. Orig
bndg, vg ex-lib.   *(Xerxes)* **$75 [≈ £46]**

Pavy, F.W.
- The Physiology of Carbohydrates; their
Application as Food and Relation to Diabetes.
London: J. & A. Churchill, 1894. 1st edn.
8vo. x,280 pp. Text figs. Orig cloth.
(Bickersteth) **£40 [≈ $64]**

Payne, George
- Elements of Mental and Moral Science
Intended to Exhibit the Original
Susceptibilities of the Mind ... London: for
B.J. Holdsworth, 1828. 8vo. Orig bndg.
(Gach) **$85 [≈ £53]**

Payne, John
- Tables for Valuing Labor and Stores, by
Weight or Number ... London: W.
Winchester & Son, 1811. Folio. Thumb
indexes along leading edge. Contemp
polished calf, upper jnt v sl cracked.
(Spelman) **£200 [≈ $320]**

Peabody, Selim H.
- Cecil's Books of Natural History ... Part I [-
III]. New York: John B. Alden, 1885. Sm
8vo. xiii,[9]-215; [11]-234; [11]-228 pp.
Divisional half-titles. Num ills. Dark yellow
morocco grained cloth gilt.
(Blackwell's) **£30 [≈ $48]**

Peach, B.N., et al.
- The Silurian Rocks of Britain, Vol 1.
Scotland. Glasgow: Geol Survey, 1899. Roy
8vo. xviii,749 pp. Fldg map (few sm tears), 27
plates, 121 figs. Orig cloth, good ex-lib, spine
faded, inner jnts strengthened.
(Wheldon & Wesley) **£65 [≈ $104]**

Pearce, E.K.
- Typical Flies. A Photographic Atlas of
Diptera, including Aphaniptera. Cambridge:
1915. Lge 8vo. xi pp text, 47 pp of photo ills.
Orig ptd bds, buckram spine. Author's pres
inscrptn & ALS.   *(Bickersteth)* **£25 [≈ $40]**

Pearsall, W.H.
- Mountains and Moorlands. London: Collins
New Naturalist, 1950. 1st edn. 8vo. Orig
cloth. Dw.   *(Wheldon & Wesley)* **£20 [≈ $32]**
- Mountains and Moorlands. London: Collins,
New Naturalist Series, 1950. 1st edn. Cold &
plain plates. Orig cloth. Dw.
(Egglishaw) **£25 [≈ $40]**

Pearson, H.J.
- Three Summers Among the Birds of Russian
Lapland. London: R.H. Porter, 1904. 1st
edn. 8vo. xvi,216 pp. Fldg map, 68 photo
plates. Orig green cloth gilt.
(Gough) **£110 [≈ $176]**

Pearson, Henry C.
- Rubber Machinery. An Encyclopaedia of
Machines ... New York: The India Rubber
World, 1915. 1st edn. Roy 8vo. 420,[34 illust
advt] pp. 428 ills. Orig cloth.
(Duck) **£35 [≈ $56]**

Pearson, Karl
- The Life, Letters and Labours of Francis
Galton. Cambridge: UP, 1914-1930. 3 vols in
4. Imperial 8vo. 5 + 2 pedigrees in 2 pockets,
190 plates & ports, num text ills. Orig dec
buckram, cvrs spotted & faded.
(Bickersteth) **£425 [≈ $680]**

Pearson, Leonard & Warren, B.H.
- Diseases and Enemies of Poultry.
Pennsylvania: Busch, 1897. Thick 4to.
116,749 pp. Lge fldg chart, num cold plates,

(Xerxes) $90 [≈ £56]

Pease, A.
- The Book of the Lion. London: 1914. 293 pp. Ills. Orig bndg.
(Trophy Room Books) $200 [≈ £125]

Pechey or Peachey, John
- A Plain and Short Treatise of an Apoplexy, Convulsions, Colick .... and several other Dangerous and Violent Diseases ... Second Edition with Additions ... London: for the author, 1708. Sm 8vo. [vi],45,[3] pp. Cut close with sl loss. Orig mrbld wraps.
(Bickersteth) £200 [≈ $320]

Peck, W.
- A Popular Handbook and Atlas of Astronomy. London: Gall & Inglis, 1890. Lge 4to. 176 pp. 54 plates. Orig dec bndg, edges rubbed, new hinges.
(Book House) £50 [≈ $80]

Peckham, George W. & Elizabeth G.
- Wasps Social and Solitary. With an Introduction by John Burroughs. Westminster: Constable, 1905. 8vo. xv,311 pp. Num ills by James H. Emerton. Orig cloth.
(Bickersteth) £28 [≈ $44]

Peele, R. (ed.)
- Mining Engineers' Handbooks. New York: John Wiley, 1941. 3rd edn. 2 vols. 8vo. 2442 pp. Num text figs, tables, graphs. Flexible cloth.
(Gemmary) $60 [≈ £37]

Peile, H.D.
- A Guide to Collecting Butterflies of India. London: 1937. xiv,312 pp. 25 plates (1 cold). Orig cloth.
(Wheldon & Wesley) £45 [≈ $72]

Peirce, Benjamin
- Ideality in the Physical Sciences. Boston: Little, Brown, 1881. 1st edn. 8vo. Orig ptd green cloth, cvrs flecked & shelfworn.
(Gach) $125 [≈ £78]

Pemberton, Christopher Robert
- A Practical Treatise on Various Diseases of the Abdominal Viscera. Second Edition, corrected and enlarged. London: W. Bulmer for C.W. Nicol, 1807. 8vo. xv,[1],201 pp. 2 plates. Half calf.
(Hemlock) $250 [≈ £156]

Pemberton, Henry
- A View of Sir Isaac Newton's Philosophy. London: ptd by S. Palmer, 1728. 1st edn. 4to. [50],407 pp. 12 fldg plates, engvd decs by John Pine. Title-page sl discold. Contemp panelled calf, rebacked, crnrs worn. Cowper family b'plate.
(Pickering) $800 [≈ £500]

- A View of Sir Isaac Newton's Philosophy. London: 1728. 1st edn. 4to. 24 ff,407 pp. 12 fldg plates, 12 vignette & tailpieces. Contemp vellum, rather badly dust stained.
(Whitehart) £300 [≈ $480]

- A View of Sir Isaac Newton's Philosophy. Dublin: reptd for John Hyde, & John Smith & William Bruce, 1728. 1st Dublin edn. 8vo. [xl], 333,[i] pp. 12 fldg plates. Orig calf, lacks label, spine sl rubbed & sm chip at hd.
(Bickersteth) £260 [≈ $416]

Pemberton, Max
- The Amateur Motorist. London: Hutchinson, 1907. 1st edn. xii,328,[4 advt] pp. Frontis port, 31 plates, 18 text ills. Orig red cloth, t.e.g., sl marked & faded.
(Duck) £65 [≈ $104]

Pennant, Thomas
- Arctic Zoology. London: 1784-85. 2 title, vignettes, 24 plates. [With] Supplement. London: 1787. 2 maps. 3 vols in 2. 4to. Contemp polished calf.
(Wheldon & Wesley) £450 [≈ $720]

- British Zoology. A New Edition. London: for Wilkie & Robinson ..., 1812. 5th edn. 4 vols. 8vo. 4 engvd titles, 293 (of 294, lacks plate 40) plates. Lacks 3 ff in vol 4 prelims. Light offsetting from some plates. Orig half leather, spines relaid.
(Schoyer) $300 [≈ £187]

- British Zoology. New Edition. London: 1812. 4 vols. 8vo. 4 addtnl engvd titles, 293 plates. Calf, rebacked.
(Henly) £195 [≈ $312]

- Genera of Birds. London: for B. White, 1781. 2nd edn (1st 4to edn). 4to. xxvi,68,[2] pp. Engvd title with vignette, 16 plates. Unobtrusive lib stamp on verso of plates. New calf backed bds.
(Claude Cox) £55 [≈ $88]

Penny Cyclopaedia ...
- Penny Cyclopaedia of the Society for the Diffusion of Useful Knowledge. London: Charles Knight, 1833. 27 vols. Lge 4to. Sl spotting. Orig cloth, spine labels, spines faded, a few sm splits to hinges.
(Hollett) £75 [≈ $120]

- Penny Encyclopaedia of the Society for the Diffusion of Useful Knowledge. London: Charles Knight, 1833-58. 27 vols in 14, plus 3 vols of Supplements, 30 vols in all. Num w'cuts. Orig cloth, mildly ex-lib, some vols with loose cvrs & tears at jnts.
(Elgen) $850 [≈ £531]

(Elgen) $850 [≈ £531]

Pepper, John H.
- The Playbook of Metals ... New Edition. London: Routledge, [ca 1870]. Cr 8vo. viii, 502,[2] pp. Frontis, 265 ills. Orig cloth gilt, a.e.g., inside hinges cracked.
(Fenning) £35 [≈ $56]

Percival, M.
- Chats on Old Jewellery and Trinkets. London: Fisher Unwin, 1912. 8vo. 384 pp. Frontis, 38 other plates, 56 drawings. Some foxing. Orig cloth. *(Gemmary)* $55 [≈ £34]

Percy, John
- Metallurgy: The Art of Extracting Metals from their Ores. London: Murray, 1875. Rvsd edn. 8vo. 616 pp. 9 fldg plates, 112 text figs. Orig cloth, worn, cvr detached.
(Gemmary) $20 [≈ £12]

Pering, Richard
- A Brief Inquiry into the Causes of Premature Decay, in our Wooden Bulwarks, with an Examination of the Means, best calculated to Prolong their Duration. Plymouth-Dock: 1812. 1st edn. 8vo. Half-title (soiled), title, 78 pp, errata leaf. Mod bds, unopened.
(Bickersteth) £145 [≈ $232]
- A Treatise on the Anchor ... Plymouth Dock: 1819. Sole edn (?). Roy 8vo. i-viii, 9-98, [4] pp. 3 fldg plates. Contemp tree calf, rebacked.
(Duck) £195 [≈ $312]

Perrin, Mrs Henry
- British Flowering Plants. New and revised Edition. Edited by A.K. Jackson. London: Ward Lock, 1939. Roy 4to. 667 pp. 260 cold plates. Orig cloth. *(Egglishaw)* £28 [≈ $44]

Perrin, Mrs Henry & Boulger, G.S.
- British Flowering Plants. London: 1914. One of 1000. 4 vols. 4to. 300 cold plates. Orig buckram. *(Wheldon & Wesley)* £150 [≈ $240]
- British Flowering Plants. London: Quaritch, 1914. One of 1000. 4 vols. 4to. 300 cold plates. Light blind stamps on plate margs. Orig cream buckram gilt, t.e.g.
(Hollett) £220 [≈ $352]

Perrin, Jean
- Atoms. Translated by D. Ll. Hammick. London: Constable, 1920. 2nd imp. [xvi],212 pp. Orig dark red cloth. Arthur W. Chapman's copy, with prize label.
(Sklaroff) £40 [≈ $64]

Perring, F.H. & Walters, S.M.
- Atlas of the British Flora. London: BSBI & Thomas Nelson, 1962. 1st edn, 2nd imp. Lge 4to. xxiv,432 pp. 6 overlays in rear pocket. Orig cloth gilt. Dw.   *(Hollett)* £60 [≈ $96]

Person, David
- Varieties: or, a Surveigh of Rare and Excellent Matters ... London: 1635. 4to in 8s. Sep titles to the 5 books. Blank ff between Books 2 & 3, 3 & 4. Aa1 cancelled & replaced by (Aa2). Contemp calf, old reback, crnrs worn, jnts cracked & tender. STC 19781.
(Clark) £300 [≈ $480]

Petersen, William
- Man Weather Sun. Springfield, Ill.: Thomas, 1947. 1st edn. Sm 4to. Orig bndg.
(Xerxes) $90 [≈ £56]

Pettigrew, J.B.
- Design in Nature, illustrated by Spiral and other Arrangements in the Inorganic and Organic Kingdoms. London: 1908. 3 vols. 4to. 3 ports, 182 plates, 581 text figs. Vol 1 trifle foxed. Cloth.
(Wheldon & Wesley) £85 [≈ $136]

Pettigrew, Thomas Joseph
- On Superstitions connected with the History and Practice of Medicine and Surgery. Philadelphia: Barrington & Haswell, 1844. 1st Amer edn, 1st printing. 12mo. 213,[3] pp. Orig embossed cloth.   *(Gach)* $65 [≈ £40]

Pettus, J.
- Fleta Minor. The Laws of Art and Nature, in Knowing, Judging, Assaying, Fining, Refining and Inlarging the Bodies of Confin'd Metals ... London: Stephen Bateman, 1686. Folio. [42],345,[1],[141] pp. Red & black title. 44 ills. Leather.
(Gemmary) $3,000 [≈ £1,875]

Pfeffer, W.
- The Physiology of Plants. A Treatise upon the Metabolism and Sources of Energy in Plants. Second Edition. Translated by A.J. Ewart. London: 1900-06. 3 vols. Roy 8vo. Text figs. Half mor.
(Wheldon & Wesley) £50 [≈ $80]

Phalon, Edward
- Treatise on the Hair. New York: Phalon, 1847. 12 pp. Ills. Orig bndg.
(Xerxes) $125 [≈ £78]

Pharmacopoeia ...
- British Pharmacopoeia, 1867. London: Spottiswoode, 1867. 2nd edn. xxiv,434 pp. Orig cloth, gilt cvr medallion, ex-lib, sm tears at spine ends, sl blistering where number removed from spine.   *(Elgen)* **$75 [≈£46]**
- The Pharmacopoeia of the United States of America. 1820. By the Authority of the Medical Societies and Colleges. Boston: Wells & Lilly, for Charles Ewer, Dec. 1820. 1st edn. 8vo. 272 pp. Few lib stamps. Contemp calf, rebacked.
      (M & S Rare Books) **$400 [≈£250]**

Pharmacopoeia Pauperum ...
- See Banyer, Henry.

Phillips, E.C.
- The Birds of Breconshire. Brecon: 1899. 8vo. 192 pp. 2 plates. A few ownership stamps. Orig cloth. *(Wheldon & Wesley)* **£25 [≈$40]**

Phillips, H.J.
- Engineering Chemistry. A Practical Treatise for Analytical Chemists, Engineers, Ironmasters, Ironfounders, & others. London: Crosby Lockwood, 1894. 398 pp. Ills. Orig bndg.   *(Book House)* **£15 [≈$24]**

Phillips, Henry
- Floral Emblems. London: 1825. 8vo. xvi,350 pp. Cold vignette, 19 cold plates. Contemp half calf, rebacked.
      (Wheldon & Wesley) **£200 [≈$320]**
- Floral Emblems. London: Saunders & Otley, 1825. 1st edn. 8vo. List of plates. 24 hand cold plates, inc frontis, addtnl engvd title, & dedic leaf. Contemp half calf, crnrs worn.
      (Hannas) **£140 [≈$224]**
- History of Cultivated Vegetables, comprising their Botanical, Medical, Edible and Chemical Qualities, Natural History, and relation to Art, Science and Commerce. London: 1822. 2nd edn. 2 vols. Roy 8vo. New bds, uncut.
      (Wheldon & Wesley) **£100 [≈$160]**
- Sylva Florifera: the Shrubbery Historically and Botanically Treated ... London: Longman, 1823. 1st edn. 2 vols. 8vo. Half-title in vol 2. Subscribers' list. Occas sl spotting. Orig grey bds, uncut, rebacked.
      (Spelman) **£95 [≈$152]**

Phillips, John, 1800-1874
- Geology of Oxford and the Valley of the Thames. Oxford: Clarendon Press, 1871. 8vo. 6 plates, num text ills. Orig cloth, sm snag in 1 jnt. *(Waterfield's)* **£70 [≈$112]**

- Illustrations of the Geology of Yorkshire ... Part 1 [with Part 2: The Mountain Limestone District]. London: 1829. 1st edn. 2 parts in one vol. 4to. xvi,192; xx,253 pp. 2 maps (1 cold), 11 plates of sections, 36 plates of fossils. Plates sl foxed. New buckram.
      (Baldwin) **£495 [≈$792]**
- Illustrations of the Geology of Yorkshire ... Part 1. The Yorkshire Coast ... Edited by R. Etheridge. London: Murray, 1875. 3rd edn. 4to. x,354 pp. Lge hand cold fldg map, 28 litho plates. A few spots. Orig cloth gilt, jnts a little tender.   *(Hollett)* **£140 [≈$224]**
- The Rivers, Mountains and Sea-Coasts of Yorkshire ... London: Murray, 1855. 2nd edn. 8vo. xv,316,20 advt pp. 34 plates & maps (1 cold). Orig cloth gilt, spine & edges faded, spine ends sl chipped. Pres inscrptn on title.
      (Hollett) **£65 [≈$104]**

Phillips, T.E.R. & Steavenson, W.H. (eds.)
- Hutchinson's Splendour of the Heavens, A Popular Authoritative Astronomy. London: Hutchinson, 1923. 2 vols. 4to. 2 cold frontis, 23 cold plates, over 1000 ills. Orig pict gilt cloth, some shelf wear, inner hinges started.
      (Elgen) **$125 [≈£78]**

Phillips, William W.A.
- Manual of the Mammals of Ceylon. Ceylon & London: Ceylon Journal of Science, 1935. Only edn. 4to. xxviii,373 pp. Fldg map, 38 plates, 55 ills. Orig cloth, 1 crnr sl bumped. Author's pres inscrptn.
      (Fenning) **£85 [≈$136]**

Phillips, William, F.R.S.
- Eight Familiar Lectures on Astronomy, intended as an Introduction to the Science, for the Use of Young Persons ... London: William Phillips, 1817. 1st edn. Lge 12mo. [10], 254,[12] pp. Lacks half-title. 2 fldg plates, fldg table. Contemp calf, sides rubbed.
      (Fenning) **£24.50 [≈$40]**
- An Elementary Introduction to Mineralogy. London: 1852. 12mo. xii,700 pp. 647 text figs. Orig cloth, rebacked, new endpapers.
      (Henly) **£16 [≈$25]**
- An Outline of Mineralogy and Geology, intended for the use ... especially of Young Persons. London: William Phillips, 1815. 1st edn. 8vo. [xii],193 pp. 4 plates (2 hand cold). Contemp lilac half calf, faded & a bit scuffed, hd of spine sl worn. *(Burmester)* **£85 [≈$136]**

Phillott, D.C. (translator)
- The Baz-Nama-Yi Nasiri. A Persian Treatise on Falconry. London: 1908. One of 500. Roy 8vo. xxiv,195 pp. 25 ills. Orig cloth gilt.

(Wheldon & Wesley) £85 [≃ $136]

Philosophy in Sport ...
- Philosophy in Sport made Science in Earnest ... see Paris, J.A.

Physical Theory of Another Life ...
- See Taylor, Isaac.

Piaget, Jean
- The Child's Conception of Physical Causality ... London: Kegan Paul ... New York: Harcourt, Brace, 1930. 1st issue in English. 8vo. viii,309,[3] pp, inserted advts dated 1929. Orig blue-green cloth.
  (Gach) $100 [≃ £62]
- Judgment and Reasoning in the Child. Translated by M. Warden. New York: Harcourt Brace, 1928. 8vo. 260 pp. Ex-lib copy with stamp on end page & rear pocket only. Orig bndg. *(Xerxes)* $65 [≃ £40]

Pickering, C.
- The Geographical Distribution of Animals and Plants. Boston & London: US Exploring Expedition 1838-42 Vol 15 Part 1, 1854. 4to. 168,[44] pp. Half cloth.
  (Wheldon & Wesley) £180 [≃ $288]

Pickering, George
- Creative Malady: Illness in the Lives and Minds of Charles Darwin, Florence Nightingale, Mary Baker Eddy, Sigmund Freud, Marcel Proust, Elizabeth Barret Browning. New York: OUP, 1974. 1st Amer edn. Sm 8vo. 327,[1] pp. Orig blue cloth, dw sl worn. *(Gach)* $25 [≃ £15]

Pidgeon, E.
- The Fossil Remains of the Animal Kingdom. London: Whittaker, 1827-32. 8vo. 536 pp. 49 litho plates (many fldg). Lacks table of contents. Sl foxing. Qtr leather.
  (Gemmary) $75 [≃ £46]

Piesse, George W.S.
- Chymical, Natural, and Physical Magic ... for the Instruction and Entertainment of Juveniles ... Third Edition. London: Longman ..., 1865. Sm 8vo. xix,267,[32 advt dated July 1864] pp. Magic frontis port, 38 text ills. Orig dec cloth gilt, inside jnts cracked. *(Fenning)* £145 [≃ $232]

Pike, O.
- The Nightingale, Its Story and Song. London: 1932. 1st edn. 8vo. 208 pp. 24 plates. Sl foxed. Orig cloth, spine sl faded.

(Henly) £14 [≃ $22]

Pinel, Philippe
- A Treatise on Insanity ... Translated from the French, by D.D. Davis ... Sheffield: 1806. 1st edn in English. 8vo. lv,288 pp. No half-title. 2 plates. Text sl foxed. Crnr torn from title. Orig bds, cloth spine defective.
  (Bickersteth) £350 [≃ $560]

Pinhey, E.
- The Dragonflies of Southern Africa. Pretoria: 1951. 4to. xv,334 pp. 146 plates. Orig cloth, crnrs trifle bumped.
  (Wheldon & Wesley) £45 [≃ $72]

Pinnock, William
- A Catechism of Mineralogy; or, an Introduction to the Knowledge of the Mineral Kingdom. Adapted to the Capacities of Youth ... By a Friend to Youth. London: [ca 1820]. 1st edn. 12mo. 72 pp. Frontis. Orig ptd paper wraps, worn & stained & sl defective.
  (Fenning) £35 [≃ $56]

Pirie, William Robinson
- An Inquiry into the Constitution, Powers, and Processes of the Human Mind with a View to the Determination of the Fundamental Principles of Religious, Moral and Political Science. Aberdeen: A. Brown, 1858. 8vo. Orig black cloth.
  (Waterfield's) £80 [≃ $128]
- An Inquiry into the Constitution, Powers, and Processes of the Human Mind ... Aberdeen, Edinburgh & London: 1858. 1st edn. 8vo. Orig embossed cloth, cvrs rubbed, hinges cracked, some dampstaining.
  (Gach) $85 [≃ £53]

Pitcairn, Archibald
- The Philosophical and Mathematical Elements of Physick, in Two Books. London: for Andrew Bell ..., 1718. 1st edn in English. xxxii,368 pp. Title-page crudely rehinged. Contemp panelled calf, rebacked, new endpapers. *(Elgen)* $550 [≃ £343]

Pitkin, Thomas
- A Statistical View of the Commerce of the United States of America: its Connection with Agriculture and Manufactures ... Hartford: Charles Hosmer, 1816. 1st edn. 8vo. xii,407, xix pp. Usual spotting. Half calf, rebacked, new endpapers.
  (Young's) £90 [≃ $144]

Pitt, William, 1749-1823
- General View of the Agriculture of the

County of Stafford ... London: for G. Nichol, 1796. 8vo. xvi,241,[vi] pp. Fldg map, 3 fldg plans, 11 plates. Orig grey wraps, buff paper spine, uncut, spine worn at ft.
(Bickersteth) **£85 [≈ $136]**

Planting and Rural Ornament ...
- See Marshall, William.

Platt, Hugh
- The Garden of Eden. Or an accurate description of all Flowers and Fruits now growing in England ... Fifth Edition. London: William Leake, 1660. 2 parts in one. 1st edn of 2nd part. 12mo. 175,[i]; [16],159, [i] pp. Some browning. Contemp calf, crnrs reprd. Wing 2387A *(Spelman)* **£420 [≈ $672]**
- The Jewel House of Art and Nature: containing divers Rare and Profitable Inventions ... London: Bernard Alsop, 1653. 2nd edn. Sm 4to. [8],232 pp. Num w'cut ills. Early 19th c contemp style calf. Wing P.2390.
(Spelman) **£650 [≈ $1,040]**

Plattes, Gabriel
- A Discovery of Subterranean Treasure; (viz.) Of All Manner of Mines and Minerals, from the Gold to the Coal ... Finding ... Melting, Refining and Assaying ... London: for Peter Parker, 1679. 3rd edn. Sm 4to. [4], 24 pp. Occas sl foxing. Mod half mor. Wing P.2411.
(Reese) **$850 [≈ £531]**

Pliny, the elder
- The Historie of the World. Commonly called, The Naturall Historie of C. Plinius Secundus. Translated ... by Philemon Holland. London: 1601. 1st edn in English. 2 vols in one. Folio. Lacks 1 index leaf, a few reprs. 17th c calf, sometime rebacked. STC 20029.   *(Sotheran)* **£385 [≈ $616]**
- The Historie of the World: commonly called, The Naturall Historie of C. Plinius Secundus. Translated into English by Philemon Holland. London: 1634-35. 2nd edn. 2 vols in one. Folio. Blank crnr of title renewed, a few sl marg dampstains. Old calf, reprd.   *(Wheldon & Wesley)* **£380 [≈ $608]**

Plues, Margaret
- British Ferns ... London: Reeve, 1866. 8vo. x,281 pp. 16 hand cold plates. Orig cloth.
(Egglishaw) **£30 [≈ $48]**
- British Grasses: an Introduction to the Study of the Gramineae of Great Britain and Ireland. London: Reeve, 1867. 8vo. viii,307 pp. 16 cold plates (sl foxed), w'engvs in text. Orig green cloth gilt, short split in jnt, back bd damp stained. *(Blackwell's)* **£21 [≈ $33]**

- Rambles in Search of Flowerless Plants. London: 1864. 8vo. vii,317 pp. 20 plates (8 hand cold). Sm blind stamp on title & stamp on verso. Qtr mor.   *(Henly)* **£35 [≈ $56]**
- Rambles in Search of Flowerless Plants. London: 1865. 2nd edn. 8vo. viii,318 pp. 20 cold plates. Orig cloth.
(Wheldon & Wesley) **£20 [≈ $32]**
- Rambles in Search of Wild Flowers. London: 1892. 4th edn. 8vo. xv,368 pp. 16 cold plates. Orig cloth, trifle used.
(Wheldon & Wesley) **£20 [≈ $32]**

Plumbe, Samuel
- A Practical Treatise on the Diseases of the Skin. From the last London edition. Philadelphia: Haswell, Barrington & Haswell, 1837. 1st Amer edn. 396,8 advt pp. 4 plates (3 hand cold). Sl foxing. Contemp calf, front jnt tender.   *(Elgen)* **$95 [≈ £59]**

Podolsky, Edward
- The Doctor prescribes Colors. The Influence of Colors on Health and Personality. New York: National Library, 1938. 1st edn. 8vo. 106 pp. Orig bndg, spine sl faded.
(Xerxes) **$65 [≈ £40]**

The Poetry of Birds ...
- See Rathbone, Hannah Maria.

Poinsett, Joel R.
- Discourse, on the Objects and Importance of the National Institution, for the Promotion of Science, established at Washington, 1840, delivered at the First Anniversary. Washington: P. Force, 1841. 1st edn. 8vo. 52 pp. Sl foxing. Rec wraps.   *(Fenning)* **£75 [≈ $120]**

Pole, Thomas
- The Anatomical Instructor; or, an Illustration of the Modern and most Approved Methods of preparing and preserving the different parts of the Human Body and of Quadrupeds ... London: 1790. 1st edn. 8vo. lxxx,304 pp,8ff. 10 plates. Contemp calf, hinges weak.
(Hemlock) **$250 [≈ £156]**

Pollard, H.B.C.
- Game Birds. Rearing, Preservation and Shooting. London: 1929. 4to. 12 mtd cold plates by Philip Rickman. Orig bndg.
(Grayling) **£80 [≈ $128]**
- Game Birds. Rearing, Preservation and Shooting. London: Eyre & Spottiswoode, 1929. 4to. xi,186 pp. 12 cold plates by Philip Rickman. Orig cloth, uncut.
(Egglishaw) **£65 [≈ $104]**

- Wildfowl and Waders. Nature and Sport in the Coastlands. London: 1928. One of 950. 4to. viii,83 pp. 16 cold mtd & 48 other plates. Sl foxing throughout, not affecting plates. Orig vellum backed bds, sl worn.
(Henly) **£28 [≈ $44]**

Polunin, N.
- Botany of the Canadian Eastern Arctic. Ottawa: Nat Mus of Canada Bulletins 92, 97, 104, 1940-48. 3 vols. 8vo. 3 fldg maps, 7 sketch maps, 133 ills. Cloth.
(Wheldon & Wesley) **£60 [≈ $96]**

Polwhele, Richard (ed.)
- Essays, by a Society of Gentlemen, at Exeter. Exeter: Trewman & Son ..., [1796]. 1st edn. 8vo. viii,573,[1],[2 errata] pp. Half-title. 5 tinted aquatint plates. Contemp mrbld bds, later calf spine, rubbed.
(Burmester) **£160 [≈ $256]**

Polyak, S.L.
- The Retina. The Anatomy and Histology of the Retina in Man, Ape and Monkey ... Chicago: (1941). 1st edn. 4to. 607 pp. Atlas of 100 plates (1 cold). Dw.
(Elgen) **$250 [≈ £156]**

Pontey, W.
- The Forest Pruner; or Timber Growers Assistant. Huddersfield: [1805]. 1st edn. 277,3 pp. 8 plates. Lower edge of plates cropped, 2 of the fldg plates cropped at the fold, folds loosely inserted. Contemp half calf, sl rubbed. John Cator's copy.
(Henly) **£135 [≈ $216]**
- The Profitable Planter: a Treatise on the Theory and Practice of Planting Forest Trees. London: Harding, 1814. 4th edn, enlgd. 8vo. viii, 267,index pp. 1 plate. Half calf, front jnt cracked. *(Egglishaw)* **£22 [≈ $35]**

Ponton, Mungo
- Earthquakes and Volcanoes, their History, Phenomena and Probable Causes. London: 1868. 8vo. 354 pp. 16 plates & text ills. Front pastedown sl damaged. Orig cloth.
(Henly) **£18 [≈ $28]**

Porta, J.B.
- Natural Magick. New York: 1958. Facs of the 1658 edn. Sm 4to. 416 pp. Slipcase sl worn.
(Book House) **£40 [≈ $64]**

Porter, C.W.
- Molecular Rearrangements. New York: Chemical Catalog Co, 1928. 1st edn. 168,4 advt pp. Orig dark blue cloth gilt, lib label.

Arthur W. Chapman's sgntr & some pencil notes. *(Sklaroff)* **£85 [≈ $136]**

Porter, Noah
- The Human Intellect. With an Introduction upon Psychology and the Soul. New York: Scribner, 1868. 1st edn. 8vo. [ii],[4], [xxviii], [5]-673, [5],[9] pp. Errata slip. Orig brown cloth, shelfworn. *(Gach)* **$75 [≈ £46]**

Portland, Duke of
- The Red Deer of Langwell and Braemore. London: 1935. 4to. 5 cold plates, num other ills. Orig bndg. *(Grayling)* **£40 [≈ $64]**

Portlock, J.E.
- Report on the Geology of the County of Londonderry, and parts of Tyrone and Fermanagh. Dublin: 1843. 8vo. xxxii,784 pp. Fldg cold geological map. 54 plates (5 cold). Orig cloth, trifle used.
(Wheldon & Wesley) **£120 [≈ $192]**
- Report on the Geology of the County of Londonderry and of parts of Tyrone and Fermanagh. Dublin: Milliken, 1843. 784 pp. Maps, plates, some cold. Orig bndg.
(Emerald Isle) **£85 [≈ $136]**

A Portrait of Geology ...
- See Murray, J.

Posepny, F.
- The Genesis of Ore-Deposits. New York: A.I.M.E., 1902. 2nd edn. 8vo. xxi,806 pp. Num figs. Some pencil underlining. Half leather, spine v sl rubbed.
(Gemmary) **$75 [≈ £46]**

Pott, Percival
- The Chirurgical Works. New Edition, with his last Corrections ... added ... Life of the Author ... by James Earle. London: for J. Johnson ..., 1790. 3 vols. Port, 19 plates. Occas sl foxing. Dampstaining in vol 1. Orig calf, needs rebinding. *(Elgen)* **$500 [≈ £312]**

Potter, Edith L.
- Pathology of the Fetus and the Newborn. Chicago: (1953). Sm 4to. 574 pp. 600 ills. Orig bndg. *(Elgen)* **$90 [≈ £56]**

Poynting, F.
- Eggs of British Birds with an Account of their Breeding Habits: Limicolae. London: 1895-96. 4to. 54 cold plates, almost free from the usual foxing. Contemp half mor, t.e.g.
(Wheldon & Wesley) **£200 [≈ $320]**

Pozzi, S.

- A Treatise on Gynaecology, Clinical and Operative. London: New Sydenham Soc, 1892-93. 1st English edn. 3 vols. Faded lib stamps on titles. Orig cloth, jnts cracked but bds firmly attached, cvrs faded.
(Elgen) **$150 [≃ £93]**
- Treatise on Gynaecology, Medical and Surgical. Translated from the French edition with additions by Brooks H. Wells. New York: Wood, 1894. 2 vols. 4to. 15 cold litho plates, 479 w'engvs. Sheep, sl scuffed, faded.
(Elgen) **$150 [≃ £93]**

Praeger, Robert Lloyd

- The Botanist in Ireland. Dublin: Hodges Figgis, 1934. 587 pp. Ills. Orig bndg.
(Emerald Isle) **£50 [≃ $80]**
- Irish Topographical Botany compiled largely from Original Material. Dublin: Royal Irish Academy Proceedings Series 3 Vol 7, 1901. 8vo. clxxxviii,410 pp. 6 cold maps. Orig cloth.   *(Wheldon & Wesley)* **£35 [≃ $56]**
- Irish Topographical Botany, compiled from original Material. Dublin: Proc of the Roy Irish Acad, 3rd series, vol 7, 1901. 8vo. clxxxviii,410 pp. 6 fldg cold maps. Orig cloth, t.e.g.   *(Fenning)* **£75 [≃ $120]**
- A Tourist's Flora of the West of Ireland. Dublin: Hodges, Figgis, 1909. 1st edn. Cr 8vo. xii,243 pp. 5 fldg maps, 27 plates. Orig cloth.   *(Fenning)* **£65 [≃ $104]**
- A Tourist's Flora of the West of Ireland. Dublin: 1909. 8vo. xii,243 pp. 5 cold maps, 27 plates, 17 text figs. Orig cloth.
(Wheldon & Wesley) **£30 [≃ $48]**
- A Tourist's Flora of the West of Ireland. Dublin: Hodges Figgis, 1909. 243 pp. Map, ills. Orig bndg. *(Emerald Isle)* **£55 [≃ $88]**
- The Way That I Went. London: 1939. 2nd edn. 8vo. xiv,394 pp. Map, 39 plates, 12 figs. Orig cloth.   *(Baldwin)* **£25 [≃ $40]**

Pratt, Anne

- The Ferns of Great Britain and their Allies the Club-Mosses, Pepperworts and Horsetails. London: n.d. 41 cold plates. Orig cloth, spine relaid. *(Henly)* **£35 [≃ $56]**
- The Flowering Plants of Great Britain and the Grasses, Sedges and Ferns of Great Britain. London: 1889. 4 vols. 8vo. 319 plates (all but 1 cold). Orig green cloth gilt.
(Henly) **£150 [≃ $240]**
- The Flowering Plants, Grasses and Ferns of Great Britain. London: n.d. 6 vols. 319 plates (all but 1 cold). Orig green cloth, 3 spines relaid.   *(Henly)* **£165 [≃ $264]**

- The Flowering Plants, Grasses, Sedges & Ferns of Great Britain, and their Allies the Club Mosses, Horsetails, &c. New Edition. Revised by Edward Step. London: Warne, 1899-1900. 4 vols. Roy 8vo. 315 cold plates. Orig blueish green cloth gilt.
(Fenning) **£145 [≃ $232]**
- The Flowering Plants, Grasses, Sedges and Ferns of Great Britain. New Edition, revised by Edward Step. London: Warne, 1899-1905. 4 vols. Roy 8vo. 5 plain & 319 cold plates. Half mor gilt, t.e.g.
(Egglishaw) **£160 [≃ $256]**
- The Flowering Plants, Grasses, Sedges & Ferns of Great Britain and their Allies the Club Mosses, Horsetails, &c. New Edition, revised by Edward Step. London: Warne, 1905. 4 vols. 8vo. 319 cold plates. Orig dec cloth gilt, edges trifle rubbed.
(Hollett) **£120 [≃ $192]**
- Wild Flowers. London: SPCK, [1853?]. 1st 4to edn. Title, contents & 96 ff, each with 2 cold ills accompanied by descriptive text. Some foxing of outer ff. Contemp half mor, a.e.g.   *(Bow Windows)* **£180 [≃ $288]**
- Wild Flowers. London: SPCK, 1857. 2 vols. Sm 8vo. iv,192; vi,195,4 advt pp. 192 chromolitho plates. Orig green cloth gilt, rebacked with most part of orig backstrips laid down.   *(Claude Cox)* **£35 [≃ $56]**
- Wild Flowers. London: SPCK, 1857. 2 vols. 12mo. 192 cold plates. Orig cloth gilt, recased.   *(Hollett)* **£120 [≃ $192]**
- Wild Flowers. Indexes by G. Egerton Warburton. London: SPCK, 1898. 1st edn. 2 vols. 96 cold plates. Orig bndg.
(Limestone Hills) **$85 [≃ £53]**

Pratt, J.H.

- The Mathematical Principles of Mechanical Philosophy, and their Application to the Theory of Universal Gravitation. Cambridge: 1836. xxvi,616 pp. 5 fldg plates. Three qtr roan, rebacked.   *(Whitehart)* **£45 [≃ $72]**

Prescott, F.C.

- Poetry and Dreams. Boston: The Four Seas, 1919. 8vo. 72 pp. Orig bndg, dw.
(Xerxes) **$55 [≃ £34]**

Prescott, George Bartlett

- History, Theory, and Practice of the Electric Telegraph. Boston: Ticknor & Fields, 1860. 1st edn. 8vo. [3],xii,468,[2] pp. Frontis, 98 w'cut text ills. Orig cloth, gilt spine, worn.
(Rootenberg) **$150 [≃ £93]**

Prescott, Henry P.
- Tobacco and its Adulterations. London: 1858. 1st edn. 8vo. Engvd title & 40 plates drawn & etched by the author. Sl marg damp stain of title & 1st few ff. Orig cloth, ft of spine nicked.   *(Robertshaw)* **£30 [≈ $48]**

Preston, T.A.
- The Flowering Plants of Wilts. London: 1888. 8vo. lxix,436 pp. Cold map. Orig cloth, cvrs sl stained, front jnt torn.
                      (Wheldon & Wesley) **£30 [≈ $48]**

Prestwich, J.
- Geology, Physical, Chemical and Stratigraphical. London: 1886. 2 vols. Roy 8vo. 1 cold geol map (only, of 2, lacks map of Europe), 3 other maps, 3 sections, 16 plates, 474 w'cuts. Orig cloth, spines relaid.
                              (Henly) **£35 [≈ $56]**
- The Water-Bearing Strata of the Country around London. London: 1851. 8vo. 240 pp. Fldg cold geol map (sm tear reprd), text figs. Orig cloth.   *(Henly)* **£25 [≈ $40]**

Prichard, James Cowles
- The Natural History of Man ... London & Paris: Bailliere, 1843. 1st edn. 8vo. xvi, 556,[1] pp, inc 2 advt pp. Hand cold frontis, 40 plates (on 39 ff, 36 hand cold), num text w'cuts. Occas sl foxing on uncold plates. Half calf, mrbld bds.   *(Rootenberg)* **$550 [≈ £343]**
- The Natural History of Man ... London: Bailliere, 1848. 3rd edn, enlgd. xvii,677 pp. 50 cold & 5 plain plates (correct, plates 18 & 19 are 2 images on 1 plate), 97 w'engvs. Orig pict red cloth gilt, recased, cvrs dulled.
                      (Box of Delights) **£120 [≈ $192]**
- The Natural History of Man ... London: Bailliere, 1848. 3rd edn, enlgd. xvii,677 pp. 50 cold & 5 plain plates, 97 w'engvs. Orig red cloth gilt, jnts reprd. [With] Six Ethnographical Maps ... London: 1843. Folio. 6 cold maps. Cloth backed bds, orig wraps preserved.
                      (Wheldon & Wesley) **£250 [≈ $400]**
- Researches into the Physical History of Man ... London: for John & Arthur Arch, 1826. 2nd edn. 2 vols. 8vo. xxxii,544; [iv],623,[1] pp. 19th c calf backed mrbld bds gilt, jnts rubbed.   *(Gach)* **$350 [≈ £218]**
- Researches into the Physical History of Mankind ... London: Houlston & Stoneman ..., [1836-47]. 3rd edn, undated iss. 5 vols. 8vo. Fldg map, 20 litho plates (sev hand cold). Orig green cloth, partly unopened, a little rubbed.   *(Pickering)* **£450 [≈ £281]**
- Researches into the Physical History of

Mankind. Third Edition. London: Sherwood, Gilbert & Piper, 1836-47. 5 vols. 8vo. Fldg map, 22 plates (9 hand cold). Contemp polished calf, contrasting labels.
                      (Bickersteth) **£450 [≈ $720]**

Priest, C.D.
- Eggs of Birds Breeding in Southern Africa. Glasgow: 1948. Orig issue. Roy 8vo. xii,180 pp. 20 cold plates. Orig buckram, crnrs bumped.   *(Wheldon & Wesley)* **£50 [≈ $80]**

Priestley, Joseph
- Directions for Impregnating Water with fixed Air; In order to communicate to it the peculiar Spirit and Virtues of Pyrmont Water ... London: J. Johnson, 1772. 22,3 advt pp. Frontis.   *(Jarndyce)* **£450 [≈ $720]**
- Disquisitions relating to Matter and Spirit ... London: for J. Johnson, 1777. 1st edn. 8vo. [xlii],356,[4] pp. Frontis. Sl foxed. Contemp calf, front bd detached. *(Gach)* **$250 [≈ £156]**
- Hartley's Theory of the Human Mind, on the Principle of the Association of Ideas; with Essays relating to the Subject of it. By Joseph Priestley. London: for J. Johnson, 1775. 1st edn. 8vo. lxii,372 pp. Calf, front bd detached.
                      (Gach) **$325 [≈ £203]**
- The History and Present State of Discoveries relating to Vision, Light, and Colours ... London: for J. Johnson ..., 1772. 1st edn. 4to. v,xvi,812,[6 index],[5],[1 advt] pp. Fldg chart frontis, 24 fldg plates. Contemp sprinkled calf, rebacked & recrnrd.
                      (Pickering) **$3,000 [≈ £1,875]**

Prime, Samuel Irenaeus
- The Life of Samuel F.B. Morse, Inventor of the Electro-Magnetic recording Telegraph. New York: Appleton, 1875. 1st edn. xii,[1],775 pp. 10 plates, text ills. Three qtr calf, v worn.   *(Elgen)* **$90 [≈ £56]**

The Principles of Mechanics ...
- See Emerson, William.

Pringle, Andrew
- Practical Photomicrography: by the Latest Methods. New York: Scoville & Adams, 1890. 1st edn. 8vo. 183,ix,[xviii] pp. 6 plates, 42 text ills. Orig cloth, lib b'plate.
                      (Charles B. Wood) **$125 [≈ £78]**
- Practical Photo-Micrography. London: Iliffe, [1893]. 1st edn. Sm 4to. 159,[1], iv,[6 advt] pp. 29 ills. Orig cloth gilt.
                      (Fenning) **£18.50 [≈ $30]**

Pringle, Sir John
- Observations on the Diseases of the Army ... With Notes, by Benjamin Rush ... Philadelphia: Edward Earle ..., 1810. 1st Amer edn. 8vo. xlvii,411 pp. Contemp calf, rubbed, hd of spine bumped.
(Hemlock) **$300 [≈ £187]**

Proctor, Percival, et al.
- See The Modern Dictionary.

Professional Anecdotes ...
- Professional Anecdotes, or Ana of Medical Literature. London: John Knight & Henry Lacey, 1825. 3 vols. 8vo. x,296; x,288; ix, 288 pp. Orig cloth, uncut.
(Hemlock) **$450 [≈ £281]**

Progress, Peter (pseud.)
- See Clarke, R.Y.

The Projection of the Sphere ...
- See Emerson, William.

Prothero, Rowland E.
- English Farming Past and Present. London: Longmans, Green, 1912. xiii,504 pp. Sl foxing, some pencil underlining. Orig cloth.
(Wreden) **$59.50 [≈ £37]**

Protheroe, E.
- The Railways of the World. London: Routledge, [1914]. 8vo. 752 pp. 16 cold plates, 419 ills. Rebound.
(Book House) **£20 [≈ $32]**

Prout, William
- Chemistry, Meteorology and the Function of Digestion considered with reference to Natural Theology. London: William Pickering, 1834. 1st edn. 8vo. [8 advt],xxiii, [5],564, [2] pp. Fldg cold map. Orig cloth, sl discold, lacks paper spine label.
(Fenning) **£45 [≈ $72]**
- Chemistry, Meteorology, and the Function of Digestion, considered with reference to Natural Theology. [Bridgewater Treatises VIII]. London: William Pickering, 1834. 1st edn. 8vo. Half-title, xxiii,[i],[iv],564 pp. Ptd table, fldg cold map. Contemp half calf, rubbed.
(Bickersteth) **£45 [≈ $72]**
- Chemistry, Meteorology and the Function of Digestion considered with reference to Natural Theology. Second Edition. William Pickering, 1834. 8vo. xxv,[i],570,[i] pp. Fldg hand cold map. Orig diced calf, gilt labels.
(Spelman) **£50 [≈ $80]**

Pryce, W.
- Mineralogia Cornubiensis. London: for the author, 1778. Folio. 17,1,xiv,[1],331 pp. Frontis port, 7 plates, 1 table. Full calf, fine.
(Gemmary) **$2,500 [≈ £1,562]**

Pryor, A.R.
- A Flora of Hertfordshire. Edited by D.B. Jackson. London: 1887. 8vo. lviii,588 pp. 3 maps. V sl foxing. Orig cloth, jnts loose.
(Wheldon & Wesley) **£45 [≈ $72]**

Pumpelly, Raphael
- Geological Researches in China, Mongolia, and Japan, 1862-1865. Washington: 1866. 1st edn. Folio. 143 pp. Cold maps, fldg plates. Orig ptd wraps. Half mor slipcase.
(Jenkins) **$200 [≈ £125]**

Purcell, F. Albert
- On Cancer, Its Allies and Other Tumours, with Special Reference to their Medical and Surgical Treatment. London: Churchill, 1881. 1st edn. 311,16 ctlg pp. Text w'cuts. Orig cloth, unopened, worn & soiled.
(Elgen) **$50 [≈ £31]**

Quain, Jones
- Elements of Anatomy. London: for Taylor & Walton, 1837. 4th edn. 8vo. xxx,910 pp. 4 plates, num text figs. Last 2 ff creased & soiled. Half calf gilt, rear hinge splitting, spine ends sl defective. *(Hollett)* **£38 [≈ $60]**

Quain, Richard (ed.)
- A Dictionary of Medicine ... London: Longmans, Green, 1882. 1st edn. 8vo. xviii, 1814 pp. Title & next leaf torn at gutter. Orig cloth backed bds.   *(Gach)* **$50 [≈ £31]**
- A Dictionary of Medicine including General Pathology, General Therapeutics, Hygiene, and the Diseases of Women and Children ... London: Longmans, Green, 1895. 2 vols. Thick 8vo. Orig buckram.
(Bickersteth) **£30 [≈ $48]**

Quebbeman, Frances
- Medicine in Territorial Arizona. Phoenix: Arizona Hist Foundation, 1966. 1st edn. Roy 8vo. 424 pp. Photo ills. Orig bndg.
(Xerxes) **$75 [≈ £46]**

The Queen of Flowers ...
- The Queen of Flowers: or, Memoirs of the Rose. London: Robert Tyas, 1840. 3rd edn. Sm 8vo. xiii,283 pp. 6 hand cold plates. Orig cloth, spine trifle faded.
(Wheldon & Wesley) **£40 [≈ $64]**

Quillet, Claude
- Callipaedia; or, An Art how to have Handsome Children ... added, Paedotrophiae; or, The Art of Nursing and Breeding Up Children ... by Monsieur St. Marthe ... London: for John Morphew, 1710. 1st edn in English. Sm 8vo. [6],263 pp. Browning. Three qtr calf. *(Elgen)* **$350 [≈£218]**

Quincy, John
- Lexicon Physico-Medicum: or, a New Medical Dictionary ... Fourth Edition, with New Improvements ... London: Osborn & Longman, 1730. 8vo. Contemp panelled calf, lower jnt cracked. *(Stewart)* **£120 [≈$192]**
- Pharmacopoeia Officinalis Extemporanea, or, A Complete English Dispensatory ... London: for E. Bell ..., 1724. 5th edn. 8vo. xvi,674,[60 advt] pp. Title sl spotted, last leaf browned & sl reprd. Mod calf gilt.
(Hollett) **£120 [≈$192]**
- Pharmacopoeia Officinalis Extemporanea, or, A Complete English Dispensatory ... London: for Thomas Longman, 1736. 10th edn. 8vo. xvi,700,lx pp. Some browning & spotting. Polished speckled calf, spine label.
(Hollett) **£150 [≈$240]**

Rachford, Benjamin Knox
- Diseases of Children. New York: 1912. 1st edn. 783 pp. 6 cold plates, 107 text ills. Orig bndg. *(Elgen)* **$75 [≈£46]**

Raffald, Elizabeth
- The Experienced English Housekeeper ... London: 1794. 11th edn. 8vo. Port, 3 plates. Page 1 sgnd by the author as usual. Contemp calf, rebacked. *(Robertshaw)* **£135 [≈$216]**
- The Experienced English Housekeeper ... New Edition. London: for Osborne & Griffin, 1794. 8vo. 398 pp. Port, 3 plates. Edge of 2 plates rubbed & worn through misfolding. Contemp calf. *(Moon)* **£165 [≈$264]**

Rafinesque, Constantin S.
- Manual of the Medical Botany of the United States ... Philadelphia: 1841. 1st edn. Sm 8vo. [iv],xii,259 pp. 52 plates ptd in green. Outer crnr of 1 p of text sl defective. Occas sl foxing. Orig sheep, rebacked.
(Charles B. Wood) **$750 [≈£468]**

Ralfe, P.G.
- The Birds of the Isle of Man. Edinburgh: David Douglas, 1905. 1st edn. Medium 8vo. Advt leaf. 2 fldg maps, gravure title, 48 tissued plates (from 51 photos). Some foxing of text. Orig gilt dec blue cloth, extremities

rubbed. Orig prospectus inserted.
(Sotheran) **£145 [≈$232]**

Ralfs, J.
- The British Desmidieae. The Drawings by Edward Jenner. London: Reeve, 1848. Roy 8vo. xxiv,226,18 advt pp. 1 plain & 34 hand cold plates. New qtr calf.
(Egglishaw) **£130 [≈$208]**

Ramsay, Sir Andrew C.
- Geological Map of England & Wales. London: Edward Stanford, Oct 2, 1879. 4th edn. Fldg single sheet 31 5/8 x 37 1/8 ins to edge of border. Hand cold, linen backed. Orig cloth on card slipcase, sl worn, v sl rubbed.
(Duck) **£85 [≈$136]**

Rand, Austin L.
- American Water and Game Birds. New York: 1956. 4to. 239 pp. 167 photos (127 cold), 35 silhouettes. Orig cloth.
(Wheldon & Wesley) **£20 [≈$32]**

Rand, Austin L. & Gilliard, E.T.
- Handbook of New Guinea Birds. London: Weidenfeld & Nicolson, 1967. 1st UK edn. Roy 8vo. x,612 pp. Map, 5 cold plates, 76 b/w plates. Orig cloth. Dw torn. This edn was publ without the index (which does appear in the Amer edn). Here supplied as a photocopy.
(Sotheran) **£180 [≈$288]**
- Handbook of New Guinea Birds. New York: (1967) 1968. Roy 8vo. x,628 pp. Index. 5 cold plates, 76 plain ills on 24 plates. Orig cloth.
(Wheldon & Wesley) **£170 [≈$272]**

Randall, John
- The Semi-Virgilian Husbandry, deduced from Various Experiments. London: 1764. 8vo. lxiii, 356,11 pp. 3 plates. Calf, jnts worn.
(Wheldon & Wesley) **£50 [≈$80]**
- The Semi-Virgilian Husbandry, deduced from Various Experiments ... London: for B. Law ..., 1764. 1st edn. 8vo. lxiv,356,11,[1] pp. 3 fldg plates. Natural flaw in M4 affecting 6 words. Contemp calf, spine ends trifle worn, upper jnt reprd.
(Burmester) **£180 [≈$288]**

Randall, P.M.
- The Quartz Operator's Handbook. New York: Van Nostrand, 1871. Cr 8vo. 175 pp. 16 text figs. Orig cloth.
(Gemmary) **$75 [≈£46]**

Raspe, Rudolph Eric
- An Account of some German Volcanos, and

their Productions ... London: Lockyer Davis, 1776. 1st edn. 8vo. xix,[1],140 pp, inc 4 advt pp. Half-title. 2 fldg plates. Contemp calf, rebacked.   *(Rootenberg)* **$525 [≈ £328]**

Rateau, A.
- Experimental Researches on the Flow of Steam through Nozzles and Orifices, to which is added a Note on the Flow of Hot Water. London: Constable, 1905. 8vo. 76 pp. 4 fldg plates. Orig bndg, used but sound.
        (Book House) **£30 [≈ $48]**

Rathbone, Hannah Maria
- The Poetry of Birds. Selected from Various Authors; with Coloured Illustrations By A Lady. Liverpool: George Smith; London: Ackermann, 1833. Sm 4to. 136 pp. 21 hand cold litho plates. Gilt dec green mor, mrbld edges, by Hatchard, sl scuffed & discold.
        (Schoyer) **$1,200 [≈ £750]**

Rauch, Frederick Augustus
- Psychology, or a View of the Human Soul, including Anthropology. New York: Dodd, 1841. 2nd edn. Foxed. Orig embossed cloth, jnts & edges chipped.   *(Gach)* **$75 [≈ £46]**

Raven, C.E.
- English Naturalists from Neckham to Ray. London: 1947. 8vo. x,379 pp. Orig cloth. Author's own copy.   *(Baldwin)* **£50 [≈ $80]**
- John Ray Naturalist. His Life and Works. London: 1942. 1st edn. 8vo. xix,502 pp. Frontis. Lib stamps. Orig cloth.
        (Baldwin) **£45 [≈ $72]**
- John Ray Naturalist. His Life and Works. London: [1950]. 2nd edn. 8vo. xix,506 pp. Frontis. Orig cloth. *(Baldwin)* **£19.50 [≈ $32]**
- John Ray Naturalist. His Life and Works. Cambridge: 1950. 2nd edn. 8vo. xx,506 pp. Port. Orig cloth.
        (Wheldon & Wesley) **£60 [≈ $96]**

Ravenscroft, E.J.
- The Pinetum Britannicum. A Descriptive Account of Hardy Coniferous Trees cultivated in Great Britain. Edinburgh & London: [1863-] 1884. 3 vols. Folio. 1 plate of maps, 48 hand cold litho plates, 4 mtd albumen prints, 643 text figs. Half mor by Zaehnsdorf, spines relaid.
        (Henly) **£3,800 [≈ $6,080]**

Rawlings, B. Burford
- A Hospital in the Making: A History of the National Hospital for the Paralysed and Epileptics (Albany Memorial) 1859-1901.

London: 1913. 1st edn. 12mo. [xvi],[272] pp. 5 plates. Sl foxed. Orig ptd straight grained cloth.   *(Gach)* **$35 [≈ £21]**

Ray, John
- The Correspondence. Edited by E. Lankester. London: 1848. 8vo. xvi,501 pp. Orig cloth, rebacked.   *(Baldwin)* **£35 [≈ $56]**
- Observations Topographical, Moral and Physiological made in a Journey through part of the Low-Countries, Germany, Italy and France ... London: 1673. 8vo. [xv],499, [vii], 115 pp. Port, 3 plates (part of 1st supplied in facs). Mod calf, antique style.
        (Wheldon & Wesley) **£220 [≈ $352]**
- Select Remains of the Learned John Ray, M.A. and F.R.S. with his Life by the late William Derham. Published by George Scott. London: Dodsley, 1760. 1st edn. 8vo. vii,336 pp. Port, 3 text engvs. Orig calf, rebacked.
        (Bickersteth) **£250 [≈ $400]**
- The Wisdom of God manifested in the Works of the Creation. Glasgow: ptd by J. Bryce & D. Paterson for W. Marshall, 1756. 13th edn, crrctd. 12mo. xxiv,25-324 pp. Mottled calf, jnts cracked but not broken.
        (Wheldon & Wesley) **£100 [≈ $160]**

Rayleigh, Lord & Ramsay, William
- Argon, a New Constituent of the Atmosphere. Washington: Smithsonian, 1896. 1st edn. 4to. [iv],43 pp. Text figs. Orig cloth.   *(Bickersteth)* **£65 [≈ $104]**

Rayner, D.H. & Hemingway, J.E. (eds.)
- The Geology and Mineral Resources of Yorkshire. Leeds: 1974. Roy 8vo. ix,405 pp. Cold frontis, fldg geol map, 6 plain plates. Orig cloth. Dw.   *(Henly)* **£49 [≈ $78]**

Rea, J.
- Flora: seu, De Florum Cultura, or, a Complete Florilege, furnished with all requisites belonging to a Florist. London: 1665. 1st edn. Folio. [xxii],240,[4] pp. 'Mind of frontis' leaf. Some wear & marks. Engvd title, 8 plates. Mod half calf, antique style.
        (Wheldon & Wesley) **£375 [≈ $600]**

Read, John
- Prelude to Chemistry. An Outline of Alchemy, its Literature and Relationships. New York: 1937. 1st edn. xxiv,328 pp. Cold frontis, 63 plates, text w'cuts. dw worn.
        (Elgen) **$65 [≈ £40]**

Reamur, R.A.F. de
- The Natural History of Ants from an

Unpublished MS. in the Archives of the Academy of Sciences of Paris. Translated and Annotated by W.M. Wheeler. New York: 1926. 8vo. xvii,280 pp. 2 ports, 2 plates. Orig cloth. *(Wheldon & Wesley)* **£35 [≈ $56]**

Rechinger, H.K.
- Flora of Lowland Iraq. London: 1964. Roy 8vo. 746 pp. Orig cloth.
  (Wheldon & Wesley) **£50 [≈ $80]**

Redding, Cyrus
- A History and Description of Modern Wines. London: Whitaker, Treacher & Arnot, 1833. 1st edn. 8vo. xxxvi,[i],407 pp. W'engvd decs. Later polished half calf, gilt spine.
  (Charles B. Wood) **$450 [≈ £281]**

Redgrove, H. Stanley
- Alchemy: Ancient and Modern ... London: Rider, 1922. 2nd edn. 8vo. Frontis, 15 plates. Orig cloth gilt, gilt dulled, extremities sl worn. *(Minster Gate)* **£24 [≈ $38]**

Reed, William
- The History of Sugar and Sugar Yielding Plants, together with an Epitome of Every Notable Process of Sugar Extraction, and Manufacture, from the Earliest Times to the Present. London: Longmans, Green, 1866. 2 fldg tables. Orig green cloth gilt, sl rubbed.
  (John Smith) **£35 [≈ $56]**

Reeve, Lovell Augustus
- The Land and Freshwater Mollusks Indigenous to, or Naturalised in the British Isles. London: 1863. Post 8vo. xx,275 pp. Map, num text figs. Lacks port. Orig cloth, rebacked. *(Wheldon & Wesley)* **£40 [≈ $64]**

Reformed Medical Journal ...
- See Beach, W. & Sheppard, W.D.

Regnault, M.V.
- Elements of Chemistry. Translated from the French by T. Forrest Betton ... Philadelphia: H.C. Baird, 1865. 2 vols. 8vo. xvi,9-671; 804 pp. 689 w'engvd ills. Orig cloth.
  (Charles B. Wood) **$125 [≈ £78]**

Regnault, Noel
- Philosophical Conversations: or, A New System of Physics ... Translated into English and illustrated with Notes by Thomas Dale. London: for W. Innys ..., 1731. 1st edn in English. 3 vols. 8vo. 89 plates (1 fldg). Contemp calf, sl rubbed, sl wear at spine ends. *(Burmester)* **£600 [≈ $960]**

Reich, Wilhelm
- The Mass Psychology of Fascism ... Translated ... by Theodore P. Wolfe. New York: Orgone Institute Press, 1946. 1st English edn. 8vo. xxiv,344 pp. Occas ink marginalia. Orig cloth. Dw sl rubbed at extremities. *(Frew Mackenzie)* **£45 [≈ $72]**

Reid, William
- The Progress of Development of the Law of Storms ... London: John Weale, 1849. Super roy 8vo. 6 fldg & other plates, & further plate tipped into endpapers (intended to be detached & v uncommon). Orig cloth, ptd label, uncut, sl worn & faded.
  (Duck) **£55 [≈ $88]**

Reik, Theodor
- Ritual. Psycho-Analytic Studies. With a Preface by Sigmund Freud. Translated from the Second German Edition by Douglas Bryan ... London: Hogarth Press, 1931. 8vo. 266 pp. Orig green cloth. Sl chipped dw.
  (Karmiole) **$75 [≈ £46]**
- The Unknown Murderer. Translated from the German by Dr. Katherine Jones. London: Hogarth Press, 1936. 8vo. 260 pp. Orig green cloth. Dw (spine sl faded).
  (Karmiole) **$65 [≈ £40]**

Remarks ...
- Remarks on the Art of Making Wine ... see MacCulloch, John.

Remington, Joseph P.
- The Practice of Pharmacy. Philadelphia: Lippincott, 1886. 1st edn. Lge 8vo. 1080 pp. Ca 500 ills. Orig cloth. *(Elgen)* **$95 [≈ £59]**

Rennie, J.
- The Domestic Habits of Birds. London: Library of Entertaining Knowledge, 1833. Sm 8vo. xvi,379 pp. Text figs. Orig cloth, trifle used. *(Wheldon & Wesley)* **£20 [≈ $32]**

Renwick, James
- Treatise on the Steam Engine. New York: Carvill, 1839. 2nd rvsd edn. 8vo. 327 pp. 10 fldg plates. Some foxing. Orig cloth, rehinged. *(Book House)* **£70 [≈ $112]**

Reyner, J.H.
- Cine-Photography for Amateurs. London: Chapman & Hall, 1931. 1st edn. viii,180,[8 advt] pp. 36 plates, text ills. Orig cloth, sl discold. *(Duck)* **£30 [≈ $48]**
- Television: Theory and Practice. London: Chapman & Hall, 1934. 1st edn. 8vo. x,[1],

196, [4 advt] pp. 13 plates, 88 other ills. Orig cloth, spine faded.   *(Fenning)* **£35 [≈ $56]**

Reynolds, Gilbert Westacott
- The Aloes of South Africa. Johannesburg: 1950. 4to. xxiv,520 pp. 77 cold plates, 572 text ills. Orig cloth, cvrs dampstained.
          (Wheldon & Wesley) **£60 [≈ $96]**
- The Aloes of South Africa. With Foreword by Field-Marshal Smuts. Johannesburg: 1950. 4to. xxiv,520 pp. 77 cold plates, num other ills. Orig green cloth, spine sunned.
          (Terramedia) **$100 [≈ £62]**
- The Aloes of Tropical Africa and Madagascar. Mbabane, Swaziland: 1966. Roy 8vo. xxii,537 pp. Map, 106 cold plates, 557 text figs. Orig cloth, cvrs somewhat stained, some sl dampstaining of edges.
          (Wheldon & Wesley) **£60 [≈ $96]**

Reynolds, Sir J. Russell
- Epilepsy: Its Symptoms, Treatment, and Relation to Other Chronic Convulsive Diseases. London: Churchill, 1861. 1st edn. xxix,360,[32 ctlg dated Oct 1861] pp. Rebound in three qtr cloth.
          (Elgen) **$300 [≈ £187]**
- Epilepsy: Its Symptoms, Treatment, and Relation to Other Chronic Convulsive Diseases. Chicago: 1981. Facs of the 1861 edn. xxix,360,[40 ctlg dated Dec 1867] pp. Orig bndg.   *(Elgen)* **$75 [≈ £46]**

Reynolds, Michael
- Continuous Railway Brakes. A Practical Treatise ... London: Crosby Lockwood, 1882. 1st edn. x,228,advt pp. 9 fldg plates, 4 fldg tables, ills. Orig olive cloth, sl soiled & rubbed.   *(Duck)* **£35 [≈ $56]**
- Locomotive-Engine Driving. A Practical Manual for Engineers in Charge of Locomotive Engines. London: Crosby Lockwood, 1877. 8vo. July 1877 ctlg. W'cut frontis. Minor pencillings in text. Orig pict gilt cloth, rather worn & shaken, loose in case.
          (Waterfield's) **£45 [≈ $72]**

Rhind, William
- A History of the Vegetable Kingdom. London: 1855. Roy 8vo. xii,720 pp. Port of Linnaeus, title vignette, 22 cold & 19 plain plates, num text figs. Somewhat foxed. New cloth.   *(Wheldon & Wesley)* **£55 [≈ $88]**
- A History of the Vegetable Kingdom. Revised Edition. London: 1868. Roy 8vo. xvi, 744 pp. Port of Linnaeus, addtnl engvd title with hand cold vignette, 22 hand cold & 23 plain plates, text figs. Occas foxing. Half mor.

          (Henly) **£78 [≈ $124]**

Ribot, Theodule A.
- Diseases of Personality. Chicago: Open Court, 1898. 4th edn, rvsd. 8vo. 163 pp. Some sl marg pencil marks. Orig bndg.
          (Xerxes) **$55 [≈ £34]**
- Heredity: a Psychological Study of its Phenomena, Laws, Causes, and Consequences. From the French ... London: Henry S. King, 1875. 1st edn in English. 8vo. x,393 pp. Orig cloth gilt.
          (Fenning) **£32.50 [≈ $52]**
- Psychology of the Emotions. London: Walter Scott, 1898. Sm 8vo. 455 pp. Orig bndg, rear crnrs sl bumped.   *(Xerxes)* **$55 [≈ £34]**

Ricardo, David
- On the Principles of Political Economy, and Taxation. London: Murray, 1817. 1st edn. 8vo. viii,589,[1 errata],[13] pp. Occas light foxing. Half Niger mor, gilt spine, mor labels, by Bernard Middleton.
          (Pickering) **$13,500 [≈ £8,437]**
- On the Principles of Political Economy, and Taxation... Third Edition. London: Murray, 1821. 3rd edn. 8vo. iii-xii,358 pp. Lacks half-title & final blank leaf. Sl foxing at start. Near contemp blindstamped mauve cloth, mor label, spine faded.
          (Pickering) **$2,500 [≈ £1,562]**
- The Works ... With a Notice of the Life and Writings of the Author, by J.R. McCulloch ... London: Murray, 1852. 2nd edn. Lge 8vo. xxxiii,[i blank],584,32 ctlg pp. Orig cloth, uncut, hinges cracking but sound, spine faded & sl rubbed.   *(Pickering)* **$650 [≈ £406]**

Rice, Nathan Lewis
- Phrenology Examined, and Shown to be Inconsistent with the Principles of Physiology, Mental and Moral Science, and the Doctrines of Christianity. Also an Examination of the Claims of Mesmerism. New York: 1849. 1st edn. 12mo. [vi],318,[20] pp. Foxed. Orig cloth.   *(Gach)* **$50 [≈ £31]**

Richard, L.C.M.
- Observations on the Structure of Fruits and Seeds; Translated from the Analyse du Fruit, with Original Notes by J. Lindley. London & Norwich: 1819. 8vo. xx,100 pp. 6 plates. Orig bds, rebacked. J.B. Balfour's copy with a few marg notes. *(Wheldon & Wesley)* **£35 [≈ $56]**

Richards, J.
- A Treatise on the Construction and Operation of Wood-Working Machines ...

London: Spon, 1872. Sm 4to. xx,283 pp. 117
plates. Orig cloth gilt, extremities worn.
(Hollett) **£220 [≈ $352]**

Richardson, Alex
- Vickers Sons and Maxim, Limited. Their
Works and Manufactures ... London: Offices
of "Engineering", 1902. Lge 4to. viii,200 pp.
70 collotype plates. Orig cloth, rear inside
hinge & ft of spine worn.
(Spelman) **£150 [≈ $240]**

Richardson, Sir B.W.
- Vita Medica: Chapters of Medical Life and
Work. London: 1897. 1st edn. 8vo. xvi,495,
[1], 16 advt pp. Rec bds with label.
(Fenning) **£45 [≈ $72]**

Richardson, Sir J., et al.
- The Museum of Natural History, being a
Popular Account of the Structure, Habits,
and Classification of the ... Animal Kingdom.
London: [1859-92]. Orig iss. 4 vols. 4to. Cold
title & 136 plates (78 hand cold). Orig dec
cloth gilt, a.e.g., inner jnts trifle loose.
(Wheldon & Wesley) **£120 [≈ $192]**
- The Museum of Natural History. London:
n.d. 4 vols. Imperial 8vo. Engvd title with
hand cold vignette, 77 hand cold & 59 plain
plates, num text figs. Orig pict cloth gilt,
a.e.g. *(Henly)* **£150 [≈ $240]**

Richardson, O.W.
- The Emission of Electricity from Hot Bodies.
London: 1921. 2nd edn. viii,320 pp. Ills. Orig
bndg, sl shelf wear. *(Elgen)* **£85 [≈ £53]**

Richerand, A.
- Elements of Physiology. Translated from the
French by G.J.M. De Lys. With Annotations
by N. Chapman ... Philadelphia: Carey &
Lea, 1825. "5th Amer from the last London
edn". 448,111 pp. Browning. Some foxing.
Orig calf, scuffed. *(Elgen)* **£85 [≈ £53]**

Rickard, George
- Practical Mining; fully and familiarly
described. London: Effingham Wilson, 1869.
72,ctlg pp. Orig green cloth.
(C.R. Johnson) **£55 [≈ $88]**

Rickman, P.
- A Bird Painter's Sketch Book. London: 1935.
Cheap edn. 4to. 150 pp. Cold frontis, 10 cold
& 23 plain plates. Title & half-title sl foxed.
Orig cloth, trifle used. AL (unsigned) by the
author inserted.
(Wheldon & Wesley) **£40 [≈ $64]**

- Bird Sketches and Some Field Observations.
London: 1938. 4to. 79 pp. Cold frontis, 32
plates. Orig cloth, trifle used.
(Wheldon & Wesley) **£35 [≈ $56]**
- Sketches and Notes from a Bird Painter's
Journal. London: 1949. 4to. 119 pp. Cold
frontis, 1 cold & 42 plain plates, num text
figs. Orig cloth.
(Wheldon & Wesley) **£28 [≈ $44]**
- Sketches and Notes from a Bird Painter's
Journal. London: 1949. 4to. x,119 pp. Cold
frontis, 1 cold & 42 plain plates, num text ills.
Orig cloth. dw. *(Henly)* **£35 [≈ $56]**

Ridgeway, William
- The Origin and Influence of the
Thoroughbred Horse. Cambridge: UP, 1905.
8vo. xvi,538,[4 advt] pp. 143 ills. Name on
half- title. Orig blue cloth, endpapers sl
discold. *(Blackwell's)* **£90 [≈ $144]**

Ridley, H.N.
- The Dispersal of Plants throughout the
World. London: 1930. Roy 8vo. xx,744 pp. 2
cold & 20 plain plates. Orig cloth.
(Henly) **£38 [≈ $60]**

Ridley, Humphrey
- The Anatomy of the Brain ... London: 1695.
1st edn. 8vo. [12],200,[24] pp. Dedic leaf.
Errata. 5 fldg plates. Lacks imprimatur leaf.
Lib stamp on title recto & verso. Full russia.
Wing R.1449.
(Rootenberg) **$6,000 [≈ £3,750]**

Riehl, G. & Zumbusch, V.
- Atlas of Diseases of the Skin. In 3 Parts.
Philadelphia: Blakiston & Co, 1925. Folio. 84
cold plates. Orig leatherette.
(Hemlock) **$475 [≈ £296]**

Rimmer, E.J.
- Boiler Explosions, Collapses and Mishaps. A
Summary of the Causes of Boiler Explosions
and Recommendations for Prevention from
Board of Trade Reports, 1882-1911 ...
London: Constable, 1912. 8vo. 135 pp. Orig
bndg. Poor dw. *(Book House)* **£15 [≈ $24]**

Rintoul, L.J. & Baxter, E.V.
- A Vertebrate Fauna of Forth. London: 1935.
8vo. lv,397 pp. Fldg cold map (mtd on linen),
16 plates. Orig cloth, crnrs sl waterstained
affecting endpapers & half-title. Authors' pres
copy. *(Wheldon & Wesley)* **£30 [≈ $48]**

Riolanus, Johannes
- A Sure Guide; or the Best and Nearest Way

to Physick and Chyrurgery ... London: 1671.
3rd edn. Some sgntrs crossed out on title.
Rebacked. Wing R.1536.
(Rittenhouse) **$1,100 [≈ £687]**

Rivers, W.H.R.
- Psychology and Ethnology. London: Kegan
Paul, 1926. 1st edn. 8vo. 324 pp. Orig bndg,
rear crnrs sl bumped. *(Xerxes)* **$55 [≈ £34]**

Riviere, Lazare
- Four Books of that Learned, and Renowned
Doctor, Lazarus Riverius, containing Five
hundred and thirteen Observations ...
Englished by Nicholas Culpeper .. London:
Peter Cole, 1658. Sm folio. [xv],417 pp. Port.
Orig sheep. Wing R.1555.
(Bickersteth) **£550 [≈ $880]**
- The Practice of Physic. In Seventeen several
books ... with the Cure of all Diseases ... by
Nicholas Culpeper ... London: Sawbridge,
1678. 6th edn. Folio. [12],'645" [≈ 517],[12],
"463" [≈ 363],[32] pp. Lacks frontis. Title
dusty. Some stains. Mod half leather.
(Hemlock) **$425 [≈ £265]**

Rizzo, T. & Kuhner, D.
- Bibliotheca De Re Metallica. The Herbert
Clark Hoover Collection of Mining &
Metallurgy. Claremont, CA: Claremont
Colleges, 1980. One of 500. Roy 8vo. xx,219
pp. Frontis, num facs. Orig olive cloth. Dw.
(Gemmary) **$150 [≈ £93]**

Roberton, John
- Observations on the Mortality and Physical
Management of Children. London: 1827. 1st
edn. 8vo. x,311 pp. Errata leaf. Orig bds,
uncut, rebacked. *(Hemlock)* **$425 [≈ £265]**

Roberts, A.
- The Mammals of South Africa. Cape Town:
1951. 4to. xlviii,700 pp. 24 cold & 54 plain
plates. Orig green cloth, spine faded, crnrs
bumped sl affecting contents, cvrs stained,
jnts trifle loose.
(Wheldon & Wesley) **£70 [≈ $112]**
- The Mammals of South Africa. Cape Town:
1951. 4to. xlviii,700 pp. 24 cold & 54 plain
plates. Orig cloth, spine partly faded.
(Wheldon & Wesley) **£95 [≈ $152]**
- The Mammals of South Africa. London:
1951. 1st edn. 4to. Num cold & other plates.
Sm tear in foredge of 3 pp. Orig bndg. Frayed
dw. *(Grayling)* **£70 [≈ $112]**
- The Mammals of South Africa. Cape Town:
1954. 2nd edn. 4to. xlviii,700 pp. 23 cold &
54 photo plates. Orig cloth, trifle used.

(Wheldon & Wesley) **£95 [≈ $152]**

Roberts, M.
- The Conchologist's Companion. London:
1834. 2nd edn. Sm 8vo. ix,210 pp. Cold
frontis. Orig cloth, reprd, ex-lib.
(Wheldon & Wesley) **£28 [≈ $44]**

Roberts, Robert
- The House Servant's Director, or a Monitor
for Private Families ... Friendly Advice for
Cooks ... Boston: 1827. 1st edn. 12mo. 180
pp. Title loose with blank piece missing from
inner marg, 2 tears reprd. Somewhat later
handmade cvrs.
(M & S Rare Books) **$1,750 [≈ £1,093]**

Roberts, T.J.
- The Mammals of Pakistan. London: 1977.
4to. 388 pp. 18 cold, 185 plain ills, 118 maps.
Orig cloth. *(Wheldon & Wesley)* **£47 [≈ $75]**

Robertson, J.H.
- The Story of the Telephone. A History of the
Telecommunications Industry of Great
Britain. London: Pitman, 1947. 8vo. 299 pp.
Orig bndg. *(Book House)* **£15 [≈ $24]**

Robertson, James
- General View of the Agriculture in the
County of Inverness ... London: for Richard
Phillips ..., 1808. lxvi,447,[4 ctlg] pp. Cold
fldg map, fldg plan of the Caledonian Canal.
Mod soft calf gilt. *(Hollett)* **£120 [≈ $192]**

Robinson, H.C.
- The Birds of the Malay Peninsula, Vol 2, The
Birds of the Hill Stations. London: 1928.
xxii,310 pp. Map, 23 cold & 2 plain plates.
Orig red cloth, somewhat used.
(Wheldon & Wesley) **£125 [≈ $200]**

Robinson, J.F.
- The Flora of the East Riding of Yorkshire ...
London: A. Brown & Sons, 1902. 1st edn.
8vo. 253,[3 advt] pp. Fldg map. Orig dec
cloth. *(Hollett)* **£40 [≈ $64]**

Robinson, Samuel
- A Course of 15 Lectures on Medical Botany,
Denominated Thomson's New Theory of
Medical Practice. Boston: Jonathan Howe,
1830. 1st edn. 8vo. 192 pp. Orig leather,
spine worn, front hinge cracked.
(Xerxes) **$175 [≈ £109]**

Robinson, Victor
- The Don Quixote of Psychiatry [biography of

Dr. S.V. Clevenger]. New York: Historico-Medical Press, 1919. 1st edn. 8vo. Photo ills. Orig bndg.   *(Xerxes)* **$85 [≈ £53]**

Robinson, William

- Alpine Flowers for Gardens: Rock, Wall, Marsh Plants, and Mountain Shrubs. Fourth Edition, revised. London: Murray, 1910. 8vo. xix,344 pp. 5 plates, num ills. Orig cloth gilt.
(Fenning) **£35 [≈ $56]**
- The English Flower Garden and Home Grounds. Design and Arrangement ... followed by a Description of the Plants, Shrubs and Trees ... Ninth Edition. London: 1905. 8vo. xii,920 pp. Num ills. Orig cloth gilt.   *(Fenning)* **£35 [≈ $56]**
- The Parks and Gardens of Paris Considered in relation to the Wants of Other Cities and of Public and Private Gardens ... London: Macmillan, 1878. 2nd, rvsd, edn. 8vo. xxiv, 548 pp. Num ills. Bds, stained.
(Bernett) **$165 [≈ £103]**
- The Wild Garden or the Naturalization and Grouping of Hardy Exotic Plants ... Fifth Edition ... London: Murray, 1895. One of 280, this copy un-numbered. 8vo. xx,304 pp. 14 plates, 76 ills. Minor foxing of endpapers. Vellum gilt, t.e.g., by Birdsall.
(Fenning) **£65 [≈ $104]**

Robson, J., et al.

- Canaries, Hybrids and British Birds in Cage and Aviary. Edited by S.H. Lewer. London: Waverley Book Co, (1911). Subscribers Edition, with the 8 extra plates. 4to. xii, 424 pp. 26 cold plates by A.F. Lydon & others. Orig half mor, t.e.g. *(Egglishaw)* **£90 [≈ $144]**

Rock, J.F.

- The Ornamental Trees of Hawaii. Honolulu: 1917. 8vo. 210 pp. 2 cold & 79 plain plates. Orig cloth, trifle used.
(Wheldon & Wesley) **£50 [≈ $80]**

Rodd, E.H.

- The Birds of Cornwall and the Scilly Islands. London: Trubner, 1880. 1st edn. 8vo. lvi,320 pp. Fldg map, port. Orig pict blue cloth gilt, hd of spine sl rubbed. *(Gough)* **£48 [≈ $76]**
- The Birds of Cornwall, and the Scilly Islands. Edited by J.E. Harting. London: 1880. 8vo. lvi,320 pp. Port, map. Orig cloth, trifle used.
(Wheldon & Wesley) **£50 [≈ $80]**

Rodway, L.

- Some Wild Flowers of Tasmania. Hobart: 1910. 8vo. viii,119 pp. 37 photo ills. Orig cloth.   *(Wheldon & Wesley)* **£30 [≈ $48]**

Roe, F.G.

- The North American Buffalo. A Critical Study of the Species in its Wild State. London: 1972. 2nd edn. 8vo. xi,991 pp. Map. Orig cloth. *(Wheldon & Wesley)* **£35 [≈ $56]**

Roget, Peter Mark

- Animal & Vegetable Physiology considered with reference to Natural Theology. London: Wm. Pickering, Bridgewater Treatise, 1834. 1st edn. 2 vols. 8vo. xxvii,593; 661 pp. Num ills. Orig cloth, paper labels (sl chipped), uncut, partly unopened, some wear.
(Hartfield) **$195 [≈ £121]**

Rohde, E.S.

- Garden-Craft in the Bible and Other Essays. London: 1927. 8vo. Ills. Orig cloth backed bds.   *(Wheldon & Wesley)* **£25 [≈ $40]**
- Gardens of Delight. London: 1934. 8vo. xii, 308 pp. 32 ills. Orig cloth.
(Wheldon & Wesley) **£30 [≈ $48]**
- The Old English Gardening Books. London: 1924. Roy 8vo. xii,144 pp. 16 plates, 8 plans. Orig linen backed bds.
(Wheldon & Wesley) **£75 [≈ $120]**
- The Old English Herbals. London: 1922. Orig edn. Roy 8vo. xii,243 pp. Cold frontis, 17 ills. Orig cloth, cvrs sl soiled.
(Wheldon & Wesley) **£75 [≈ $120]**
- Oxford's College Gardens. London: 1932. Roy 8vo. xiv,193 pp. Cold frontis, 23 cold & 8 plain plates. A little foxing, v sl inkstain on a few extreme outer margs. Orig cloth, cvr sl bubbled.   *(Wheldon & Wesley)* **£28 [≈ $44]**
- Shakespeare's Wild Flowers, Fairy Lore, Gardens, Herbs, Gatherers of Simples and Bee Lore. London: [1935]. 8vo. xi,236 pp. Cold frontis, 4 cold plates, 2 text figs. Label partly removed from endpapers. Orig cloth.
(Wheldon & Wesley) **£30 [≈ $48]**
- The Story of the Garden. London: 1932. Roy 8vo. xii,326 pp. 36 plates (5 cold). Orig cloth.   *(Wheldon & Wesley)* **£25 [≈ $40]**
- The Story of the Garden, with a Chapter on American Gardens. London: Medici Society, 1933. 8vo. xii,326 pp. 5 cold & 46 other plates. Orig cloth. Dw.
(Blackwell's) **£45 [≈ $72]**
- Vegetable Cultivation and Cookery. London: 1938. 8vo. viii,275 pp. Num ills. Orig cloth.
(Wheldon & Wesley) **£18 [≈ $28]**

Roheim, Geza

- The Riddle of the Sphinx, or Human Origins. Authorized Translation from the German by R. Money-Kyrle. With a Preface by Ernest Jones, M.D. London: Hogarth Press, 1934.

1st edn. 8vo. 304 pp. Orig green cloth.
(Karmiole) **$50 [≃ £31]**

Roll, Erich
- An Early Experiment in Industrial Organisation. Being a History of the Firm of Boulton & Watt, 1775-1805. London: Longmans, Green, 1930. 320 pp. 3 plates. Orig bndg, sl shaken.
(Book House) **£45 [≃ $72]**

Rolleston, H.D.
- Diseases of the Liver, Gall-Bladder and Bile-Ducts. Philadelphia: 1905. 1st edn. Lge 8vo. 794,16 ctlg pp. 7 cold litho plates, text ills. Orig bndg, some shelf wear.
(Elgen) **$100 [≃ £62]**

Rolleston, J.D. & Ronaldson, G.W.
- Acute Infectious Diseases. A Handbook for Practitioners and Students. Third Edition. London: 1940. 8vo. [ix],477 pp. Orig cloth.
(Bickersteth) **£24 [≃ $38]**

Rollins, William
- Notes on X-Light. Boston: 1904. 1st edn. xlii,errata, 400 pp. 152 plates. Some margs sl dampstained. Orig bndg, t.e.g.
(Elgen) **$325 [≃ £203]**

Rolt, L.T.C.
- Isambard Kingdom Brunel. London: Longman, BCA, 1972. 345 pp. Ills. Orig bndg. Dw. *(Book House)* **£15 [≃ $24]**

Romains, Jules
- Eyeless Sight. A Study of Extraretinal Vision and the Paroptic Sense. Translated by C. Ogden. New York: Putnam, 1924. 1st edn. 8vo. 251 pp. Orig bndg. *(Xerxes)* **$70 [≃ £43]**

Romanes, E.
- The Life and Letters of George John Romanes. London: 1896. 2nd edn. 8vo. x,360 pp. Port, 2 plates. Orig cloth, trifle used.
(Wheldon & Wesley) **£40 [≃ $64]**

Romanes, George John
- Darwin, and After Darwin. An Exposition of the Darwinian Theory ... London: 1892-95-97. 1st edn. 3 vols. 8vo. 3 ports, num figs. Some spots & other marks. Orig cloth.
(Bow Windows) **£125 [≃ $200]**
- Mental Evolution in Man: Origin of Human Faculty. London: Kegan Paul ..., 1888. 1st edn. 8vo. [ii],[x],452,[4] pp. Fldg frontis chart. Orig mauve cloth.
(Gach) **$175 [≃ £109]**

- Mental Evolution in Man: Origin of Human Faculty. New York: Appleton, 1889. 1st Amer edn. 8vo. [ii],[x],452,[4] pp. Orig maroon cloth. *(Gach)* **$85 [≃ £53]**

Roosevelt, Theodore
- The Deer Family. 1903. 8vo. Ills. Orig bndg.
(Grayling) **£30 [≃ $48]**

Rorschach, Hermann
- Psycho-Diagnostics. Text and Tables. Translated by Paul Lemkau. New York: Grune & Stratton, 1942. 2nd edn. 2 vols. 8vo. 238 pp. 10 mtd cold plates. Orig bndg, 1 backstrip detached. *(Xerxes)* **$140 [≃ £87]**

Roscoe, Sir Henry E.
- Spectrum Analysis ... Fourth Edition Revised and Considerably Enlarged by the Author and Arthur Schuster. London: 1885. 8vo. xvi,452 pp. Dble-page frontis, 3 cold & 2 plain plates, text ills. Orig gilt illust brown cloth, t.e.g., cold onlay on upper cvr.
(Bow Windows) **£40 [≃ $64]**

Rosen, George
- The History of Miners' Diseases. With an Introduction by Henry E. Sigerist. New York: 1943. 1st edn. 490 pp. Frontis port, ills. Dw. *(Elgen)* **$100 [≃ £62]**
- The Reception of William Beaumont's Discovery in Europe. With a Foreword by John F. Fulton. New York: 1942. One of 500. 8vo. 97 pp. Tipped in sepia port. Orig bndg, paper facs mtd on cvrs, paper label.
(Elgen) **$65 [≃ £40]**

Rothschild, W.
- The Avifauna of Laysan and the Neighbouring Islands, with a Complete History of the Birds of the Hawaiian Possessions. London: 1893-1900. One of 250. 3 parts in 2 vols. Folio. 55 hand cold plates by Keulemans, 20 collotype & 8 litho plates. Half mor.
(Wheldon & Wesley) **£2,500 [≃ $4,000]**

Rothwell, C.F. Seymour
- The Printing of Textile Fabrics ... London: Charles Griffin, 1897. 1st edn. 8vo. xii,312,ii pp. 20 ff containing 109 fabric samples. Clean tear in half-title. Orig cloth, inner hinges weak. *(Charles B. Wood)* **$175 [≃ £109]**

Rourke, Constance
- Audubon. London: 1936. 1st edn. 8vo. 342 pp. Cold plates. Orig cloth.
(Henly) **£18 [≃ $28]**

- Audubon. London: 1936. 8vo. 12 cold plates, ills. Orig cloth.
(Wheldon & Wesley) **£20 [≃ $32]**

Rouse, William

- The Doctrine of Chances, or the Theory of Gaming, made easy ... London: Gye & Balne for the author ..., [1814]. 1st edn. 8vo. Errata leaf. Engvd title, 4 fldg tables. Contemp calf, rebacked.
(Hannas) **£180 [≃ $288]**

Routh, Edward John

- A Treatise on Analytical Statics with Numerous Examples. Cambridge: 1896-1902. 2nd edn, rvsd & enlgd. 2 vols. Ills. Orig cloth.
(Elgen) **$65 [≃ £40]**

Rowe, Nicholas

- The Golden Verses of Pythagoras. Translated from the Greek. [Edinburgh]: ptd in the Year, 1740. 1st sep edn. 8vo. 12 pp. Disbound.
(Hannas) **£85 [≃ $136]**

Rowley, G.D.

- Ornithological Miscellany. London: 1876-78. 3 vols in one. 4to. Text only, without plates. Vol 1 title sl browned. Orig cloth.
(Wheldon & Wesley) **£100 [≃ $160]**

Roy, D.N.

- Entomology (Medical & Veterinary). Calcutta: Saraswaty Library, 1946. 1st edn. 4to. 358 pp. 162 ills. Orig cloth, sl worn.
(Savona) **£25 [≃ $40]**

Royal Albert Asylum

- Royal Albert Asylum for Idiots and Imbeciles. Annual Reports from 1866-1916. 6 vols. Thick 8vo. Polished red leather, rubbed.
(Gach) **$350 [≃ £218]**

Royal College of Surgeons

- Catalogue of the Hunterian Collection in the Museum. London: 1830-31. 6 vols. 4to. Sl foxing. Contemp cloth. Only fasc. 1 of Part 4 was published.
(Wheldon & Wesley) **£100 [≃ $160]**

Royal Horticultural Society

- Dictionary of Gardening. Second Edition. Edited by F.J. Chittenden. London: 1956. 5 vols, inc Supplement. 4to. Ills. Orig cloth.
(Henly) **£110 [≃ $176]**

Royal Natural History ...

- See Lydekker, R. (ed.).

Rudd, Sayer, M.D. & Man-Midwife

- The Certain Method to Know the Disease. A Lecture address'd to Students in Physic. London: for J. Roberts, 1742. Sole edn. Sm 4to. Title, viii,28 pp, final blank leaf. Title sl soiled at edges & crnrs creased. Sewed as issued.
(Bickersteth) **£180 [≃ $288]**

Ruddock, E.H.

- Stepping-Stone to Homoeopathy and Health. New American Edition, edited and enlarged ... by William Boericke. San Francisco: 1890. 256 pp. Orig bndg.
(Xerxes) **$75 [≃ £46]**

Rufinus

- The Herbal of Rufinus. Edited from the Unique Manuscript by L. Thorndike and F.S. Benjamin, Jr. Chicago: (1946) 1949. 2nd imp. 8vo. xliv,476 pp. Sm tear in lower edges of prelims. Orig cloth.
(Wheldon & Wesley) **£85 [≃ $136]**

Rumford, Sir Benjamin Thompson, Count of

- Proposals for Forming ... a Public Institution for Diffusing the Knowledge and Facilitating the General Introduction of Useful Mechanical Inventions and Improvements ... London: 1799. 50,[4] pp. Orig wraps, a.e.g., soiled. Orig subscription form inserted.
(Duck) **£385 [≃ $616]**

Rundall, L.P.

- The Ibex of Sha-Ping and Other Himalayan Studies. London: 1915. 4to. Mtd cold plates, text ills. Orig bndg, v sl rubbed.
(Grayling) **£35 [≃ $56]**

Rundell, Maria Eliza

- A New System of Domestic Cookery ... By a Lady. Sixty-Ninth Edition. Augmented and Improved ... by Miss Emma Roberts. London: Murray, 1846. 8vo. liv,517,12 advt pp. 10 engvd plates. Flyleaf replaced. Orig green cloth gilt, v clean. *(Moon)* **£50 [≃ $80]**
- A New System of Domestic Cookery ... By a Lady. Halifax: Milner & Sowerby, 1862. i,376 pp. 9 plates. Engvd frontis, addtnl pict title. Orig cloth, orig pict spine relaid, recipes written on flyleaves.
(Box of Delights) **£25 [≃ $40]**
- A New System of Domestic Cookery ... By a Lady. Halifax: Milner, [ca 1870]. i,350,14 ctlg pp. 9 plates. Orig cloth, new endpapers.
(Box of Delights) **£20 [≃ $32]**

Rusden, M.

- A Further Discovery of Bees, treating of the

Nature, Government, Generation and Preservation of the Bee ... Improvements arising from keeping them in Transparent Boxes ... London: 1679. 8vo. [xxiv],143 pp. 4 plates on 1 fldg leaf. Contemp calf, jnts cracked. *(Wheldon & Wesley)* **£300 [≈ $480]**

Rush, Benjamin
- An Inquiry into the Influence of Physical Causes upon the Moral Faculty. Delivered before the American Philosophical Society, Held at Philadelphia on the Twenty-Seventh of February, 1768. 8vo. [ii],[95]-124 pp. Sl foxed. Mod lib binder. *(Gach)* **$450 [≈ £281]**
- An Inquiry into the Various Sources of the Usual Forms of Summer & Autumnal Disease in the United States ... Philadelphia: J. Conrad & Co ..., 1805. 1st edn. 4to. 113 pp. Contemp calf, jnts rubbed, uncut.
  (Hemlock) **$300 [≈ £187]**
- Letters ... Edited by L.H. Butterfield. Princeton, NJ: Princeton UP for the Amer Philosophical Soc, 1951. 1st edn. 2 vols. 8vo. 17 half-tones. Ex-lib, sl marked. Orig bndg, vol 2 front bd sl abraded.
  (Gach) **$125 [≈ £78]**
- Medical Inquiries and Observations. Philadelphia: 1809. 3rd edn. 4 vols. Contemp leather. *(Rittenhouse)* **$750 [≈ £468]**
- Medical Inquiries and Observations. Philadelphia: ptd by Griggs & Dickinson for M. Carey, 1815. 4th edn. 4 vols in two. 8vo. 264; 273; 244; 249,[x] pp. Occas sl foxing. New cloth & bds. *(Schoyer)* **$200 [≈ £125]**
- Medical Inquiries and Observations, upon the Diseases of the Mind ... Philadelphia: John Richardson, 1818. 2nd edn. 8vo. viii, 9-367 pp. Occas foxing. V occas sl marg staining. Contemp mottled calf, jnts split.
  (Hemlock) **$350 [≈ £218]**
- Sixteen Introductory Letters to Courses of Lectures upon the Institutes and Practice of Medicine ... Philadelphia: 1811. Lib stamp on title. Rebound in leather.
  (Rittenhouse) **$600 [≈ £375]**

Rush, R.W.
- The Lancashire & Yorkshire Railway and its Locomotives, 1846-1923. London: Railway World, 1949. 1st edn. 4to. 154 pp. Fldg map, num ills. Orig qtr cloth, bd sides, minor wear to crnrs, minor bruise to top edge of cvr.
  (Duck) **£35 [≈ $56]**

Russ, K.
- The Speaking Parrots: A Scientific Manual. Translated by L. Schultze ... London: n.d. 8vo. viii,296,[32 advt dated 1887] pp. 8 cold

plates, 2 figs. Some marks. Orig illust cloth, a little rubbed & dull.
  (Bow Windows) **£85 [≈ $136]**

Russel, A.
- The Salmon. Edinburgh: Douglas, 1864. 8vo. viii, 248,advt pp. New cloth.
  (Egglishaw) **£25 [≈ $40]**

Russell, Bertrand
- The Analysis of Mind. London: Allen & Unwin, [1921]. 1st edn. 310 pp. Lib b'plate & withdrawn stamp on front pastedown. Orig cloth. *(Gach)* **$50 [≈ £31]**
- Introduction to Mathematical Philosophy. London: 1919. 1st edn. viii,208 pp. Orig cloth, sl worn, spine faded.
  (Whitehart) **£40 [≈ $64]**
- Our Knowledge of the External World. Chicago & London: The Open Court Publ Co, 1914. 1st edn. [x],[246],16 ctlg pp. Name blotted from flyleaf. Orig ptd green cloth.
  (Gach) **$175 [≈ £109]**
- The Philosophy of Bergson. With a Reply to Mr. H. Wildon Carr ... and a Rejoinder by Mr. Russell. Cambridge: for the "Heretics" by Bowes & Bowes, 1914. 1st edn. 8vo. [ii],36, [2] pp. Orig ptd wraps, sl worn & chipped. *(Frew Mackenzie)* **£50 [≈ $80]**
- The Principles of Mathematics. Vol 1 [all publ]. Cambridge: UP, 1903. 1st edn. 8vo. 29, 534,index pp. Half-title. Some pencilled notes. Lib buckram.
  (M & S Rare Books) **$200 [≈ £125]**

Russell, Kenneth
- British Anatomy, 1525-1800: A Bibliography. Melbourne: 1963. 1st edn. Slipcase. Sgnd by the author.
  (David White) **£50 [≈ $80]**

Russell, Richard
- A Dissertation concerning the Use of Sea Water in Diseases of the Glands ... Oxford: at the Theatre ..., 1753. 1st authorised edn in English. 8vo. Imprimatur & corrigenda ff. 7 plates. Contemp calf, sm sm holes in lower panel of spine. *(Bow Windows)* **£230 [≈ $368]**

Russell, W. Ritchie
- Poliomyelitis. London: [1952]. 8vo. [vii], 84 pp. Text figs. Orig cloth.
  (Bickersteth) **£20 [≈ $32]**

Russell, William Howard
- The Atlantic Telegraph. London: Day & Son, [1866]. 1st edn. Folio. Cold litho title, 24 lithotint plates, 1 chart. Orig elab gilt dec

green cloth, a.e.g. Cloth case.
(Offenbacher) **$1,250 [≈ £781]**

Russell, William Logie
- The New York Hospital: A History of the Psychiatric Service 1771-1936. New York: 1945. 1st edn. 8vo. [xviii],556,[2] pp. 8 ff of plates, text ills. Orig green cloth.
(Gach) **$85 [≈ £53]**

Ruston, Arthur G. & Witney, Denis
- Hooton Pagnell. The Agricultural Evolution of a Yorkshire Village. London: 1934. 1st edn. 8vo. 12 plates. Orig cloth, sl rubbed.
(Robertshaw) **£15 [≈ $24]**

Rutherford, Ernest
- The Newer Alchemy. Cambridge: Univ Press, 1937. 67 pp. Ills. Orig bndg.
(Book House) **£15 [≈ $24]**
- Radioactive Substances and their Radiations. London: 1913. 1st edn. 8vo. vii, 699 pp. Num figs. Orig cloth.   *(Baldwin)* **£80 [≈ $128]**

Rutherford, Ernest, Chadwick, James & Ellis, C.D.
- Radiations from Radioactive Substances. Cambridge: UP, 1930. 1st edn. 8vo. xii,588 pp. 12 pp of plates. Orig green cloth, spine faded, spine extremities sl rubbed.
(Karmiole) **$45 [≈ £28]**

Rydberg, P.A.
- Flora of Colorado. Fort Collins: 1906. 8vo. xxii,448 pp. New cloth.
(Wheldon & Wesley) **£40 [≈ $64]**

Rye, E.C.
- British Beetles: an Introduction to the Study of our Indigenous Coleoptera. London: 1866. 1st edn. 8vo. 16 hand cold plates. Orig cloth, spine faded & sl worn. *(Henly)* **£65 [≈ $104]**
- British Beetles. Second Edition, by A.C. Fowler. London: Reeve, 1890. 8vo. xii,288 pp. 16 cold plates. Orig cloth gilt.
(Egglishaw) **£20 [≈ $32]**

Ryley, J. Beresford
- Sterility in Women: Its Causes and Cure. London: H. Renshaw, 1887. 2nd edn. Sm 8vo. 80 pp. Orig cloth, inner hinge strengthened.   *(Elgen)* **$95 [≈ £59]**

S., H., Philokepos
- The Young Gard'ners Director ... see Stevenson, Henry.

Sachs, B.
- Treatise on the Nervous Diseases of Children. New York: Wood, 1905. 2nd edn, rvsd. Thick sm 4to. Ills. Orig bndg.
(Xerxes) **$55 [≈ £34]**

Sachs, J. von
- Lectures on the Physiology of Plants. Translated by H.M. Ward. Oxford: 1887. 8vo. xiv,836 pp. 455 w'cuts. Orig qtr mor, rubbed.   *(Wheldon & Wesley)* **£35 [≈ $56]**

Saint-Pierre, Jacques H. Bernardin de
- Studies of Nature. Translated by Henry Hunter ... London: for C. Dilly, 1796. 1st English edn. 5 vols. 8vo. Frontis in vol 1, 4 fldg plates. Contemp diced russia gilt.
(Sotheran) **£465 [≈ $744]**
- Studies in Nature. Translated by Henry Hunter ... Original Notes and Illustrations by Benjamin Smith Barton. Philadelphia: Abraham Small ..., 1808. 3 vols. 8vo. Frontis & map in vol 1, 3 plates vol 2. Occas foxing. Later three qtr mor, crnrs & spine ends rubbed.   *(Schoyer)* **$135 [≈ £84]**

Sajous, Charles
- Hay Fever and its Successful Treatment by Superficial Organic Alternation of the Nasal Mucous Membrane. Philadelphia: Davis, 1885. 1st edn. 8vo. 103 pp. 13 w'engvs. Orig bndg.   *(Xerxes)* **$85 [≈ £53]**

Salaman, R.N.
- The History and Social Influence of the Potato. Cambridge: 1949. 8vo. xxiv,685 pp. 33 plates, 17 text figs, 9 maps & diags. Orig cloth, trifle faded.
(Wheldon & Wesley) **£30 [≈ $48]**

Salisbury, E.J.
- Downs and Dunes, their Plant Life and Environment. London: 1952. 8vo. xiii,328 pp. Num ills. Orig cloth.
(Wheldon & Wesley) **£25 [≈ $40]**

Salisbury, William
- Hints Addressed to Proprietors of Orchards, and to Growers of Fruit, in General... State of the Apple Trees, in the Cider Countries ... London: Longman & the author ..., 1816. 1st edn. 8vo. [iv],188 pp. 2 fldg plates (sl spotted). Orig bds, spine ends worn.
(Burmester) **£75 [≈ $120]**
- Hortus Siccus Gramineus, or, a Collection of Dried Specimens of British Grasses, with their Latin and English Names, classed according to W. Curtis. London: 1802. 2 vols.

Folio. 126 mtd specimens, with MS labels.
Lib blindstamps. Contemp half calf, worn.
(Wheldon & Wesley) **£750 [≃ $1,200]**

Salmon, G.
- Lessons Introductory to the Modern Higher
Algebra. Dublin: 1866. 2nd edn. xv,296 pp.
Traces of lib stamp removal from title. Orig
cloth, spine relaid. *(Whitehart)* **£35 [≃ $56]**
- A Treatise on the Higher Plane Curves
intended as a Sequel to A Treatise on Conic
Sections. Dublin: 1873. 2nd edn. xix,379 pp.
Orig cloth, unopened, spine faded. 'From the
Author' in ink on title.
(Whitehart) **£25 [≃ $40]**

Salmon, William
- Botanologia, the English Herbal: or, History
of Plants ... London: I. Dawks, for H. Rhodes
..., 1710. 1st edn. Folio. Over 1150 text
w'cuts. Frontis, title & 1st leaf of dedic
supplied from a sl smaller copy. With the
'Index Morborum'. Rec calf, gilt spine.
(Stewart) **£1,200 [≃ $1,920]**
- The Compleat English Physician: or, the
Druggist's Shop Opened ... London:
Gilliflower & Sawbridge, 1693. Thick 8vo.
Half-title laid down. Sl browning. Old calf,
rather defective & rebacked, rubbed. Wing
S.542.   *(Hollett)* **£185 [≃ $296]**
- Select Physical and Chyrurgical
Observations: containing divers remarkable
histories of Cures ... performed by the Author
... London: 1687. 1st edn. 8vo. 8 ff, 523,[1]
pp,17 ff. Advt leaf. Port, 8 plates of
instruments. Contemp calf, rubbed, cvrs
detached. Wing S435
(Hemlock) **$550 [≃ £343]**

Salt, Henry
- Animals' Rights considered in Relation to
Social Progress. London: G. Bell, 1915. Rvsd
edn. 8vo. 124 pp. Orig bndg, sl soiled.
(Xerxes) **$45 [≃ £28]**

Salter, Frank
- Economy in the Use of Steam: A Statement of
the Principles on which a Saving of Steam can
Best be Effected. London: Spon, 1874. Post
8vo. [iv],92,16 price list pp. 17 text figs. Orig
cloth.   *(Duck)* **£20 [≃ $32]**

Salter, J.H.
- The Flowering Plants and Ferns of
Cardiganshire. Cardiff: 1935. 8vo. 182 pp.
Orig cloth. *(Wheldon & Wesley)* **£25 [≃ $40]**

Salter, J.W.
- A Catalogue of the Cambrian and Silurian
Fossils contained in the Geological Museum
of the University of Cambridge. London:
1873. 4to. xlviii,204 pp. Num figs. New
cloth.   *(Baldwin)* **£35 [≃ $56]**

Samouelle, G.
- The Entomologist's Useful Compendium ...
of British Insects ... London: Thomas Boys,
1819. 1st edn. 8vo. 496 pp. 12 plain plates.
Some pp & 6 plates browned. Contemp calf,
worn, cvrs detached. *(Egglishaw)* **£30 [≃ $48]**
- The Entomologist's Useful Compendium ...
of British Insects ... London: Thomas Boys,
1819. 1st edn. 8vo. 496 pp. 12 hand cold
plates, 1 plate of apparatus. Calf gilt, spine in
6 compartments, mrbld edges.
(Egglishaw) **£130 [≃ $208]**
- The Entomologist's Useful Compendium ...
of British Insects ... London: for Thomas
Boys, 1819. 1st edn. 8vo. 496 pp. 12 plates.
Some foxing & sl stains. Old half calf,
rebacked, reprd. *(Bow Windows)* **£50 [≃ $80]**

Sampson, G.V.
- Statistical Survey of the County of
Londonderry. Dublin: Graisberry, 1802.
Maps & plates. Calf.
(Emerald Isle) **£200 [≃ $320]**

Samuels, E.A.
- The Birds of New England and Adjacent
States. Boston: 1870. 5th edn, rvsd. Roy 8vo.
vii,591 pp. Cold frontis, 4 plates of eggs, 23
plates of birds. A little foxing. Orig cloth,
trifle loose, sl wear to spine ends.
(Wheldon & Wesley) **£50 [≃ $80]**

Sanctorius
- Medicina Statica: being the Aphorisms of
Sanctorius. Translated into English, with
Large Explanations ... By John Quincy. The
Fourth Edition. London: 1728. 8vo. viii,463,
index pp. Frontis, fldg plate. Sl browning.
Half calf.   *(Hemlock)* **$375 [≃ £234]**
- Medicina Statica: being the Aphorisms of
Sanctorius, translated into English, with
Large Explanations ... Fifth Edition. By John
Quincy. London: Longman & Newton, 1737.
8vo. viii,463,[xvii] pp. Frontis, fldg plate.
Orig calf, rubbed, jnts cracked, spine ends
worn.   *(Bickersteth)* **£135 [≃ $216]**

Sanger, William W.
- The History of Prostitution: Its Extent,
Causes and Effects throughout the World.
New York: Medical Publ Co, [1899]. 8vo. 709

pp. Orig bndg. *(Gach)* **$30 [≈ £18]**

Sargent, Eric
- The Aircraft Calendar. London: Sampson Low, October 1938. Post 8vo. xiv,144,[2 blank] pp. Num photo ills. Orig cloth.
(Duck) **£20 [≈ $32]**

Sarton, George
- Introduction to the History of Science. Baltimore: 1927-48. 1st edn, 1st iss. 3 vols in 5. Lge thick 8vo. Orig cloth, vol 2 spine ends sl frayed & inner hinges cracked.
(Elgen) **$450 [≈ £281]**

Saunders, Benjamin
- Forging, Stamping and General Smithing ... London & New York: Spon, Spon & Chamberlain, 1912. 1st edn. x,428,[2,56 advt] pp. 728 ills. Orig blue cloth.
(Duck) **£20 [≈ $32]**

Saunders, H.
- An Illustrated Manual of British Birds. London: Gurney, 1899. 2nd edn, rvsd. 8vo. xl, 776 pp. 3 cold maps, 384 ills. Orig cloth gilt.
(Egglishaw) **£25 [≈ $40]**

Saunders, John Cunningham
- A Treatise on Some Practical Points relating to Diseases of the Eye. To which is added, a Short Account of the Author's Life ... London: 1811. 1st edn. 8vo. xlii,216 pp. Port, 8 uncold plates. Rec qtr calf antique.
(Bickersteth) **£240 [≈ $384]**

Saunders, William
- A Treatise on the Structure, Economy and Diseases of the Liver ... Boston: W. Pelham, 1797. 1st Amer edn, with addtns. 12mo. xx,231 pp. Contemp calf, lower jnt starting.
(Elgen) **$95 [≈ £59]**

Saunderson, Nicholas
- The Method of Fluxions applied to a Select Number of Useful Problems ... London: for A. Millar ..., 1756. 1st edn. xxiv,309,[1 errata] pp. 12 fldg plates. Some marg browning. Rebound in calf backed bds.
(Pickering) **$850 [≈ £531]**

Saville-Kent, W.
- The Great Barrier Reef of Australia; its Products and Potentialities. London: [1893]. Imperial 4to. xviii,387 pp. Fldg map, 64 plates (16 cold). Orig cloth, trifle used, refixed. *(Wheldon & Wesley)* **£150 [≈ $240]**
- A Manual of the Infusoria. London: 1881-82.

3 vols in 2. Roy 8vo. 52 plates (numbered 1-51, 48a). Occas sl foxing. Green half mor, spines faded.
(Wheldon & Wesley) **£160 [≈ $256]**
- The Naturalist in Australia. London: 1897. 4to. xv,302 pp. Port, 9 cold & 50 collotype plates, num ills. 1 plate taped in & with short marg. Orig cloth, trifle used, inner jnts reprd.
(Wheldon & Wesley) **£150 [≈ $240]**

Savory, T.H.
- The Spider's Web. London: 1952. Post 8vo. 27 plates (8 cold). Orig cloth.
(Wheldon & Wesley) **£28 [≈ $44]**
- The Spiders and Allied Orders of the British Isles. London: Warne, Wayside & Woodland Series, 1945. 2nd edn. Sm 8vo. 224 pp. 95 plates, 70 text figs. Orig cloth gilt. Dw.
(Egglishaw) **£23 [≈ $36]**

Sawyer, A.R.
- Accidents in Mines in the North Staffordshire Coalfield, from Falls of Roofs & Sides ... Hanley: 1886. 8vo. 101 pp. Map, 302 plates, many cold, some fldg. Mod cloth.
(Book House) **£65 [≈ $104]**

Scarpa, Antonio
- See Knox, Robert.

Schacht, H.
- The Microscope and its Application to Vegetable Anatomy and Physiology. London: S. Highley, 1855. 2nd edn, enlgd. 8vo. 202 pp. Orig cloth, discold, crnrs bumped, sl wear to spine & bd edges. *(Savona)* **£25 [≈ $40]**

Scheele, Karl-Wilhelm
- The Chemical Essays of Charles-William Scheele. Translated ... with Additions ... With a Sketch of the Life ... by John Geddes McIntosh. London: Scott, Greenwood, 1901. 2nd edn. [xxx],294,36 ctlg pp. Orig cloth, spine sl darkened & rubbed at ends.
(Sklaroff) **£90 [≈ $144]**

Schellberg, O. Boto
- Colonic Irrigation in the Treatment of Disease. New York: Oboschell, 1923. 1st edn. 8vo. 202 pp. Orig bndg. *(Xerxes)* **$50 [≈ £31]**

Schimper, A.E.W.
- Plant Geography upon a Physiological Basis. English Translation by W.R. Fisher. Oxford: 1903. Roy 8vo. xxx,839 pp. 4 maps, port, 502 plates & ills. Binder's cloth.
(Wheldon & Wesley) **£60 [≈ $96]**

Schmucker, Samuel
- Psychology ... New York: Harper, 1842. 1st edn. 8vo. 227,[5],[24 ctlg dated 1842] pp. Sl foxed. Contemp calf.   *(Gach)* **$125 [≃ £78]**

Scholl, William
- The Human Foot. Anatomy, Deformities and Treatment. Chicago: Foot Specialist, 1920. Sm 4to. 415 pp. Num photo ills. Orig bndg.
(Xerxes) **$55 [≃ £34]**

Schrenck-Notzing, A.
- Therapeutic Suggestion in Psychopathia Sexualis (Pathological Manifestations of the Sexual Sense) ... Authorised Translation from the German by C.G. Chaddock. Phila: F.A. Davis, 1895. xix,320 pp. Orig cloth.
(Box of Delights) **£30 [≃ $48]**

Schrotter, L. & C.
- Coloured Vade-Mecum to the Alpine Flora. Zurich: [1906]. 11th edn. 8vo. Text in English, French & German. 24 cold & 2 plain plates. Sm lib stamp on title. Orig pict cloth.
(Henly) **£35 [≃ $56]**

Schubert, H.R.
- History of the British Iron and Steel Industry from c. 450 B.C. to a.d. 1775. London: Routledge, [1957]. 1st edn. Tall 8vo. 445 pp. Ills. Orig blue cloth, label.
(Terramedia) **$52 [≃ £32]**

Schuckard, W.E.
- British Bees ... London: Reeve, 1866. 1st edn. 8vo. xvi,371 pp. 16 hand cold plates. Orig cloth, spine & part of cvr faded.
(Egglishaw) **£34 [≃ $54]**

Schuster, Arthur
- An Introduction to the Theory of Optics. London: 1904. 8vo. xv,340,6 advt pp. 2 plates, num text ills. Orig blue buckram.
(Hemlock) **$175 [≃ £109]**

Schwann, Theodor
- Microscopical Researches into the Accordance in the Structure and Growth of Animals and Plants. Translated by Henry Smith. London: Sydenham Society, 1847. 1st English edn. xx,268 pp. 6 plates. Orig cloth, t.e.g., spine relaid.   *(Elgen)* **$200 [≃ £125]**

Schwarz, Herbert F.
- Stingless Bees (Meliponidae) of the Western Hemisphere ... New York: Amer Museum of Nat Hist Bulletin Vol 90, 1948. 4to. xvii, 546 pp. 8 plates, 87 text figs. Buckram. Author's

inscrptn.   *(Bickersteth)* **£85 [≃ $136]**

Sclater, P.L.
- Challenger Voyage 1873-76. Report on the Birds. London: Challenger Reports, Zoology Vol 2 Part 8, 1881. 4to. [iii],166 pp. 30 hand cold plates by J. Smit. Title sl creased. Half calf.   *(Wheldon & Wesley)* **£525 [≃ $840]**
- A Monograph of the Jacamars and Puff-Birds, or Families Galbulidae and Bucconidae. London: [1879-82]. One of 250. Roy 4to. liii,171 pp. 55 hand cold plates. Title sl creased. Red half mor.
(Wheldon & Wesley) **£2,000 [≃ $3,200]**

Sclater, P.L. & Salvin, O.
- Exotic Ornithology ... New or Rare Specimens of American Birds. London: [1866-] 1869. [One of 150]. Demy folio. vi,204 pp. 100 hand cold plates by J. Smit, 10 text figs. Red half mor.
(Wheldon & Wesley) **£4,200 [≃ $6,720]**

Sclater, W.L.
- A History of the Birds of Colorado. London: 1912. 8vo. xxiv,576 pp. Port, map, 16 plates. Orig cloth. *(Wheldon & Wesley)* **£50 [≃ $80]**

Scloppetaria ...
- See Beaufoy, Henry.

Scott, H.H.
- Some Notable Epidemics. London: Edward Arnold, 1934. 1st edn. xix,272 pp. Orig cloth gilt. Dw spine defective. *(Hollett)* **£35 [≃ $56]**

Scott, J.D.
- Siemens Brothers 1858-1958. An Essay in the History of Industry. London: Weidenfeld, 1958. 279 pp. Ills. Orig bndg.
(Book House) **£14 [≃ $22]**

Scott, J.F.
- A History of Mathematics. London: 1960. 2nd edn. viii,[2],266 pp. Frontis, 6 plates, 23 text figs. Orig cloth. Dw.
(Whitehart) **£35 [≃ $56]**

Scott, Peter
- The Swans. London: Michael Joseph, 1972. 1st edn. 4to. x,242 pp. Cold frontis, 48 plates. Orig cloth gilt. Dw. Mint.
(Hollett) **£35 [≃ $56]**

Scott, William Henry
- See Lawrence, John, "William Henry Scott".

Scrope, G. Poulett

- The Geology and Extinct Volcanoes of Central France. London: 1858. 2nd edn. 8vo. xvii,258 pp. 2 fldg geological maps in pockets at end, 17 plates (some fldg), text w'cuts. Orig cloth, refixed. *(Henly)* **£98 [≃ $156]**

- Volcanos. The Character of their Phenomena ... Descriptive Catalogue of all known Volcanos and Volcanic Formations. London: 1862. 2nd edn, rvsd & enlgd. 8vo. xi,[i],490 pp. Cold frontis, fldg map, ills. Orig pict cloth gilt, spine relaid. Author's inscrptn. *(Bickersteth)* **£130 [≃ $208]**

Scrope, W.

- The Art of Deer Stalking. London: 1839. 2nd edn. 8vo. Tinted litho plates, some heavily foxed. Orig cloth, spine dulled with sl snags. *(Grayling)* **£150 [≃ $240]**

Searle, A.B.

- The Clayworker's Handbook: A Manual for all engaged in the Manufacture of Articles from Clay. London: Griffin, 1911. 8vo. 416 pp. Ills. Orig bndg, cvrs marked. *(Book House)* **£15 [≃ $24]**

Secret Remedies ...

- Secret Remedies. What They Cost and What They Contain. Based on Analyses made for the British Medical Association. [With] More Secret Remedies. London: BMA, 1909-12. Sole edns (vol 1 "105th thousand'). 2 vols. 8vo. vii,195; vii,282 pp. Orig cloth cvrd bds. *(Bickersteth)* **£35 [≃ $56]**

Sedgwick, Adam

- Life and Letters ... see Clark, J.W. & Hughes, T.M.

See, T.J.J.

- Researches on the Evolution of the Stellar Systems. Lynn, MA: 1896-1910. 2 vols. Lge 4to. 95 charts & photos. Orig cloth, vol 1 spine ends sl frayed. *(Elgen)* **$225 [≃ £140]**

Seebohm, Henry

- The Birds of Siberia. London: Murray, 1901. xix,512 pp. Map, text ills. Orig dec cloth, t.e.g., spine ends worn, upper jnt split. *(High Latitude)* **$60 [≃ £37]**

- The Birds of Siberia. London: Murray, 1901. 1st edn. Thick 8vo. xix,512 pp. Ills. Orig pict cloth gilt, recased. *(Hollett)* **£60 [≃ $96]**

- Coloured Figures of the Eggs of British Birds. Edited by R. Bowdler Sharpe. London: 1896. Roy 8vo. xxiv,304 pp. Port, 60 cold plates. Some mainly marg foxing. Orig cloth, trifle

used. *(Wheldon & Wesley)* **£40 [≃ $64]**

- Coloured Figures of the Eggs of British Birds. Edited by R. Bowdler Sharpe. Sheffield: Pawson & Brailsford, 1896. Roy 8vo. xxiv,304 pp. Port, 60 cold plates. Orig cloth. *(Egglishaw)* **£45 [≃ $72]**

- The Geographical Distribution of the Family Charadriidae, or the Plovers, Sandpipers, Snipes and their Allies. London: [1887]. 1st iss. Thick 4to. xxix,524 pp. 24 cold plates by Keulemans. Text ills. Orig cloth gilt, trifle rubbed. *(Hollett)* **£750 [≃ $1,200]**

- The Geographical Distribution of the Family Charadriidae, or the Plovers, Sand-pipers, Snipes and their Allies. London: [1888]. 1st iss, with fine impressions, well cold. Roy 4to. xxix,524 pp. 21 hand cold plates by Keulemans, num text figs. Orig green cloth. *(Wheldon & Wesley)* **£750 [≃ $1,200]**

- The Geographical Distribution of the Family Charadriidae, or the Plovers, Sandpipers, Snipes and their Allies. London: [1888]. 2nd iss, but superior colouring. 21 hand cold plates by Keulemans. Inscrptn on title. Red half mor, front cvr trifle faded. *(Wheldon & Wesley)* **£400 [≃ $640]**

Seeley, H.G.

- The Fresh-Water Fishes of Europe. London: 1886. Roy 8vo. x,444 pp. 214 ills. Orig cloth, sl used ex-lib. *(Wheldon & Wesley)* **£40 [≃ $64]**

Seeley, H.G. & Etheridge, R.

- Phillips' Manual of Geology. London: 1885. 2 vols. 8vo. xiv,546,34 advt; xxiv,712 pp. Cold frontis, cold geol map, 36 plates (numbered 1-32, 23a, 23b, 23c, 32a), 147 text figs. Orig cloth, internal jnts becoming weak. *(Henly)* **£35 [≃ $56]**

Seeman, B.

- Popular History of the Palms and their Allies. London: Reeve, 1856. Sm 8vo. xvi,359 pp. 20 cold litho plates. New qtr calf gilt. *(Egglishaw)* **£90 [≃ $144]**

Seigne, J.W.

- A Bird Watcher's Note Book. Studies of Woodcock, Snipe and Other Birds. London: 1930. 8vo. Cold frontis, other plates by Philip Rickman. Orig cloth. *(Grayling)* **£22 [≃ $35]**

Selby, Prideaux John

- Gallinaceous Birds. Vol 5. Pigeons. Edinburgh: Jardine's Naturalist's Library, 1835. 8vo. 30 hand cold plates. Orig cloth, shabby. *(Moorhead)* **£45 [≃ $72]**

- A History of British Forest Trees, Indigenous and Introduced. London: Van Voorst, 1842. 8vo. xx,540 pp. 4 pp July 1839 ctlg tipped in at end. Engvs. Orig dark green cloth gilt, spine faded.   *(Blackwell's)* **£60 [≈ $96]**

- The Natural History of Pigeons. Edinburgh: Jardine's Naturalist's Library, 1835. 1st edn. Cr 8vo. 228 pp. Port, cold vignette title, 30 hand cold plates by Edward Lear. New half calf.   *(Egglishaw)* **£85 [≈ $136]**

- Parrots. Edinburgh: Jardine's Naturalist's Library, 1836. 1st edn. Cr 8vo. 187 pp. Port, cold vignette title, 27 (of30) hand cold plates. 19th c half mor, lettered Ornithology Vol 6.
        (Egglishaw) **£55 [≈ $88]**

Sellew, William H.
- Steel Rails: Their History, Properties, Strength and Manufacture ... London: Constable, 1913. 1st English edn. Thick imperial 8vo. 33 fldg plates, other ills. Occas dustiness. Orig blue cloth, worn & soiled, front hinge cracked but firm.
        (Duck) **£75 [≈ $120]**

Selous, E.
- Realities of Bird Life. London: 1927. 8vo. xvi,351 pp. Orig cloth, faded.
        (Wheldon & Wesley) **£25 [≈ $40]**

Selwyn-Brown, Arthur
- The Physician throughout the Ages. A Record of the Doctor from the Earliest Historical Period ... New York: 1928. 2 vols. Lge 4to. Num ills & ports. Orig dec dark brown fabricoid gilt. *(Karmiole)* **$100 [≈ £62]**

Sendivogius, Michael
- A New Light of Alchymie ... translated ... by J. F[rench]. London: 1650. 1st edn in English. Sm 4to. [xvi],147,[v], [viii],145,50 pp. Sep titles for Paracelsus & Dictionary. Sm marg worm holes. Old ink underlining. Orig calf, reprd. Wing S.2505.
        (Bickersteth) **£750 [≈ $1,200]**

A Series of Essays on Agriculture ...
- See Jeffreys, George Washington.

Serven, James E.
- Colt Firearms, 1836-1954. Santa Ana, Calif.: the author, 1954. Roy 4to. x,385 pp. Num ills. Orig olive cloth, minor marking to top extremities of cvrs.   *(Duck)* **£110 [≈ $176]**

Seward, A.C.
- Plant Life through the Ages, a Geological and Botanical Retrospect. Cambridge: 1931. 8vo.

xxi,601 pp. 140 figs. Orig cloth. Henry N. Ridley's b'plate.   *(Henly)* **£42 [≈ $67]**

Seward, A.C. (ed.)
- Darwin and Modern Science. Essays in Commemoration of the Centenary of the Birth of Charles Darwin ... Cambridge: 1909. 8vo. xvii,595 pp. Port, 1 cold & 3 other plates. Buckram.   *(Egglishaw)* **£24 [≈ $38]**

Seward, Anna
- Memoirs of the Life of Dr. [Erasmus] Darwin, chiefly during his residence at Lichfield, with Anecdotes of his Friends and Criticisms of his Writings ... London: 1804. 8vo. xiv,430,[2] pp. Contemp calf.
        (Wheldon & Wesley) **£85 [≈ $136]**

Sharp, Granville
- The Gilbart Prize Essay on the Adaptation of Recent Discoveries and Inventions in Science and Art to the Purposes of Practical Banking. Third English Edition ... London: 1854. Thick 8vo. viii,356 pp. 90 numbered "plates" + 14 bis plates. Rec half mor, t.e.g.
        (Charles B. Wood) **$1,500 [≈ £937]**

Sharpe, J.
- An Analytical Index to the Works of the late John Gould, F.R.S. London: 1893. One of 250. Imperial 8vo. xlviii,375 pp. Port. A few lib stamps. Orig cloth, trifle used.
        (Wheldon & Wesley) **£350 [≈ $560]**

Sharpe, R.B.
- British Museum Catalogue of Birds. Vol 12 Fringilliformes. Part 3 Fringillidae. London: 1888. 8vo. xv,871 pp. 16 hand cold plates. Lib stamps on title & plate versos. Orig cloth, reprd.   *(Wheldon & Wesley)* **£75 [≈ $120]**

Shaw, Sir Eyre Massey
- Fire Protection. A Complete Manual of ... the Fire Brigade of London. London: Layton, 1876. 1st edn. xvi,332,iv,60 illust advt pp. 2 plates, num ills. Orig cloth, uncut, largely unopened, cvrs sl marked, front hinge cracked.   *(Duck)* **£150 [≈ $240]**

Shaw, Peter
- An Enquiry into the Contents, Virtues and Uses, of the Scarborough Spaw-Waters ... London: for the author, 1734. 1st edn. 8vo. viii,[2],166 pp, advt leaf depicting trade labels. Contemp calf, rebacked, crnrs reprd.
        (Spelman) **£110 [≈ $176]**

Shaw, S. Parsons
- Odontologia, commonly called Tooth-Ache;

its Causes, Prevention, and Cure. Manchester: Palmer & Howe, 1868. 1st edn. xi,258 pp. Engvs, diags. Orig cloth, endpapers grubby.
(Box of Delights) **£20 [≃ $32]**

Shaw, T.R.
- Machine Tools - Their Design and Construction. For Planing, Shaping, Slotting, Drilling, Boring, Milling, Wheel Cutting ... Manchester: [ca 1900]. 8vo. 684 pp. 467 ills. Lacks fly leaf. Orig bndg, cvrs sl marked.
(Book House) **£20 [≃ $32]**
- The Mechanisms of Machine Tools. London: Frowde, 1923. 4to. 351 pp. Ills. Lacks fly leaf. Orig bndg, cvrs marked.
(Book House) **£18 [≃ $28]**

Shaw, Thomas
- The Study of Breeds in America: Cattle, Sheep and Swine. New York & Chicago: Orange Judd Co, 1905. xvi,371,12 advt pp. Frontis, plate, ills. Orig cloth.
(Wreden) **$59.50 [≃ £37]**

Sheehan, J. Eastman
- Plastic Surgery of the Nose. New York: 1925. 1st edn. 4to. xix,249 pp. Plates, ills. Orig cloth, ex-lib, crnrs worn.
(Elgen) **$125 [≃ £78]**
- Plastic Surgery of the Nose. New York: 1936. 2nd edn, rewritten. Sm 4to. xvii,186 pp. Frontis, plates, ills. Orig bndg.
(Elgen) **$75 [≃ £46]**

Sheen, James Richmond
- Wines and other Fermented Liquors; from the Earliest Ages to the Present Time. London: Hardwicke, [preface dated 1864]. xii, 292,7 advt pp. Orig gilt dec maroon cloth. Author's pres inscrptn.
(Box of Delights) **£150 [≃ $240]**

Shelley, G.E.
- A Handbook of the Birds of Egypt. London: 1872. Roy 8vo. ix,342 pp. 14 hand cold plates by Keulemans. Occas browning in text, a few sm wormholes in lower marg. New half calf.
(Henly) **£280 [≃ $448]**
- Handbook of the Birds of Egypt. London: Van Voorst, 1872. Roy 8vo. ix,342 pp. 14 hand cold plates by Keulemans. Sl foxing, mainly to text but sl affecting 2 plates. Contemp half calf, sl rubbed, upper hinge beginning to crack but firm.
(Sotheran) **£375 [≃ $600]**

The Shepherd's Kalendar ...
- The Shepherd's Kalendar, or the Citizen's and Country Man's Daily Companion ... being above Forty Years Study and Experience of a Learned Shepherd. London: [?1710]. 4th edn. 818mo. [vi],157,[pp3] pp. Frontis, w'cuts. New calf, antique style.
(Wheldon & Wesley) **£150 [≃ $240]**

Sheppard, T.
- Geological Rambles in East Yorkshire. London: n.d. 8vo. x,235 pp. Frontis, cold fldg map, 53 text ills. Orig cloth, soiled.
(Henly) **£17 [≃ $27]**
- Geological Rambles in East Yorkshire. London: [1903]. 8vo. xi,235 pp. Maps & 53 figs. Orig cloth.
(Wheldon & Wesley) **£28 [≃ $44]**

Sherrin, R.A.A.
- Handbook of the Fishes of New Zealand. Auckland: 1886. 8vo. 307,iv pp. Fldg map. Orig limp cloth, spine faded.
(Wheldon & Wesley) **£30 [≃ $48]**

Sherrington, Sir Charles Scott
- Goethe On Nature & On Science. Cambridge: 1924. 1st edn. 12mo. 32 pp. Lib perf stamp on title. Orig ptd wraps. *(Elgen)* **$50 [≃ £31]**
- Man on His Nature. The Gifford Lectures Edinburgh, 1937-38. Cambridge: UP, 1940. 1st edn, 1st printing. 8vo. [xii],413,[3] pp. 7 plate, s. Orig green cloth.
(Gach) **$65 [≃ £40]**
- Man on His Nature. Gifford Lectures Edinburgh, 1937-38. New York: Macmillan, 1941. 1st Amer & edn. 8vo. [x],413,[1] pp. Orig brown cloth.
(Gach) **$50 [≃ £31]**

Sherwin, Henry (ed.)
- Mathematical Tables ... Logarithms ... Sines, Tangents, and Secants ... London: for Thomas Page ..., 1726. 8vo. [viii[x],64,[2 errata], [304], 39,[1] pp. Fldg plate. Contemp calf, rebacked & reprd.
(Burmester) **£110 [≃ $176]**

Shew, Joel
- Children, their Hydropathic Management in Health and Disease; a Descriptive & Practical Work, designed as a Guide for Families and Physicians ... New York: Fowler & Wells, 1852. 8vo. 432,4 advt pp. Ills. Orig cloth.
(Hemlock) **pp$150 [≃ £93]**

Short, Thomas
- The Natural, Experimental, and Medicinal

History of the Mineral Waters of Derbyshire, Lincolnshire, and Yorkshire ... [With] An Essay ... Mineral Waters ... London: 1734; Sheffield: 1740. 1st edn. 2 vols. 4to. 5 plates. Vol 2 sl browned. Calf.
(Rootenberg) **$475 [≈£296]**

Shortridge, G.C.
- The Mammals of South West Africa: a Biological Account of the Forms occurring in that Region. London: 1934. 2 vols. Roy 8vo. Orig cloth, a working copy, bndg sl worn & stained, contents sound though rather foxed, some pencil annotations.
(Wheldon & Wesley) **£60 [≈$96]**

Shuckard, W.E.
- British Bees: an Introduction to the Study of thean Natural History and Economy of the Bees indigenous to the British Isles. London: 1866. Sole edn. 158vo. xvi,371 pp. S16 hand cold plates. Orig cloth, spine faded.
(Bickersteth) **qn qb£45 [≈$72]**

Shuldham, E.B.
- Chronic Sore Throat, or, Follicular Disease of the Pharynx ... With Special Chapters on the Hygiene of the Voice. London: 1881. 90 pp. Orig bndg.
(Moorhead) **£15 [≈$24]**

Sibly, Ebenezer
- A New and Complete Illustration of the Celestial Science of Astrology ... In Four Parts ... London: for the proprietor, & sold by W. Nicoll, M. Sibly & E. Sibly, 1784 [-88]. 1st edn. 4to. 1126,[iv] pp. Frontis, 29 plates. Some marks & marg reprs. Half calf antique.
(Finch) **£200 [≈$320]**

Sidgwick, N.V.
- The Chemical Elements and their Compounds. OUP: 1950. 2 vols. 1703 pp. Orig bndg.
(Book House) **£20 [≈$32]**

Sidis, Boris
- Psycho-Pathological Researches, Studies in Mental Dissociation. New York: Stechert, 1902. 1st edn. Sm 4to. 10 fldg plates. Sl water stained. Orig bndg, hinges cracked.
(Xerxes) **$140 [≈£87]**

Sigerist, Henry E.
- A History of Medicine. New York: 1951-61. 1st edn. 2 vols (all publ). 48 plates, num ills. Orig bndg. dws.
(Elgen) **$135 [≈£84]**

Sigmond, G.
- Tea; its Effects, Medicinal and Moral. London: Longman ..., 1839. 16mo.

viii,144,16 ctlg dated March 1840 pp. Orig cloth, unopened, v sl hinge split at hd.
(de Beaumont) **£68 [≈$108]**

Silvertop, C.
- A Geological Sketch of the Tertiary Formation in the Provinces of Granada and Murcia, Spain, with Notices respecting Primary, Secondary, and Volcanic Rocks in the same District and Sections. London: 1836. 8vo. 236 pp. Frontis, 7 plates. Contemp half calf gilt.
(Wheldon & Wesley) **£75 [≈$120]**

Sim, G.
- The Vertebrate Fauna of Dee. Aberdeen: 1903. 8vo. 295 pp. 8 ills. Orig cloth, t.e.g., spine sl frayed.
(Egglishaw) **£28 [≈$44]**

Sim, T.R.
- Tree Planting in South Africa. Pietermaritzburg: 1927. 8vo. 452 pp. 40 plates. Orig cloth.
(Wheldon & Wesley) **£38 [≈$60]**

Simmonds, J.H.
- Trees from other Lands for Shelter and Timber in New Zealand. Auckland: 1927. 4to. xviii,164 pp. 76 botanical & 28 other plates. Orig cloth, sl used.
(Wheldon & Wesley) **£50 [≈$80]**

Simms, Frederick W.
- A Treatise on the Principles and Practice of Levelling ... Railway Engineering ... Roads ... London: Weale, 1856. 4th edn, rvsd, with addtns. viii,215,[8,16 advt] pp. 7 plates, a few ills. Orig cloth, uncut, sl rubbed, front hinge cracked but firm.
(Duck) **£30 [≈$48]**
- A Treatise on the Principles and Practice of Levelling ... Railway Engineering ... Roads ... London: Lockwood, 1866. 5th edn, rvsd, with addtns. vii,215 pp. Plates, w'cuts. Orig cloth, new endpapers.
(Lamb) **£40 [≈$64]**

Simon, Sir John
- General Pathology, as Conducive to the Establishment of Rational Principles for the Diagnosis and Treatment of Diseases; A Course of Lectures ... London: H. Renshaw, 1850. 1st edn. 288 pp. Orig cloth, spine relaid.
(Elgen) **$75 [≈£46]**
- General Pathology ... Philadelphia: Blanchard & Lea, 1852. 1st Amer edn. 211 pp. Some browning. Sheep.
(Elgen) **$50 [≈£31]**

Simonds, Herbert R.
- Industrial Plastics. Third Edition. London

[US ptd]: Pitman, 1947. x,396 pp. Num ills. Orig cloth. Dw worn & soiled.
(Duck) **£15 [≈ $24]**

Simonds, Herbert R. & Bigelow, M.H.
- The New Plastics. Third Printing. New York: Van Nostrand, January 1946. xii,320 pp. Cold frontis, 46 other ills. Top crnr of a few ff sl bumped. Orig cloth, spine ends sl rubbed.
(Duck) **£15 [≈ $24]**

Simpson, G.G.
- Tempo and Mode in Evolution. New York: 1944. 8vo. xviii,237 pp. 36 text figs. Orig cloth, trifle used.
(Wheldon & Wesley) **£25 [≈ $40]**

Simpson, James Y.
- Acupressure a New Method of Arresting Surgical Haemorrhage and of accelerating the Healing of Wounds. Edinburgh: A. & C. Black, 1864. 1st edn. 8vo. [5],xiv,580 pp. Half- title. Over 40 text w'cuts. Orig cloth, uncut.
(Rootenberg) **$350 [≈ £218]**

Simpson, N.D.
- A Bibliographical Index of the British Flora. Bournemouth: privately ptd, 1960. One of 750. Imperial 8vo. xix,429 pp. Orig cloth, spine faded. *(Wheldon & Wesley)* **£48 [≈ $76]**

Simpson, Thomas
- Mathematical Dissertations on a Variety of Physical and Analytical Subjects ... London: for T. Woodward ..., 1743. 1st edn. 4to. viii, 168 pp. Some foxing. Rec half calf.
(Pickering) **$800 [≈ £500]**
- Select Exercises for Young Proficients in the Mathematics ... New Edition ... Account of the Life and Writings of the Author, by Charles Hutton. London: for E. Wingrave, 1792. 8vo. [4],iv,xxiii,[1],252 pp. Ink splashes on 4 pp. Contemp calf. *(Fenning)* **£45 [≈ $72]**

Simson, Robert
- The Elements of Euclid ... Edinburgh: for E. Wingrove ..., 1799. 10th edn. 8vo. 520 pp. 2 fldg plates, num text diags. Contemp calf, jnts cracked but holding. *(Young's)* **£30 [≈ $48]**

Sinclair, George
- Hortus Gramineus Woburnensis. London: 1824. 2nd edn. Roy 8vo. xx,438 pp. 60 plain plates. Cloth, used, contents good.
(Wheldon & Wesley) **£35 [≈ $56]**

Sinclair, Sir John
- The Code of Agriculture; including

Observations on Gardens, Orchards, Woods and Plantations. London: Sherwood, 1817. 1st edn. 8vo. viii,492,96 pp. Port frontis, plate, 9 plates in the Appendix. Browned. Half calf, sl rubbed.
(Blackwell's) **£110 [≈ $176]**

Sindall, R.W.
- Bamboo for Papermaking. London: Marchant, Singer, 1909. 1st edn. 60,[iv] pp. 1 ill. Illust advts. Orig cloth backed bds. *(Charles B. Wood)* **$75 [≈ £46]**

Singer, Charles
- A Short History of Biology. Oxford: 1931. 194 ills. Cloth, sl used.
(Wheldon & Wesley) **£25 [≈ $40]**
- Studies in the History and Method of Science. Oxford: 1917-21. 2 vols. Imperial 8vo. 96 plates (12 cold). Buckram.
(Wheldon & Wesley) **£80 [≈ $128]**
- Festschrift ... see Underwood, E. Ashworth (ed.).

Singer, Edgar Arthur
- Mind as Behavior and Studies in Empirical Idealism. Columbus, Ohio: Adams, 1924. 1st edn. 8vo. 293 pp. Orig bndg.
(Xerxes) **$45 [≈ £28]**

Sitwell, Sacheverell & Blunt, W.
- Great Flower Books, 1700-1900. With a Bibliography ... London: 1956. One of 295. Roy folio. 104 pp. 20 cold & 16 plain plates. Half mor. Slipcase.
(Wheldon & Wesley) **£500 [≈ $800]**

Sitz, A.
- Macrolepidoptera of the World, Vol 3, Palearctic Noctuae. Stuttgart: 1914. 2 vols. 4to. iii,511 pp. Interleaved. 75 cold plates. Half mor, trifle rubbed.
(Wheldon & Wesley) **£200 [≈ $320]**
- Macrolepidoptera of the World, Vol 4, Palearctic Geometrae. Stuttgart: 1912. 2 vols. 4to. 479 pp. Interleaved. 25 cold plates. Half mor, trifle rubbed.
(Wheldon & Wesley) **£160 [≈ $256]**

A Six Weeks Tour ...
- A Six Weeks Tour through the Southern Counties of England and Wales ... see Young, Arthur.

Skaife, S.H.
- African Insect Life. London: [1954]. Roy 8vo. viii,387 pp. 75 plates (5 cold). Orig cloth.
(Wheldon & Wesley) **£35 [≈ $56]**

Skinner, Burrhus Frederick
- The Behavior of Organisms ... New York &
London: Appleton, [1938]. 1st edn, 1st iss
(black bndg). 8vo. [x],458 pp. Name stamp on
flyleaf. Orig black cloth, shaken. Sgnd pres
inscrptn at time of publication.
(Gach) **$650 [≈ £406]**
- Science and Human Behavior. New York:
Macmillan, [1953]. 1st edn. 8vo. x,461,[1] pp.
Orig grey cloth. Carroll Pratt's copy.
(Gach) **$50 [≈ £31]**

Sladen, F.W.L.
- The Humble-Bee. Its Life-History, and how
to domesticate it. With Descriptions of all the
British Species of Bombus and Psithyrus.
London: 1912. 8vo. xiii,283 pp. 7 plates, 34
text figs. Orig cloth. *(Bickersteth)* **£25 [≈ $40]**

Slare, Frederick
- Experiments and Observations upon Oriental
and Other Bezoar-Stones, which Prove them
to be of no Use in Physick ... with Further
Discoveries and Remarks. London: 1715.
Sole edn. 8vo. [iv],v,xviii, [viii],47, [x],64 pp,
advt leaf. Orig calf, jnt ends cracked.
(Bickersteth) **£285 [≈ $456]**

Sloane, Hans
- A Voyage to the Islands Madera, Barbados,
Nieves, S. Christophers and Jamaica, with the
Natural History ... London: 1707-25. 4 vols.
Folio. Lge fldg map, 284 dble-page plates. 4
plates restored with xerox. 2 blank crnrs
reprd. Contemp calf, sl rubbed.
(Wheldon & Wesley) **£2,400 [≈ $3,840]**

Small, A.E.
- Manual   of   Homeopathic   Practice.
Philadelphia: Boericke, 1863. 10th edn,
enlgd. 8vo. 831 pp. Orig bndg, leather
backstrip worn, hinges cracked.
(Xerxes) **$75 [≈ £46]**

Smedley, John
- Practical Hydropathy, including Plans of
Baths, and Remarks on Diet, Clothing and
Habits of Life. Seventh Edition. London:
1864. 8vo. W'engvs in text. Orig cloth, spine
ends sl snagged.   *(Robertshaw)* **£10 [≈ $16]**

Smee, Alfred
- Instinct and Reason: Deduced from Electro-
Biology. London: Reeve & Benham, 1850. 1st
edn. 8vo. xxxiv,320 pp. 10 litho plates (7
cold), 65 text w'cuts. Orig ochre cloth.
(Gach) **$250 [≈ £156]**
- The Mind of Man: Being a Natural System of

Mental Philosophy. London: George Bell,
1875. 1st edn. 8vo. xx,262,2 pp. 3 engvs, text
ills. Some pencil notes. Orig dec red cloth
gilt, some fraying to jnts. Inscrbd copy.
(Gach) **$85 [≈ £53]**

Smellie, William
- The   Philosophy   of   Natural   History.
Edinburgh: 1790-99. 1st edn. 2 vols. 4to.
[xvi], [548]; xii,[516] pp. Orig drab bds,
untrimmed, rebacked with cloth, some wear
to bds.   *(Gach)* **$475 [≈ £296]**

Smiles, Samuel
- Industrial Biography: Iron Workers and Tool
Makers. Fifteenth Thousand. London: 1863.
8vo. xiv,342,[4 advt] pp. Sl spotting. Orig
cloth.   *(Bow Windows)* **£40 [≈ $64]**
- The Life of George Stephenson, Railway
Engineer.L: Murray, 1858. 5th rvsd edn. 8vo.
Port. Contemp polished calf, gilt panelled
spine, labels.   *(Stewart)* **£30 [≈ $48]**
- Lives of the Engineers ... London: 1861- 62.
1st edn. 3 vols. 8vo. Frontis, 4 other plates,
265 text ills & maps. Orig embossed cloth.
(Argosy) **£150 [≈ $156]**
- Lives of the Engineers. A New and Revised
Edition. London: Murray, 1874-77. 5 vols.
Post 8vo. Ca 2000 pp. Half-title vol 1 only as
correct. 9 ports, 342 other ills, maps, plans.
Occas minor marking. Contemp polished calf
gilt extra, jnts & edges sl rubbed.
(Duck) **£425 [≈ $680]**

Smit, Pieter
- History of the Life Sciences: An Annotated
Bibliography. Amsterdam: Asher, 1974. Lge
8vo. xiv,1072 pp. Orig cloth.
(Gach) **$75 [≈ £46]**

Smith, A.L.
- A Monograph of the British Lichens. A
Descriptive Catalogue of the Species in the
Department of Botany, British Museum.
London: 1918-26. 2nd edn. 2 vols. 8vo. 134
plates. Orig cloth, good ex-lib.
(Wheldon & Wesley) **£75 [≈ $120]**

Smith, Adam
- Essays on Philosophical Subjects ... prefixed,
An Account of the Life and Writings of the
Author; by Dugald Stewart. Dublin: 1795.
1st Dublin edn. 8vo. cxxiii,[i blank], 332 pp.
Sm piece cut from title marg. 1 sm repr.
Contemp sheep, rebacked, crnrs rubbed.
(Pickering) **$2,500 [≈ £1,562]**
- An Inquiry into the Nature and Causes of the
Wealth of Nations ... Dublin: 1776. 1st

Dublin edn. 3 vols. 8vo. [viii],391; [vii],[i blank],524,[3 advt]; [iv],412 pp. Half-title in vol 1 only, correctly. A few sl marks. Mod half calf over mrbld bds.
(Pickering) **$6,000 [≈ £3,750]**

- An Inquiry into the Nature and Causes of the Wealth of Nations ... Eleventh Edition; with Notes ... Life ... by William Playfair. London: 1805. 11th edn. 3 vols. 8vo. xl,515; viii, 567; viii,590,[2 advt] pp. Lacks half- titles vols 2 & 3. Contemp calf gilt, reprd.
(Pickering) **$1,000 [≈ £625]**

- An Inquiry into the Nature and Causes of the Wealth of Nations ... With a Life of the Author ... Also, a View of the Doctrine of Smith ... London: for William Allason, 1819. 3 vols. 8vo. lxv,360; vi,514;v,499 pp. Contemp calf gilt, lacks labels, sl bumped.
(Hollett) **£120 [≈ $192]**

Smith, C.
- Birds of Guernsey ... London: R.H. Porter, 1879. 1st edn. 8vo. xix,223 pp. Orig green cloth, uncut & unopened.
(Gough) **£60 [≈ $96]**
- The Birds of Somersetshire. London: 1969. Cr 8vo. xi,643 pp. Orig cloth, trifle used.
(Wheldon & Wesley) **£20 [≈ $32]**

Smith, C.H.
- An Introduction to the Mammalia. Edinburgh: Jardine's Naturalist's Library, 1842. Sm 8vo. 313 pp. Port, vignette title, 31 hand cold plates. Orig cloth.
(Egglishaw) **£40 [≈ $64]**
- The Natural History of Dogs. Edinburgh: Jardine's Naturalist's Library, 1839-40. 1st edn. 2 vols. Sm 8vo. 2 ports, 2 vignette titles, 66 hand cold & 5 plain plates. Contemp mor, gilt spines, a.e.g., jnts & crnrs sl rubbed.
(Egglishaw) **£130 [≈ $208]**

Smith, C.S.
- A History of Metallography. Chicago: Univ Press, 1960. Roy 8vo. xxi,291 pp. 100 ills. Orig cloth. Dw.   *(Gemmary)* **$75 [≈ £46]**

Smith, David
- The English Dyer ... Manchester: Palmer & Howe, 1882. Only edn. 8vo. xxxii,340 pp. 500 mtd samples. Rec cloth, orig spine laid down.   *(Charles B. Wood)* **$500 [≈ £312]**

Smith, Edgar C.
- A Short History of Naval and Marine Engineering. Cambridge: 1937. 8vo. xix,376 pp. 16 plates, 46 text figs. Orig cloth, spine faded.   *(Bickersteth)* **£40 [≈ $64]**

- A Short History of Naval and Marine Engineering ... Cambridge: for Babcock & Wilcox at the UP, 1937. 1st edn. Post 8vo. xx,376 pp. 16 plates, 46 ills. Orig cherry cloth, uncut, spine faded as usual, minor bumping.   *(Duck)* **£35 [≈ $56]**

Smith, Edward
- Foods. Second Edition. London: 1873. 8vo. xvi,485 pp. Text figs & charts. Sm ink stain at ft of title & prelims. Orig cloth.
(Bickersteth) **£24 [≈ $38]**

Smith, Eliza
- The Compleat Housewife: or Accomplish'd Gentlewoman's Companion ... Eighth Edition, with very Large Additions ... London: Pemberton, 1737. 8vo. [xviii],354,xv,[3 advt] pp. Frontis, 6 fldg plates. Rec qtr calf.
(Burmester) **£275 [≈ $440]**

Smith, F.
- Catalogue of Hymenopterous Insects in the Collection of the British Museum. London: 1853-59. 7 parts. Parts 1-6 bound in 3 vols. 12mo. 47 plates. Lib stamp on titles. 3 vols in half calf, spine ends worn, Part 7 in wraps.
(Egglishaw) **£65 [≈ $104]**

Smith, G.
- The Ferns of Derbyshire, Illustrated from Nature. London: [1877]. New edn. 8vo. xxiv,24 pp. 25 cold plates. Without the cold title- page found in some copies. Orig cloth, trifle used. *(Wheldon & Wesley)* **£25 [≈ $40]**

Smith, G. Elliot & Dawson, Warren R.
- Egyptian Mummies. New York: Dial Press, 1924. 4to. 189 pp. cold frontis, 71 plates. Orig pict blue cloth, spine sl sunned.
(Terramedia) **$250 [≈ £156]**

Smith, Godfrey
- The Laboratory; or, School of Arts: containing a large Collection of Valuable Secrets, Experiments, and Manual Operations in Arts and Manufactures ... Sixth Edition ... London: 1799. 2 vols. 8vo. 40 plates. Orig calf, rubbed, spine ends sl worn, Two jnts cracked.
(Bickersteth) **£125 [≈ $200]**
- The Laboratory; or, School of Arts: containing a large Collection of Valuable Secrets, Experiments, and Manual Operations, in Arts and Manufactures ... Seventh Edition. London: Law & Gilbert, 1810. 2 vols. 8vo. 32 plates. Contemp calf, rebacked.   *(Spelman)* **£240 [≈ $384]**

Smith, J. Bucknall
- A Treatise upon Wire, its Manufacture and Uses ... London: "Engineering'; New York: Wiley, 1891. 1st edn. Post 4to. xxii,[ii], 348, [2,xviii advt] pp. 2 fldg plates, fldg table, num ills. Occas thumbing or soiling. Orig cloth, sl scuffed, marked & soiled, recased.
(Duck) £110 [≈ $176]

Smith, J.C.
- The Manufacture of Paint. A Practical Handbook for Manufacturers, Merchants, and Painters. London: Scott Greenwood, 1915. 2nd rvsd edn. 271 pp. 80 ills. Orig bndg. *(Book House)* £15 [≈ $24]

Smith, James
- The Panorama of Science and Art; embracing the Sciences of Aerostation, Agriculture ... Building, Brewing ... Japanning ... Staining Glass ... Liverpool: for Nuttall, Fisher, 1815. 1st edn. 2 vols. 8vo. x,626; xii,862, [2] pp. Frontis, 49 plates. Rec bds.
(Fenning) £95 [≈ $152]

Smith, Sir James Edward
- Compendium Florae Britannicae. London: 1825. 4th edn. 8vo. viii,288 pp. Sm lib stamp on title. Half calf. *(Henly)* £25 [≈ $40]
- A Grammar of Botany ... added, a reduction of all the Genera contained in the Catalogue of North American Plants ... New York: Seaman, 1822. 8vo. xvi,17-284 pp. 21 hand cold litho plates. Rec calf in orig style.
(Charles B. Wood) £850 [≈ £531]
- A Grammar of Botany ... Second Edition. London: Longman ..., 1826. 8vo. xxii,240 pp. 21 plates (somewhat spotted). Contemp half green calf gilt. *(Gough)* £32 [≈ $51]
- A Grammar of Botany ... Second Edition. London: Longman ..., 1826. 8vo. xxii,240 pp. 21 hand cold plates showing 277 figs. Orig pink bds, uncut, paper label & backstrip defective, sides held on 2 cords.
(Claude Cox) £65 [≈ $104]
- An Introduction to Physiological and Systematical Botany. Third Edition. London: Longman ..., 1814. 8vo. xxiv,407 pp. 15 plates. Contemp gilt panelled calf.
(Gough) £48 [≈ $76]
- An Introduction to Physiological and Systematical Botany. Fifth Edition, corrected and enlarged. London: Longman, 1825. 8vo. xxi, 435,[1 advt] pp. 15 plates. Contemp half calf, elab gilt spine, by John Pritchard of Caernarvon. *(Spelman)* £60 [≈ $96]
- An Introduction to the Study of Botany. Seventh Edition, Corrected ... London:

Longman ..., 1833. Lge 8vo. xx,504,16 ctlg pp. 15 steel-engvd plates plus supplement of 21 plates. Orig green cloth, gilt dec spine.
(Schoyer) $75 [≈ £46]

Smith, James, of Jordanhill
- Researches in New Pliocene and Post-Tertiary Geology. Glasgow: 1862. 1st edn. 8vo. xi,191 pp. 4 tinted litho plates, text figs. Cancelled lib blind stamp on title. Orig cloth, rebacked. *(Bickersteth)* £35 [≈ $56]

Smith, John, 1630-1679
- [Greek title, then] King Solomon's Portraiture of Old Age ... Secret Anatomy ... Account of the Infirmities of Age ... London: 1666. 1st edn. 8vo. [16],266,[6] pp. Tables (1 on a fldg plate). Browning. Crnr of 1 leaf torn. Lacks A1. Old calf, rebacked. Wing S.4114.
(Hemlock) $650 [≈ £406]

Smith, John, of Kew
- Ferns: British & Foreign ... With a Treatise on their Cultivation. London: 1879. New & enlgd edn. 8vo. xv,450 pp. Plate, text figs. Orig cloth. *(Wheldon & Wesley)* £15 [≈ $24]

Smith, M.
- The British Amphibians and Reptiles. London: Collins, New Naturalist Series, 1951. 1st edn. Cold & plain plates. Orig cloth. Dw in protective cvr taped down.
(Egglishaw) £30 [≈ $48]

Smith, Peter
- The Indian Doctor's Dispensatory, being Father Smith's Advice respecting Diseases and their Cure ... Cincinnati: 1813. 108,[4] pp. Some staining. Old calf, worn & rubbed, hinges cracking, cords sound.
(Reese) $3,500 [≈ £2,187]

Smith, Robert
- A Compleat System of Opticks in Four Books ... Cambridge: for the author, 1738. 1st edn. 2 vols. 4to. [4],vi,280; [2],281-455,171,[12] pp. 83 fldg plates. Contemp calf, worn, jnts cracked. *(Offenbacher)* $950 [≈ £593]
- A Compleat System of Opticks in Four Books ... Cambridge: for the author ..., 1738. 1st edn. 2 vols. 4to. 83 fldg plates. Contemp half calf, gilt spines, entirely uncut, sl worn, later endpapers. *(Ximenes)* $3,000 [≈ £1,875]
- Harmonics, or the Philosophy of Musical Sounds. Cambridge: ptd by J. Bentham, 1749. 1st edn. 8vo. xv,[i], 292,[xii] pp, advt leaf. 3 ptd tables, 25 fldg engvd plates. Orig calf, upper jnt cracked but bndg firm.
(Bickersteth) £380 [≈ $608]

Smith, S.B.
- British Waders in their Haunts. London: 1950. Roy 8vo. 79 photos & 3 plates. Orig cloth.   *(Wheldon & Wesley)* £20 [≈ $32]

Smith, Samuel Stanhope
- An Essay on the Causes of the Variety of Complexion and Figure in the Human Species. New Brunswick: J. Simpson, 1810. 2nd edn, enlgd. 411 pp. Usual foxing. Contemp calf, scuffed. *(Elgen)* $140 [≈ £87]
- An Essay on the Causes of the Variety of Complexion and Figure in the Human Species ... New Brunswick: 1810. 2nd edn, enlgd. 8vo. 411 pp. Contemp tree calf, rubbed, upper cvr detached.
            (Hemlock) $225 [≈ £140]

Smith, Thomas
- The Naturalist's Cabinet: containing Interesting Sketches of Animal History. London: 1806-07. 6 vols. 8vo. 5 (of 6) cold engvd titles, 47 (of 60) hand cold plates. Contemp calf, jnts weak.
            (Wheldon & Wesley) £110 [≈ $176]

Smith, William
- A Synopsis of the British Diatomaceae. London: 1853-56. 2 vols. Roy 8vo. 69 plates (7 cold) (somewhat foxed). Ex-lib. Half mor, reprd.   *(Wheldon & Wesley)* £175 [≈ $280]

Smyth, C. Piazzi
- Madeira Meteorologic. Being a Paper on the Above Subject Read before the Royal Society, Edinburgh, on the 1st of May 1882. Edinburgh: David Douglas, 1882. 1st edn. Cr 4to. viii,83 pp. 2 plates, ills & tables. Occas sl thumbing. Orig cloth, t.e.g., uncut.
            (Duck) £25 [≈ $40]

Smyth, H.D.
- Atomic Energy. A General Account of the Development of Methods of Using Atomic Energy for Military Purposes under the Auspices of the United States Government. London: HMSO, 1945. 1st edn. iv,144 pp. Wrappers, sl dust stained.
            (Whitehart) £18 [≈ $28]
- A General Account of the Development of Methods of Using Atomic Energy for Military Purposes under the Auspices of the United States Government. London: HMSO, 1945. 1st English edn. Tall 8vo. 144 pp. Orig ptd wraps.   *(Hemlock)* $200 [≈ £125]

Smythe, F.S.
- The Valley of Flowers. London: 1947. 2nd edn. 8vo. 318 pp. 2 maps, 16 cold plates. Orig cloth.   *(Henly)* £35 [≈ $56]

Smythies, B.E.
- The Birds of Borneo. London: 1960. 1st edn. 8vo. xvi,562 pp. Cold frontis, 49 photo plates (2 cold), 4 plain & 45 cold plates of birds. Orig cloth. dw.   *(Henly)* £75 [≈ $120]
- Birds of Burma. London: 1953. 2nd edn. 8vo. xliii,668 pp. Map, 31 cold plates. Orig cloth, dampstained, lower cvr & map have sm worm holes. Dw.   *(Henly)* £85 [≈ $136]
- The Birds of Burma. London: 1953. 2nd edn. Roy 8vo. xliii,688 pp. Fldg map, 31 cold plates. Orig cloth. Dw.
            (Wheldon & Wesley) £90 [≈ $144]

Snapper, I.
- Chinese Lessons to Western Medicine. A Contribution to Geographical Medicine from the Clinics of Peking. New York: Interscience, 1941. 1st edn. 8vo. 308 pp. Photo ills. Orig bndg, sm spine chip, ends worn, cvr sl soiled. *(Xerxes)* $75 [≈ £46]

The Society for Psychical Research
- Catalogue of the Library for Psychical Research. London & Boston: G.K. Hall, 1976. Only edn. Folio. [viii],[342] pp. Orig buckram.   *(Gach)* $50 [≈ £31]

Soddy, Frederick
- The Interpretation of Radium. Being the Substance of Six Free Popular Experimental Lectures delivered at the University of Glasgow, 1908. London: Murray, 1909. 1st edn. 8vo. xviii,256 pp, advt leaf. Orig cloth, spotted, spine faded & sl worn at hd.
            (Bickersteth) £65 [≈ $104]
- The Interpretation of the Atom. London: 1932. 8vo. xviii,355 pp. 2 fldg tables, 20 plates, 32 text figs. Orig cloth.
            (Wheldon & Wesley) £25 [≈ $40]

Solms-Laubach, H.
- Fossil Botany. Oxford: 1891. 8vo. xii,401 pp. 49 text figs. Qtr mor, rubbed.
            (Henly) £40 [≈ $64]

Somervell, John
- Water-Power Mills of South Westmorland, on the Kent, Bela and Gilpin and their Tributaries. Kendal: Titus Wilson, 1930. 1st edn. 4 plates. Orig cloth gilt.
            (Hollett) £80 [≈ $128]

Somerville, Mary
- On the Connexion of the Physical Sciences.

London: Murray, 1836. 3rd edn. 12mo. xv,475 pp. 5 plates. Prize mor gilt.
(Schoyer) **$65 [≈ £40]**

Sommerfeld, A.
- Atomic Structure and Spectral Lines. Translated by H.L. Brose. London: 1923. 1st English edn. 8vo. xiii,626 pp. 125 figs. Occas sl foxing. Front blank torn out. Orig cloth, worn & sl dust stained.
(Whitehart) **£25 [≈ $40]**

"A Son of the Marshes" (Denham Jordan)
- Drift from Longshore. London: 1898. 8vo. Frontis. Orig cloth, sl faded.
(Grayling) **£20 [≈ $32]**
- From Spring to Fall, or When Life Stirs. London: 1894. 8vo. Orig cloth, sl rubbed.
(Grayling) **£15 [≈ $24]**
- The Wild-Fowl and Sea-Fowl of Great Britain. London: 1895. 8vo. Ills. Frontis sl foxed. Orig cloth, sl rubbed.
(Grayling) **£22 [≈ $35]**

Sotheran, Henry
- Bibliotheca   Chemico-Mathematica. Compiled and Annotated by Heinrich Zeitlinger. London: 1921-52. 1st edn. Vols 1 & 2, 1921, plus 3 supplementary vols bound in 4, 1932-52. 6 vols. Num plates. Minor ex-lib marks, some vols with perf stamps on titles.
(Elgen) **$900 [≈ £562]**
- Bibliotheca   Chemico-Mathematica: Catalogue of Works in Many Tongues on Exact and Applied Science, with a Subject-Index. London: Henry Sotheran & Co, 1921. 2 vols. 8vo. xii,964 pp. 127 plates. Red cloth, spines faded, extremities sl rubbed.
(Karmiole) **$250 [≈ £156]**

Sothern, J.W.
- The Marine Steam Turbine. A Practical Description of the Parsons Marine Turbine ... London: Crosby Lockwood, 1909. 3rd edn, rewritten & enlgd. Roy 8vo. xvi,337 pp. 45 plates, text ills. Orig cloth, uncut, spine sl marked.
(Duck) **£25 [≈ $40]**

South, John Flint
- Memorials of the Craft of Surgery in England ... Edited by D'Arcy Power ... London: 1886. 1st edn. 8vo. xxx,[i],412 pp. 6 plates. Orig cloth.
(Bickersteth) **£60 [≈ $96]**

Southgate, F. & Pollard, H.B.C.
- Wildfowl and Waders. Nature and Sport in the Coastlands. London: 1928. Large Paper. One of 950. 4to. viii,83 pp. 16 mtd cold plates, 48 other ills. Orig half parchment.
(Wheldon & Wesley) **£120 [≈ $192]**

Sowerby, A. de Carle
- A Naturalist's Holiday by the Sea, being a Collection of Essays on Marine, Littoral, and Shore-Line Life of the Cornish Peninsula ... London: Routledge, 1923. 8vo. xvi,262 pp. 21 plates, 21 text ills. Orig blue cloth gilt.
(Blackwell's) **£25 [≈ $40]**
- A Naturalist's Note-Book in China. Shanghai: 1925. Roy 8vo. 270 pp. 20 plates, num text figs. Sl foxing. Orig cloth.
(Wheldon & Wesley) **£50 [≈ $80]**
- A Naturalist's Note-Book in China. Shanghai: 1925. [xii],270 pp. Errata tipped in. 20 plates, num text ills. Orig dec cream cloth.
(Lyon) **£150 [≈ $240]**

Sowerby, G.B.
- A Conchological Manual. London: 1839. 1st edn. 8vo. x,[1],130 pp. Cold frontis, 23 plates, fldg table, 530 figs. Blind stamp on title. New cloth.
(Wheldon & Wesley) **£45 [≈ $72]**
- A Conchological Manual. London: Bohn, 1842. 2nd edn. 8vo. vii,313,[12 advt] pp. 27 hand cold plates. Interleaved with MS notes. Orig gilt dec cloth. *(Egglishaw)* **£180 [≈ $288]**
- Conchological Manual. London: Bohn, 1842. 2nd enlgd edn. Tall 8vo. vi,[i],313 pp, advt leaf. 27 uncold plates. 2 plates loose. Orig cloth gilt, trifle faded. *(Hollett)* **£40 [≈ $64]**
- A Conchological Manual. London: 1852. 4th edn. 8vo. vii,337 pp. 2 fldg tables, 28 hand cold & 1 plain plates. Sl foxing. Orig cloth.
(Henly) **£120 [≈ $192]**
- Illustrated Index of British Shells, containing Figures of all the Recent Species. London: Simpkin, Marshall, 1859. 1st edn. Roy 8vo. xv,(48) pp. 4 advt ff bound in. 24 hand cold plates. Orig cloth, a.e.g.
(Egglishaw) **£195 [≈ $312]**
- Illustrated Index of British Shells, containing Figures of all the Recent Species with Names and other Information. London: 1867. 2nd edn. 4to. 26 hand cold plates. Strip cut from blank fore marg of title & 2 ff. Orig red cloth, sl worn, spine faded. *(Henly)* **£135 [≈ $216]**
- Popular British Conchology. London: 1854. Sq post 8vo. xii,304 pp. 20 hand cold plates. Blind stamps on plates. Orig cloth, rebacked, ex-lib.   *(Wheldon & Wesley)* **£50 [≈ $80]**
- Popular History of the Aquarium of Marine and Fresh-Water Animals and Plants. London: Lovell Reeve, 1857. 1st edn. Sm 8vo. xvi,327 pp. 20 cold plates. Orig cloth gilt.   *(Egglishaw)* **£40 [≈ $64]**

Sowerby, James

- English Botany ... 3rd edn. 12 vols. 1938 hand cold plates. [With] Supplement to Vols 1 to 4 ... 2 plain & 13 hand cold plates. London: 1866-92. Together 13 vols. Roy 8vo. Num contemp annotations. Occas foxing. Contemp half calf, some jnts beginning to crack. *(Wheldon & Wesley)* **£625 [≈$1,000]**

Sowerby, John Edward

- An Illustrated Key to the Natural orders of British Wild Flowers. London: Van Voorst, 1865. 8vo. 42 pp. 9 hand cold plates. Occas sl foxing. Some marg ink-notes. Orig cloth, a.e.g. *(Egglishaw)* **£32 [≈$51]**

Sowerby, John Edward & Johnson, Charles

- The Ferns of Great Britain. London: 1855. 1st edn. 87 pp. 49 hand cold plates. [Bound with] A Supplement to The Ferns of Great Britain. London: 1856. 1st edn. 52 pp. 31 hand cold plates. 8vo. Half mor, spine relaid. *(Henly)* **£80 [≈$128]**
- The Ferns of Great Britain. London: 1855. 1st edn. 87 pp. 49 partly hand cold plates. Orig cloth, spine sl faded. *(Henly)* **£55 [≈$88]**

Sowerby, John Edward & Johnson, Charles Pierpoint

- British Wild Flowers. London: 1863. Re-issue with Supplement. 8vo. xlix,186 pp. 89 hand cold & 2 plain plates. Contemp half mor. *(Wheldon & Wesley)* **£120 [≈$192]**
- British Wild Flowers. London: 1863. Re-issue with Supplement by J.W. Salter. Sm 4to. 90 hand cold plates. Orig cloth gilt, rather rubbed. *(Hollett)* **£85 [≈$136]**

Soyer, Alexis

- The Modern Housewife or Menagere, comprising nearly 1000 Receipts ... London: Simpkin Marshall, 1851. 25th thousand. xvi, 454, [22 advt, 16 ctlg] pp. Port frontis (foxed). Orig cloth, pict spine darkened, headband sl chipped, sm split at top jnt. *(Box of Delights)* **£50 [≈$80]**
- The Modern Housewife or Menagere ... Thirtieth Thousand. London: Simpkin Marshall, 1853. 8vo. xvi,508,[4,8 advt] pp. 6 plates, text ills. Orig green cloth, dec spine, recased. *(Moon)* **£120 [≈$192]**

Soysa, E. (ed.)

- Orchid Culture in Ceylon. Colombo: 1943. Ltd edn. Roy 8vo. 176 pp. 11 cold & 55 other ills. Half mor. *(Wheldon & Wesley)* **£65 [≈$104]**

Spence, W.

- An Essay on the Theory of the Various Orders of Logarithmic Transcendents ... London: 1809. xiv,128 pp. Sm lib stamp on title. Occas sl foxing & discolouration. Paper-cvrd spine & bds, rather worn & marked. *(Whitehart)* **£40 [≈$64]**

Spencer, Herbert

- Essays: Scientific, Political, and Speculative. Vol III. London & Edinburgh: Williams & Norgate, 1874. 1st British edn. 8vo. [xii],341,[3] pp, 2 ff of front advts. Orig cloth, jnts frayed. *(Gach)* **$75 [≈£46]**
- The Principles of Psychology. London: Williams & Norgate, 1870. 2nd edn. 8vo. Orig cloth. *(Gach)* **$75 [≈£46]**
- The Principles of Psychology. New York: Appleton, 1897. 1st complete edn, 1st Amer printing. 5 vols. 8vo. Leather backed mrbld bds, sl shelfworn. *(Gach)* **$85 [≈£53]**

Spencer, John

- A Discourse concerning Prodigies ... Second Edition corrected and inlarged ... London: J. Field, for Will. Graves, 1665. 2 parts in one vol. 8vo. 3 index ff at end, lacks final blank. Marg wormholes at end. Contemp calf, jnt cracking. Wing S.4948 (inc S.4949). *(Hannas)* **£120 [≈$192]**
- A Discourse concerning Prodigies ... Second Edition corrected and inlarged ... added a short Treatise concerning Vulgar prophecies ... London: J. Field for Will. Graves, 1665. 8vo. [xxxii],408,[viii], 136, [vi] pp. 18th c panelled calf, rebacked. Wing S.4948. *(Finch)* **£130 [≈$208]**

Spencer, K.G.

- The Lapwing in Britain. Some Account of its Distribution and Behaviour, and of its Role in Dialect, Folk-Lore and Literature. Hull: 1953. 8vo. 12 plates. Orig cloth bds. *(Wheldon & Wesley)* **£28 [≈$44]**

Spencer, L.J.

- The Worlds Minerals. New York: Stokes, 1911. 8vo. xi,272 pp. 40 cold plates, 21 text ills. Orig cloth. *(Gemmary)* **$45 [≈£28]**

Spitta, E.J.

- Microscopy (The Construction, Theory and Use of the Microscope). London: Murray, 1909. 2nd edn. 8vo. xxii,498 pp. 16 plates, 243 text figs. Lib stamps on title. Orig cloth, lib label on front pastedown. *(Savona)* **£35 [≈$56]**

The Sportsman's Calendar ...
- See Lawrence, John, "William Henry Scott".

Sprat, Thomas
- The History of the Royal-Society of London, for the Improvement of Natural Knowledge ... London: T.R. for J. Martyn ..., 1667. 1st edn. 4to. Dampstained. Frontis sl defective. Working copy only. Contemp calf, jnts cracked. Wing S.5032. *(Falkner)* £30 [≈$48]
- The History of the Royal Society of London, for the Improving of Natural Knowledge. The Third Edition, Corrected. London: J. Knapton ..., 1722. 4to. [xvi],438 pp. Armorial frontis, 2 fldg plates. Contemp calf, rebacked, old label laid down, crnrs sl worn.
(Clark) £200 [≈$320]

Sprigg, Christopher
- The Airship. Its Design, History, Operation and Future. London: Sampson Low, [not after 1930]. Sole edn (?). viii,248 pp. 48 plates. Orig cloth, uncut, cvrs marked, removed shelfmark on spine. Author's own copy. *(Duck)* £65 [≈$104]

Spriggs, A.O.
- Champion Restorative Art. Springfield, Ohio: Champion, 1934. 1st edn. 8vo. 125 pp. Ills. Orig bndg. *(Xerxes)* $75 [≈£46]

Spruce, Richard
- Notes of a Botanist on the Amazon & Andes being Records of Travel on the Amazon and Its Tributaries ... Edited by Alfred R. Wallace. London: Macmillan, 1908. 1st edn. 2 vols. Frontis port, 7 maps, 71 ills. Orig green cloth, t.e.g., spines recased. Boxed.
(Jenkins) $325 [≈£203]

Spry, W. & Shuckard, W.E.
- The British Coleoptera Delineated. London: Crofts, 1840. 1st edn. 8vo. vii,83 pp. 94 plates. Sl spotting. Half mor.
(Egglishaw) £30 [≈$48]

Spurzheim, J.G.
- Outlines of Phrenology; being also a Manual of Reference for the Marked Busts ... London: 1827. 1st edn. 8vo. x,[4],100,[2] pp. Frontis. Orig bds. *(Hemlock)* $100 [≈£62]
- Phrenology, in Connexion with the Study of Physiognomy. To which is prefixed a Biography of the Author ... Boston: Marsh, Capen & Lyon, 1833. 1st Amer edn. Lge 8vo. Errata tipped in. Frontis, 34 litho plates. Orig cloth, extremities worn.
(Elgen) $200 [≈£125]

Stackhouse, John
- Nereis Britannica: Containing All the Species of Fuci, Natives of the British Coasts ... Bath: for the author, [1795-] 1801. Folio. Port, cold vignettes on titles (Lat & Engl), 17 cold plates. Orig bds, rebacked. Without appendix plates sometimes found.
(W. Thomas Taylor) $1,000 [≈£625]

Stainton, H.T.
- British Butterflies and Moths: an Introduction to the Study of Our Native Lepidoptera. London: Reeve & Co, 1867. 1st edn. 8vo. 16 hand cold engvd plates. Contemp maroon calf, gilt dec spine, sl rubbed & scuffed, gilt faded.
(Clark) £35 [≈$56]
- A Manual of British Butterflies and Moths. London: 1857-59. 2 vols. Cr 8vo. Text figs. Half calf, trifle rubbed, lacks labels.
(Wheldon & Wesley) £20 [≈$32]

Stamp, L. Dudley
- Britain's Structure and Scenery. London: Collins, New Naturalist Series, 1946. 1st edn. Cold & plain plates, diags & maps. Orig cloth. Dw. *(Egglishaw)* £14 [≈$22]
- The Land of Britain, its Use and Misuse. London: Longmans, 1948. 4to. viii,507 pp. 237 text figs (inc distribution maps). Orig green cloth gilt. *(Blackwell's)* £60 [≈$96]
- Man and the Land. London: New Naturalist, 1955. 1st edn. 8vo. xvi,272 pp. 56 plates (24 cold). Orig cloth. Sl frayed dw.
(Wheldon & Wesley) £25 [≈$40]

Stark, A. & Sclater, W.L.
- The Birds of South Africa. Vol 3. Picarians, Parrots, Owls and Hawks. London: 1903. 8vo. xvii,416 pp. Text figs. New cloth.
(Wheldon & Wesley) £35 [≈$56]

Stark, R.M.
- A Popular History of British Mosses. London: Reeve, 1854. 1st edn. Sm 8vo. xvi,322 pp. 20 hand cold plates. Orig cloth, spine worn. *(Egglishaw)* £16 [≈$25]
- A Popular History of British Mosses. London: 1860. 2nd edn. 8vo. xx,348 pp. 20 hand cold plates. Orig cloth, trifle used.
(Wheldon & Wesley) £20 [≈$32]

Staveley, E.F.
- British Insects. London: [1871]. Cr 8vo. xvi,392 pp. 16 hand cold plates. Orig cloth, rebacked. *(Wheldon & Wesley)* £25 [≈$40]
- British Spiders ... London: Reeve, 1866. Cr 8vo. xvi,280 pp. 2 plain & 14 hand cold

plates. Orig cloth gilt, dampstained along outer marg.   *(Egglishaw)* £23 [≈ $36]

Stead, E.F.

- The Life Histories of New Zealand Birds. London: 1932. Roy 8vo. xvi,162 pp. 93 photo plates. Orig cloth, trifle used.
(Wheldon & Wesley) £40 [≈ $64]

Stearns, S.

- The American Herbal or Materia Medica ... Walpole: (David Carlisle for Thomas & Thomas & the author), 1801. 12mo. 360 pp. Upper part waterstained. Browned. 3 ff have sm pieces missing from margs. Contemp calf, reprd.   *(Wheldon & Wesley)* £250 [≈ $400]

Stebbing, Edward Percy

- Indian Forest Insects of Economic Importance. Coleoptera. London: 1914. 4to. xvi,648 pp. 64 plates (7 cold), 401 text figs. Orig cloth. Author's pres inscrptn.
(Bickersteth) £75 [≈ $120]
- Stalks in the Himalaya. Jottings of a Sportsman Naturalist. London: Bodley Head, 1912. 8vo. xxviii,321,18 advt pp. Frontis, num ills. Occas sl foxing. Orig green cloth.
(Bates & Hindmarch) £80 [≈ $128]

Steers, J.A.

- The Coastline of England and Wales. Cambridge: 1946. 8vo. xix,644 pp. 117 photos (2 cold), 114 text figs. Orig cloth, front free endpaper partly stuck down.
(Wheldon & Wesley) £30 [≈ $48]
- The Sea Coast. London: Collins, New Naturalist Series, 1953. 1st edn. 8vo. 10 cold & 24 plain plates, maps & diags. Orig cloth. Dw.   *(Egglishaw)* £25 [≈ $40]

Steinbeck, John & Ricketts, E.E.

- Sea of Cortez, a Journal of Travel and Research with Materials for a Source Book on the Marine Animals of the Panamic Faunal Province. New York: 1941. 1st edn. Roy 8vo. x, 598 pp. 40 plates (8 cold). Inscrptns on title & endpapers. Orig cloth.
(Wheldon & Wesley) £80 [≈ $128]

Steinmetz, Charles Proteus

- Four Lectures on Relativity and Space. New York: 1923. 1st edn. x,126 pp. 7 stereoscopic photos in pocket. Ills. Orig cloth, spine ends sl frayed.   *(Elgen)* $75 [≈ £46]

Stejneger, Leonhard

- Georg Wilhelm Steller. The Pioneer of Alaskan Natural History. Cambridge:

Harvard Univ Press, 1936. 8vo. xxiv,623 pp. Cold frontis, 29 plates inc fldg map. Orig cloth. Chipped & torn dw.
(High Latitude) $60 [≈ £37]

Step, E.

- Bees, Wasps, Ants and Allied Insects of the British Isles. London: Warne, 1932. Sm 8vo. 44 cold & 67 plain plates, 64 wing maps, text figs. Orig cloth gilt. Dw.
(Egglishaw) £22 [≈ $35]
- Wayside and Woodland Blossoms. London: 1948. New edn. 3 vols. 8vo. Orig cloth.
(Wheldon & Wesley) £20 [≈ $32]

Stephens, George

- The Practical Irrigator and Drainer. Edinburgh: Blackwood ..., 1831. 8vo. viii,196 pp, directions to the binder. 11 full page plans, 2 text ills. Orig blue bds, uncut, remains of label, jnts cracked but firm, spine chipped.   *(Rankin)* £55 [≈ $88]

Stephens, Henry

- The Book of the Farm ... Edinburgh: Blackwood, 1849-52. 2nd edn. 2 vols. Roy 8vo. Num text ills. Endpapers foxed. Orig cloth, pict spines gilt, colour of cloth not uniform, reprd.   *(Stewart)* £75 [≈ $120]
- The Book of the Farm ... Edinburgh: 1850. 2nd edn. 2 vols. Lge 8vo. 14 steel engvs (damp stained), 589 text ills. Lacks vol 1 half-title. Rec cloth.   *(Blackwell's)* £85 [≈ $136]

Stephens, J.F.

- Illustrations of British Entomology. London: Baldwin & Cradock, 1828-46. 11 vols, plus Supplement bound in 5 vols. Roy 8vo. 95 hand cold plates. New qtr calf, raised bands, mrbld bds.   *(Egglishaw)* £845 [≈ $1,352]
- Illustrations of British Entomology ... Haustellata. London: Baldwin & Cradock, 1828-35. 4 vols. Roy 8vo. 41 hand cold plates. Cloth backed bds. *(Egglishaw)* £240 [≈ $384]

Stephenson, J.

- The Oligochaeta. Oxford: 1930. 1st edn. Roy 8vo. xvi,978 pp. 242 text figs. Orig cloth.
(Egglishaw) £34 [≈ $54]

Stephenson, T.A.

- The British Sea Anemones. London: Ray Society, 1928-35. 2 vols. 8vo. 19 cold & 14 plain plates, 108 text figs. Orig cloth gilt.
(Egglishaw) £75 [≈ $120]

Stern, Louis William

- General Psychology from the Personalistic

Standpoint. New York: Macmillan, 1938. 1st edn in English, 1st printing. 8vo. xxii, [590] pp. Orig cloth, sl soiled.. *(Gach)* **$65 [≈ £40]**

Sternberg, George M.
- A Text-Book of Bacteriology. New York: Wood, 1896. 1st edn. 693 pp. 9 plates (4 chromolithos), 200 text ills (some cold). Orig bndg.   *(Elgen)* **$75 [≈ £46]**

Steuart, Sir Henry
- The Planter's Guide; or, a Practical Essay on the Best Method of Giving Immediate Effect to Wood, by the removal of large Trees and Underwood. Edinburgh: 1828. 1st edn. 8vo. [vi], xxii,47,8 advt pp. Frontis & 4 plates. Orig cloth backed bds, sl worn.
        (Rankin) **£65 [≈ $104]**
- The Planter's Guide ... Originally intended for the Climate of Scotland. Second Edition, greatly improved and enlarged. Edinburgh: 1828. 8vo. [6],xxxvii,[i],527 pp. 6 litho plates. Some marg dusting. Orig linen backed bds, uncut, paper label, evenly soiled.
        (Spelman) **£120 [≈ $192]**

Stevenson, David
- A Treatise on the Application of Marine Surveying & Hydrometry to the Practice of Civil Engineering. Edinburgh: A. & C. Black ..., 1842. 1st edn. Roy 8vo. xiii,173,[1],[2 blank] pp. 14 plates, 16 other ills. Orig cloth, spine worn but holding.
        (Fenning) **£75 [≈ $120]**

Stevenson, Henry
- The Young Gard'ners Director. Furnishing him with Instructions for Planting and Sowing ... By H.S. Philokepos, Master of the Free- School in East Retford. London: for Anthony Barker, 1716. 1st edn. 12mo. Frontis, 1 plate. Contemp calf, rebacked.
        (Ximenes) **$475 [≈ £296]**

Stevenson-Hamilton, J.
- Animal Life in Africa. London: 1912. 538 pp. Ills. Orig bndg.
        (Trophy Room Books) **$200 [≈ £125]**

Stewart, S.A. & Corry, T.H.
- A Flora of the North East of Ireland. Cambridge: 1888. One of a few copies ptd on heavy paper. Interleaved. Black leather gilt, uncut, beginning to wear.
        (Emerald Isle) **£75 [≈ $120]**

Stewart, S.A. & Wright, Joseph
- Systematic Lists of the Flora, Fauna, Palaeontology and Archaeology of the North

of Ireland, by Members of the Belfast Field Naturalists Field Club. Vol I [all published]. Belfast: [1886]. 342 pp. 27 plates. 21 papers. Half leather.   *(Emerald Isle)* **£65 [≈ $104]**

Stewart, S.A., Corry, T.H. & Praeger, R.L.
- A Flora of the North East of Ireland. Belfast: Quota, 1938. Supplement tipped in.
        (Emerald Isle) **£35 [≈ $56]**

Stigand, C.H.
- The Game of British East Africa. Second Edition. London: 1913. 4to. xii,310 pp. 76 ills. Some foxing, a few short marg tears. Orig cloth, a little shaken, spine ends v sl fingered.
        (Bow Windows) **£205 [≈ $328]**

Still, George Frederic
- Common Disorders and Diseases of Childhood. London: 1909. 1st edn. 731 pp. Ills. Orig bndg, sl shelf wear.
        (Elgen) **$100 [≈ £62]**

Stillingfleet, Benjamin
- Miscellaneous Tracts relating to Natural History. Third Edition. London: 1775. 8vo. xxxii,391 pp. 11 plates. 1st 3 plates water stained. Half calf.
        (Wheldon & Wesley) **£80 [≈ $128]**

Stillman, J.D.B.
- The Horse in Motion as shown by Instantaneous Photography [by Edward Muybridge]. Edited by L. Stanford. Boston: 1882. 4to. viii,127 pp. 107 plates (9 cold). Orig dec cloth, edges sl worn.
        (Wheldon & Wesley) **£300 [≈ $480]**
- The Horse in Motion as shown by Instantaneous Photography with a Study on Animal Mechanics ... Edited by L. Stanford. Boston: 1882. Lge 4to. 127 pp. Frontis, 107 plates (some cold). Orig pict cloth gilt, spine worn, sl shaken.   *(Jenkins)* **$625 [≈ £390]**

Stimpson, W.
- Shells of New England: A Revision of the Synonymy of the Testaceous Mollusks. Boston: 1851. 8vo. 56,[2] pp. 2 plates. Blind stamp on title. New bds.
        (Wheldon & Wesley) **£35 [≈ $56]**

Stix, Hugh & Marguerite
- The Shell. Five Hundred Million Years of Inspired Design. New York: Abrams, 1968. Oblong 4to. 188 photo ills, inc 82 cold, 15 text figs. 1 gathering bound in upside down. Orig blue cloth gilt. Dw.
        (Blackwell's) **£50 [≈ $80]**

Stoddard, H.L.
- The Bobwhite Quail. Its Habits, Preservation and Increase. New York: (1931) 1946. Roy 8vo. xxix,559 pp. 6 cold & 63 plain plates. Sgntr on title. Orig buckram, jnts trifle loose.
(Wheldon & Wesley) **£35 [≈ $56]**

Stoddart, Thomas
- Tables for Computing the Solid Contents of Timber ... Leith: for the author, 1818. Thin 8vo. Sl soiled, notes on endpapers. Contemp calf, crnrs worn. *(Stewart)* **£45 [≈ $72]**

Stoerck, Anthony
- An Essay on the Use and Effects of the Root of the Colchicum Autumnale, or Meadow Saffron ... Translated from the Latin ... London: Becket & De Hondt, 1764. 1st edn in English. 8vo. 47 pp. Fldg frontis. Title margs sl spotted. Later mor backed bds, uncut.
(Burmester) **£170 [≈ $272]**

Stokers and Pokers ...
- See Head, Sir Francis Bond.

Stokes, E.
- Birds of the Atlantic Ocean. London: 1968. 4to. 156 pp. 38 cold plates by K. Shackleton. Orig cloth. *(Wheldon & Wesley)* **£30 [≈ $48]**

Stokes, R.S.G.
- Mines and Minerals of the British Empire. London: 1908. 8vo. xx,403,8 pp. 40 plates. Sm stamp on title. Orig cloth, sl soiled.
(Henly) **£18 [≈ $28]**

Stokes, William
- The Diseases of the Heart and the Aorta ... Dublin: Hodges & Smith, 1854. 1st edn. 8vo. xvi,689 pp. Orig black cloth, uncut, spine reprd. *(Rootenberg)* **$850 [≈ £531]**

Stout, G.D. (ed.)
- The Shorebirds of North America. Text by P. Matthiesen and R.S. Palmer. Paintings by R.V. Clem. New York: 1968. Imperial 4to. 270 pp. 32 cold plates. Orig buckram. Dw.
(Wheldon & Wesley) **£90 [≈ $144]**

Stratton, George Malcolm
- Theophrastus and the Greek Physiological Psychology before Aristotle. London: George Allen; New York: Macmillan, [1917]. 1st edn. 8vo. [228] pp. Orig ptd red cloth.
(Gach) **£65 [≈ $40]**

Strickland, H.E. & Melville, A.G.
- The Dodo and its Kindred; or the History,

Affinities and Osteology of the Dodo, Solitaire and other Extinct Birds ... London: 1848. 4to. 141 pp. 18 plates (2 cold). A few light pencil marks & notes. Orig cloth, trifle worn & soiled. Authors' pres inscrptn.
(Wheldon & Wesley) **£350 [≈ $560]**

Strutt, J.G.
- Sylva Britannica; or Portraits of Forest Trees. London: [1830]. 4to. xvi,151 pp. Engvd title, 49 plates on india paper. Some foxing throughout. Half calf, rebacked, rubbed.
(Henly) **£110 [≈ $176]**

Stuart, Charles E. & Pratt, J.G. (eds.)
- A Handbook for Testing Extra-Sensory Perception. New York & Toronto: Farrar & Rinehart, [1937]. 1st edn, 1st printing. 12mo. [3]-96,[2] pp. Orig cloth, cvrs dull.
(Gach) **$37.50 [≈ £23]**

Stuart, Robert
- A Descriptive History of the Steam Engine. London: Knight & Lacey, 1824. 8vo. 228 pp. 47 plates. Orig bds, uncut, rebacked, new endpapers. *(Book House)* **£85 [≈ $136]**

The Student and Intellectual Observer ...
- The Student and Intellectual Observer of Science, Literature and Art ... London: Groombridge, 1868-71. 5 vols, complete run. Tall 8vo. Num chromolitho plates, num text ills. Green cloth, gilt spines, sl rubbed, 2 vols with some staining to cvrs.
(Karmiole) **$175 [≈ £109]**

Studer, Paul & Evans, Joan
- Anglo-Norman Lapidaries. Paris: 1924. 1st edn. xx,404 pp. Fldg table. Orig ptd wraps, uncut. *(Elgen)* **$95 [≈ £59]**

Stukeley, William
- The Healing of Diseases, a Character of the Messiah. Being the Anniversary Sermon preached before the Royal College of Physicians, London, on September 20, 1750 ... London: Corbet, 1750. 1st edn. 4to. [vi], 21 pp. Half-title. Disbound.
(Burmester) **£35 [≈ $56]**
- A Letter ... to Mr. Macpherson, on his Publication of Fingal and Temora. With a Print of Catmor's Shield. London: ptd by Richard Hett; sold by J. Baillie, 1763. 1st edn. 4to. 19,[1] pp. Hand cold frontis. Disbound. *(Burmester)* **£110 [≈ $176]**
- Palaeographia Sacra. Or Discourses on Sacred Subjects. London: Richard Hett, sold by J. Baillie, 1763. 1st edn. 4to. [xii], 133, [7] pp. Half-title. Hand cold engvd frontis of a

dandelion head. Old wraps.
                      (Burmester) **£30 [≃ $48]**

Sully, James
- Sensation   and   Intuition: Studies   in
Psychology and Aesthetics. London: Henry
King, 1874. 1st edn. 8vo. 372 pp. Orig bndg,
hinges cracked.   *(Xerxes)* **$60 [≃ £37]**

A Summary of Material Heads ...
- See Lovell, Archibald.

Sutcliffe, Thomas
- The Earthquake of Juan Fernandez, as it
occurred in the year 1835. Authenticated by
the Retired Governor of that Island ...
Manchester: 1839. 32,4 pp. 5 plates & maps.
Half mor.   *(Jenkins)* **$225 [≃ £140]**

Sutton, M.J.
- Permanent   and   Temporary   Pastures.
London: Hamilton, Adams, 1887. 2nd edn.
4to. 23 chromolithos. Orig cloth gilt, trifle
worn.   *(Hollett)* **£40 [≃ $64]**

Swainson, W.
- The Natural History of Fishes, Amphibians
and Reptiles. London: 1838-39. 2 vols. Post
8vo. 2 vignettes & 235 text figs. Vol 1 orig
cloth, vol 2 half leather (trifle used).
                (Wheldon & Wesley) **£35 [≃ $56]**
- A Preliminary Discourse on the Study of
Natural History. London: 1834. Post 8vo.
viii, 462 pp. Engvd title (rather foxed). New
cloth.   *(Wheldon & Wesley)* **£40 [≃ $64]**
- Taxidermy. With the Biography of Zoologists
and Notices of their Works. London: 1840.
Post 8vo. 392 pp. Engvd title. Calf (used).
                (Wheldon & Wesley) **£40 [≃ $64]**
- A   Treatise   on   the   Geography   and
Classification of Animals. London: Lardner's
Cabinet Cyclopaedia, 1835. Sm 8vo. vii,367
pp. New cloth.
                (Wheldon & Wesley) **£45 [≃ $72]**

Swainson, William
- A Preliminary Discourse on the Study of
Natural   History.   London: Cabinet
Cyclopaedia, 1834. Sm 8vo. Orig cloth, ptd
label worn, soiled.   *(Farahar)* **£20 [≃ $32]**

Swammerdam, J.
- The Book of Nature; or, the History of
Insects ... translated by Thomas Flloyd.
Revised ... by John Hill. London: for C.G.
Seyffert, 1758. Folio. Frontis, 53 plates.
Some spotting. Contemp calf, spine relaid,
crnrs relaid.   *(Stewart)* **£550 [≃ $880]**

- The Book of Nature or the History of Insects,
translated from the Dutch and Latin original
by T. Flloyd, revised ... by J. Hill. London:
1758. Folio. [iv],xx,[viii], 236,153, lxiii,[12]
pp. 53 plates. Minor foxing, 1 marg stain, 1
sm tear. Contemp russia, rebacked.
                (Wheldon & Wesley) **£600 [≃ $960]**

Swayer, Arthur Robert
- Miscellaneous Accidents in Mines, with
Special Reference to the North Staffordshire
Coalfield ... Manchester & London: John
Heywood, 1889. Super roy 8vo. 278,[2] pp.
Fldg cold table, 55 plates. No fig 85a,
probably a misnumbering. Orig red cloth, sl
worn.   *(Duck)* **£95 [≃ $152]**

Swayne, George
- Gramina Pascua: or, a Collection of
Specimens of the Common Pasture Grasses ...
Bristol: the author by S. Bonner, sold by W.
Richardson, 1790. Folio. 6 plates displaying
19 mtd specimens, each with leaf of text. Orig
bds, rebacked, stained, crnrs worn.
                (Blackwell's) **£750 [≃ $1,200]**
- Gramina Pascua: or, a Collection of
Specimens of the Common Pasture Grasses ...
Bristol: for the author, 1790. 1st edn. Lge
folio. 6 plates with 19 mtd specimens of actual
grasses, each with 2 labels. Orig bds, a little
worn.
        (W. Thomas Taylor) **$3,000 [≃ £1,875]**

Swaysland, W.
- Familiar Wild Birds. London: Cassell, 1883.
4 vols. 8vo. 160 cold plates, mostly by
Thorburn. A little spotting. Half calf gilt,
hinges rubbed.   *(Hollett)* **£150 [≃ $240]**
- Familiar   Wild   Birds.   London: Cassell,
1883-88. 1st edn. 4 vols. Sm 8vo. 160 tissued
chromolitho plates, mostly by Thorburn,
num w'engvs. Half calf, red labels.
                (Egglishaw) **£130 [≃ $208]**
- Familiar Wild Birds. London: 1901. 4 vols in
2. 8vo. 160 plates by Thorburn & others. Half
calf.   *(Henly)* **£35 [≃ $56]**

Swedenborg, Emanuel
- The   Generative   Organs,   considered
Anatomically,   Physically,   and
Philosophically. A Posthumous Work ...
Translated   from   the   Latin   by   J.J.G.
Wilkinson. London: 1852. 8vo. x,327,4,4
advt pp. Orig cloth.   *(Hollett)* **£85 [≃ $136]**

Sweet, R.
- Hortus Britannicus: or, A Catalogue of Plants
... London: James Ridgway, 1827(-26). 1st
edn. 8vo. [viii],492,xxv pp. Some spotting,

mostly light. Contemp green half calf gilt.
(Gough) **£32 [≈ $51]**

- The Hot-House and Greenhouse Manual ...
Disposed under the Generic Names of the
Plants. London: 1825. 2nd edn. 8vo. 574 pp.
New cloth. *(Wheldon & Wesley)* **£25 [≈ $40]**

Swinton, A.H.
- Insect Variety: its Propagation and
Distribution. London: [1881]. 8vo. x,326 pp.
8 plates (some cold). Orig cloth.
(Wheldon & Wesley) **£15 [≈ $24]**

Switzer, Stephen
- The Nobleman, Gentleman, and Gardener's
Recreation: or an Introduction to Gardening,
Planting, Agriculture ... London: B. Barker,
1715. 1st edn. 8vo. [viii],xxxiv,266, [xvi] pp.
Frontis. Mod calf antique.
(Charles B. Wood) **$900 [≈ £562]**

Sydenham, Thomas
- Dr. Sydenham's Compleat Method of Curing
almost all Diseases and Descriptions of their
Symptoms ... Seventh Edition. London: for J.
Hodges, 1737. 12mo. [viii],202,[6] pp. 18th c
'school-book' cloth. *(Burmester)* **£45 [≈ $72]**
- Works on Acute and Chronic Disease ... with
Notes by Benjamin Rush. Philadelphia: 1809.
1st edn. 8vo. xl,473 pp,6 ff (inc final blank).
Minimal browning. Contemp calf, rubbed,
hinges cracked but holding.
(Hemlock) **$250 [≈ £156]**

Syme, Patrick
- A Treatise on British Song-Birds ...
Edinburgh: John Anderson ..., 1823. 12mo.
vi, 231 pp. 15 hand cold engvd plates.
Contents grubby around the edges. Sl
offsetting of plates. Contemp half calf, spine
relaid. *(Sotheran)* **£225 [≈ $360]**

Symonds, John Addington, senior
- Sleep and Dreams: Two Lectures delivered at
the Bristol Literary and Philosophical
Institution. London: Murray, 1851. Orig
cloth, respined. *(C.R. Johnson)* **£38 [≈ $60]**

A Table of Redemption ...
- A Table of Redemption ... by T.W., F.R.S.
... see Watkins, Thomas.

Tait, Peter Guthrie
- An Elementary Treatise on Quaternions.
Oxford: 1873. 2nd edn, enlgd. xx,296 pp.
Orig cloth, inner hinge cracked but bndg
firm, spine sl torn at hd.
(Whitehart) **£35 [≈ $56]**

- Lectures on Some Recent Advances in
Physical Science. London: 1876. 1st edn. 8vo.
xii,337 pp. Orig cloth, inner hinge sl cracked.
(Whitehart) **£18 [≈ $28]**
- Lectures on Some Recent Advances in
Physical Science ... Second Edition, revised.
London: Macmillan, 1876. 8vo. xx,363,errata
slip, 24 advt pp. Lib stamp on title. Orig
brown cloth. *(Pickering)* **$75 [≈ £46]**
- Life and Scientific Work ... Supplementing
the Two Volumes of Scientific Papers ... By
Cargill Gilston Knott ... Cambridge: Univ
Press, 1911. 1st edn. 4to. x,379 pp. 5 ports.
Orig dark blue cloth. Pres inscrptn from the
author. *(Pickering)* **$200 [≈ £125]**
- Light. Edinburgh: 1884. 1st edn. viii,276 pp.
47 text figs. Orig cloth, dull, spine v sl worn.
Prize label sgnd by Tait.
(Whitehart) **£18 [≈ $28]**
- Light. Edinburgh: A. & C. Black, 1884. 1st
edn. 8vo. viii,276 pp. Orig red cloth.
(Pickering) **$100 [≈ £62]**
- Newton's Laws of Motion. London: A. & C.
Black, 1899. 1st edn. viii,53,2 advt pp. Orig
red cloth. *(Pickering)* **$75 [≈ £46]**
- Properties of Matter. Edinburgh: A. & C.
Black, 1885. 1st edn. viii,320,errata slip, [1
advt] pp. Orig red cloth.
(Pickering) **$100 [≈ £62]**
- Scientific Papers ... Cambridge: UP,
1898-1900. 1st edn. 2 vols. 4to. xiv,498; xiv,
500 pp. 19 plates, text ills. Orig blue cloth,
partly unopened, spine ends a little worn.
(Pickering) **$350 [≈ £218]**
- Sketch of Thermodynamics. Edinburgh:
1868. 1st edn. 8vo. viii,128 pp. Title sl dust
stained. Orig cloth, new endpapers.
(Whitehart) **£25 [≈ $40]**
- Sketch of Thermodynamics ... Second
Edition. Revised and Extended. Edinburgh:
Douglas, 1877. 8vo. xx,162 pp. Ex-lib copy.
Buckram. *(Pickering)* **$75 [≈ £46]**

Tait, Peter Guthrie & Steele, William John
- A Treatise on the Dynamics of a Particle ...
Third Edition ... augmented. London:
Macmillan, 1871. 8vo. xvi,428,32 advt pp.
Orig green cloth, worn.
(Pickering) **$75 [≈ £46]**

Tait, Peter Guthrie & Thomson, W.
- Elements of Natural Philosophy. Part I [all
published]. Oxford: Clarendon Press, 1873.
1st edn. 8vo. viii,279,16 advts. Pencil
marginalia. Orig plum cloth, shaken, spine
frayed. *(Pickering)* **$300 [≈ £187]**

Talbot, Frederick A.
- Motor-Cars and Their Story. London: Cassell, 1912. 1st edn. Thick super roy 8vo. xvi,368 pp. Cold frontis, 84 b/w plates. Occas soiling. Orig gilt dec cloth with pict onlay, t.e.g., sl rubbed & marked, hinges shade slack, minor straining or cracking to sewing.
(Duck) **£135 [≈ $216]**

Talbot, H. Fox
- The Process of Calotype Photogenic Drawing, Communicated to the Royal Society, June 10th, 1841. [Caption title]. [London: 1841?]. 1st sep edn. Offprint. Sq 8vo. 4 pp leaflet. Mod bds.
(M & S Rare Books) **$1,200 [≈ £750]**

Tallis's Crystal Palace ...
- Tallis's The Crystal Palace and The Exhibition of the World's Industry. Described and Illustrated by Beautiful Engravings chiefly from Daguerrotypes by Beard, Myall, &c. London: Tallis, 1851. 3 vols. 4to. 640 pp. Cold frontis. Half leather, worn.
(Book House) **£160 [≈ $256]**

Tallis's Illustrated London ...
- See under Gaspey, William.

Tanner, John
- The Hidden Treasures of the Art of Physick. Fully discovered, in Four Books ... London: for George Sawbridge, 1659. 1st edn. Sm 8vo. [16],543,[25] pp. Browned. Lacks 3 ff (pp 13/14,209-212). Pp 5-12 torn with loss of 4 lines. Lacks title. Half calf, rebacked.
(Hemlock) **$375 [≈ £234]**

Tansley, A.G.
- Britain's Green Mantle. Past Present and Future. London: 1949. 1st edn. 8vo. xii,294 pp. 69 plates. Orig cloth. Dw.
(Henly) **£15 [≈ $24]**
- The British Islands and their Vegetation. London: 1939. 1st edn. Roy 8vo. xxxviii,930 pp. 162 plates, 179 text figs. Orig buckram, spine sl faded, top edge of lower cvr dampstained. *(Henly)* **£36 [≈ $57]**

Taplin, William
- The Gentleman's Stable Directory; or, Modern System of Farriery, to which is added a Supplement. Dublin: P. Wogan, 1800. 14th edn. 184 pp. Orig bds, ptd paper label, ink name on upper cvr, edges sl rubbed.
(Moon) **£55 [≈ $88]**

Tate, R.
- A Plain and Easy Account of the Land and Fresh Water Mollusks of Great Britain. London: 1866. 8vo. viii,244 pp. 11 hand cold plates, 28 text figs. Some foxing & marg staining. Orig cloth, sl used, rear inner jnt taped.   *(Wheldon & Wesley)* **£20 [≈ $32]**

Tate, Thomas
- On the Strength of Materials; containing Various Original and Useful Formulae, specially applied to Tubular Bridges, Wrought Iron and Cast Iron Beams ... London: Longman, Brown, 1850. 8vo. x,96,32 advt pp. 22 figs. Orig cloth, spine splitting & sl defective. *(Hollett)* **£40 [≈ $64]**

Taussif, Frank William
- International Trade ... New York: Macmillan, 1927. 1st edn. 8vo. xxi,[i blank], 425 pp. Orig blue cloth gilt.
(Pickering) **$150 [≈ £93]**

Taverner, P.A.
- Birds of Canada. London: 1938. Roy 8vo. 445 pp. 87 cold plates, 488 text figs. Orig cloth, spine faded, cvrs trifle soiled.
(Wheldon & Wesley) **£30 [≈ $48]**
- Birds of Eastern Canada. Ottawa: 1922. 2nd edn. Roy 8vo. 380 pp. 50 cold plates, 68 text figs. Orig cloth, trifle used.
(Wheldon & Wesley) **£20 [≈ $32]**
- Birds of Western Canada. Ottawa: 1926. Roy 8vo. 380 pp. 84 cold plates, 315 text figs. Orig cloth, sl used.
(Wheldon & Wesley) **£25 [≈ $40]**

Tavistock, Marquess of
- Parrots and Parrot-like Birds in Aviculture. London: [1929]. Roy 8vo. 298 pp. 8 cold plates. Orig cloth, sl used, front jnt rather loose, good ex-lib.
(Wheldon & Wesley) **£40 [≈ $64]**
- Parrots and Parrot-Like Birds. London: [1929]. 1st edn. Cr 4to. 298 pp. 8 cold plates. Occas sl foxing of text. Orig cloth.
(Henly) **£28 [≈ $44]**

Taylor, Alfred
- Birds of a County Palatine. London: 'Wild Life', 1913. 1st edn. 4to. 148 pp. 30 mtd b/w plates, num other ills. Orig pict cloth gilt.
(Gough) **£30 [≈ $48]**

Taylor, Charles Fayette
- Theory and Practice of the Movement-Cure, or, the Treatment of Lateral Curvature of the Spine ... by the Swedish System of Localized

Movements. Philadelphia: Lindsay & Blaikiston, (1860) 1861. 8vo. 295 pp. Ills. 1 sgntr loosening. Orig bndg.
(Xerxes) **$70 [≈ £43]**

Taylor, Fanny M.
- Eastern Hospitals and English Nurses; the Narrative of Twelve Months' Experience in the Hospitals of Koulari and Scutari. By a Lady Volunteer. Third Edition, Revised. London: Hurst & Blackett, 1857. xii,356 pp. 2 lithotints. Orig pict cloth gilt, v sl soiled.
(Duck) **£135 [≈ $216]**

Taylor, Frederick Winslow
- The Principles of Scientific Management ... New York & London: Harper, 1911. 1st edn. 8vo. 144 pp, inc half-title. Orig red cloth, spine v sl sunned. *(Pickering)* **$500 [≈ £312]**

Taylor, H.V.
- The Apples of England. London: (1946) 1947. 3rd edn. Roy 8vo. 206 pp. 36 cold plates. Orig cloth, trifle warped.
(Wheldon & Wesley) **£25 [≈ $40]**
- The Plums of England. London: 1949. Roy 8vo. 160 pp. 32 cold plates. Orig cloth.
(Wheldon & Wesley) **£20 [≈ $32]**

Taylor, Isaac
- Physical Theory of Another Life. London: William Pickering, 1836. 1st edn. 8vo. Contemp calf, ruled in gilt.
(Robertshaw) **£32 [≈ $51]**

Taylor, J.
- Geological Essays and Sketch of the Geology of Manchester. London: 1864. 8vo. ix, 282 pp. Ills. Orig cloth. *(Baldwin)* **£22 [≈ $35]**

Taylor, J.E.
- The Aquarium: its Inhabitants, Structure and Management. London: Hardwicke & Bogue, 1876. 8vo. xv,316,[4 advt] pp. 239 ills. Orig cloth gilt, a.e.g. *(Lamb)* **£15 [≈ $24]**
- Flowers, their Origin, Shapes, Perfumes and Colours. Illustrated with 32 Coloured Figures by Sowerby and 161 Woodcuts. Fourth Edition. London: W.H. Allen, n.d. 8vo. xxiv, 347 pp. Orig cloth. *(Lamb)* **£25 [≈ $40]**
- Half-Hours in the Green Lanes: A Book for a Country Stroll. London: Hardwicke, 1873. 8vo. viii,328 pp. 262 w'engvs. Orig green cloth gilt. *(Lamb)* **£12 [≈ $19]**
- Notes on Collecting and Preserving Natural History Objects. London: Hardwicke & Bogue, 1876. 8vo. viii,215,32 advt pp. Ills. Orig green cloth. *(Lamb)* **£12 [≈ $19]**

- Our Common British Fossils and Where to Find Them. A Handbook for Students. London: Chatto & Windus, 1885. 8vo. xii,331,32 advt pp. 331 ills. Orig maroon cloth gilt, uncut. *(Lamb)* **£15 [≈ $24]**
- Our Common British Fossils and Where to Find Them ... London: 1885. 1st edn. 8vo. xii, 331 pp. 331 figs. Orig cloth.
(Bow Windows) **£30 [≈ $48]**

Taylor, John
- African Rifles and Cartridges. London: 1948. 1st edn. 8vo. Ills. Orig bndg. Dw.
(Grayling) **£150 [≈ $240]**

Taylor, Joseph
- The Complete Weather Guide; a Collection of Practical Observations for prognosticating the Weather ... Second Edition. London: for Gale, Curtis & Fenner, 1814. 160 pp. Half-title. Fldg frontis. A little foxing. Cloth, uncut. *(Wreden)* **$40 [≈ £25]**

Teale, Thomas Pridgin
- Dangers to Health; a Pictorial Guide to Domestic Sanitary Defects. London: Churchill, 1878. 1st edn. 8vo. 55 plates. Orig cloth, spine extremities sl rubbed.
(de Beaumont) **£68 [≈ $108]**
- A Treatise on Neuralgic Diseases, dependent upon Irritation of the Spinal Marrow and Ganglia of the Sympathetic Nerve. Philadelphia: 1830. 1st Amer edn. 120 pp. Usual browning. Foxing. Orig cloth backed bds, worn. *(Elgen)* **$100 [≈ £62]**

Teall, J.J.H.
- British Petrography, with Special Reference to Igneous Rocks. London: 1888. Imperial 8vo. viii,469 pp. 47 plates (46 cold), 45 key plates. Orig cloth, trifle loose, tear in spine reprd. *(Wheldon & Wesley)* **£70 [≈ $112]**

Tegetmeier, W.B.
- Pheasants. For Coverts and Aviaries ... London: Horace Cox, 1873. Folio. viii,124 pp. Num ills. Orig green cloth gilt, sl rubbed, extremities sl frayed, sl discold by water damage at outer cvr margs.
(Karmiole) **$125 [≈ £78]**
- Pheasants: their Natural History and Practical Management.. London: Horace Cox, 1881. 2nd edn, enlgd. 4to. v,142,advt pp. 13 plates, text figs. Orig pict gilt cloth, a.e.g. *(Egglishaw)* **£48 [≈ $76]**
- Pheasants; their Natural History and Practical Management. London: 1904. 4th edn, enlgd. 8vo. 6 cold & 16 plain plates. Orig

cloth gilt, lacks front free endpaper.
(Henly) **£55 [≈ $88]**
- Pheasants, their Natural History and Practical Management. London: 1904. 8vo. Cold & other plates, text ills. Title & prelims foxed. Orig cloth, spine v sl snagged.
(Grayling) **£20 [≈ $32]**

Tegg, Thomas
- The Young Man's Book of Knowledge ... The Works of Nature, Logic, Eloquence ... Matter and Motion, Magnetism, Mechanical Powers, Hydrostatics, Hydraulics ... American History ... New York:1834. 1st Amer edn. 8vo. 396 pp. Frontis. Sl spotting. Cloth, uncut, soiled.   *(Young's)* **£30 [≈ $48]**

Telescope, Tom
- See Newbery, J.

Television ...
- Television. A Monthly Magazine. The World's First Television Journal. London: The Television Press Ltd, 1928-31. Nos 1-36, March 1928 - February 1931. 36 issues in orig wraps as issued. 4to. Edges a trifle dusty.
(Hollett) **£150 [≈ $240]**

Thackrah, Charles Turner
- An Inquiry into the Nature and Properties of the Blood ... New and Enlarged Edition; arranged and revised by T.C. Wright ... London: 1834. 8vo. [4],246 pp. Contemp mor gilt, somewhat worn & scuffed.
(Hemlock) **$225 [≈ £140]**

Thayer, Gerald H.
- Concealing-Coloration in the Animal Kingdom ... Laws of Disguise through Color and Pattern: being a Summary of Abbott H. Thayer's Discoveries ... New York: 1909. 1st edn. 4to. xix,260 pp. 16 cold & 58 other plates. Blind stamp on title. Orig cloth, a.e.g. hd of spine sl worn.
(Bickersteth) **£85 [≈ $136]**
- Concealing-Coloration in the Animal Kingdom ... New York: 1918. New edn, with new preface. Cr 4to. [xx],260 pp. 16 cold tissued plates, over 140 other ills. Orig cloth.
(Bow Windows) **£145 [≈ $232]**

Thearle, S.J.P.
- The Modern Practice of Shipbuilding in Iron and Steel. London: Collins, 1886. 2nd rvsd edn. 2 vols. 8vo & 4to (text & atlas). 341 pp. 34 plates. Orig bndg, a little soiled.
(Book House) **£30 [≈ $48]**

Thelmar, E.
- The Maniac. A Realistic Study of Madness from the Maniac's Point of View. Introduction by Hereward Carrington. New York: Amer Psychical Institute, [ca 1932]. 12mo. 259 pp. Orig bndg. Dw.
(Xerxes) **$60 [≈ £37]**

Theobald, F.V.
- A Monograph of the Culicidae or Mosquitoes ... London: BM, 1901-10. 6 vols inc atlas vol. 8vo. 39 plates, plus atlas 37 cold & 5 plain plates. Lib stamp on titles. Cloth.
(Egglishaw) **£200 [≈ $320]**

Thomas, H.S.
- The Rod in India, being Hints how to Obtain Sport with Remarks on the Natural History of Fish ... London: 1881. 2nd edn. Roy 8vo. xxvi,436 pp. 25 litho plates (8 cold). Orig dec cloth gilt, sm tears to hd of spine.
(Egglishaw) **£75 [≈ $120]**

Thomas, John J.
- The American Fruit Culturist ... New York: William Wood, 1875. 8th edn, rvsd. 8vo. vi-576 pp. Chromolitho frontis, 508 text ills. Orig pict green cloth gilt, sl rubbed.
(de Beaumont) **£48 [≈ $76]**

Thompson, C.J.S.
- The Mystic Mandrake. London: Rider, (1934). 253 pp. 8 plates, 19 text ills. Orig blue cloth.   *(Karmiole)* **$35 [≈ £21]**
- The Quacks of Old London. Philadelphia: Lippincott, 1929. 1st Amer edn. 8vo. 356 pp. Plates, text ills. Orig cloth.
(Hemlock) **$100 [≈ £62]**

Thompson, D'A.W.
- On Growth and Form. Cambridge, (1942) 1968. 2nd edn, rvsd & enlgd. 2 vols. 8vo. 1116 pp. 2 plates, 554 text figs. Orig cloth.
(Wheldon & Wesley) **£98 [≈ $156]**

Thompson, Henry
- Modern Cremation. Cremation: Its History and Practice to the Present Date ... Third Edition, revised and much enlarged. London: Smith, Elder, 1899. xii,187 pp. Frontis, plates, ills. Three qtr mor, spine & crnr worn. Author's pres inscrptn. *(Wreden)* **$65 [≈ £40]**

Thompson, Robert
- The Gardener's Assistant: Practical and Scientific ... London: Blackie & Son, 1846. 1st edn. Sm 4to. xv,774 pp. 12 cold plates. Mod half mor silvered. *(Hollett)* **£60 [≈ $96]**

- The Gardener's Assistant: Practical and Scientific ... London: Blackie & Son, n.d. Sm 4to. xv,774 pp. 12 cold plates. A little fingering, a few ff at end chipped & soiled. Old half calf gilt, rather soiled & worn.
(Hollett) **£60 [≈ $96]**
- The Gardener's Assistant ... London: Blackie & Son, [ca 1859]. 1st edn (?). 4to. 4,xv,774,[8 advt] pp. Hand cold frontis, 11 plates. Occas minor foxing. Orig green cloth gilt-lettered, extremities sl rubbed.
(Heritage) **$175 [≈ £109]**
- The Gardener's Assistant ... London: [1859]. 8vo. xv,774 pp. 12 hand cold plates. Sl foxing throughout, lower marg of last 12 pp dampstained. Half calf, a little worn.
(Henly) **£140 [≈ $224]**
- The Gardener's Assistant: Practical and Scientific ... London: Blackie & Son, 1859. 1st edn. Lge thick 8vo. xv,774,[8 advt] pp. 12 cold plates. Orig green cloth gilt, lower bd sl damped.
(Hollett) **£140 [≈ $224]**
- The Gardener's Assistant: Practical and Scientific. New Edition by T. Moore. London: 1881. Roy 8vo. 956 pp. 32 plates (12 cold). Cold plates rather foxed. Orig cloth, refixed & reprd.
(Wheldon & Wesley) **£60 [≈ $96]**

Thompson, Silvanus P.
- Cantor Lectures on Dynamo-Electric Machinery. London: William Trounce, 1883. 1st edn. 8vo. 54 pp. Text ills & diags. Lacks wraps.
(Rootenberg) **$150 [≈ £93]**

Thompson, W.
- The English Flower Garden: a Monthly Magazine of Hardy and Half-Hardy Plants. London: 1852-53. Vols 1 & 2. 2 vols in one. Roy 8vo. 23 hand cold plates. Mod half mor. Vol 3 part 1 (with 1 cold plate) was also publ but is not included.
(Wheldon & Wesley) **£275 [≈ $440]**

Thompson, William
- Sickness. A Poem. In Three Books. London: for R. Dodsley, sold by M. Cooper, 1745-46. 1st edn. 3 parts in one vol. 4to. viii,47, [1]; [ii],51-104; [ii],107-159 pp. Half- title. Rec bds.
(Burmester) **£50 [≈ $80]**

Thomson, A.L.
- Britain's Birds and their Nests. London: [1910]. Sm 4to. xxviii,340 pp. 132 cold plates. Minor foxing. Orig dec cloth.
(Wheldon & Wesley) **£18 [≈ $28]**

Thomson, Adam
- Time and Timekeepers. London: T. & W. Boone, 1842. 1st edn. Sm 8vo. xii,195 pp. 54 text ills. Occas sl browning. Orig cloth gilt.
(Fenning) **£55 [≈ $88]**

Thomson, Anthony Todd
- The Domestic Management of the Sick-Room ... London: Longman ..., 1841. 8vo. Orig cloth, spine faded & worn at hd.
(Stewart) **£100 [≈ $160]**

Thomson, H. Campbell
- Acute Dilation of the Stomach. New York: William Wood, 1902. 1st edn. 8vo. 54 pp. Photo ills. Some pencil & ink notes. Orig bndg.
(Xerxes) **$50 [≈ £31]**

Thomson, J.A.
- Darwinism and Human Life. The South African Lectures for 1909. London: 1909. 1st edn. 8vo. xii,245 pp. Port. Orig cloth.
(Bow Windows) **£25 [≈ $40]**

Thomson, Sir J.J.
- The Discharge of Electricity through Gases. London: 1898. 1st edn. 8vo. x,203 pp. 41 text diags. Orig cloth.   *(Whitehart)* **£40 [≈ $64]**
- Rays of Positive Electricity and their Applications to Chemical Analyses. London: Longmans, Green, Monographs on Physics Series, 1921. 2nd edn, enlgd. [x],238 pp. 9 pp plates, text ills. Orig blue cloth, spine sl dull. Arthur W. Chapman's sgntr.
(Sklaroff) **£65 [≈ $104]**

Thomson, John
- Lectures on Inflammation ... a View of the General Doctrines, Pathological and Practical of Medical Surgery. Edinburgh: 1813. 1st edn. 8vo. xii,649 pp. Lib stamp on title. Contemp half calf, worn, lacks front cvr, lower hinge split.   *(Hemlock)* **$175 [≈ £109]**

Thomson, Spencer
- Wanderings among the Wild Flowers ... With Two Chapters on the Economical and Medicinal Uses of Native Plants. Second Edition. London: Groombridge, 1854. Sm 8vo. xix,318,[2 advt] pp. 171 ills. Orig cloth gilt.   *(Fenning)* **£24.50 [≈ $40]**

Thomson, W.
- A Practical Treatise on the Cultivation of the Grape Vine. Tenth Edition. London: William Blackwood, 1890. 8vo. viii,105,[1]32 advt pp. Text figs. Sev pp carelessly opened at hd. Orig green cloth gilt. *(Spelman)* **£20 [≈ $32]**

Thonner, F.
- The Flowering Plants of Africa. An Analytical Key to the Genera of African Phanerogams. London: (1916) 1962. xvi,647 pp. Map, 150 plates.
  (Wheldon & Wesley) **£60 [≈ $96]**

Thorburn, Archibald
- British Birds. London: 1925-26. New edn. 4 vols. 8vo. 192 cold plates. Sgntr on title, some ink annotations & extra material pasted in. Orig red cloth, spines trifle faded.
  (Wheldon & Wesley) **£65 [≈ $104]**
- British Birds. London: 1925-26. New edn. 4 vols. 8vo. 192 cold plates. Orig cloth, spines faded. *(Henly)* **£120 [≈ $192]**
- British Birds. London: 1925-26. New edn. 4 vols. 8vo. 192 cold plates. Orig red cloth, trifle used. *(Wheldon & Wesley)* **£85 [≈ $136]**
- British Mammals. London: Longmans, Green, 1920. 1st edn. 2 vols in one. 4to. 50 cold plates on thick card. Orig blue cloth gilt, bds a little marked. *(Gough)* **£550 [≈ $880]**
- British Mammals. London: 1920-21. 2 vols. 4to. 50 cold plates. New cloth.
  (Wheldon & Wesley) **£350 [≈ $560]**
- Game Birds and Wild Fowl of Great Britain and Ireland. London: 1923. Imperial 4to. vi, 78 pp. 30 cold plates. Orig red buckram.
  (Wheldon & Wesley) **£500 [≈ $800]**

Thorley, J.
- [In Greek: Melisselogia], or, the Female Monarchy, being an Enquiry into the Nature, Order and Government of Bees. London: 1744. 1st edn. 8vo. xlvi,206,[2] pp. Frontis & 4 plates. Contemp calf, rebacked.
  (Wheldon & Wesley) **£200 [≈ $320]**
- An Enquiry into the Nature, Order and Government of Bees. London: 1774. 4th edn. 8vo. 158 pp. 2 plates. Sl staining. Contemp calf, worn. *(Wheldon & Wesley)* **£75 [≈ $120]**

Thorndike, Edward Lee
- Animal Intelligence: Experimental Studies. New York: Macmillan, 1911. 1st edn. 8vo. [x], 297, [13] pp. Linen. *(Gach)* **$125 [≈ £78]**
- The Elements of Psychology. New York: A.G. Seiler, 1905. 1st edn. Sm 8vo. [iv],[xx], [352] pp. Russet cloth.   *(Gach)* **$50 [≈ £31]**
- The Fundamentals of Learning. New York: Bureau of Publications Teachers College, Columbia Univ, 1932. 1st edn, 1st printing. 8vo. [xviii],638 pp. Orig cloth. Chipped dw.
  (Gach) **$60 [≈ £37]**
- The Psychology of Wants, Interests, and Attitudes. New York: Appleton, [1935]. 1st edn. 8vo. [ii],x,301,[3] pp. Black cloth, jnts worn. Inscrbd copy. *(Gach)* **$85 [≈ £53]**
- The Psychology of Wants, Interests, and Attitudes. New York: Appleton, [1935]. 1st edn. 1st printing. 8vo. [ii],x,301,[3] pp. Cloth. Lightly worn dw. *(Gach)* **$60 [≈ £37]**

Thorndike, Lynn
- A History of Magic and Experimental Science. New York: Columbia UP, 1923-58. Only edn, later printing. 8 vols. 8vo. Orig buckram.   *(Gach)* **$250 [≈ £156]**

Thorne, W. Beazly
- The Schott Method of Treatment of Chronic Diseases of the Heart. With an Account of the Nauheim Baths, and of the Therapeutic Exercises. Second Edition. London: 1896. 8vo. 83 pp. 9 plates, text figs. Orig cloth.
  (Bickersteth) **£25 [≈ $40]**

Thornthwaite, William Henry
- Hints on Reflecting & Refracting Telescopes and their Accessories. Fifth Edition. London: Horne, Thornthwaite & Wood, 1890. 8vo. 84 pp. 24 ills. Orig ptd paper wraps.
  (Fenning) **£18.50 [≈ $30]**

Thornton, Phineas
- The Southern Gardener and Receipt Book: Containing Directions for Gardening; a Collection of Valuable Receipts for Cookery ... [Camden, S.C.]: for the author, 1840. 1st edn. 12mo. 12,330 pp. Lacks front endpaper. Orig cloth, sl shaken.
  (M & S Rare Books) **$1,250 [≈ £781]**

Thornton, Robert John
- Elements of Botany. London: 1812. 2 vols in one. Roy 8vo. viii,90; 73 pp. 171 engvd plates (most copies seem to have fewer than the 176 called for). Foredge of title strengthened, sm section cut from endpaper. Mod half calf.
  (Henly) **£115 [≈ $184]**
- Elements of Botany. Part 1 Classification. Part 2 Terms of the Science. London: 1812. 2 vols. Roy 8vo. 173 engvd plates (misbound). Contemp cloth, uncut.
  (Wheldon & Wesley) **£48 [≈ $76]**
- A Grammar of Botany ... for the Use of Schools. New York: 1818. 12mo. 317 pp. 45 plates. Some foxing & blind stamps. New cloth. *(Wheldon & Wesley)* **£30 [≈ $48]**
- A New Family Herbal, or Popular Account of the Natures and Properties of the various Plants used in Medicine. London: 1814. 2nd edn. 8vo. xxviii,901 pp. Port, 258 engvs by Thomas Bewick, cold by hand. Contemp

green half calf.
(Wheldon & Wesley) **£250 [≈ $400]**

- Temple of Flora ... London: 1812. 'Lottery' edn. Imperial 4to. Cold engvd frontis, engvd title on 2 ff, 28 cold & 2 plain plates. Sl marg soiling & foxing. Mod half green levant mor, t.e.g. Inc Persian Cyclamen plate not mentioned by Dunthorne.
(Wheldon & Wesley) **£3,500 [≈ $5,600]**

- Temple of Flora. Edited by Geoffrey Grigson. London: 1951. Folio. 24 pp. 12 cold & 25 [sic] collotype plates. Orig buckram backed bds. Dw.
(Wheldon & Wesley) **£75 [≈ $120]**

- Temple of Flora. Edited by Geoffrey Grigson. London: Collins, 1951. 1st edn thus. Folio. viii,18 pp. 12 cold & 24 [sic] collotype plates. Orig half buckram. Slipcase.
(Gough) **£110 [≈ $176]**

- See also Grigson, Geoffrey & Buchanan, Handasyde.

A Thousand Notable Things ...
- See Lupton, Thomas.

Ticehurst, Claud Buchanan
- A History of the Birds of Suffolk. London: 1932. 8vo. xi,502 pp. Map, 18 plates. Orig cloth.   *(Wheldon & Wesley)* **£38 [≈ $60]**
- A Systematic Review of the Genus Phylloscopus (Willow-Warblers or Leaf-Warblers). London: 1938. Cr 4to. viii,193 pp. 8 maps, 2 cold plates. Orig cloth, crnrs trifle bumped.   *(Wheldon & Wesley)* **£50 [≈ $80]**

Ticehurst, Norman Frederic
- A History of the Birds of Kent. London: Witherby, 1909. 1st edn. 8vo. lvi,568 pp. Fldg map, 24 b/w plates. Orig blue cloth, bds somewhat damp marked.
(Gough) **£60 [≈ $96]**
- The Mute Swan in England and the Ancient Custom of Swan-Keeping. London: 1957. Roy 8vo. xiii,133 pp. 31 plates (4 cold). Orig cloth.   *(Wheldon & Wesley)* **£30 [≈ $48]**

Tilden, John
- Etiology of Typhoid Fever. Denver, Co.: Tilden, 1909. 1st edn. Sm 8vo. 135,3,index pp. Orig bndg.   *(Xerxes)* **$45 [≈ £28]**

Timbs, John
- Hints for the Table: or, The Economy of Good Living. With a Few Words on Wines. London: Kent & Co ..., 1859. 1st edn. 8vo. W'engvd frontis & title-vignette. Orig salmon cloth gilt, spine faded.
(Sanders) **£100 [≈ $160]**

Tinbergen, N.
- The Herring Gull's World. London: Collins, New Naturalist Monograph, 1953. 1st edn. 8vo. xvi,255 pp. 109 ills. Orig cloth. Dw torn.   *(Gough)* **£35 [≈ $56]**
- The Study of Instinct. Oxford: (1951) 1958. Roy 8vo. 242 pp. 2 plates, 138 text figs. Orig cloth.   *(Wheldon & Wesley)* **£20 [≈ $32]**

Tinterow, Maurice M.
- Foundations of Hypnosis from Mesmer to Freud. Springfield, IL: Charles C. Thomas, [1971]. 8vo. [xvi],606,[2] pp. Orig bndg. Dw.
(Gach) **$87.50 [≈ £55]**

Tissot, Samuel A.A.D.
- An Essay on the Disorders of People of Fashion. Translated from the French by Francis Bacon Lee. London: for Richardson & Urquhart ..., [1771]. 1st edn in English. 8vo. xvi,163 pp. Disbound.
(Bickersteth) **£120 [≈ $192]**

Titcomb, Sara
- Mind-Cure on a Material Basis. Boston: Cupples, 1885. 1st edn. 8vo. 288 pp. Orig bndg.   *(Xerxes)* **$55 [≈ £34]**

Todd, Sereno Edwards
- The American Wheat Culturist. A Practical Treatise on the Culture of Wheat ... New York: Taintor Bros, 1868. 1st edn. 432 pp. Ills. Orig cloth, front jnt splitting.
(Wreden) **$50 [≈ £31]**

The Toilet of Flora ...
- See Buc'hoz, Pierre Joseph.

Tollemache, The Hon. Stanhope
- British Trees with Illustrations. London: Sampson Low, 1901. Lge 8vo. 98 pp. 61 plates. Orig cloth, sl bubbled on front cvr.
(Lamb) **£18 [≈ $28]**

Tomlinson, C.
- Cyclopaedia of Useful Arts, Mechanical and Chemical, Manufacturing, Mining and Engineering. London: George Virtue, 1854. 1st edn. 2 vols in the orig 9 parts. 4to. xvi, clx, 832; iv,1052,[8 advt] pp. Addtnl engvd titles, 42 plates, 2399 ills. Orig cloth gilt.
(Fenning) **£125 [≈ $200]**

Tomlinson, W.W.
- The North Eastern Railway; Its Rise and Development. Newcastle: Andrew Reid, 1914. Lge thick 8vo. xvi,820 pp. 40 plates, num photo ills. Mod half calf gilt.

(Hollett) **£120 [≈ $192]**

Tonks, Eric S.
- The Ironstone Railways & Tramways of the Midlands. Second Impression (Revised). London: Locomotive Publ Co, 1961. [viii],316 pp. 8 fldg maps & plans, 34 plates, plans in text. Orig cloth. Dw sl rubbed.
(Duck) **£75 [≈ $120]**

Topsell, E.
- The History of Four-Footed Beasts and Serpents ... whereunto is now added The Theater of Insects ... by T. Muffet. Introduction by W. Ley. New York: 1967. Facs reprint of 1658 edn. 3 vols. Folio. Num ills. Orig dec bds.
(Wheldon & Wesley) **£100 [≈ $160]**

Tovey, Charles
- Wine and Wine Countries; a Record Manual for Wine Merchants and Wine Consumers. London: Hamilton Adams, 1862. 1st edn. xiv, 359 pp. Fldg & other tables. Orig embossed cloth.
(Box of Delights) **£200 [≈ $320]**

Townsend, John, 1757-1826
- Memoirs of the Rev. John Townsend, Founder of the Asylum for the Deaf and Dumb, and of the Congregational School. Boston: Crocker & Brewster; New York: J. Leavitt, 1831. 1st Amer edn. 12mo. Frontis port. Sl foxed. Orig cloth backed bds, paper label, chipped.   *(Gach)* **$75 [≈ £46]**

Townsend, R.
- Chapters on the Modern Geometry of the Point, Line, and Circle ... Dublin: 1863-65. 2 vols. xx,300; xx,400 pp. Figs. Traces of lib stamp erasure. New cloth.
(Whitehart) **£40 [≈ $64]**

Trade catalogues
- Bristol Wagon Works Co: Illustrated Catalogue. Cart & Wagon Builders, Agricultural Implement Makers, Wheelwrights, Ironfounders, and Millwrights, Railway Carriage and Wagon Builders. Bristol: July, 1877. 38 pp. Ills. Title sl frayed. Lacks wraps, shaken.
(Ambra) **£32 [≈ $51]**
- Clark, Hunt & Co, Shoreditch: General Catalogue No. 1138, of Builders & Electricians Hardware etc. London: [1930s]. 4to. 477 pp. Ills. Orig bndg. Priced.
(Book House) **£25 [≈ $40]**
- Cooper, William: Catalogue of William Cooper, Portable Building Manufacturers ...

London: 761, Old Kent Road, [ca 1900]. 638 pp. Num ills. Orig cloth.
(Harrison) **£19 [≈ $30]**
- Deakin, James: Silver, Electro Plate, Britannia Metal and Cutlery. Sheffield: Hames Deakin, Sidney Works, [ca 1903]. 4to. 307 pp. Num ills. Sm snippets cut from 16 pp. Bndg rather worn, esp hd of spine.
(Harrison) **£55 [≈ $88]**
- Gibson Co: Illustrated Catalogue of Surgical Instruments and Surgical Supplies. Washington: [ca 1920]. 416pp, 30 pp price list in pocket. Num ills. Orig cloth.
(Elgen) **£100 [≈ £62]**
- Long, Joseph: Sundries Catalogue "B" [tools & instruments]. London: Joseph Long, Eastcheap, E.C., 1914. Tall 8vo. 80 pp. Num ills. Staining & fingermarking, well creased crnrs at bottom.   *(Harrison)* **£19 [≈ $30]**
- Macfarlane, Walter, & Co: Illustrated Catalogue of Macfarlane's Castings. Glasgow: [1890s]. 6th edn. 2 vols. 696 pp. Num ills. Orig bndg.   *(Book House)* **£80 [≈ $128]**
- Mather & Platt: Catalogue of Mather & Platt, Salford Ironworks, Makers of Textile Machinery of all Kinds. Salford: 1910. 4to. 243 pp. Text in French, German & Russian only. Ills. Orig cloth gilt, spine faded.
(Harrison) **£29 [≈ $46]**
- Morris, Herbert, Ltd: Modern Conveyors (Book 187). Catalogue of Roller, Belt, Chain, Overhead and other Conveyors and Elevators. Loughborough: 1931. 4to. 437 pp. Ills. Orig bndg.   *(Book House)* **£15 [≈ $24]**
- Morris, Herbert, Ltd: Modern Lifting (Book 50). Catalogue of Cranes, Conveyors, Pulley Blocks, etc. Loughborough: 1910. Lge 4to. 472 pp. Ills. Orig bndg, cvrs poor, shaken.
(Book House) **£20 [≈ $32]**
- Morris, Herbert, Ltd: Modern Lifting (Book 56). Loughborough: 1933. Sm 4to. 1124 pp. Ills. Orig bndg.   *(Book House)* **£18 [≈ $28]**
- Newman, Hender & Co: Catalogue 59 of Brass, Gun-Metal, Cast-Iron and Steel Valves, Cocks, Fittings, etc. Woodchester, Glos.: 1939. Sm 4to. 392 pp. Ills. Orig bndg.
(Book House) **£16 [≈ $25]**
- Rowe Bros & Co Ltd: Baths, Sinks, Lavatories, Closets & General Plumbing Fittings. Liverpool: [ca 1910]. 4to. 215 pp. Num ills. Orig blue cloth.
(Harrison) **£16 [≈ $25]**
- Rowe Bros & Co: Catalogue of Sanitary Ware, Fireplaces, Cookers, Tiles, & similar Builders Merchandise. Bristol: [ca 1930]. 4to. 604 pp. Ills. Orig bndg, cvrs washed.
(Book House) **£20 [≈ $32]**
- Siebe, Gorman & Co: Manual for Divers, and

Illustrated Catalogue. Siebe, Gorman & Co.'s Description of Diving Apparatus ... London: [ca 1886-90]. Cr 4to. iv,[iv],60 pp. Tinted litho frontis, 23 litho plates. Soiled & used. Orig cloth backed card wraps.
(Duck) **£275 [≈ $440]**

- Simon, Henry, Ltd: Modern Flour Mill Machinery. Manchester: 1898. 4to. 133 pp. Fldg map, ills. Orig bndg.
(Book House) **£60 [≈ $96]**

- Slingsby, H.C.: Catalogue of Trucks and Ladders. [Ca 1950s]. 280 pp. Num ills. Orig silver lettered cloth. *(Harrison)* **£18 [≈ $28]**

- Stone, J., & Co: Catalogue A1: Valves and Fittings for Water Supply. Hydrants, Boxes, Water-Cranes, etc. Deptford: 1903. 4to. 129 pp. Ills. Well used. Orig bndg.
(Book House) **£20 [≈ $32]**

- Stone, J., & Co: Catalogue C: Ship Work, Parts 1-5. Deptford: 1903. 4to. 198 pp. Ills. Well used. Orig bndg, shaken, cvrs marked.
(Book House) **£30 [≈ $48]**

- F.H. Thomas Co: Illustrated Catalogue. Boston: n.d. 2nd edn. 576,xvi pp. Num ills. Orig cloth. *(Elgen)* **$100 [≈ £62]**

- Veritys Ltd, Birmingham: Priced Catalogue of Electrical Plant and Sundries. Birmingham, [ca 1909). 2 vols. 373; 386 pp. Ills. Orig bndgs, back cvrs vol 2 a little marked. *(Book House)* **£45 [≈ $72]**

- Watson, & Sons Ltd, W.: Microscopes & Accessories for Special Purposes. London: W. Watson & Sons Ltd, High Holborn, [ca 1910]. 40 pp. Num ills. Prices advanced by overprint. Some creasing. Sm piece gone from blank edge of title. Silver lettered front cvr.
(Harrison) **£36 [≈ $57]**

- Woolley, James, Sons & Co: Catalogue of Surgical Instruments and Appliances. Manchester: 1931. Num ills.
(David White) **£22 [≈ $35]**

- Wyleys Ltd: Pharmaceutical Specialities. A Priced Catalogue. Coventry: 1899. 185 pp. Ills. Card cvrs. *(Harrison)* **£32 [≈ $51]**

Trail, W.

- Account of the Life and Writings of Robert Simson, M.D. Late Professor of Mathematics in the University of Glasgow. London: 1812. vii, 192 pp. 1 fldg plate. Sm repr to dedic, sm stain on pp 21-33. New cloth.
(Whitehart) **£135 [≈ $216]**

Traill, C.P.

- Canadian Wild Flowers Painted & Lithographed by Agnes Fitzgibbon with Botanical Descriptions by C.P. Traill. Fourth Edition. Toronto: William Briggs, 1895. Ltd

edn, sgnd by the artist. Folio. 88 pp. Litho frontis, 10 hand cold plates. Orig dec cloth gilt. *(Lamb)* **£150 [≈ $240]**

- Studies of Plant Life in Canada. Toronto: 1906. New & rvsd edn. 8vo. xvii,227 pp. Port, 8 cold & 12 plain plates. Orig cloth.
(Wheldon & Wesley) **£35 [≈ $56]**

Trall, Russell Thacher

- The Hydropathic Encyclopedia: A System of Hydropathy and Hygiene ... New York: Fowler & Wells, 1853. 2 vols. 12mo. Num ills. Prelims foxed. Orig cloth, extremities rubbed, 2 sgntrs sprung.
(Heritage) **$150 [≈ £93]**

- Sexual Physiology; a Scientific and Popular Exposition of the Fundamental Problems in Sociology. New York: 1866. xiv, 312 pp. Diags. Orig cloth.
(Box of Delights) **£20 [≈ $32]**

Treatise ...

- A Treatise of Diseases of the Head, Brain, and Nerves ... By a Physician. The Fourth Edition, Corrected. London: 1721. 8vo. [vi], 74 pp. Crnrs sl creased, title sl dusty. Orig mrbld wraps. *(Bickersteth)* **£350 [≈ $560]**

- A Treatise on the Progressive Improvement and Present State of the Manufactures in Metal ... see Holland, John.

- A Treatise on the Steam Engine ... By the Artizan Club ... see Bourne, John.

Tredgold, Thomas

- The Principles of Warming and Ventilating Public Buildings, Dwelling Houses, Manufactories, Hospitals ... Third Edition ... Appendix ... London: M. Taylor, 1836. 8vo. xvi,347 pp. 12 plates, text figs. Orig cloth, faded, stained, lacks paper label.
(Bickersteth) **£150 [≈ $240]**

Trimmer, Kirby

- Flora of Norfolk. London: 1866. Cr 8vo. xxxvi,[4],195 pp. Orig cloth, trifle used.
(Wheldon & Wesley) **£45 [≈ $72]**

- Flora of Norfolk: A Catalogue of Plants ... London: Hamilton Adams, 1866. 1st edn. 12mo. xxxvi,[iv],195 pp. Orig cloth, red edges. *(Lamb)* **£40 [≈ $64]**

Trimmer, Mary (pseud.?)

- A Natural History of the Most Remarkable Quadrupeds, Birds, Fishes, Serpents, Reptiles and Insects. Chiswick: Whittingham, 1825. Sm 8vo. vi,224; viii,256 pp. Num sm w'cuts. Contemp half calf, mor labels, sl rubbed. *(Lamb)* **£65 [≈ $104]**

Tripp, F.E.
- British Mosses, their Homes, Aspects, Structure and Uses. London: 1874. 2 vols. Roy 8vo. 39 cold plates. Orig cloth.
(Henly) **£120 [≃ $192]**
- British Mosses ... London: 1874. 2 vols. Roy 8vo. 2 dec titles, 39 hand cold plates. Orig cloth.   *(Wheldon & Wesley)* **£100 [≃ $160]**
- British Mosses ... London: 1888. New edn. 2 vols. Text & plates bound separately, with a typed leaf of description to each plate. Roy 8vo. 37 cold plates. Half cloth.
(Wheldon & Wesley) **£65 [≃ $104]**

Triquet, R.
- The Planters Manual ... see Cotton, Charles.

Trismosin, Solomon
- Splendor Solis. Alchemical Treatises ... Introduction ... Notes by J.K. London: Kegan Paul, [1920]. 104 pp. Plates. Orig cloth, faded, spine ends chipped.
(Wreden) **$50 [≃ £31]**

Tristram, H.B.
- The Survey of Western Palestine: The Fauna and Flora of Palestine. London: Palestine Exploration Fund, 1884. 1st edn. Lge 4to. 455 pp. 13 litho plates (only, of 20). Perf lib stamp on title, plates stamped on reverse. Orig cloth, hd of spine worn, inner hinge broken.   *(Terramedia)* **$250 [≃ £156]**

Trotter, Thomas
- Medicina Nautica: an Essay on the Diseases of Seamen ... Vols 1 and 2. London: 1797-99. 1st edns of vols 1 & 2. 8vo. viii,487; x,475, 4 advt pp. Sm lib stamps. Contemp calf, rubbed, jnts splitting. A 3rd volume was published in 1803. *(Hemlock)* **$875 [≃ £546]**

A True and Particular Relation ...
- A True and Particular Relation of the Dreadful Earthquake which happened at Lima ... see Lozano, Pedro.

True, F.W.
- The Whalebone Whales of the Western North Atlantic. Washington: Smithsonian Contribs to Knowledge Vol 33, 1904. 4to. 332 pp. 40 plates. Orig cloth.
(Wheldon & Wesley) **£75 [≃ $120]**

Tryon, Thomas
- A New Art of Brewing Beer, Ale, and other sorts of Liquors ... added, The Art of Making Mault ... London: Tho. Salusbury, 1691. 3rd edn, enlgd. 12mo. [6],138 pp. 1st 3 & last ff

chipped & soiled without loss of text. Old calf, rebacked. Wing T.3189.
(Heritage) **$550 [≃ £343]**

Tulasne, L.R. & C.
- Selecta Fungorum Carpologia. Edited by A.H.R. Buller and C.L. Shear. Oxford: 1931. 3 vols. 4to. 886 pp. 61 plates. Orig cloth, edges v sl worn. Signed by the translator.
(Wheldon & Wesley) **£100 [≃ $160]**

Tull, Jethro
- The Horse-Hoing Husbandry ... London: for the author, 1733. 1st edn. 6 plates. 2 sm rubber stamps & a perf lib mark on title. [Bound with] A Supplement on the Essay on Horse-Hoing Husbandry ... London: the author, 1736. 1st edn. Plate. Some foxing. Old style half calf.
(Blackwell's) **£950 [≃ $1,520]**
- The Horse-Hoeing Husbandry ... Dublin: A. Rhames, 1733. 1st Dublin edn. 8vo. xvii,417, [4] pp. 6 fldg plates. Sl foxing at ends. New qtr calf over orig pink mrbld bds, red leather label.   *(Blackwell's)* **£250 [≃ $400]**
- The Horse-Hoeing Husbandry: or, an Essay on the Principles of Tillage and Vegetation. Dublin: 1733. 2nd edn. 8vo. xvii,417,[4] pp. 6 plates. Contemp calf.
(Wheldon & Wesley) **£250 [≃ $400]**
- Horse-Hoeing Husbandry ... London: for A. Millar, 1762. 4th edn. 8vo. xvi,432 pp. 7 fldg engvd plates. New half calf.
(Egglishaw) **£240 [≃ $384]**

Tunnicliffe, C.F.
- Shorelands Summer Diary. London: 1952. 4to. 160 pp. 16 cold plates, 180 ills. Endpapers sl foxed. Orig cloth. Torn dw.
(Wheldon & Wesley) **£50 [≃ $80]**

Turner, Charles C.
- Aerial Navigation of To-Day. A Popular Account of the Evolution of Aeronautics. London: Seeley, 1910. 1st edn. 328,[8,16 advt] pp. Addendum slip at p xiii. 24 plates, 41 text ills. Orig cloth, uncut, sl soiled & marked.   *(Duck)* **£25 [≃ $40]**

Turner, Daniel, 1667-1741
- De Morbis Cutaneis. A Treatise of Diseases incident to the Skin. In Two Parts ... Third Edition, Revised and very much Enlarged. London: 1726. 8vo. [xvi],534 pp. Port. Contemp calf, showing age.
(Hemlock) **$375 [≃ £234]**
- De Morbis Cutaneis. A Treatise of Diseases incident to the Skin. In Two Parts ... Third

Edition, Revised and very much Enlarged. London: 1726. 8vo. [xvi],524 pp. Port. Orig calf, v worn, front jnt split, rear jnt rubbed. Cloth case. *(Schoyer)* **$150 [≈ £93]**

Turner, Nicholas
- An Essay on Draining and Improving Peat Bogs ... London: ptd for R. Baldwin, & J. Bew; sold by W. Wilson, Dublin, 1784. 1st edn. 8vo. [ii],x,86,[6 errata & blanks] pp. Half-title. Rec cloth.
(Burmester) **£110 [≈ $176]**

Turner, Robert
- Botanologia. The British Physician: or The Nature and Vertues of English Plants ... London: Obadiah Blagrave, 1687. 8vo. [2],363, [24],8 ctlg pp. Lacks frontis. Browning & sl staining in margs. 2 ff defective. Rec leather.
(Hemlock) **$400 [≈ £250]**

Turner, Sharon
- The Sacred History of the World as displayed in the Creation and Subsequent Events to the Deluge. London: 1834. 5th edn. 8vo. xvi,569 pp. New buckram. *(Baldwin)* **£45 [≈ $72]**

Turner, W.
- Turner on Birds. Cambridge: UP, 1903. 1st edn thus. 8vo. xviii,223 pp. Orig green cloth.
(Gough) **£30 [≈ $48]**

Turner, William
- Sound Anatomized, in a Philosophical Essay on Musick. Wherein is explained the Nature of Sound ... London: William Pearson ..., 1724. 4to. 79,7 pp. 1 fldg plate, num music examples. Mrbld wraps, edges dusty.
(W. Thomas Taylor) **$450 [≈ £281]**

Turnor, Edmund
- Collections for the History of the Town and Soke of Grantham. Containing Authentic Memoirs of Sir Isaac Newton ... London: 1806. 1st edn. Lge 4to. xvi,200 pp. Map, port, 1 plate, 2 text engvs. Without the 8 extra plates sometimes found. Half calf, spine chipped at ft. *(Burmester)* **£150 [≈ $240]**

Turton, William
- Conchylia Insularum Britannicarum, the Shells of the British Islands, systematically arranged. Exeter: 1822. 1st edn. xlviii,279 pp. 20 hand cold plates. Lib blind stamp on title causing localized browning a few pp further on. *(Baldwin)* **£260 [≈ $416]**
- Conchylia Insularum Britannicarum: the Shells of the British Islands, systematically arranged. Exeter: 1822. 1st edn. 4to. xlvii,

279 pp. 20 hand cold plates. Sl spotting of 1st few ff. Sm blind stamps on plates. New half calf. *(Wheldon & Wesley)* **£275 [≈ $440]**
- A Manual of the Land and Fresh-Water Shells of the British Islands. London: 1831. Cr 8vo. viii,152,16 pp. 10 hand cold plates. Orig cloth, sound ex-lib.
(Wheldon & Wesley) **£35 [≈ $56]**
- A Manual of the Land and Fresh-Water Shells of the British Islands. London: 1840. 8vo. ix,[i],324 pp. 12 cold plates. Orig cloth.
(Baldwin) **£45 [≈ $72]**
- A Manual of the Land and Fresh-Water Shells of the British Islands .. New Edition, revised and enlarged by J.E. Gray. London: Longman, 1840. Cr 8vo. ix,(i),324,[16 advt] pp. 12 hand cold plates. Orig cloth, spine sl torn. *(Egglishaw)* **£45 [≈ $72]**
- A General System of Nature ... see also Linnaeus, C.

Tusser, Thomas
- Five Hundred Points of Good Husbandry ... New Edition with Notes ... by William Mavor. London: Lackington, Allen ..., 1812. Large Paper. 2 parts in one vol. 4to. Ptd in red & black throughout. Contemp calf, rebacked & crnrd in roan.
(Hannas) **£120 [≈ $192]**
- Five Hundred Points of Good Husbandry ... New Edition, with Notes ... by William Mavor. London: Lackington, Allen ..., 1812. Large Paper (9 1/4 x 7 1/2 ins.). 4to. [2],36,xl, 338 pp. Addtnl title. Later 19th c half mor, t.e.g., rubbed & sl worn. Editor's pres copy.
(Claude Cox) **£110 [≈ $176]**
- Five Hundred Points of Good Husbandry. With an Introduction by Sir Walter Scott and Benediction by Rudyard Kipling incorporated in a Foreword by E.V. Lucas. London: 1931. One of 500. Cr 4to. xiv,336 pp. Hermitage calf.
(Wheldon & Wesley) **£80 [≈ $128]**
- Some of the Five Hundred Points of Good Husbandry ... now newly corrected and edited ... by H.M.W. Oxford: 1848. Sm 8vo. 116,45 pp. Orig cloth.
(Wheldon & Wesley) **£25 [≈ $40]**
- Tusser's Husbandry reprinted verbatim from the Original Edition of 1557. To which is added The Life of Thomas Tusser and a Glossary. Great Totham: 1834. One of 100. Sm 4to. 31,[1] pp. Repr to title. Orig wraps, back wrapper defective.
(Wheldon & Wesley) **£40 [≈ $64]**

Tutt, J.W.
- British Moths. London: Routledge, 1896. 1st

edn. Sm 8vo. xii,368 pp. 12 cold plates, num text figs. Orig cloth. *(Egglishaw)* **£15 [≃ $24]**

Twamley, Louisa Anne
- See Meredith, Louisa Anne, nee Twamley.

Tweeddale, Arthur, Marquis of
- Ornithological Works. Reprinted from the Originals by desire of his Widow. Edited and Revised by R.C.W. Ramsey. London: privately ptd, 1881. 4to. lxiv,760 pp. Orig cloth.   *(Wheldon & Wesley)* **£75 [≃ $120]**

Twenhoffel, W.H.
- Treatise on Sedimentation. Baltimore: 1926. 1st edn. 8vo. Plates, text figs. Orig cloth.
(Henly) **£32 [≃ $51]**

Tyas, T.
- Favourite Field Flowers. Series I and II. London: 1848-50. 2 vols. Sm 8vo. 24 hand cold plates by James Andrews. Orig cloth, vol 1 trifle faded, vol 2 rather used.
(Wheldon & Wesley) **£65 [≃ $104]**

Tyndall, John
- Essays on the Use and Limit of the Imagination in Science. London: 1871. 2nd edn. 72,24 ctlg pp. Orig purple cloth, spine sunned & sl frayed at ends.
(Elgen) **$85 [≃ £53]**
- The Forms of Water in Clouds, Rivers, Ice, Glaciers. London: 1872. 1st edn. Orig cloth, somewhat marked.
(Wheldon & Wesley) **£20 [≃ $32]**
- Fragments of Science for Unscientific People ... New York: Appleton, 1871. 1st edn (the Amer edn is the correct 1st edn). 8vo. [ii],422,[12] pp. Orig dec cloth.
(Gach) **$85 [≃ £53]**
- Sound: A Course of Eight Lectures Delivered at the Royal Institution of Great Britain. London: Longmans, Green, 1869. 2nd edn. Post 8vo. xvi,341 pp. Frontis, 169 ills. Blind stamp on title. Contemp half mor, crnrs & edges of cvrs sl rubbed, endpapers spotty.
(Duck) **£35 [≃ $56]**

Tyrrell, Frederick
- A Practical Work on the Diseases of the Eye, and Their Treatment, Medically, Topically, and by Operation. London: Churchill, 1840. Sole edn. 2 vols. 8vo. lviii,533; xii,566 pp. 8 cold & 1 uncold plates. Orig half calf, a little rubbed.   *(Bickersteth)* **£380 [≃ $608]**

Tyson, E.
- Orang-outang, sive Homo Sylvestris, or the

Anatomy of a Pygmie, compared with that of a Monkey, an Ape and a Man. London: 1966. Facs of the 1699 edn. 4to. 8 fldg plates. Half mor.   *(Wheldon & Wesley)* **£85 [≃ $136]**

Ukers, William H.
- All About Coffee. New York: The Tea and Coffee Trade Journal Co, 1922. 1st edn. 4to. xxx,796 pp. Fldg map, 17 cold plates, num ills. Orig brown cloth gilt, sl rubbed. Sgnd by Ukers on endpaper.
(Karmiole) **$200 [≃ £125]**

Underhill, Frank P.
- The Physiology of the Amino Acids. New Haven: 1915. 1st edn. Sm 8vo. 169 pp. Frontis, plates, ills. Orig bndg.
(Elgen) **$50 [≃ £31]**

Underwood, E. Ashworth (ed.)
- Science, Medicine and History. Essays on the Evolution of Scientific Thought and Medical Practice written in honour of Charles Singer. Collected and Edited by E. Ashworth Underwood. OUP: 1953. 2 vols. Lge 8vo. 106 plates. Orig buckram gilt. Dws.
(Hollett) **£145 [≃ $232]**

Underwood, Michael
- A Treatise on the Diseases of Children, and Management of Infants from the Birth ... 2nd American, from the 6th London Edition. Boston: David West, 1806. 3 vols in one. 8vo. xx,476 pp. Contemp calf, hinges rubbed.
(Hemlock) **$225 [≃ £140]**

Upham, Thomas C.
- Elements of Mental Philosophy Embracing the Two Departments of the Intellect and the Sensibilities. Portland, ME: Shirley & Hyde, 1828. 2nd edn. 8vo. [ii],576,[2] pp. Contemp calf, lightly shelfworn. *(Gach)* **$125 [≃ £78]**
- Mental Philosophy Embracing the Three Departments of the Intellect, Sensibilities, and Will. New York: 1869. Rvsd edn. 12mo. [ii], 561,[5]; [ii],705,[5] pp. Orig cloth.
(Gach) **$50 [≃ £31]**
- Outlines of Disordered and Imperfect Mental Action. New York: Harper, 1840. 1st edn. 16mo. xvi,[2],[17]-[400] pp. Lightly foxed. Leather backed mrbld bds.
(Gach) **$150 [≃ £93]**
- A Philosophical and Practical Treatise on the Will. Portland, ME: published by William H. Hyde, for Z. Hyde, 1834. 1st edn. 8vo. 400 pp. Sl foxed. Orig cloth, jnts worn, remains of tape markings on spine. *(Gach)* **$150 [≃ £93]**

Ure, Andrew
- A Dictionary of Arts, Manufactures, and Mines ... Reprinted from the Fourth English Edition, corrected and greatly enlarged. New York: Appleton, 1853. 2 vols. Thick 8vo. xiv, 1118; [iii],998 pp. 1588 w'engvd ills. Orig cloth.   *(Charles B. Wood)* **$180 [≃£112]**

Vancouver, Charles
- General View of the Agriculture of the County of Devon ... London: for Richard Phillips ..., 1808. xii,479,[2 advt],[2 ctlg of seeds],[errata] pp. Hand cold fldg map, 28 plates. Old polished calf gilt, edges & spine rubbed, hinges cracked, upper bd loose.
(Hollett) **£140 [≃$224]**

Veblen, O. & Young, J.W.
- Projective Geometry. Boston: 1910-18. 2 vols. x,342; xii,511 pp. Orig cloth, sl marked, bottom page edges sl grubby.
(Whitehart) **£25 [≃$40]**

Veitch, James Herbert
- Hortus Veitchii. A History of the Rise and Progress of the Nurseries of Messrs. James Veitch and Sons ... London: James Veitch & Sons, for private circulation, 1906. 4to. 542 pp. 50 plates. Orig cloth, jnts sl rubbed.
(Spelman) **£95 [≃$152]**
- A Manual of the Coniferae ... London: 1881. Roy 8vo. 342 pp. W'cut plates & text ills. Orig cloth gilt, sl worn. Publisher's pres copy.
(Egglishaw) **£40 [≃$64]**
- A Manual of the Coniferae ... see also Kent, A.H.; Wooster, D.
- A Traveller's Notes or Notes of a Tour through India, Malaysia, Japan, Corea, the Australian Colonies and New Zealand during the Years 1891-1893. Chelsea: Royal Exotic Nursery, privately ptd, 1896. 219 pp. Fldg map, plates, ills. Orig dec blue cloth.
(Lyon) **£195 [≃$312]**

Venable, F.P.
- The Development of the Periodic Law. Easton: 1896. 1st edn. 12mo. viii,321,2 ctlg pp. 2 fldg plates, ills. Orig cloth, worn.
(Elgen) **$150 [≃£93]**

Venables, L.S.V. & U.M.
- Birds and Mammals of Shetland. London: 1955. 8vo. xii,391 pp. 3 maps, 8 plates. Orig cloth.   *(Wheldon & Wesley)* **£48 [≃$76]**

Venette, Nicolas
- The Art of Pruning Fruit-Trees, with an Explanation of some Words which Gardiners make us of in speaking of Trees ... London: for Tho. Basset, 1685. 1st edn in English. 8vo. 7 w'cut text ills. Contemp calf, some rubbing, spine a bit worn. Wing V.187.
(Ximenes) **$850 [≃£531]**

Venn, John
- Symbolic Logic ... Second Edition, Revised and Rewritten. London: Macmillan, 1894. 2nd, rvsd, edn. 8vo. [ii],xxxviii,540 pp. Orig red cloth.   *(Pickering)* **$500 [≃£312]**

Vernon-Harcourt, L.F.
- Civil Engineering as applied in Construction. London: Longmans, 1902. 624 pp. Ills. Orig bndg, cvrs a bit rubbed.
(Book House) **£25 [≃$40]**

Verrall, G.H.
- British Flies, Vol 5. Stratiomyidae and succeeding Families of the Diptera Brachycera of Great Britain. London: 1909. Roy 8vo. [viii], 780,34 pp. Port, 407 text figs. Half mor. *(Wheldon & Wesley)* **£70 [≃$112]**
- British Flies, Vol 8. Platypezidae, Pipunculidae and Syrphidae. London: 1901. Roy 8vo. 691,121 pp. Port, 458 text figs. Minor foxing. Half mor. Port of the author pasted in at front.
(Wheldon & Wesley) **£70 [≃$112]**

Vesey-Fitzgerald, Brian & La Monte, F. (eds.)
- Game Fish of the World. London: [1949]. Roy 8vo. xvii,446 pp. 81 cold plates. Orig cloth, faded, trifle warped.
(Wheldon & Wesley) **£25 [≃$40]**

Vestiges of the Natural History of Creation ...
- See Chambers, Robert.

Vieck, J.H.v.
- The Theory of Electric and Magnetic Susceptibilities. Oxford: 1932. 1st edn. xii, 384 pp. 16 text diags. Orig cloth, dull. Sgnd by the author.   *(Whitehart)* **£35 [≃$56]**

Vince, Samuel
- The Elements of the Conic Sections, adapted to the Use of Students in Philosophy. Cambridge: ptd by J. Burges ..., 1800. 2nd edn, enlgd. 8vo. [ii],60 pp. 3 fldg plates. Disbound.   *(Burmester)* **£35 [≃$56]**
- A Treatise on Practical Astronomy. Cambridge: J. Archdeacon, 1790. 4to. [vi],204 pp. 8 fldg plates. Period half green roan gilt, mrbld sides, spine faded, spine &

crnrs rubbed. Marquis Townshend's b'plate.
(Rankin) **£165 [≃ $264]**

- A Treatise on Plain and Spherical Trigonometry ... Adapted to the Use of Students in Philosophy. Cambridge: at the University Press ..., 1805. 2nd edn, crrctd. 8vo. [ii],148 pp. 2 fldg plates. Disbound.
(Burmester) **£25 [≃ $40]**

Virgil
- Bucolica - The Eclogues ... with an English Translation and Notes by John Martyn. New Edition [by Richard Duppa]. London: 1813. Lge 8vo. 270,[x] pp. 37 cold engvs by Sowerby & others. Contemp green half mor gilt, a.e.g., spine faded & sl rubbed.
(Blackwell's) **£200 [≃ $320]**

Vivian, E. Charles
- A History of Aeronautics ... London: [1921]. Sole edn. 8vo. x,521 pp. 56 photo plates, text figs. Orig cloth.   *(Bickersteth)* **£40 [≃ $64]**

Vogel, Hermann
- The Chemistry of Light and Photography in their Application to Art, Science, and Industry. London: Kegan Paul, International Scientific Library, 1883. 4th edn. 8vo. viii, 280,47 ctlg pp. 6 plates inc 3 mtd photos, 100 text ills. Orig cloth, stained & rubbed.
(de Beaumont) **£55 [≃ $88]**

Voltaire, F.M.A. de
- The Elements of Sir Isaac Newton's Philosophy ... Translated from the French. Revised and Corrected by John Hanna ... London: for Stephen Austen ..., 1738. 1st edn in English. 8vo. xvi,363,3 advt pp. 10 fldg plates. Title dust soiled. Mod half calf.
(Pickering) **$1,200 [≃ £750]**

Von Hagen, Victor Wolfgang
- The Aztec and Maya Papermakers. With an Introduction by Dard Hunter. New York: J.J. Augustin, [1943]. One of 220. Folio. 116 pp. 32 plates, inc 7 orig paper samples & 3 fldg plates (2 maps). Orig beige cloth, leather label.   *(Karmiole)* **$450 [≃ £281]**

Von Neumann, John & Morgenstern, Oscar
- Theory of Games and Economic Behaviour ... Princeton: Univ Press, 1944. 1st edn. 8vo. xviii,625,[1] pp. Corrigenda laid in. Orig cloth.   *(Rootenberg)* **$400 [≃ £250]**

Voronoff, Serge
- Rejuvenation by Grafting. Translation edited by Fred F. Imianitoff. New York: (1925). 1st

edn in English. 224 pp. Ports, plates. Orig bndg.   *(Elgen)* **$75 [≃ £46]**

- Sources of Life. Boston: Bruce Humphrie, 1943. 1st edn. Sm 4to. 240 pp. Photo ills. Orig bndg, unmarked ex-lib.
(Xerxes) **$90 [≃ £56]**

Vries, H. de
- The Mutation Theory. Experiments and Observations on the Origin of Species in the Vegetable Kingdom. London: 1910-11. 2 vols. Roy 8vo. 10 cold plates, ills. Orig cloth, trifle used, good ex-lib.
(Wheldon & Wesley) **£120 [≃ $192]**

Vyner, Robert T.
- Notitia Venatica: a Treatise on Fox- Hunting, embracing the General Management of Hounds. A New Edition ... by William C. Blew. London: Nimmo, 1892. Roy 8vo. xxii,406 pp. 12 hand cold ills by Alken & others. Paper v sl discold. Mod three qtr gilt dec mor, t.e.g.   *(Blackwell's)* **£120 [≃ $192]**

W., J., Gent.
- Systema Agriculturae ... see Worlidge, John.

Wadd, William
- Comments on Corpulency, Lineaments of Leanness. Mems. on Diet and Dietetics. London: J. Ebers, 1829. Enlgd edn. Sm 8vo. 170 pp. Frontis, 5 plates. Half leather, mrbld bds sl rubbed. Sgnd pres copy.
(Elgen) **$150 [≃ £93]**

Wade, F.B.
- Diamonds. A Study of the Factors that Govern their Value. New York & London: Putnam's, 1916. 8vo. ix,150 pp. Orig cloth.
(Gemmary) **$27.50 [≃ £17]**

Wade, Walter
- Plantae Rariores in Hibernia Inventae, or Habitats of some Plants rather Scarce and Valuable found in Ireland, with Concise Remarks of the Properties and Uses of many of them. Dublin: Graisberry, 1804. Cold plates. Half calf, rebacked.
(Emerald Isle) **£85 [≃ $136]**

Wagner, M.
- The Darwinian Theory and the Law of the Migration of Organisms. Translated by J.L. Laird. London: 1873. 8vo. 79 pp. Mod cloth, orig wraps bound in.
(Wheldon & Wesley) **£20 [≃ $32]**

Wainewright, Jeremiah
- An Anatomical Treatise of the Liver, with the Diseases incident to it. By a Member of the College of Physicians. London: Lacy & Clarke, 1722. Sole edn. 8vo. Title, 100 pp, advt leaf. Title sl soiled. Sewn as issued.
(Bickersteth) **£250 [≈ $400]**

Wait, W.E.
- Manual of the Birds of Ceylon. Colombo: 1931. 2nd edn. Roy 8vo. xxxiii,494 pp. Map. Orig cloth, trifle worn.
(Wheldon & Wesley) **£30 [≈ $48]**

Waite, A.E.
- The Secret Tradition in Alchemy. London: Kegan Paul, 1926. 1st edn. 8vo. xxii,415 pp. Some foxing. Orig cloth gilt, sl dampstained.
(Minster Gate) **£68 [≈ $108]**

Wakefield, Priscilla
- An Introduction to Botany in a Series of Familiar Letters... Second Edition. London: E. Newberry, Darton & Harvey, Vernor & Hood, 1798. 12mo. 200 pp. 11 hand cold plates (1 fldg with fldg key). Mod half roan.
(Claude Cox) **£60 [≈ $96]**
- An Introduction to Botany, in a Series of Familiar Letters. Dirst American Edition from the Fifth London Edition. Boston: 1811. 8vo. xii, 216 pp. 2 tables, 12 plates. Some marg staining at ends. Contemp calf, crudely rebacked, lib b'plates.
(Wheldon & Wesley) **£20 [≈ $32]**
- An Introduction to the Natural History and Classification of Insects, in a Series of Familiar Letters. London: Darton, 1816. Cr 8vo. x,192 pp. 12 hand cold plates. Sl spotting. Contemp calf gilt.
(Egglishaw) **£54 [≈ $86]**

Wakeman, Geoffrey
- Victorian Book Illustration. The Technical Revolution. Newton Abbot: David & Charles, 1973. 1st edn. 8vo. 182 pp. Num ills. Orig buckram. Dw.   *(de Beaumont)* **£48 [≈ $76]**

Waksmundzaka, M.
- Pneumatic Therapy (Balancing the Vasomotor System) (Junod's Haemospasia) (Bier's Hyperemia). New York: Benedict Lust, 1936. 2nd edn. 8vo. 112 pp. Photo ills. Orig bndg. Sgnd pres copy.
(Xerxes) **$85 [≈ £53]**

Walcott, M.V.
- North American Wild Flowers. Washington: Smithsonian Institution, 1925. 5 vols. Cr folio. 400 cold plates. Orig cloth portfolios.
(Wheldon & Wesley) **£300 [≈ $480]**

Walker, Mrs A.
- Female Beauty, as preserved and improved by Regimen, Cleanliness and Dress ... London: Thomas Hurst, 1837. 1st edn. 8vo. xxxvi,432 pp. 10 hand cold costume plates, each with hand cold overlay, 1 b/w plate. Contemp green mor, gilt dentelles, a.e.g., jnt sl rubbed.
(Spelman) **£550 [≈ $880]**

Walker, Adam
- A System of Familiar Philosophy: in Twelve Lectures ... Chemical Properties of Matter ... Mechanics ... Electricity ... London: for the author ..., 1799. 1st edn. 4to. xviii,571 pp. 48 fldg plates on 47 ff. Orig half calf, sl rubbed, hd of spine sl worn.
(Bickersteth) **£185 [≈ $296]**
- A System of Familiar Philosophy: in Twelve Lectures ... Chemical Properties of Matter ... Mechanics ... Electricity ... London: for the author ..., 1802. 2nd edn. 2 vols. 4to. xvi, 354; iv,251,[25] pp. 49 fldg plates. Occas spotting. Contemp half calf, rebacked
(Young's) **£220 [≈ $352]**

Walker, F.
- List of the Specimens of Dipterous Insects in the British Museum. London: 1848-55. 7 parts in 3 vols. 12mo. 1172,775 pp. New cloth.   *(Egglishaw)* **£125 [≈ $200]**
- List of the Specimens of Homopterous Insects in the British Museum. London: 1850-58. 4 parts & Supplement, bound in 3 vols. 12mo. 1188,369 pp. 8 litho plates. Lib stamp on titles. Half calf, jnts & crnrs rubbed.
(Egglishaw) **£90 [≈ $144]**

Walker, Frederick
- Aerial Navigation. A Practical Handbook on the Construction of Dirigible Balloons, Aerostats, Aeroplanes, and Aeromoters. London: Crosby Lockwood, 1902. 1st edn. Post 8vo. xvi,152,advt pp. Frontis & 104 ills. Orig pict dec cloth, sl rubbed & marked.
(Duck) **£75 [≈ $120]**

Walker, John
- The Oculist's Vade-Mecum. A Complete Practical System of Ophthalmic Surgery. London: 1857. 2nd edn. 405 pp.
(Rittenhouse) **$60 [≈ £37]**

Walker, W.B.
- Cyclical Deluges: An Explanation of the Chief Geological Phenomena of the Globe by Proofs of the Periodical Changes of the

Earth's Axis. London: 1871. 8vo. 142 pp.
Frontis. Orig cloth. Author's inscrptn.
(Henly) **£15 [≃ $24]**

Walkingame, Francis
- The Tutor's Assistant. Being a Compendium
of Arithmetic, and a Complete Question
Book. The Twentieth Edition. London: J.
Scatcherd, 1784. 8vo. 12,180 pp. MS list of
works inside front bd. Contemp calf, jnts
cracked but attached, lacks front endpaper.
(Spelman) **£18 [≃ $28]**
- The Tutor's Assistant; being a Compendium
of Arithmetic ... New Edition, corrected, and
every question worked aneew by T. Crosby,
Mathematician. York: T. Wilson, 1835. 8vo.
199, [1] pp. Fldg plates of tables. Mozley's 4
pp ctlg at front. Orig sheep.
(Claude Cox) **£15 [≃ $24]**

Walkington, Thomas
- The Optick Glasse of Humours or the
Touchstone of a Golden Temperature: Or the
Philosophers Stone to make a golden Temper
... London: 1664. Sm 8vo. [22],168 pp. 1
plate. Lacks title & frontis. Contemp calf, dull
& worn. STC 24967.
(Hemlock) **$200 [≃ £125]**

Wall, E.J.
- The History of Three Color Photography.
Boston: Amer Photo Publ Co, 1925. 1st edn.
8vo. [x],747 pp. 203 text ills. Orig cloth.
(Charles B. Wood) **$125 [≃ £78]**

Wall, John
- Experiments and Observations on the
Malvern Waters. The Second Edition, with
an Appendix ... London: sold by W. Sandby
... & S. Mountfort & R. Lewis in Worcester,
[1757]. 8vo. 77 pp. Disbound.
(Bickersteth) **£180 [≃ $288]**

Wallace, Alfred Russell
- Contributions to the Theory of Natural
Selection. London: Macmillan, 1870. 1st edn.
8vo. xvi,384 pp. Title soiled with lib stamp
recto & verso. Rec bds, new endpapers.
(Rootenberg) **$250 [≃ £156]**
- Darwinism, an Exposition of the Theory of
Natural Selection, with some of its
Applications ... London: Macmillan, 1889.
1st edn. 8vo. xvi,494,[2 advt] pp. Map, ills.
Orig green cloth gilt, backstrip dulled, sm
snag in jnt, upper hinge strained.
(Blackwell's) **£55 [≃ $88]**
- Darwinism. London: 1889. 1st edn, 3rd iss.
8vo. xvi,494 pp. Map, port, 37 text figs. Orig
cloth. *(Wheldon & Wesley)* **£60 [≃ $96]**

- Darwinism ... London: Macmillan, 1889.
8vo. Orig cloth, sl shaken.
(Waterfield's) **£75 [≃ $120]**
- Darwinism. London: 1890. 2nd edn. 8vo. xvi,
494 pp. Map, port, 37 text figs. Mor, sl
rubbed. *(Wheldon & Wesley)* **£35 [≃ $56]**
- Darwinism ... London: Macmillan, 1890. 2nd
edn. 8vo. [5],494,[2] pp. Frontis port, text ills.
Orig green cloth, recased. ALS by Wallace
tipped in. *(Rootenberg)* **$400 [≃ £250]**
- Darwinism. London: 1897. 8vo. xvi,494,[2
advt] pp. Map, port, figs. Sl spotting, 1 or 2
finger marks. Orig cloth, lower tips of crnrs
rubbed. *(Bow Windows)* **£45 [≃ $72]**
- The Geographical Distribution of Animals ...
London: Macmillan, 1876. 1st edn. 2 vols.
8vo. xxi,[2],503; viii,[4],607 pp. 7 cold maps,
text ills. Orig cloth.
(Rootenberg) **$750 [≃ £468]**
- Island Life: or, the Phenomena and Causes of
Insular Faunas and Floras. London: 1880. 1st
edn. 8vo. xix,526 pp. 26 maps & ills. Orig
cloth, inner jnts weak.
(Wheldon & Wesley) **£150 [≃ $240]**
- Man's Place in the Universe. London: 1904.
4th edn. 8vo. Title foxed. Orig cloth, spine
faded. *(Wheldon & Wesley)* **£20 [≃ $32]**
- Natural Selection and Tropical Nature.
Essays on Descriptive and Theoretical
Biology. New Edition, with corrections and
additions. London: Macmillan, 1891. 8vo.
xii, 492 pp. Orig green cloth gilt, spine v sl
dulled. *(Blackwell's)* **£50 [≃ $80]**
- Tropical Nature, and Other Essays. London:
Macmillan, 1878. 1st edn. 8vo. xiii,356 pp.
advt leaf. 1 sgntr sl loose. Orig green cloth
blocked in black, gilt lettered spine.
(Bickersteth) **£125 [≃ $200]**
- The World of Life. London: 1911. 2nd edn.
8vo. xvi,408 pp. 110 ills. Orig cloth, front
inner jnt & frontis taped.
(Wheldon & Wesley) **£25 [≃ $40]**

Wallace, R.L.
- The Canary Book. London: [1893]. 3rd edn.
8vo. viii,429 pp. 6 cold & 16 plain plates.
Orig cloth, used, lacks front endpaper.
(Wheldon & Wesley) **£28 [≃ $44]**

Wallace, W.
- The Laws which regulate the Deposition of
Lead Ore in Veins: illustrated by an
Examination of the Geological Structure of
the Mining Districts of Alston Moor.
London: 1861. 8vo. xx,258,2 pp. 19 hand
cold sections, 2 plates, map in pocket. Orig
cloth, spine relaid. *(Henly)* **£60 [≃ $96]**

Wallis, George

- The Art of Preventing Diseases, and Restoring Health. New York: S. Campbell, (1794). 1st Amer edn. [3],vi-xv, [17]-571 pp. Usual foxing. Minor tear in 1 f affecting 1 word. Contemp calf. *(Elgen)* **$175 [≈ £109]**

Wallis, John

- A Brief Letter from a Young Oxonian to One of his late Fellow-Pupils upon the Subject of Magnetism. London: for S. Keble, 1697. Sole edn. 8vo. Title, 14 pp. 1 w'cut text fig. Title sl creased & frayed. Old wraps. Wing W.562.
(Bickersteth) **£660 [≈ $1,056]**
- Cono-Cuneus: or, the Shipwright's Circular Wedge ... Geometrically Considered. London: John Playford, for Richard Davis ..., 1684. 4to. [iv],17,[1] pp. 7 fldg plates, num tables in text. Disbound, stabbed & tied at inner marg. Cloth case. Wing W.565.
(Karmiole) **$350 [≈ £218]**

Walpole-Bond, J.

- A History of Sussex Birds. London: 1938. 3 vols. Roy 8vo. 53 cold plates by Philip Rickman. Orig brown buckram. Dws.
(Henly) **£220 [≈ $352]**

Walsh, David

- The Roentgen Rays in Medical Work ... London: 1902. 3rd edn. 316 pp. Plates, ills. Orig cloth, hd of spine sl worn.
(Elgen) **$95 [≈ £59]**

Walsh, J.H.

- A Manual of Domestic Economy suited to Families spending from £100 to £1000 a Year ... Management of the Nursery and Sick Room ... Domestic Remedies. London: Routledge ..., 1861. New edn. xvi,736 pp. Num ills. Orig brown cloth, simulated leather spine sl chipped.
(Box of Delights) **£44 [≈ $70]**

Walsh, James

- Psychotherapy. New York: Appleton, (1912) 1913. Sm 4to. 806 pp. Orig bndg.
(Xerxes) **$65 [≈ £40]**

Walshe, F.M.R.

- Critical Studies in Neurology. Baltimore: William & Wilkins, 1948. 8vo. 256 pp. Ills. Orig bndg. *(Xerxes)* **$40 [≈ £25]**

Walton, W.

- Problems in Illustration of the Principles of Plane Coordinate Geometry. Cambridge: 1851. viii,429 pp. Orig cloth, dust stained,

tear in spine reprd, top of spine sl dfefective.
(Whitehart) **£25 [≈ $40]**

Wang, Chung Yu

- Antimony: its History, Chemistry, Mineralogy, Geology ... London: 1909. 1st edn. 8vo. x,217 pp. 1 plate, tables, figs. Orig cloth. *(Bow Windows)* **£45 [≈ $72]**

Ward, The Hon Mrs Mary

- The Microscope or Descriptions of Various Objects of Especial Interest and Beauty. London: Groombridge, 1869. 3rd edn. Fcap 8vo. 154 pp. 8 hand cold plates, 25 text figs. Leather, gilt dec spine, prize label.
(Savona) **£25 [≈ $40]**

Ward, Rowland

- Records of Big Game. London: 1896. 2nd edn. 8vo. Ills. Orig cloth, considerably faded, some rubbing. *(Grayling)* **£130 [≈ $208]**
- Records of Big Game. London: 1907. 5th edn. 8vo. Ills. Orig buckram, sl rubbed & discold. *(Grayling)* **£100 [≈ $160]**
- Records of Big Game. London: 1910. 6th edn. 8vo. Ills. Some foxing throughout. New buckram, endpapers preserved.
(Grayling) **£60 [≈ $96]**
- Records of Big Game. London: 1910. 6th edn. 8vo. Ills. Orig cloth, rather rubbed.
(Grayling) **£90 [≈ $144]**
- Records of Big Game. London: 1922. 8th edn. 8vo. Ills. Orig cloth. Reprd dw.
(Grayling) **£250 [≈ $400]**
- Records of Big Game. London: 1928. 9th edn. 8vo. Ills. Orig blue buckram.
(Grayling) **£280 [≈ $448]**
- Records of Big Game. London: 1928. 9th edn. 8vo. Ills. Orig blue buckram, spine sl rubbed & faded. *(Grayling)* **£240 [≈ $384]**
- Records of Big Game. London: 1935. 10th edn. 8vo. Ills. Prelims sl foxed. Orig bndg.
(Grayling) **£320 [≈ $512]**

Warder, J.

- The True Amazons: or, The Monarchy of Bees. London: 1716. 3rd edn. 8vo. xiii,[ii], 120 pp. Port. Trifle foxed, a few annotations. Contemp calf.
(Wheldon & Wesley) **£125 [≈ $200]**
- The True Amazons: or, The Monarchy of Bees. London: 1726. 6th edn. 8vo. xxiv,112 pp. Port. Old inscrptns on title & port. Unlettered cloth.
(Wheldon & Wesley) **£70 [≈ $112]**
- The True Amazons: or, The Monarchy of Bees. London: 1765. 9th edn. 8vo. 164 pp.

Port. Wraps, spine worn.
(Wheldon & Wesley) **£60 [≈ $96]**

Warming, E. & Vahl, M.
- Oecology of Plants. An Introduction to the Study of Plant Communities ... Oxford: 1909. Roy 8vo. xii,422 pp. Endpapers foxed. Orig cloth.   *(Wheldon & Wesley)* **£25 [≈ $40]**

Warner, Sir Frank
- The Silk Industry of the United Kingdom. Its Origin and Development. London: Drane, [ca 1921]. 4to. 664 pp. Ills. Orig bndg, spine a little worn.   *(Book House)* **£90 [≈ $144]**

Warnes, John
- On the Cultivation of Flax; the Fattening of Cattle with Native Produce; Box-Feeding; and Summer Grazing ... London: W. Clowes, 1846. 1st edn. 8vo. xv,321 pp. 7 plates & text ills. Orig fine reeded cloth, uncut.
        (Claude Cox) **£35 [≈ $56]**
- On the Cultivation of Flax; the Fattening of Cattle with Native Produce; Box-Feeding; and Summer Grazing. Second Edition. London: James Ridgway, , 1847. 8vo. xxii,362 [inc 2 advt] pp. W'cut ills. Orig gilt dec cloth, spine ends sl worn.
        (Rankin) **£50 [≈ $80]**

Warren, Ina Russelle (ed.)
- The Doctor's Window. Poems by the Doctor, for the Doctor and about the Doctor. Buffalo, New York: Wharles Wells Moulton, (1897) 1898. 8vo. 288 pp. Photogravure plates. Orig bndg, gilt stamped, uncut, lower cvr crnrs bumped.   *(Xerxes)* **$55 [≈ £34]**

Warren, J.
- The Conchologist. Boston: 1834. Sm 4to. 204 pp. Frontis, 16 plates. Blind stamp on title & plates. New half mor.
        (Wheldon & Wesley) **£165 [≈ $264]**

Warren, John Collin
- Etherization; with Surgical Remarks. Boston: Ticknor, 1848. 1st edn. 8vo. [2],v, [3],100, 4,[4 advt dated October 1, 1847] pp. Orig brown cloth.   *(Rootenberg)* **$700 [≈ £437]**

Warren, Samuel
- Passages from the Diary of a Late Physician. With Notes and Illustrations by the Editor. Fourth Edition. Edinburgh: 1835. 2 vols. 12mo. Contemp half calf, endpapers spotted.
        (Robertshaw) **£12.50 [≈ $20]**

Wasson, R. Gordon
- The Wondrous Mushroom: Mycolatry in Mesoamerica. New York: 1980. 1st edn. One of 501, sgnd. 4to. 247 pp. 139 ills, 54 cold. Qtr green levant, dec buckram bds, slipcase.
        (Argosy) **$300 [≈ £187]**

Wasson, R. Gordon, et al., (ed.)
- Maria Sabina and Her Mazatec Mushroom Velada. New York & London: 1974. 1st edn. One of 250. 4to. 281 pp. 27 plates, 10 mtd cold. Qtr navy levant, dec cloth bds, cloth case, with 4 LP record set & accompanying musical score.   *(Argosy)* **$500 [≈ £312]**

Wasson, Valentina Pavlovna & R. Gordon
- Mushrooms, Russia and History. New York: Pantheon Books, (1957). One of 512. 2 vols. Lge 4to. xx,434 pp. 82 cold plates. Orig green cloth, t.e.g. Slipcase.
        (Karmiole) **$2,250 [≈ £1,406]**

Waterhouse, Benjamin
- The Rise, Progress, and Present State of Medicine. A Discourse ... July 6th, 1791. Boston: 1792. 1st edn. Sm 8vo. 12,31 pp. Last 2ff reprd without loss. New wraps, good ex lib. Clamshell case.
        (M & S Rare Books) **$600 [≈ £375]**

Waterhouse, G.R.
- The Natural History of Marsupalia or Pouched Animals. Edinburgh: Jardine's Naturalist's Library, 1841. 1st edn. Sm 8vo. 323 pp. Port, vignette title, 35 cold plates. Contemp mor, gilt spines, a.e.g., jnts sl rubbed.   *(Egglishaw)* **£80 [≈ $128]**

Waters, David
- The Art of Navigation in England in Elizabethan and Early Stuart Times. London: Hollis & Carter, (1958). Roy 8vo. 87 plates, 43 diags. Orig cloth.   *(Stewart)* **£45 [≈ $72]**

Waterton, Charles
- Letters of Charles Waterton of Walton Hall. Edited by R.A. Irwin. London: 1955. 8vo. 172 pp. 6 plates. Cloth.
        (Wheldon & Wesley) **£18 [≈ $28]**
- Natural History. Essays. Edited with a Life of the Author by N. Moore. London: [1870]. 8vo. viii,631 pp. 5 ills. Title & port foxed. Calf gilt, mrbld edges.
        (Wheldon & Wesley) **£25 [≈ $40]**

Watkins, Thomas
- A Table of Redemption. Shewing at one view in what time the principal and interest of any

debt ... may be discharged ... By T.W. F.R.S. London: W. Wilkins ... sold by J. Roberts, 1717. 1st edn. Folio. [8],[iv] pp. Disbound. *(Pickering)* **$1,000 [≈ £625]**

Watmough, W.

- The Cult of the Budgerigar. London: "Cage Birds", [1935]. 8vo. viii,292,[4] pp. 6 cold & 11 b/w plates, 4 line drawings. Half dark green calf gilt. *(Spelman)* **£40 [≈ $64]**

Watson, H.C.

- Remarks on the Geographical Distribution of British Plants ... London: 1835. Post 8vo. xvi,288 pp. Orig cloth. *(Wheldon & Wesley)* **£25 [≈ $40]**
- Topographical Botany, being Local and Personal Records towards shewing the Distribution of British Plants. London: for private distribution, 1873-74. 2 vols. 8vo. Map. Orig cloth. *(Wheldon & Wesley)* **£28 [≈ $44]**
- Topographical Botany ... London: 1883. 2nd edn. 8vo. xlvii,612 pp. Fldg cold map. Orig cloth, unopened. *(Henly)* **£20 [≈ $32]**

Watson, James & Crick, F.H.C.

- The Structure of DNA. [In] Cold Spring Harbour Symposia on Quantitative Biology, Vol XVII Viruses ... New York: 1953. 1st edn. 4to. The paper occupies pp 123-131. Orig cloth. *(Pickering)* **$650 [≈ £406]**

Watson, James D.

- The Double Helix. New York: 1968. 1st edn. 8vo. Orig cloth. Dw. *(Argosy)* **$75 [≈ £46]**

Watson, John B.

- Behavior: An Introduction to Comparative Psychology. New York: Holt, 1914. 1st edn. 1st iss, with 1914 on title-page. Sm 8vo. [ii], [xiv],[440] pp. Orig green cloth. *(Gach)* **$250 [≈ £156]**
- Behaviorism. New York: The People's Institute Publishing Company Incorporated, 1924. 1st edn, 1st iss. The correct 1st appearance in book form. 12 pamphlets loose in ptd paper cvrd green bds as issued. Orig cardboard slipcase (spine & edges of case taped). *(Gach)* **$175 [≈ £109]**

Watson, John Selby

- The Reasoning Power in Animals. London: Reeve & Co, 1867. 1st edn. Cr 8vo. [2],viii, 471,24 advt pp. Orig cloth gilt, remains of lib label on upper bd. *(Fenning)* **£35 [≈ $56]**

Watson, John, 1847-1939

- Comte, Mill, and Spencer: An Outline of Philosophy. Glasgow: Maclehose, 1895. 1st edn. 12mo. xx,302,[2] pp. Orig maroon cloth, edges rubbed. *(Gach)* **$50 [≈ £31]**

Watson, M.

- Observations on Human and Comparative Anatomy. Edinburgh: 1874. 8vo. 78 pp. 2 plates. Orig bds, spine worn. *(Wheldon & Wesley)* **£20 [≈ $32]**

Watson, P.W.

- Dendrologia Britannica, or Trees and Shrubs that will live in the Open Air of Britain ... London: 1825. 2 vols. Roy 8vo. 172 hand cold plates. Occas spotting in text. Mod green half mor, uncut. *(Henly)* **£980 [≈ $1,568]**

Watson, Richard, Bishop of Llandaff

- Chemical Essays. Cambridge: J. Archdeacon for T. & J. Merrill ..., 1781-81-83-86. 1st edn vols 1 & 2, 2nd edn vols 3 & 4. 4 vols. 8vo. [ii],[x],349; [iv],368; [iv],ii,376; [xxiv], 354,[1] pp. Rec half calf antique. *(Charles B. Wood)* **$350 [≈ £218]**
- Chemical Essays. Seventh Edition. London: 1800. 5 vols. Sm 8vo. Half-title in vol 3 only (probably all called for). Fldg ptd table in vol 1. Orig speckled calf, partly unopened, sl wear, 1 jnt cracked but firm. *(Bickersteth)* **£140 [≈ $224]**

Watson, Thomas

- Lectures on the Principles and Practice of Physic. Revised with additions by D. Francis Condie. Philadelphia: Blanchard & Lea, 1854. 3rd Amer edn. Thick sm 4to. 1040 pp. Leather. *(Xerxes)* **$60 [≈ £37]**

Watson, W.

- Orchids: Their Culture and Management. With Descriptions of all the kinds in General Cultivation ... Second Edition, Revised. London: Upcott Gill, [ca 1900]. Thick 8vo. xi, 554,[2].12 advt pp. 8 chromolitho plates, num w'engvs. Dec half parchment & bds, sl worn. *(Heritage)* **$250 [≈ £156]**
- Orchids: Their Culture and Management ... London: 1903. New (last) edn. 8vo. xi,559 pp. 20 cold plates, num ills. A few red ink notes. Orig cloth gilt. *(Wheldon & Wesley)* **£55 [≈ $88]**

Watson, William

- Experiments and Observations tending to Illustrate the Nature and Properties of Electricity ... Second Edition. London: for C.

Davies, 1746. 8vo. [ii],viii,3-59 pp. Disbound. *(Bow Windows)* **£95 [≃ $152]**

Watt, Alexander
- The Art of Soapmaking: a Practical Handbook of Hard and Soft Soaps, Toilet Soaps, etc. ... Appendix on Candle-Making. London: Crosby Lockwood, 1896. 5th edn, rvsd. xiii,310,[48,15 ctlg] pp. Tables, diags. Orig cloth.   *(Box of Delights)* **£25 [≃ $40]**

Watters, John J.
- The Natural History of the Birds of Ireland, Indigenous and Migratory. Dublin: McGlashan, 1853. 299 pp. Orig bndg.
    (Emerald Isle) **£38 [≃ $60]**

Watts, Henry
- A Dictionary of Chemistry and the Allied Branches of other Sciences. London: Longmans, Green, 1874. New edn. 7 vols. Thick 8vo. Text ills. Half calf gilt, hd of spines sl chipped.   *(Hollett)* **£65 [≃ $104]**

Webster, A.D.
- British Orchids. London: 1898. 2nd enlgd edn. 8vo. xii,132,3 pp. Frontis & 39 text ills. Orig cloth.   *(Henly)* **£36 [≃ $57]**

Webster, Mrs Alfred
- Dancing, as a means of Physical Education: with remarks on Deformities, and their Prevention and Cure. London: David Bogue ..., 1851. 1st edn. 4to. Orig blue cloth, a.e.g. Author's pres inscrptn.
    (Ximenes) **$325 [≃ £203]**

Webster, Charles
- Facts tending to show the Connection of the Stomach with Life, Disease and Recovery. London: for J. Murray ..., 1793. 1st edn. 8vo. [iv],59 pp. Rec cloth.
    (Burmester) **£70 [≃ $112]**

Webster, John
- Metallographia; or, an History of Metals ... London: Walter Kettilby, 1671. 1st (& only) edn. 8 ff,388,2 ctlg pp. 1st 3 ff reinforced at outer edges. New contemp style calf by Bliss.
    (Elgen) **$1,000 [≃ £625]**

Wechselmann, Wilhelm
- The Treatment of Syphilis with Salvarsan ... Only Authorized Translation by Abr. L. Wolbarst. New York: 1911. 1st edn in English. 175 pp. 16 chromolitho plates. Orig cloth, leather spine label sl chipped.
    (Elgen) **$135 [≃ £84]**

Weeks, Nora
- Medical Discoveries of Edward Bach, Physician. What the Flowers do for the Human Body. Ashingdon: Daniel, 1963. Sm 8vo. 141 pp. Some marg pen lines. Orig bndg.
    (Xerxes) **$65 [≃ £40]**

Wegmann, Edward
- The Water-Supply of the City of New York, 1658-1895. New York: John Wiley; London: Chapman & Hall, 1896. 1st edn. Roy 4to. xii, 316,9 advt pp. 148 plates, 73 text ills. A few sl marks, 1 minor repr. New buckram.
    (Duck) **£300 [≃ $480]**

Weismann, August
- Essays upon Heredity and Kindred Biological Problems. Edited by E.B. Poulton, S. Schonland and A.E. Shipley. Oxford: 1889. 8vo. xii,455 pp. Orig cloth, sound ex-lib.
    (Wheldon & Wesley) **£35 [≃ $56]**
- The Evolution Theory. Translated by J.A. and M.R. Thomson. London: 1904. 2 vols. Roy 8vo. 3 cold plates, 131 text figs. Orig cloth.   *(Wheldon & Wesley)* **£60 [≃ $96]**
- The Germ-Plasm: a Theory of Heredity. Translated by W.N. Parker and H. Ronnfeldt. London: 1893. 8vo. xxiii,477 pp. 24 text figs. Orig cloth, sl used.
    (Wheldon & Wesley) **£45 [≃ $72]**

Weld, Charles
- A History of the Royal Society, with Memoirs of the Presidents. Compiled from Authentic Documents ... London: Parker, 1848. 1st edn. 2 vols. 8vo. xx,527,8 advt; viii, 612, 4 advt pp. 14 plates, text ills. Frontises foxed. Orig cloth, spines & bd edges sl frayed.
    (Pickering) **$600 [≃ £375]**

Wells, Edward
- The Young Gentleman's Astronomy, Chronology, and Dialling ... London: Knapton, 1712. 1st edn. 8vo. General title, [iv],149 pp, 2 advt ff, [viii],87 pp, [viii],56 pp. 25 plates. Sl waterstaining. Orig calf, sl worn.   *(Bickersteth)* **£285 [≃ $456]**
- The Young Gentleman's Astronomy, Chronology, and Dialling ... Fourth Edition Revised and Corrected, with Additions. London: Knapton, 1736. 8vo. [viii], 148, [viii], 86, [viii], 54 pp. 25 plates (some browning). Period gilt panelled calf, rebacked.   *(Rankin)* **£65 [≃ $104]**

Wells, Horace
- Horace Wells, Dentist, father of Surgical Anesthesia. Proceedings of Centenary

Commemoration of Wells' Discovery in 1844. ADA: 1948. 1st edn. Lge 8vo. 415 pp. Plates, ills. Orig bndg. *(Elgen)* **$100 [≈ £62]**

Wells, William Charles
- An Essay on Dew, and several Appearances connected with it. London: 1815. 2nd edn. 8vo. 150 pp. Orig bds, uncut, spine worn.
    (Wheldon & Wesley) **£48 [≈ $76]**
- Two Essays: One upon Single Vision with Two Eyes; the Other on Dew ... London: Constable, 1818. 1st edn of his coll writings. 8vo. lxxiv,[2],439,[1] pp. Contemp half calf.
    (Rootenberg) **£650 [≈ £406]**
- Two Essays: One upon Single Vision with 2 Eyes: the Other on Dew ... London: 1818. New lib buckram.
    (Rittenhouse) **£350 [≈ £218]**

Weltmer, Professor Sidney
- Healing Hand. Nevada, Missouri: Weltmer Institute of Suggestive Therapeutics, 1922. 8vo. 248 pp. Orig bndg. *(Xerxes)* **$40 [≈ £25]**

Wentworth, Lady
- The Authentic Arabian Horse and His Descendants. London: Allen & Unwin, 1945. 1st edn. 4to. 388 pp. 287 plates (24 cold). Orig cloth gilt. *(Hollett)* **£240 [≈ $384]**

Wenyon, C.M.
- Protozoology. New York: 1926. 2 vols. Roy 8vo. 1589 pp. 20 cold plates, 565 figs. Orig cloth, rebacked.
    (Wheldon & Wesley) **£75 [≈ $120]**

Westermarck, Edward
- Ethical Relativity. London: 1932. 1st edn in English. 8vo. xviii,301 pp. Orig cloth, cvrs marked. *(Bow Windows)* **£60 [≈ $96]**
- History of Human Marriage. New York: Allerton, 1922. 5th edn, rewritten. 3 vols. 8vo. 571; 595; 587 pp. Orig bndg.
    (Xerxes) **$55 [≈ £34]**
- The History of Human Marriage. Fifth Edition ... London: 1925. 3 vols. 8vo. Orig cloth, tiny nick in 2 lower jnts.
    (Bow Windows) **£140 [≈ $224]**
- The Origin and Development of the Moral Ideas. Second Edition. London: 1924. 2 vols. 8vo. Orig cloth, spine ends just a trifle frayed.
    (Bow Windows) **£105 [≈ $168]**

Westropp, Hodder & Wake, C.
- Ancient Symbol Worship. Influence of the Phallic Idea in the Religions of Antiquity. With Introduction, Notes and Appendix by Alexander Wilder. New York: Bouton, 1875.

2nd edn. 8vo. 98 pp. Ills. Orig bndg.
    (Xerxes) **$45 [≈ £28]**

Westwood, J.O.
- An Introduction to the Modern Classification of Insects ... London: Longman ..., 1839-40. 1st edn. 2 vols. 8vo. xii,[i], 462; xi,587,158 pp. 1 hand cold plate, text figs. Orig cloth, partly unopened, spines faded.
    (Bickersteth) **£70 [≈ $112]**

Weyl, H.
- Space -Time - Matter. Translated by H.L. Brose. London: 1922. 1st English edn. xi,330 pp. 15 diags. Orig cloth, sl worn, jnt cracked but bndg firm. *(Whitehart)* **£40 [≈ $64]**

Whately, Thomas
- Observations on Modern Gardening, illustrated by Descriptions. London: 1770. 2nd edn. 8vo. [viii],257 pp. Mod bds.
    (Wheldon & Wesley) **£120 [≈ $192]**
- Observations on Modern Gardening. Illustrated by Descriptions. Fifth Edition. London: T. Payne, 1793. 8vo. [8],263 pp. Orig glazed green cloth, uncut, some edge wear, paper label sl indistinct.
    (Spelman) **£95 [≈ $152]**

Wheeler Gift
- Catalogue of the Books, Pamphlets and Periodicals in the Library of the American Institute of Electrical Engineers. Edited by William D. Weaver ... New York: 1909. 1st edn. 2 vols. Frontis ports, num plates. Orig cloth, sl soiled. *(Elgen)* **$225 [≈ £140]**

Wheeler, A.
- The Fishes of the British Isles and North-West Europe. London: 1969. Roy 8vo. xvii,613 pp. 10 cold & 6 plain plates, num maps & text figs. Orig cloth.
    (Wheldon & Wesley) **£28 [≈ $44]**

Wheeler, W.H.
- A Practical Manual of Tides and Waves. London: Longmans, Green, 1906. 8vo. 201 pp. Orig cloth gilt, partly uncut.
    (Moon) **£35 [≈ $56]**

Wheeler, William Morton
- Demons of the Dust, a Study in Insect Behaviour. New York: [1930]. 8vo. xviii, 378 pp. Frontis, 49 text figs. Orig cloth.
    (Wheldon & Wesley) **£35 [≈ $56]**
- The Social Insects. Their Origin and Evolution. London: 1928. 1st edn. 8vo. xviii, 378 pp. 48 plates. Orig cloth.

(Bickersteth) **£28 [≃ $44]**

Wheldon, J.A. & Wilson, A.
- The Flora of West Lancashire. Privately printed: 1907. 1st edn. 8vo. 511 pp. Fldg map, 15 plates. Orig cloth gilt, extremities trifle rubbed, inner jnt strengthened.
(Hollett) **£45 [≃ $72]**

Whewell, William
- Astronomy and general Physics considered with reference to Natural Theology. London: William Pickering, Bridgewater Treatises, 1834. 3rd edn. 8vo. xv,381[1] pp, advt leaf. 2 advts tipped in at front. Orig cloth, paper spine label sl rubbed.
(Fenning) **£32.50 [≃ $52]**
- Astronomy and general Physics considered with reference to Natural Theology. London: William Pickering, Bridgewater Treatises, 1834. 3rd edn. 8vo. xv,381[1] pp, advt leaf. Lacks half-title. Contemp half calf, a little rubbed.
(Fenning) **£21.50 [≃ $35]**
- History of the Inductive Sciences, From the Earliest to the Present Time ... London: John W. Parker, 1857. 3rd rvsd & enlgd edn, 1st printing. 3 vols. 12mo. Orig red cloth, 1st vol recased, vol 3 hinges cracked.
(Gach) **$175 [≃ £109]**
- History of the Inductive Sciences, From the Earliest to the Present Time ... London: John W. Parker, 1857. 3rd rvsd & enlgd edn, 1st printing. 3 vols. Sm 8vo. Orig red cloth, cloth darkened.
(Gach) **$250 [≃ £156]**
- The Mechanics of Engineering ... Cambridge: 1841. xii,216 pp. Diags. Half roan, rebacked in cloth with orig spine laid on, edges worn.
(Whitehart) **£40 [≃ $64]**
- A Treatise on Dynamics containing a considerable Collection of Mechanical Problems. Cambridge: 1823. xvi,403 pp. 6 fldg plates. Three qtr calf, worn, hinges cracked but bndg firm, lacks label, spine sl defective.
(Whitehart) **£25 [≃ $40]**

Whidborne, G.F.
- The Devonian Fauna of the South of England. London: 1889-1907. 3 vols in 12 parts, complete. 94 plates. 11 parts disbound.
(Baldwin) **£100 [≃ $160]**

Whiston, William
- Memoirs of the Life and Writings of Mr. William Whiston ... written by himself. London: for the author, 1749 [-50]. 3 parts in one vol. 8vo. 18th c calf, spine relaid. Part III contains his Lectures on Meteors, dated 1750.
(Waterfield's) **£200 [≃ $320]**

- A New Theory of the Earth ... London: R. Roberts for Benj. Tooke, 1696. 1st edn. 8vo. [iv],388 pp, advt leaf, errata leaf. Frontis, 7 plates, 5 text figs. Orig calf. Wing W.1696.
(Bickersteth) **£425 [≃ $680]**

Whitcombe, Charles Edward
- The Canadian Farmer's Manual of Agriculture ... Counsel to the Immigrant-Settler ... Toronto: 1879. 1st edn. 8vo. 571 pp. Sev text ills. Lacks table of contents. Orig cloth, rubbed.
(Young's) **£120 [≃ $192]**

White, A.
- A Popular History of Birds. London: Lovell Reeve, 1855. 1st edn. Sq 8vo. viii,347 pp. 20 hand cold litho plates. New cloth.
(Egglishaw) **£42 [≃ $67]**

White, Andrew Dickson
- A History of the Warfare of Science with Theology in Christendom. New York: Appleton, 1896. 1st edn. 2 vols. 8vo. 2nd title sl jagged at edge. Orig red cloth, hinges cracked.
(Gach) **$85 [≃ £53]**

White, B.
- Gold: Its Place in the Economy of Mankind. London: Pitman, 1920. Cr 8vo. 145 pp. Ills. Orig cloth.
(Gemmary) **$35 [≃ £21]**
- Silver: Its Intimate Associations with the Daily Life of Man. London: Pitman, 1920. Cr 8vo. 160 pp. Fldg table, num ills. Orig cloth.
(Gemmary) **$35 [≃ £21]**

White, Charles
- An Account of the Regular Gradation in Man, and in Different Animals and Vegetables; and from the Former to the Latter. London: 1799. 1st edn. 4to. xii,138, cxxxix-clxvi, 139-146 pp. 4 plates. Some offsetting. 2 sm blind stamps. 19th c half mor, trifle rubbed.
(Wheldon & Wesley) **£330 [≃ $528]**
- Cases in Surgery, with Remarks. Part the First [all published]. To which is added, an Essay on the Ligature of Arteries. By James Aikin ... London: for W. Johnson, 1770. Sole edn. 8vo. xv,198,[ii] pp, errata leaf. 7 fldg plates. Early 19th c half calf, rebacked.
(Bickersteth) **£180 [≃ $288]**

White, F.
- Forest Flora of Northern Rhodesia. With the Assistance of A. Angus. OUP: 1962. 1st edn. Tall 8vo (slim 4to). 455 pp. Fldg map. 73 ills. Cloth. Dw.
(Terramedia) **$100 [≃ £62]**

White, F.B.W.

- The Flora of Perthshire. Edited by J.W.H. Trail. Edinburgh: 1898. One of 500. 8vo. lix, 407 pp. Port, cold map. Orig cloth.
(Wheldon & Wesley) **£50 [≈ $80]**

White, Francis Sellon

- A History of Inventions and Discoveries. London: Rivington, 1827. 1st edn. iv,547 pp. Occas sl foxing. Orig bds, unopened, spine split. *(Elgen)* **$95 [≈ £59]**

White, Gilbert

- The Natural History and Antiquities of Selborne. London: 1789. 1st edn. 4to. v, title-page with vignette,468,[2] pp, errata sheet. 6 plates. Later red mor. Qtr mor case.
(Baldwin) **£900 [≈ $1,440]**
- The Natural History and Antiquities of Selborne. London: 1789; reptd Scolar Press, 1970. Facs of the 1st edn. 4to. Plates. Red three qtr Levant mor gilt, t.e.g.
(Bow Windows) **£155 [≈ $248]**
- The Natural History and Antiquities of Selborne. New Edition, to which are added The Naturalist's Calendar; Observations ... Poems. London: 1813. Mitford's edn. 4to. x, title vignettes, 585 pp. 9 plates (1 hand cold). 1 plate a little spotted. Later calf gilt.
(Baldwin) **£175 [≈ $280]**
- The Natural History of Selborne, to which are added the Naturalists Calendar, Miscellaneous Observations, and Poems. London: 1822. 2 vols. 8vo. 4 plates (1 cold). Buckram, t.e.g. *(Henly)* **£120 [≈ $192]**
- The Natural History of Selborne ... London: 1835. Captain T. Brown's 5th edn. 12mo. xii,356 pp. Frontis, 1 plate, 2 ills. Some v light marg staining & foxing. Orig bds, new spine label. *(Baldwin)* **£25 [≈ $40]**
- The Natural History of Selborne ... London: 1837. New edn by E.T. Bennett. 8vo. xxiv,640 pp. 46 engvs. Name on title. Orig cloth. *(Baldwin)* **£40 [≈ $64]**
- The Natural History of Selborne ... London: 1844. Harting's 3rd edn. 8vo. xxii, 568 pp. Engvs by Bewick & others. Minor spotting. Contemp prize tree calf, hinges reprd.
(Baldwin) **£40 [≈ $64]**
- Natural History of Selborne. London: 1853. Brown's 9th edn. 12mo. xii,348 pp. Addtnl title with hand cold vignette, 14 hand cold plates. Orig bndg, a.e.g. *(Henly)* **£25 [≈ $40]**
- The Natural History of Selborne. Edited by J.E. Harting. London: 1875. 1st Harting edn. 8vo. xx,532 pp. 60 ills. Orig cloth gilt, a.e.g. *(Henly)* **£40 [≈ $64]**
- The Natural History of Selborne ... London:

1876. The standard edn of E.T. Bennett rvsd by J.E. Harting. 2nd edn. 8vo. xxii,532 pp. Bewick engvs. Contemp prize calf, 1 hinge reprd, 1 cracked but sound.
(Baldwin) **£40 [≈ $64]**
- The Natural History of Selborne ... London: 1877. Edited by T. Bell. 2 vols. 8vo. Ills. Orig cloth, sl worn & faded.
(Baldwin) **£95 [≈ $152]**
- Natural History and Antiquities of Selborne. London: 1883. 6th Buckland edn. 8vo. xxix,480,32 pp. Num ills. Orig cloth, partly unopened. *(Henly)* **£20 [≈ $32]**
- The Natural History of Selborne. Edited by J.E. Harting. London: 1887. 5th Harting edn. 8vo. xxii,568 pp. 60 ills. Orig pict cloth gilt, a.e.g. *(Henly)* **£30 [≈ $48]**
- The Natural History of Selborne. Edited by J.E. Harting. London: 1890. 8th Harting edn. 8vo. xxii,568 pp. 60 ills. Orig pict cloth gilt, a.e.g. Thomas Ashton's b'plate.
(Henly) **£35 [≈ $56]**
- The Natural History of Selborne. London: 1898. Jesse & Jardine's edn. Cr 8vo. xxiv, 416, 32 pp. Frontis, 40 hand cold plates. Orig cloth. *(Henly)* **£30 [≈ $48]**
- The Natural History and Antiquities of Selborne ... Edited by R. Bowdler Sharpe. London: 1900. One of 160 Large Paper, signed. 2 vols. Thick 4to. Num ills by Keulemans, Railton & Sullivan. Some tissues in vol 1 browned. Vellum gilt, t.e.g., vol 1 spine dulled. *(Hollett)* **£375 [≈ $600]**
- The Natural History and Antiquities of Selborne. Edited by R. Bowdler Sharpe. London: 1900. 2 vols. 8vo. 99 plates by Keulemans, Railton & Sullivan, 69 text ills, inc 2 facs letters. Half green mor gilt.
(Wheldon & Wesley) **£180 [≈ $288]**
- The Natural History and Antiquities of Selborne. Edited by R. Bowdler Sharpe. London: 1900. 2 vols. 8vo. 99 plates by Keulemans, Railton & Sullivan, 69 text ills, inc 2 facs letters. Orig cloth, t.e.g., 2 faint stains on vol 2 spine.
(Wheldon & Wesley) **£120 [≈ $192]**
- The Natural History of Selborne. London: 1902. 1st Kearton edn. Roy 8vo. xvi,294 pp. 124 ills. Orig green cloth gilt.
(Henly) **£25 [≈ $40]**
- A Naturalist's Calendar, with Observations in Various Branches of Natural History ... London: 1795. 1st edn. 8vo. 170,[3] pp. Hand cold plate. Orig mrbld bds.
(Baldwin) **£90 [≈ $144]**
- The Works in Natural History ... Natural History of Selborne; The Naturalist's Calendar ... added A Calendar and

Observations by W. Markwick. London: 1802. 2 vols. 8vo. 4 plates (2 cold). Contemp calf. The 2nd edn of White's Selborne.
(Wheldon & Wesley) **£200 [≃ $320]**

- The Writings of Gilbert White of Selborne. Selected and Edited by H.J. Massingham. Nonesuch Press: 1938. 2 vols. W'engvs by Eric Ravilious. Orig slipcase.
(Baldwin) **£425 [≃ $680]**

White, J.W.
- The Flora of Bristol. Bristol: 1912. 8vo. x,722 pp. Map, 3 plates. Orig cloth.
(Wheldon & Wesley) **£45 [≃ $72]**

White, James
- A Treatise on Veterinary Medicine. London: 1807. 8th edn, enlgd. 8vo. 400,[14] pp. 14 plates (4 hand cold). Mod half mor gilt.
(Hollett) **£75 [≃ $120]**

White, Stephen, 1697-1773
- Collateral Bee-Boxes; or, a New, Easy, and Advantageous Method of Managing Bees ... Part of the Honey is taken away ... without Destroying ... the Bees ... London: 1764. 3rd edn. 8vo. ix,47 pp. Plate. Wraps.
(Wheldon & Wesley) **£50 [≃ $80]**
- Collateral Bee-Boxes; or, a New, Easy, and Advantageous Method of Managing Bees ... London: Davis & Reyners, 1764. 3rd edn. 8vo. ix, 47 pp. Marg reprs to plate & last leaf. Mod qtr mor, uncut.
(Wheldon & Wesley) **£85 [≃ $136]**

White, William
- The Story of a Great Delusion. London: Allen, 1885. 1st edn. 8vo. 627 pp. 2 w'engvs. Orig bndg.   *(Xerxes)* **$150 [≃ £93]**

Whitehead, Alfred North
- The Aims of Education and Other Essays. New York: Macmillan, 1929. 1st Amer edn, 1st printing. Sm 8vo. [viii],[248] pp. Orig red cloth. Dw lightly worn. *(Gach)* **$50 [≃ £31]**
- The Concept of Nature ... Cambridge: Univ Press, 1920. 1st edn. 8vo. [ii],[x],202,[2] pp. Orig blue cloth.   *(Gach)* **$75 [≃ £46]**
- An Enquiry concerning the Principles of Natural Knowledge. Cambridge: Univ Press, 1919. 1st edn. xii,200 pp. Orig blue cloth.
(Gach) **$75 [≃ £46]**
- The Function of Reason. Princeton: Univ Press, 1929. 1st edn. Sm 8vo. [viii],72 pp. Orig brown cloth.   *(Gach)* **$40 [≃ £25]**
- Process and Reality: An Essay in Cosmology. New York: Macmillan, 1929. 1st edn. xii,545,[3] pp. Orig blue cloth, spine dull (as

usual).   *(Gach)* **$50 [≃ £31]**
- Science and the Modern World. Lowell Lectures 1925. New York: Macmillan, 1925. 1st edn. [xii],296,[4] pp. Orig blue cloth, dull.
(Gach) **$50 [≃ £31]**

Whitehead, Alfred North & Russell, Bertrand
- Principia Mathematica ... Second Edition. Cambridge: Univ Press, 1925-27-27. 2nd edn. 3 vols. Lge 8vo. xlvi,674; xxxi,742; viii,491 pp. Orig blue cloth, sl worn.
(Pickering) **$2,500 [≃ £1,562]**

Whitehead, G.K.
- The Deer of Great Britain and Ireland. London: 1964. 8vo. Ills. Orig bndg. Dw.
(Grayling) **£70 [≃ $112]**

Whitehurst, John
- An Enquiry into the Original State and Formation of the Earth ... London: for the author by J. Cooper, 1778. 1st edn. 4to. [xvi], iv,199 pp, inc subscribers. 9 plates on 4 ff. 19th c port inserted. Half calf, old mrbld bds.
(Bickersteth) **£380 [≃ $608]**
- An Enquiry into the Original State and Formation of the Earth. London: 1786. 2nd edn. 4to. [x],283 pp. Port (trifle water stained & offset on title), 7 plates. Mod half calf.
(Wheldon & Wesley) **£150 [≃ $240]**

Whitmill, Benjamin
- Kalendarium Universale: or the Gardner's Universal Calendar ... London: for J. Clarke, 1757. 6th edn, 'adopted to the new style'. 12mo. iv,262,[7 index],[4 advt] pp. Sm marg wormhole at end. Contemp sheep, some insect damage.   *(Burmester)* **£60 [≃ $96]**

Whitworth, W.A.
- Trilinear Coordinates and other Methods of Modern Analytical Geometry of Two Dimensions. Cambridge: 1866. xxxvi,506 pp. Few diags in text. Orig cloth, spine sl marked, lib stamps on front inner cvr.
(Whitehart) **£25 [≃ $40]**

Whymper, Charles
- Egyptian Birds. London: A. & C. Black Colour Books, 1909. One of 100, sgnd by Whymper. 51 mtd cold plates, 11 text ills. Orig dec cloth, t.e.g., signs of ageing with occas darkening, some wear to ft of spine, damp stain on top half inch of front endpapers. *(Old Cathay)* **£195 [≃ $312]**
- Egyptian Birds. London: A. & C. Black Colour Books, Twenty Shilling Series, 1909.

1st edn. 51 cold plates, 11 text ills. Orig dec cloth, t.e.g., rear hinge tender.
(Old Cathay) £85 [≈ $136]

Whytt, Robert

- Observations on the Nature, Causes, and Cure of those Disorders which have commonly been called Nervous Hypochondriac, or Hysteric ... Edinburgh & London: 1765. 1st edn. 8vo. [4],viii,[8],520,[2] pp. Some sl browning. Rec bds. Half mor slipcase.
(Rootenberg) $1,500 [≈ £937]

Wickham, Louis & Degrais, Paul

- Radiumtherapy. Translated by S. Ernest Dore ... New York: 1912. 396 pp. 20 cold litho plates. Orig cloth, worn, sl shaken.
(Elgen) $60 [≈ £37]

Wildman, T.

- A Complete Guide for the Management of Bees throughout the Year. London: 1780. 3rd edn. 8vo. 48 pp. 2 plates. Trifle soiled. New bds. *(Wheldon & Wesley)* £50 [≈ $80]
- A Complete Guide for the Management of Bees throughout the Year. London: 1792. 10th edn. Sm 4to. 48 pp. 2 plates. New bds.
(Wheldon & Wesley) £45 [≈ $72]
- A Treatise on the Management of Bees ... added the Natural History of Wasps and Hornets. London: 1768. 1st edn. 4to. xx,169, [7] pp. 3 fldg plates (some offsetting). Contemp calf, rebacked.
(Wheldon & Wesley) £250 [≈ $400]
- A Treatise on the Management of Bees. London: 1770. 2nd edn. 8vo. xx,311,[8],16 pp. 3 plates. New half calf, antique style.
(Wheldon & Wesley) £120 [≈ $192]
- A Treatise on the Management of Bees. London: 1778. 3rd edn. 8vo. xx,318,vii,16 pp. 3 plates. Sm stamp erased from title. New bds. *(Wheldon & Wesley)* £75 [≈ $120]

Wile, Ira

- Handedness Right or Left. Boston: Lothrop, Lee, 1934. 1st edn. 8vo. 439 pp. Orig bndg, hd of spine sl worn. *(Xerxes)* $40 [≈ £25]

Wilkins, John

- A Discovery of a New World ... Habitable World in the Moon ... Fifth Edition, Corrected and Amended. London: J. Rawlins for John Gellibrand, 1684. 8vo. Lacks engvd title. Sl staining on title & next leaf. Contemp calf, rebacked. Wing W.2187.
(Waterfield's) £175 [≈ $280]
- The Mathematical and Philosophical Works

... Author's Life ... London: for J. Nicholson ..., 1708 (1707, 1707). 1st edn. 8vo. viii,vi,274, x,90,viii,184 pp. Port frontis & engvd title to 'Discovery'. Text ills. Old lib stamps. Rec calf, sewn rather tight.
(Pickering) $1,500 [≈ £937]
- Mathematical Magick ... Third Edition ... London: for Edw. Gellibrand ..., 1680. 8vo. [xvi], 295 pp inc engvd port. Text ills. Engv pasted in on p 88. Contemporary calf, gilt spine, rubbed, jnt ends starting to split. Wing W.2200. *(Pickering)* $1,250 [≈ £781]
- Mercury: or the Secret and Swift Messenger ... communicate ... Thoughts to a Friend at any distance ... Second Edition. London: 1694. 8vo. [xvi],172,[4] pp. Port. 4 advt pp at end. W'cut ills. Title darkened, occas foxing & browning. Rec qtr calf. Wing W.2203.
(Clark) £185 [≈ $296]

Wilkinson, Charles Henry

- Elements of Galvanism, in Theory and Practice ... London: Murray, 1804. 1st edn. 2 vols. 8vo. xvi,468; xii,[2],472,[40] pp. Frontis, 12 engvd plates (1 hand cold). Occas browning. Sl marg dampstain on plates of vol 1. Orig calf, gilt spines.
(Rootenberg) $850 [≈ £531]

Wilkinson, George

- Experiments and Observations on the Cortex Salicis Latifoliae or Broad-Leafed Willow-Bark ... Newcastle upon Tyne: for the author ... [1803]. 1st edn. 8vo. [14],118 pp. Hand cold engvd frontis by Thomas Bewick. Minor browning. Contemp half calf, gilt spin
(Rootenberg) $400 [≈ £250]

Wilkinson, Oscar

- Strabismus. Its Etiology and Treatment. Boston: Meador, 1943. 2nd edn. 8vo. 369 pp. Photo ills. Orig bndg, ex-lib, rear hinge cracked. *(Xerxes)* $35 [≈ £21]

Williams, A.F.

- The Genesis of the Diamond. London: Ernest Benn, 1932. 2 vols. Roy 8vo. xv,352,vi; xii, 353-636,iv pp. 152 cold & 221 b/w plates, num text figs. Orig cloth.
(Gemmary) $400 [≈ £250]
- Some Dreams Come True. Cape Town: Howard B. Timmins, 1948. Roy 8vo. 14,590 pp. Frontis port, 3 cold plates, num other ills. Orig cloth. Dw. *(Gemmary)* $60 [≈ £37]

Williams, B.S.

- The Orchid Grower's Manual. London: (1894) 1961. 7th edn. 815 pp. Over 300 ills. Orig cloth. *(Wheldon & Wesley)* £50 [≈ $80]

Williams, C.B.
- Insect Migration. London: New Naturalist, 1958. 1st edn. 8vo. 24 plates (8 cold). Orig cloth. Dw. *(Wheldon & Wesley)* £25 [≈ $40]
- The Migrations of Butterflies. London: 1930. 8vo. xi,473 pp. 71 text figs. Orig cloth. *(Wheldon & Wesley)* £35 [≈ $56]

Williams, Charles J.B.
- Authentic Narrative of the Case of the late Earl St. Maur. With the Short-Hand Notes of the Retraction and Apology. London: 1870. 1st edn. 8vo. 44 pp. Disbound. *(Bickersteth)* £25 [≈ $40]
- The Pathology and Diagnosis of Diseases of the Chest ... Fourth Edition, much enlarged ... London: John Churchill, 1840. 8vo. x, [ii], 331,16 ctlg pp. 2 engvd plates, num text figs. Contemp half calf. *(Bickersteth)* £58 [≈ $92]

Williams, Frederick S.
- The Midland Railway: Its Rise and Progress. London & Derby: Bemrose, 1887. 3rd edn. xvi,682,[6 advt] pp. Port, 6 maps, ills. Orig cloth gilt, sl used, front hinge cracked but firm. *(Duck)* £50 [≈ $80]
- Our Iron Roads. London: Bemrose, 1884. 5th edn. 8vo. 514 pp. Ills. Orig bndg. *(Book House)* £30 [≈ $48]
- Our Iron Roads: Their History, Construction, and Administration. Seventh Edition. London & Derby: Bemrose, 1885. 7th edn, rvsd. xvi,514,515-520 advt pp. Frontis, ills. Orig dec cloth, spine sl faded, endpapers sl spotted. *(Duck)* £50 [≈ $80]

Williams, Herbert
- A Scrapbook of Dental Informalities. NJ: Williams, 1938. 8vo. 172 pp. Ills. Orig bndg. Author's pres copy. *(Xerxes)* $50 [≈ £31]

Williams, John, mining engineer
- An Account of Some Remarkable Ancient Ruins, lately discovered in the Highlands, and Northern Parts of Scotland ... Edinburgh: Creech, 1777. 1st edn. 8vo. [7],[1],83,[1] pp. Fldg frontis. Half calf, mrbld bds. *(Rootenberg)* $400 [≈ £250]
- The Natural History of the Mineral Kingdom. In Three Parts ... Edinburgh: for the author, by Thomas Ruddiman, 1789. 1st edn. 2 vols. 8vo. [ii],iii, [ii]-lxii, 450, [2]; 531,[1],iv pp. Contemp calf, gilt spines, mor labels, upper jnts reprd. *(Burmester)* £275 [≈ $440]

Williamson, George
- Memorials of the Lineage, Early Life,

Education and Development of the Genius of James Watt. London: Constable for the Watt Club, 1856. 4to. 262 pp. Ills. Lacks map. Some worming not affecting text. Orig bndg, rebacked, rubbed, shaken. *(Book House)* £28 [≈ $44]

Willich, A.F.M.
- Lectures on Diet and Regimen ... calculated chiefly for the Use of Families ... Fourth Edition, revised, corrected, and improved. London: for Longman ..., 1809. 8vo. xxiii,448 pp. Contemp half calf, rebacked, new label. *(Bickersteth)* £80 [≈ $128]

Willis, Robert
- Principles of Mechanism. Designed for the Use of Students in the Universities and for Engineering Students generally. London: 1841. 8vo. xxxi,446 pp. Figs & tables. Prize calf gilt, a.e.g., sl stained. *(Dillons)* £35 [≈ $56]

Willison, John
- Afflicted Man's Companion: or, A Directory for Families and Persons Afflicted with Sickness or other Duress with Directions to the Sick ... With a Collection of Dying Words. Glasgow: P. Tait, 1777. 12mo. 304,[4] pp. Stained. Leather, crudely reprd. *(Xerxes)* $175 [≈ £109]

Willmott, E.
- The Genus Rosa. London: [1910-] 1914. 2 vols. Folio. 132 colour-ptd plates, 1 full-page line drawing, 81 plates, text figs, after Alfred Parsons. A few sl stains on inner blank margs vol 1. New half mor. *(Wheldon & Wesley)* £775 [≈ $1,240]

Willocks, W.
- Egyptian Irrigation. London: Spon; New York: Spon & Chamberlain, 1899. 2nd edn. Sm 4to. xxviii,485 pp. Fldg plans, num plates, text ills. Orig cloth, edges a little rubbed, hd of spine sl frayed. *(Worldwide Antiqu'n)* $150 [≈ £93]

Wills, L.J.
- Concealed Coalfields. London: 1956. 8vo. xiii,208 pp. 2 fldg maps, figs. Orig cloth. Dw. *(Baldwin)* £25 [≈ $40]

Wilson, A.
- The Illustrated Natural History, being a Systematic Arrangement of Descriptive Zoology, edited by James Wylde. London: London Printing & Publishing Co, n.d. 4to. 847 pp. 36 cold plates, 2236 engvs. Half

leather, gilt dec spine, a.e.g.
(Egglishaw) £75 [≈ $120]

Wilson, Alexander & Bonaparte, Charles Lucien

- American Ornithology; or the Natural History of the Birds of the United States. Edited by Rbt. Jameson. London: Constable's Miscellany, 1831. 4 vols. 12mo. Addtnl engvd titles, port. Orig green linen bds, paper labels, upper cvr vol 3 waterstained, 1 label defective. *(Claude Cox)* £25 [≈ $40]
- American Ornithology; or, the Natural History of the Birds of the United States. London: Chatto & Windus, 1876. 3 vols. 8vo. Port, 103 colour-printed plates. Text foxed at ends, a few pp carelessly opened. Orig half mor, spine ends somewhat worn.
  (Wheldon & Wesley) £200 [≈ $320]

Wilson, C.

- The History of Unilever. London: Cassell, 1954. 2 vols. 335; 480 pp. Charts. Orig bndgs. Dws rubbed.
  (Book House) £20 [≈ $32]

Wilson, E.

- Birds of the Antarctic. Edited by B. Roberts. London: 1967. 4to. 60 cold & 42 plain plates. Orig cloth. *(Wheldon & Wesley)* £30 [≈ $48]

Wilson, Sir Erasmus

- On Diseases of the Skin. Philadelphia: 1847-63. 2nd Amer edn, plate vol 5th edn. 2 vols. 8vo. 20 cold litho plates. Ex-lib. Orig embossed cloth. *(Argosy)* $125 [≈ £78]

Wilson, George

- Religio Chemici. Essays. London: 1862. 1st edn. 12mo. viii,386 pp. Title vignette. Orig gilt dec cloth, spine ends worn, rear inner hinge open. *(Elgen)* $85 [≈ £53]

Wilson, J.O.

- Birds of Westmorland and the Northern Pennines. London: [1933]. 8vo. 319 pp. 153 ills. Orig cloth.
  (Wheldon & Wesley) £20 [≈ $32]

Wilson, John, d.1751

- A Synopsis of British Plants, in Mr. Ray's Method ... Newcastle upon Tyne: ptd by John Gooding, 1744. 1st edn. 8vo. 2 plates. Contemp calf, re-cvrd at an early date in reversed calf, cvrs panelled in blind & stitched inside, trifle rubbed.
  (Ximenes) $750 [≈ £468]

Wilson, L.G.

- Charles Lyell. The Years to 1841: The Revolution in Geology. London: 1972. xiii,553 pp. 17 maps, 52 ills. Orig cloth. dw.
  (Baldwin) £48 [≈ $76]

Wilson, Netta

- Alfred Owre, Dentistry's Militant Educator. Minneapolis: Univ of Minnesota, 1937. 1st edn. 8vo. 331 pp. Photo ills. Orig bndg, front bd spotted. *(Xerxes)* $75 [≈ £46]

Wilson, O.S.

- The Larvae of British Lepidoptera and their Food Plants ... London: Reeve, 1880. 4to. xxix,367 pp. 40 cold plates. Orig cloth.
  (Egglishaw) £140 [≈ $224]

Wilson, Samuel Alexander Kinnier

- Neurology ... Edited by A. Ninian Bruce. Baltimore: Williams & Wilkins, 1941. 1st edn. 2 vols. Tall 8vo. Ex-lib, margs dampstained. Orig cloth. *(Gach)* $75 [≈ £46]

Wilson, William

- Bryologia Britannica; containing the Mosses of Great Britain and Ireland. London: 1855. 8vo. xx,445 pp. 61 plain plates. Binder's cloth. A new (3rd) edn of Hooker & Taylor's Muscologia Britannica.
  (Wheldon & Wesley) £70 [≈ $112]

Wimmer, August

- Psychiatric Neurologic Examination Methods. Translated by Andrew Hoisholt. St. Louis, Mo.: Mosby, 1919. 1st edn. 8vo. 177 pp. Orig bndg. *(Xerxes)* $60 [≈ £37]

Winans, W.

- The Sporting Rifle. The Shooting of Big and Little Game, together with a Description of the Principal Classes of Sporting Weapons. London: 1908. Sm folio. Num ills. Orig cloth, sl rubbed. *(Grayling)* £100 [≈ $160]

Wingate, Edmund

- Mr Wingate's Arithmetick ... carefully Revis'd ... By John Kersey ... Fifteenth Edition. London: 1726. 8vo. viii,448 pp. Orig panelled calf, upper jnt cracked, lacks label, sl rubbed. *(Bickersteth)* £45 [≈ $72]

Winkworth, Susanna

- The Life and Letters of Barthold George Niebuhr, and Selections from his Minor Writings ... London: Chapman & Hall, 1852. 2nd edn. 3 vols. Contemp mor gilt, by Leighton, crnrs sl rubbed.

(John Smith) **£65 [≈ $104]**

Winslow, Forbes
- The Anatomy of Suicide. London: 1840. 1st edn. 8vo. Frontis (foxed). Orig cloth, sl worn, rear inner hinge broken.
(Robertshaw) **£100 [≈ $160]**

Winthrop, John
- Relation of a Voyage from Boston to Newfoundland, for the Observation of The Transit of Venus, June 6, 1761. Boston, N.E.: Edes & Gill, 1761. 24 pp. 1 diag.
(Jarndyce) **£380 [≈ $608]**

Winton, John G. & Millar, W.J.
- Modern Steam Practice and Engineering: A Guide to Approved Methods of Construction ... London: Blackie, 1883. 1st edn. One vol in 2. Roy 8vo. xiii,[2],529; [1],530-1120 pp. 14 plates, ca 800 ills. Minor foxing. Orig half calf, rubbed & worn but v strong.
(Fenning) **£75 [≈ $120]**

Wistar, Caspar
- A System of Anatomy. Philadelphia: T. Dobson, 1811-17. Vol 1 1st edn, vol 2 1st edn 2nd printing. 2 vols. Fldg table. Browning. Orig calf, vol 2 v worn. *(Elgen)* **$400 [≈ £250]**

Witherby, H.F., et al.
- The Handbook of British Birds. London: Witherby, 1938. 1st edn. 5 vols. 8vo. Plates. Orig cloth gilt.   *(Hollett)* **£95 [≈ $152]**
- The Handbook of British Birds. London: Witherby, 1938-48. 5 vols. Vol 5 contains Supplement of additions. 8vo. 157 cold & other plates. Occas sl foxing vol 1, a torn page reprd. Cloth.   *(Egglishaw)* **£52 [≈ $83]**
- The Handbook of British Birds. London: 1938-41. 1st edn. 5 vols. 8vo. 157 plates, mostly cold, maps & text figs. Orig cloth. The 52 pp Supplement loosely inserted.
(Wheldon & Wesley) **£125 [≈ $200]**
- The Handbook of British Birds. London: 1943-44. 2nd (rvsd) impression. 5 vols. 8vo. 157 plates, maps, text figs. A little minor foxing. Orig cloth. Dws.
(Wheldon & Wesley) **£120 [≈ $192]**
- The Handbook of British Birds. London: 1946. 5 vols. 8vo. 157 cold & other plates. Orig cloth. *(Wheldon & Wesley)* **£90 [≈ $144]**
- The Handbook of British Birds. London: Witherby, 1948. 5th imp. 5 vols. 8vo. 147 plates, mostly cold. Orig dark blue cloth.
(Gough) **£85 [≈ $136]**
- The Handbook of British Birds. London: 1948. 5 vols. 8vo. 157 cold & other plates.

Orig cloth. *(Wheldon & Wesley)* **£90 [≈ $144]**

Witney, Caspar, et al.
- Musk-Ox, Bison, Sheep & Goat. New York & London: Macmillan, American Sportsman's Library, 1904. 1st edn. 8vo. 284,advt pp. Frontis & other plates. Orig green cloth, sl worn.   *(Terramedia)* **$150 [≈ £93]**

Wittels, Fritz
- Sigmund Freud: His Personality, His Teaching, & His School. New York: Dodd, Mead, 1924. 1st Amer edn, ptd on British sheets. 8vo. 287,[1] pp. Frontis. Orig cloth.
(Gach) **$50 [≈ £31]**

Wittie, Robert
- Scarbrough Spaw, or, a Description of the Nature and Vertues of the Spaw at Scarborough in Yorkshire ... London: Charles Tyus, 1660. 1st edn. Cr 8vo. Errata leaf. 4 ff misbound but present. Contemp sheep, rubbed. Wing W.3231.
(Stewart) **£295 [≈ $472]**

Wodarch, C.
- Introduction to the Study of Conchology. Fourth Edition, by J. Mawe. London: 1831. 8vo. xii,149 pp. 7 hand cold plates. Blind stamps on title & plates. New bds.
(Wheldon & Wesley) **£45 [≈ $72]**

Wodzicki, K.A.
- Introduced Mammals of New Zealand. An Ecological and Economic Survey. London: 1950. 8vo. Maps, ills. Orig bndg. Dw.
(Grayling) **£25 [≈ $40]**

Wolberg, Lewis Robert
- The Psychology of Eating. New York: McBride, 1936. 1st edn. 8vo. 321 pp. Orig bndg, rear hinge cracked.
(Xerxes) **$20 [≈ £12]**

Wolf, Joseph
- The Life and Habits of Wild Animals ... see Elliot, Daniel Giraud.

Wolley-Dod, A.H.
- Flora of Sussex. Hastings: 1937. 1st edn. 8vo. lxxiii,571 pp. 2 maps (1 in pocket), 6 plates. Orig cloth.   *(Henly)* **£35 [≈ $56]**

Wood, Alexander
- Homoeopathy Unmasked; being an Exposure of its Principal Absurdities and Contradictions: with an Estimate of its Recorded Cures. Edinburgh: 1844. 1st sep

edn. 8vo. Orig cloth.
(Robertshaw) £18 [≈ $28]

Wood, Casey A.
- An Introduction to the Literature of Vertebrate Zoology. London: 1931. 1st edn. 4to. 643 pp. Cold frontis. Orig bndg.
(Elgen) $225 [≈ £140]
- An Introduction to the Literature of Vertebrate Zoology. (Oxford, 1931), reprint, Hildesheim, 1974. Roy 8vo. xix,643 pp. Orig cloth. *(Wheldon & Wesley)* £48 [≈ $76]

Wood, H.C.
- A Contribution to the History of the Fresh-Water Algae of North America. Washington: Smithsonian Contributions to Knowledge, 1872. 4to. x,262 pp. 21 plates (19 cold). Stamps on title & front wrapper which is bound in. Buckram.
(Wheldon & Wesley) £85 [≈ $136]

Wood, James
- The Elements of Algebra designed for the use of Students in the University ... Eighth Edition. Cambridge: J. Smith, 1825. 8vo. Contemp diced calf, mor label sl chipped. J. Cator's sgntr on endpaper.
(Waterfield's) £50 [≈ $80]

Wood, John George
- The Common Moths of England. London: Routledge, [1870]. 1st edn. Sm sq 8vo. [iv], 186,[ii] pp. 12 hand cold plates, text engvs. Frontis spotted. Orig plum cloth gilt, a.e.g.
(de Beaumont) £48 [≈ $76]
- The Common Moths of England. London: [189-]. Cr 8vo. iv,188 pp. 12 cold plates, 12 figs. Orig cloth.
(Wheldon & Wesley) £18 [≈ $28]
- Common Objects of the Microscope. London: Routledge, [ca 1870?]. Sm 8vo. iv,132 pp. 12 composite engvd plates & text ills. Orig cold pict paper bds, some rubbing.
(de Beaumont) £30 [≈ $48]
- The Handy Natural History. London: RTS, 1886. 8vo. xvi,367 pp. 226 engvs. Orig dec cloth gilt, a.e.g. *(Egglishaw)* £11 [≈ $17]
- Homes Without Hands, being a Description of the Habitations of Animals. London: Longmans, 1866. 8vo. xix,632 pp. Num engvs. Calf gilt. *(Egglishaw)* £26 [≈ $41]
- The Illustrated Natural History, with New designs by Wolf, Zwecker, Weir, Coleman, Harvey etc. engraved by the Brothers Dalziel. London: Routledge, 1862-63. 3 vols. Roy 8vo. Half calf gilt, red label, mrbld edges.
(Egglishaw) £45 [≈ $72]

- Insects Abroad. London: Longmans, Green, 1874. 1st edn. 8vo. xii,780,48 ctlg pp. 600 ills. Orig green cloth gilt, short splits at hd of hinges. *(Claude Cox)* £18 [≈ $28]
- Insects Abroad. New Edition. London: Longmans, 1883. 8vo. xii,780 pp. 20 plates, num ills. Prize mor gilt, a.e.g., sl scuffed.
(Fenning) £28.50 [≈ $46]
- Insects at Home. London: Longmans, Green, 1872. 1st edn. 8vo. xx,670,24 ctlg pp. Chromolitho frontis, over 700 w'engvs. Orig green cloth gilt, a little rubbed & shaken.
(Claude Cox) £18 [≈ $28]
- Insects at Home. London: Longmans, Green, 1883. New edn. 8vo. xx,670,24 ctlg pp. Cold frontis, over 700 figs. Orig brown cloth gilt, a.e.g., jnts sl split, hd of spine frayed.
(Blackwell's) £45 [≈ $72]
- Insects at Home. London: 1883. New edn. 8vo. xx,670 pp. Cold frontis (trifle stained), 20 plates, num text figs. Orig cloth gilt, cvrs trifle stained.
(Wheldon & Wesley) £20 [≈ $32]

Wood, N.A.
- The Birds of Michigan. Ann Arbor: 1951. 8vo. 559 pp. 16 plates. A few sm lib stamps. Mod qtr mor.
(Wheldon & Wesley) £25 [≈ $40]

Wood, S.
- The Tree Planter and Plant Propagator. London: Crosby Lockwood, 1880. 1st edn. 8vo. xii,188,xii,153,48 advt pp. 107 w'engvs in text. Orig pict qtr straight-grain mor gilt.
(Gough) £25 [≈ $40]

Wood, Samuel
- The Bulb Garden or How to Cultivate Bulbous and Tuberous-Rooted Flowering Plants to Perfection ... London: 1878. 1st edn. Sm 8vo. viii,128,48 advt pp. 8 cold plates, 22 ills. Minor spots. Orig dec cloth gilt. *(Bow Windows)* £78 [≈ $124]

Wood, William
- General Conchology: or, a Description of Shells arranged according to the Linnean System. Vol 1. London: 1815. All published. 8vo. iv,lxi,7,246 pp. 60 hand cold plates (numbered 1-59, 4*). Blind stamp on title, sm blind stamp on plates. New buckram.
(Wheldon & Wesley) £200 [≈ $320]
- Illustrations of the Linnean Genera of Insects. London: 1821. 3 vols in one. Sm 8vo. 86 hand cold plates. Edges sl shaved, affecting some plate numerals. 1 leaf of text torn without loss. Contemp half leather.

(Wheldon & Wesley) £185 [≈ $296]
- Index Entomologicus. A Complete Illustrated Catalogue of the Lepidopterous Insects. New and Revised Edition, with a Supplement by J.O. Westwood. London: 1854. Roy 8vo. viii,21,298 pp. 59 hand cold plates. Qtr mor, lib emblem on front cvr.
(Henly) £180 [≈ $288]

Woodcock, H.B. & Stearn, W.T.
- Lilies of the World. London: Country Life, 1949. 1st edn. Roy 8vo. 132 plates. Orig cloth, spine faded.   *(Hollett)* £40 [≈ $64]

Woodhouse, L.G.O.
- The Butterfly Fauna of Ceylon. Colombo: 1950. 2nd (abridged) edn. 4to. xvi,133 pp. Map, 37 cold & 12 plain plates. Tiny stain in lower marg of a few plates, 2 tissues reprd. Orig cloth, crnr bumped.
(Wheldon & Wesley) £50 [≈ $80]
- The Butterfly Fauna of Ceylon. Colombo: 1951. 2nd complete edn. 4to. xxxii,231 pp. Map, 37 cold & 19 plain plates. Orig cloth.
(Wheldon & Wesley) £75 [≈ $120]

Woodruff, Charles Edward
- Expansion of Races. New York: Rebman, 1909. 1st edn. 8vo. 495 pp. Orig bndg.
(Xerxes) $70 [≈ £43]

Woodward, A.S.
- The Fossil Fishes of the English Chalk. London: Pal. Soc. Monograph, 1902-12. 7 parts, complete. viii,264 pp. 54 plates. Disbound.   *(Baldwin)* £50 [≈ $80]

Woodward, Augustus B.
- Considerations on the Substance of the Sun. Washington: Way & Groff, Sept. 1801. 1st edn.89,[1] pp. Interleaved with blanks. Minimal foxing & light transparent stains to ff at ends. Contemp calf, somewhat chipped & worn, inner hinges strengthened.
(Reese) $950 [≈ £593]

Woodward, H.B.
- History of the Geological Society of London. London: 1907. 8vo. xix,336 pp. Ills. Orig cloth.   *(Baldwin)* £40 [≈ $64]

Woodward, John
- An Essay towards a Natural History of the Earth, and Terrestrial Bodyes, Especially Minerals ... Third Edition. London: 1723. 8vo. [xii],304,[1 advt,3 blank] pp. Minor marks. Contemp gilt ruled calf, mor label.
(Bow Windows) £275 [≈ $440]

- The Natural History of the Earth, illustrated, inlarged and defended. Written originally in Latin and now first made English by Benj. Holloway ... London: Tho. Edlin, 1726. 2 vols in one. 8vo. Contemp calf, sl rubbed.
(Falkner) £240 [≈ $384]
- The State of Physick: and of Diseases; With an Inquiry into the Causes in the late Increase of them: But more particularly of the Small-Pox ... London: for T. Horne ..., 1718. 5 ff, 274 pp,5 ff,2 pp. Contemp calf, cvr detaching.   *(Elgen)* $300 [≈ £187]

Woodward, R.B. & J.D.S.
- Natal Birds, including the Species belonging to Natal and the Eastern Districts of the Cape Colony. Pietermaritzburg: 1899. 8vo. vi,215,ii,v pp. Cold plate. A few annotations. Orig cloth, back cvr affected by damp.
(Wheldon & Wesley) £85 [≈ $136]

Woodward, S.
- An Outline of the Geology of Norfolk. London: 1833. 1st edn. 8vo. 54,[iv] pp. Cold geological map, 6 plates. Orig cloth.
(Baldwin) £45 [≈ $72]

Woolward, Florence H.
- The Genus Masdevallia. Issued by the Marquess of Lothian chiefly from Plants in his Collection at Newbattle Abbey. London: 1890-96. One of 250. Folio. Cold map, 87 hand cold plates. New half mor.
(Wheldon & Wesley) £6,000 [≈ $9,600]

Wooster, David
- Alpine Plants ... Second Series. London: 1874. 1st edn. 8vo. [iv],140 pp. 54 cold plates. Tear in flyleaf reprd, b'plate removed. Orig gilt illust blue cloth.
(Bow Windows) £145 [≈ $232]
- A Manual of the Coniferae, containing a General Review of the Order ... [Compiled by D. Wooster] ... London: James Veitch and Sons, 1881. 1st edn. 8vo. [iv],350 pp. 3 fldg plates, num other ills. Some foxing. Orig dec cloth gilt, extremities rubbed.
(Finch) £90 [≈ $144]

Worden, E.C.
- Nitrocellulose Industry. A Compendium of the History, Manufactures, Commercial Application and Analysis ... Peaceful Arts ... Chapter on Gun Cotton ... Explosive Cellulose Nitrates. London: Constable, 1911. 2 vols. 4to. 1239 pp. Ills. Orig bndg, vol 2 cvrs marked.   *(Book House)* £30 [≈ $48]

Worlidge, John
- The Second Parts of Systema Agriculturae, or
the Mystery of Husbandry, and Vinetum
Britannicum, or a Treatise of Cider. London:
1689. 8vo. xlix,[iii],191 pp. Engvd title & 5
plates. Licence leaf before title. Contemp
sheep, jnts cracked.
(Wheldon & Wesley) **£180 [≈$288]**
- Systema Agriculturae; The Mystery of
Husbandry Discovered ... Kalendarium
Rusticum ... Dictionarium Rusticum ...
Second Edition ... Additions ... By J.W. Gent.
London: 1675. 3 parts in 1 vol. Sm folio.
[32],324,[4] pp. Frontis, ills. Old sheep,
worn.   *(M & S Rare Books)* **$350 [≈£218]**
- Systema Agriculturae. The Mystery of
Husbandry Discouvered. To which is added
'Kalendarium Rusticum' and 'Dictionarium
Rusticum'. Los Angeles: Sherwin & Freutel,
1970. Facs reprint of 2nd edn, London, 1675.
Roy 8vo. Frontis. Orig cloth.
(Stewart) **£30 [≈$48]**
- Systema Agriculturae; the Mystery of
Husbandry Discovered ... to which is added
Kalendarium Rusticum ... and Dictionarium
Rusticum ... London: 1681. 3rd edn. Folio.
[xxviii],326,[6] pp. Irregular pagn. Frontis
(with leaf of explanation). Contemp calf,
rebacked. *(Wheldon & Wesley)* **£325 [≈$520]**

Worster, Benjamin
- A Compendious and Methodical Account of
the Principles of Natural Philosophy ...
London: for the author, & sold by W. & J.
Innys ..., 1722. 1st edn. 8vo. viii,[iv], 239,[1
advt] pp. W'cut text ills. Contemp panelled
calf, new label, jnts just cracking.
(Pickering) **$850 [≈£531]**
- A Compendious and Methodical Account of
the Principles of Natural Philosophy ...
Second Edition. Revis'd and Corrected ...
London: for Stephen Austen, 1730. 8vo. xviii,
[ii],269,[3] pp. Intl advt leaf. Some marg
worm, sl waterstaining. Contemp calf,
rubbed.   *(Burmester)* **£70 [≈$112]**

Worth, Claud
- Squint. Its Causes, Pathology and Treatment.
London: 1906. 234 pp. Ills. Orig bndg.
(Moorhead) **£18 [≈$28]**

Worthington, A.M.
- A Study of Splashes. With 197 Illustrations
from Instantaneous Photographs. London:
Longmans, Green, 1908. 1st edn. Super roy
8vo. xii,130,[2] pp. Ills. Orig gilt dec cloth,
spine ends & crnrs v sl rubbed.
(Duck) **£60 [≈$96]**

Wredden, J.H.
- The Microscope - Its Theory and
Applications. London: Churchill, 1947. 1st
edn. 8vo. xxiv,296 pp. Photo ills, text figs.
Orig cloth, spine rubbed, crnrs bumped.
(Savona) **£16 [≈$25]**

Wright, E. Perceval
- Cassell's Concise Natural History, being a
Complete Series of Descriptions of Animal
Life. London: Cassell, [1891]. [vi],618,16
advt pp. Num steel engvs, other ills. Orig
bright blue cloth gilt over bevelled bds, hd of
spine sl pulled.   *(Sklaroff)* **£25 [≈$40]**

Wright, J.
- The Flower Grower's Guide. London: [1897].
3 vols in 6. 4to. 46 cold plates, 519 text ills.
Orig dec cloth.
(Wheldon & Wesley) **£60 [≈$96]**
- The Fruit Grower's Guide. London: 1924.
New & rvsd edn. 2 vols. Imperial 8vo. 24 cold
plates, num text figs. A little minor foxing.
Orig cloth, trifle used.
(Wheldon & Wesley) **£65 [≈$104]**

Wright, L.
- Light. A Course of Experimental Optics
chiefly with the Lantern. London: 1892. 2nd
edn, rvsd & enlgd. xv,391 pp. 9 plates, 207
text figs. Orig bndg. *(Whitehart)* **£35 [≈$56]**

Wright, Lewis
- The New Book of Poultry. London: [1902].
4to. viii,600 pp. 46 plates (30 cold) by J.W.
Ludlow. Marg foxing to a few plates, sm marg
reprs in 2 plates, 1 plate sl trimmed. Half
mor.   *(Wheldon & Wesley)* **£180 [≈$288]**
- The Practical Poultry Keeper. London:
Cassell, 1912. New edn. Cr 8vo. viii,315 pp.
8 cold plates, other ills. Orig green cloth.
(Blackwell's) **£35 [≈$56]**

Wright, R.G. & Dewar, D.
- The Ducks of India. London: 1925. 8vo. 231
pp. 22 cold plates. Orig buckram, spine faded,
new front endpaper.   *(Henly)* **£70 [≈$112]**

Wright, R.P. (ed.)
- The Standard Cyclopaedia of Modern
Agriculture and Rural Economy ... London:
Gresham Publishing ..., 1908-11. 12 vols.
8vo. 33 cold & 188 other plates, num text ills,
3 cold anatomical models with overlays. Sl
used. Orig pict cloth, sl used.
(Bow Windows) **£75 [≈$120]**

Wright, W.H.
- The Black Bear. London: [ca 1900]. 8vo. Ills. Orig bndg, sl rubbed. *(Grayling)* £30 [≃ $48]

Wright, W.P.
- Alpine Flowers and Rock Gardens. London: n.d. 1st edn. 8vo. 292 pp. 40 mtd cold & 15 plain plates. Sl marg foxing. Orig pict cloth. *(Henly)* £48 [≃ $76]

Wundt, Wilhelm Max
- Elements of Folk Psychology: Outlines of a Psychological History of the Development of Mankind. Translated by Edward Leroy Schaub. London: Allen & Unwin; New York: Macmillan, [1921]. 1st edn in English, 2nd printing. 8vo. Orig cloth, some scratching to cvrs. *(Gach)* $125 [≃ £78]

Wyld, R.S.
- The Physics and Philosophy of the Senses; or, the Mental and the Physical in their Mutual Relation. London: Henry S. King, 1875. 1st edn. 8vo. [xvi],562,[2] pp. Orig mauve cloth. *(Gach)* $75 [≃ £46]

Wylde, James (ed.)
- The Circle of the Sciences. With an Introductory Discourse ... by Henry, Lord Brougham. London: London Printing & Publishing Co, [1862]. 2 vols. 4to. 63 plates, num w'cut vignettes. Sl spotting. Contemp half calf, gilt spines, rubbed, spines sl frayed at ft. *(Clark)* £45 [≃ $72]
- The Illustrated Natural History ... see Wilson, A.

Yarrell, William
- A History of British Birds. London: Van Voorst, 1843. 1st edn. 3 vols. 8vo. 520 text w'cuts. Mod three qtr mor gilt. *(Hollett)* £150 [≃ $240]
- A History of British Birds. Illustrated with 520 Wood-Engravings. London: Van Voorst, 1843. 1st edn in book form. 3 vols. 8vo. Contemp polished leather gilt, vol 1 spine relaid, spines & jnts sl rubbed, spine labels chipped. Yorkshire Club b'plate. *(Schoyer)* $225 [≃ £140]
- A History of British Birds. Illustrated by 520 Wood-Engravings. London: 1843. 3 vols. 8vo. Some foxing & other marks. Contemp half calf, spines relaid. *(Bow Windows)* £105 [≃ $168]
- A History of British Birds. London: Van Voorst, 1845. 2nd edn. 3 vols. 8vo. 535 text w'cuts. Mod cloth gilt. *(Hollett)* £95 [≃ $152]
- A History of British Birds. Illustrated by 550

Wood Engravings. Third Edition with many additions. London: Van Voorst, 1856. 3 vols. 8vo. xxxvi,614; [4],702; [4],679 pp. Contemp calf gilt extra, unusual red & tan mottled sides, a.e.g., extremities rubbed. *(Claude Cox)* £100 [≃ $160]
- A History of British Birds. London: Van Voorst, 1865. 2nd edn. 3 vols. 8vo. 535 w'engvs. Contemp half calf. *(Egglishaw)* £68 [≃ $108]
- A History of British Birds. Fourth Edition, by A. Newton and H. Saunders. London: 1871-85. 4 vols. 8vo. 564 w'engvs. Polished calf, jnts of 1 vol reprd, other jnts a little worn. *(Wheldon & Wesley)* £85 [≃ $136]
- A History of British Fishes. London: 1836. Large Paper. 2 vols. Roy 8vo. Nearly 400 w'cuts. Calf, trifle rubbed. *(Wheldon & Wesley)* £75 [≃ $120]
- A History of British Fishes. London: Van Voorst, 1836. 1st edn. 2 vols. 8vo. Nearly 400 w'engvs. Later half calf, sl rubbed. *(Egglishaw)* £78 [≃ $124]
- A History of British Fishes. London: 1859. 3rd edn. 2 vols. 8vo. 522 w'engvs. Sm lib blind stamps on title & in text. Buckram. *(Henly)* £65 [≃ $104]
- A History of British Fishes. London: 1859. 3rd edn. 2 vols. 8vo. Port, num w'cuts. Orig cloth, trifle used. *(Wheldon & Wesley)* £55 [≃ $88]
- A History of British Fishes. Edited by Sir John Richardson. London: Van Voorst, 1859. 3rd edn. 2 vols. 8vo. Port, 522 w'engvs. Orig cloth. *(Egglishaw)* £46 [≃ $73]
- A History of British Fishes. Edited by Sir John Richardson. London: Van Voorst, 1859. 3rd edn. 2 vols. 8vo. Port, 522 w'engvs. Orig cloth, stained. *(Egglishaw)* £26 [≃ $41]
- On the Growth of the Salmon in Fresh Water. London: 1839. Oblong folio. 3 pp. 3 hand cold plates. Sm hole in plate 3, plate 2 sl waterstained. Orig wraps, rebacked, creased in the middle, tear in back cvr reprd, margs sl frayed. *(Claude Cox)* £80 [≃ $128]

Yates, William & Maclean, Charles
- A View of the Science of Life, on the Principles ... of ... John Brown ... Philadelphia: William Young, 1797. 1st Amer edn. 8vo. 232 pp. 3 subsidiary title-pages. Old mottled calf, ex-lib. *(Argosy)* $175 [≃ £109]

Yeaton, Charles
- Manual of the Alden Type-Setting and Distributing Machine ... New York: Francis Hart & Co, 1865. 1st edn. "Stockholder's

Copy". One of 100. Folio. [8],245,[3] pp. Num ills. Orig three qtr mor, rubbed, sm piece gone from ft of spine, orig mor gilt pres label. *(M & S Rare Books)* **$2,500 [≈£1,562]**

Yerkes, Robert Mearns & Ada W.
- The Great Apes: A Study of Anthropoid Life. New Haven: Yale UP, 1929. 1st edn. Lge 8vo. [ii],xx,652,[4] pp. 172 text ills. Lib stamps at front, inscrptn on flyleaf. Orig cloth, faint spine stamp. *(Gach)* **$85 [≈£53]**

Yerkes, Robert Mearns (ed.)
- Psychological Examining in the United States Army ... Washington, DC: GPO, 1921. 1st edn. 4to. [ii],vi,890,[2] pp. Ex-lib, sl marked. Mauve cloth. *(Gach)* **$100 [≈£62]**

Yonge, C.M.
- The Sea Shore. London: Collins, New Naturalist Series, 1949. 1st edn. Cold & plain plates. Orig cloth. Dw.
(Egglishaw) **£12 [≈$19]**

Youatt, William
- Cattle: Their Breeds, Management, and Diseases. London: SPCK ..., 1843. 8vo. viii, 600 pp. Text ills. A few pp soiled. Diced calf gilt, front jnt cracking. *(Hollett)* **£75 [≈$120]**
- Cattle: Their Breeds, Management, and Diseases. London: Simpkin, Marshall, 1876. 8vo. viii,600 pp. Text ills. Mod half levant mor gilt. *(Hollett)* **£45 [≈$72]**

Young, Arthur
- The Farmer's Guide to Hiring and Stocking Farms ... London: Strahan, 1770. 2 vols. 8vo. 10 engvs (4 fldg). Sl discold. Contemp tan calf, gilt spines, jnts rubbed.
(Blackwell's) **£175 [≈$280]**
- The Farmer's Kalendar, or a Monthly Directory for all sorts of Country Business ... London: for Robinson & Roberts ... & J. Knox, 1771. 1st edn. 8vo. [xxxii],399 pp, inc half-title. Contemp sheep, upper jnt reprd.
(Pickering) **$450 [≈£281]**
- The Farmer's Calendar. London: 1804. New edn. 8vo. x,604 pp. 2 plates. Calf, rubbed.
(Wheldon & Wesley) **£50 [≈$80]**
- The Farmer's Letters to the People of England ... added Sylvae: or Occasional Tracts on Husbandry and Rural Oeconomics. The Second Edition, corrected and enlarged. London: W. Nicoll, 1768. 8vo. [6],482 pp. Early 19th c half calf, upper jnt v sl cracked. *(Spelman)* **£90 [≈$144]**
- The Farmer's Tour through the East of England ... London: Strahan, 1771. 1st edn.

4 vols. 8vo. Half-title in vol 1. 28 plates (17 fldg), fldg table. Some sl discolouration. Contemp tan calf, some jnts cracked & some spine ends chipped.
(Blackwell's) **£210 [≈$336]**
- The Farmer's Tour through the East of England. London: 1771. 4 vols. 8vo. 29 plates, table. Calf, trifle used.
(Wheldon & Wesley) **£160 [≈$256]**
- General View of the Agriculture of the County of Lincoln. London: 1799. 1st edn. 8vo. 3 maps, 11 plates. Contemp half calf. *(Robertshaw)* **£100 [≈$160]**
- General View of the Agriculture of the County of Norfolk; drawn up for the Consideration of the Board of Agriculture ... London: Richard Phillips, 1804. 1st edn. 8vo. xx, 532,2 pp. Cold fldg map, 7 plates (sl foxed) & diags. New qtr calf, untrimmed.
(Blackwell's) **£100 [≈$160]**
- General View of the Agriculture of the County of Suffolk; drawn up for the consideration of the Board of Agriculture and Internal Improvement. London: Bulmer, 1804. 3rd edn. Cold fldg map, 2 plates (browned). Final ff spotted. New grey bds.
(Stewart) **£60 [≈$96]**
- Letters concerning the Present State of the French Nation ... London: W. Nicoll, 1769. 1st edn. 8vo. iv,[ix],497 pp. Contemp sprinkled calf, sl rubbed, jnts cracking, lacks label. *(Blackwell's)* **£175 [≈$280]**
- A Six Weeks Tour through the Southern Counties of England and Wales ... London: for W. Nicoll, 1768. 1st edn. 8vo. [iv],284 pp. Some foxing in the 1st gathering. Contemp calf gilt, red mor label.
(Pickering) **$470 [≈£293]**
- A Six Weeks Tour through the Southern Counties of England and Wales ... By the Author of The Farmer's Letters. London: for W. Nicholl, 1768. 8vo. Mod qtr calf.
(Waterfield's) **£250 [≈$400]**
- Travels during the Years 1787, 1788, & 1789 ... Kingdom of France ... London: 1794. 2nd edn. 2 vols in one. Lge thick 4to. 629, [3]; 336,[4] pp. 3 fldg maps (1 hand cold). Contemp calf, spine gilt extra, gilt dentelles, outer hinges worn, rear broken & loose.
(Reese) **$850 [≈£531]**

Young, Charles Frederic T.
- The Economy of Steam Power on Common Roads ... With its History and Practice in Great Britain. London: Atchley & Co, [1861]. 8vo. 7 plates (1 fldg). Orig gilt dec brown cloth, spine faded.
(Waterfield's) **£165 [≈$264]**

Young, Edward
- The Ferns of Wales. Neath: 1856. 4to. v,29 pp. Cold pict title (sl foxed), 28 ff of 34 mtd specimens. Tissue guards. A few sm pieces of frond missing. Orig cloth gilt, sm splits in jnts, spine ends sl defective.
(Egglishaw) £130 [≈ $208]
- The Ferns of Wales. Neath: Thomas Thomas, 1856. Only edn. Folio. v,29 pp. Addtnl tinted litho title, 34 actual specimens on 28 plates. 1 tissue guard creased. Orig cloth gilt, 2 v sm nicks in backstrip. Orig prospectus tipped-in. *(Fenning)* £265 [≈ $424]

Young, Francis C.
- Every Man His Own Mechanic. A Complete and Comprehensive Guide to Every Description of Constructive and Decorative Work ... Tenth Edition. London: Ward Lock ..., 1893. 8vo. x, 924,[16 advt] pp. 3 fldg plates, 850 ills. Orig cloth gilt, inner hinges weak. *(Fenning)* £45 [≈ $72]

Young, George
- A Geological Survey of the Yorkshire Coast ... Whitby: R. Kirby, 1828. 2nd edn, enlgd. 4to. iv,356 pp. Frontis (spotted), hand cold map, dble-page plate of sections, 17 hand cold plates. Title sl browned. Minor loss from last plate. Half calf gilt, sl rubbed.
(Hollett) £250 [≈ $400]

Young, George & Bird, J.
- A Geological Survey of the Yorkshire Coast. London: 1822. 1st edn. 4to. iv,332 pp. Cold geological map & sections, 17 hand cold plates (16 of fossils). Offsetting from frontis to title, sm ink mark on 1 page. Half calf, mrbld bds. *(Baldwin)* £200 [≈ $320]
- A Geological Survey of the Yorkshire Coast. London: 1828. 2nd edn. 4to. iv,367 pp. Cold geological map & sections, 17 hand cold plates of fossils. Frontis & title sl stained. Mrbld bds, new buckram spine.
(Baldwin) £150 [≈ $240]

Young, Thomas
- Miscellaneous Works ... Edited by George

Peacock ... [With] Life of Thomas Young ... London: Murray, 1855. 1st edns. 4 vols (Works 3 vols, Life 1 vol). 8vo. 2 frontis, 20 plates. Orig uniform brown cloth, spine ends v sl frayed. *(Pickering)* $1,200 [≈ £750]

Zeeman, P.
- Researches in Magneto-Optics ... London: 1913. 1st edn. xvi,219 pp. 8 plates, 74 diags. Orig cloth, sl stained, edge of spine sl marked.
(Whitehart) £45 [≈ $72]

Zeitlinger, Heinrich
- Bibliotheca Chemico-Mathematica ... see Sotheran, Henry.

Zeller, H.
- Wild Flowers of the Holy Land. London: Nisbet, 1877. Roy 8vo. xiv pp. 54 cold plates. New cloth, a.e.g.
(Egglishaw) £80 [≈ $128]
- Wild Flowers of the Holy Land. London: 1883. 3rd edn. Roy 8vo. xiv pp. 54 cold plates. Occas v sl foxing. Front free endpaper removed. Orig cloth.
(Wheldon & Wesley) £85 [≈ $136]

Zeuner, F.E.
- The Pleistocene Period. Its Climate, Chronology and Faunal Successions. London: Ray Society, 1945. [One of 600]. 8vo. xii,322 pp. Text figs. Orig cloth.
(Wheldon & Wesley) £35 [≈ $56]

Ziemssen, H. Von (ed.)
- Handbook of Diseases of the Skin. New York: Wood, 1885. 1st Amer edn. 4to. 658 pp. Ills. Orig bndg. *(Elgen)* $75 [≈ £46]

Zimmer, George Frederick
- The Mechanical Handling & Storing of Material ... London: Crosby Lockwood, 1916. 1st edn. Post 4to. xiv,752 pp. Cold frontis, over 1000 ills, many fldg. 2 minor misfoldings. Orig cloth, endpapers spotty.
(Duck) £45 [≈ $72]

Catalogue Booksellers Contributing to IRBP

The booksellers who have provided catalogues during 1989 specifically for the purpose of compiling the various titles in the *IRBP* series, and from whose catalogues books have been selected, are listed below in alphabetical order of the abbreviation employed for each. This listing is therefore a complete key to the booksellers contributing to the series as a whole; only a proportion of the listed names is represented in this particular subject volume.

The majority of these booksellers issue periodic catalogues free, on request, to potential customers. Sufficient indication of the type of book handled by each bookseller can be gleaned from the individual book entries set out in the main body of this work and in the companion titles in the series.

Adelson	=	Richard H. Adelson, Antiquarian Bookseller, North Pomfret, Vermont 05053, U.S.A. (802 457 2608)
Ambra	=	Ambra Books, 22 West Shrubbery, Redland, Bristol BS6 6TA, England (0272 741962)
Ampersand	=	Ampersand Books, P.O. Box 674, Cooper Station, New York City 10276, U.S.A. (212 674 6795)
Antic Hay	=	Antic Hay Rare Books, P.O. Box 2185, Asbury Park, NJ 07712, U.S.A. (201 774 4590)
Argosy	=	Argosy Book Store, Inc., 116 East 59th Street, New York, NY 10022, U.S.A. (212 753 4455)
Ars Libri	=	Ars Libri, Ltd., 560 Harrison Avenue, Boston, Massachusetts 02118, U.S.A. (617 357 5212)
Astoria	=	Astoria Books & Prints, 1801 Lawrence Street, Denver, CO 80202, U.S.A. (303 292 4122)
Baldwin	=	Baldwin's Books, Fossil Hall, Boars Tye Road, Silver End, Witham, Essex CM8 3QA, England (0376 83502)
Bates & Hindmarch	=	Bates and Hindmarch, Antiquarian Bookseller, Fishergate, Boroughbridge, North Yorkshire Y05 9AL, England (0423 324258)
Berkelouw	=	Messrs. Berkelouw, Antiquarian Booksellers, 830 N. Highland Ave., Los Angeles, CA 90038, U.S.A. (213 466 3321)
Bernett	=	F.A. Bernett Inc., 2001 Palmer Avenue, Larchmont, N.Y. 10538, U.S.A. (914 834 3026)
Beyer	=	Preston C. Beyer, 752A Pontiac Lane, Stratford, Connecticut 06497, U.S.A. (203 375 9073)
Bickersteth	=	David Bickersteth, 4 South End, Bassingbourn, Royston, Hertfordshire SG8 5NG, England ((0763) 45619)
Blackwell's	=	Blackwell's Rare Books, B.H. Blackwell Ltd., Fyfield Manor, Fyfield, Abingdon, Oxon OX13 5LR, England (0865 390692)
Black Sun Books	=	Black Sun Books, P.O. Box 7916 - F.D.R. Sta., New York, New York 10150-1915, U.S.A. (212 688 6622)
Bookmark	=	Bookmark, Children's Books, Fortnight, Wick Down, Broad Hinton, Swindon, Wiltshire SN4 9NR, England (0793 73693)
Book Block	=	The Book Block, 8 Loughlin Avenue, Cos Cob, Connecticut 06807, U.S.A. (203 629 2990)
Book House	=	The Book House, Grey Garth, Ravenstonedale, Kirkby Stephen, Cumbria CA17 4NQ, England (058 73634)
Bow Windows	=	Bow Windows Book Shop, 128 High Street, Lewes, East Sussex BN7 1XL, England (0273 480780)
Box of Delights	=	The Box of Delights, 25 Otley Street, Skipton, North Yorkshire BD23 1DY, England (0756 60111)

Bromer	=	Bromer Booksellers, 607 Boylston Street, at Copley Square, Boston, MA 02116, U.S.A. (617 247 2818)
Buckley	=	Brian & Margaret Buckley, 11 Convent Close, Kenilworth, Warwickshire CV8 2FQ, England (0926 55223)
Burmester	=	James Burmester, Manor House Farmhouse, North Stoke, Bath BA1 9AT, England (0272 327265)
Castle	=	The Castle Bookshop, 37 North Hill, Colchester, Essex CO1 1QR, England (0206 577520)
Chapel Hill	=	Chapel Hill Rare Books, P.O. Box 456, Carrboro, NC 27510, U.S.A. (919 929 8351)
City Spirit	=	City Spirit Books, 1434 Blake Street, Denver, Colorado 80202, U.S.A. (303 595 0434)
Clark	=	Robert Clark, 24 Sidney Street, Oxford OX4 3AG, England (0865 243406)
Collected Works	=	Collected Works, 3 Melbourne Terrace, Melbourne Grove, London SE22 8RE, England (01 299 4195)
Coombes	=	A.J. Coombes, Bookseller, 24 Horsham Road, Dorking, Surrey RH4 2JA, England (0306 880736)
Claude Cox	=	Claude Cox, The White House, Kelsale, Saxmundham, Suffolk IP17 2PQ, England (0728 602786)
Dalian	=	Dalian Books, David P. Williams, 81 Albion Drive, London Fields, London E8 4LT, England (01 249 1587)
de Beaumont	=	Robin de Beaumont, 25 Park Walk, Chelsea, London SW10 0AJ, England (01 352 3440)
Dillons	=	Dillons, 82 Gower Street, London WC1E 6EQ, England (01 636 1577)
Doyle	=	Mrs. Assia Doyle, Teffont, Salisbury SP3 5QF, England
Duck	=	William Duck, The Glebe House, Brightling, East Sussex TN32 5HE, England (042 482 295)
Edrich	=	I.D. Edrich, 17 Selsdon Road, London E11 2QF, England (01 989 9541)
Egglishaw	=	H.J. Egglishaw, Bruach Mhor, 54 West Moulin Road, Pitlochry, Perthshire PH16 5EQ, Scotland (0796 2084)
Egret	=	Egret Books, 6 Priory Place, Wells, Somerset BA5 1SP, England (0749 79312)
Elgen	=	Elgen Books, 336 DeMott Avenue, Rockville Centre, New York 11570, U.S.A. (516 536 6276)
Emerald Isle	=	Emerald Isle Books, 539 Antrim Road, Belfast BT15 3BU, Northern Ireland (0232 370798)
Falkner	=	Falkner Greirson & Co Ltd, Glenaglogh, Tallow, Co. Waterford, Ireland, Eire (058 56349)
Farahar	=	Clive Farahar, XIV The Green, Calne, Wiltshire SN11 8DG, England (0249 816793)
Fenning	=	James Fenning, 12 Glenview, Rochestown Avenue, Dun Laoghaire, County Dublin, Eire (01 857855)
Finch	=	Simon Finch Rare Books, Clifford Chambers, 10 New Bond Street, London W1Y 9PF, England (01 499 0799)
Firsts & Company	=	Firsts & Company, 1066 Madison Avenue, New York, New York 10028, U.S.A. (212 249 4122)
First Issues	=	First Issues Ltd, 7 York Terrace, Dorchester, Dorset DT1 2DP, England (0305 65583)
Frew Mackenzie	=	Frew Mackenzie plc, 106 Great Russell Street, London WC1B 3NA, England (01 580 2311)
Gach	=	John Gach Books, 5620 Waterloo Road, Columbia, Md. 21045, U.S.A. (301 465 9023)
Gekoski	=	R.A. Gekoski, 33B Chalcot Square, London NW1 8YA, England (01 722 9037)

Gemmary	=	The Gemmary, Inc, PO Box 816, Redondo Beach, CA 90277, U.S.A. (213 372 5969)
Gough	=	Simon Gough Books, 5 Fish Hill, Holt, Norfolk, England (026371 2650)
Grayling	=	David A.H. Grayling, Lyvennet, Crosby Ravensworth, Penrith, Cumbria CA10 3JP, England (09315 282)
Green Meadow Books	=	Green Meadow Books, Kinoulton, Notts NG12 3EN, England (0949 81723)
Greyne	=	Greyne House, Marshfield, Chippenham, Wiltshire SN14 8LU, England (0225 891279)
Handy Book	=	Handy Book, 1762 Avenue Road, Toronto, Ontario M5M 3Y9, Canada (416 781 4139)
Hannas	=	Torgrim Hannas, 29a Canon Street, Winchester, Hampshire SO23 9JJ, England (0962 62730)
Harrison	=	T. Harrison (Books), 25 Clayfields, Wentworth, nr Rotherham, Yorkshire S62 7TD, England (0226 742097)
Hartfield	=	Hartfield, Fine and Rare Books, 117 Dixboro Road, Ann Arbor, MI 48105, U.S.A. (313 662 6035)
Hartley	=	Hartley Moorhouse Books, 142 Petersham Road, Richmond, Surrey TW10 6UX, England (01 948 7742)
Hawthorn	=	Hawthorn Books, 7 College Park Drive, Westbury-on-Trym, Bristol BS10 7AN, England (0272 509175)
Hemlock	=	Hemlock Books, 170 Beach 145th Street, Neponsit, New York 11694, U.S.A. (718 318 0737)
Henly	=	John Henly, Bookseller, Brooklands, Walderton, Chichester, West Sussex PO18 9EE, England (0705 631426)
Heritage	=	Heritage Book Shop, Inc., 8540 Melrose Avenue, Los Angeles, California 90069, U.S.A. (213 659 3674)
High Latitude	=	High Latitude, P.O. Box 11254, Bainbridge Island, WA 98110, U.S.A. (206 598 3454)
Hollett	=	R.F.G. Hollett and Son, 6 Finkle Street, Sedbergh, Cumbria LA10 5BZ, England (05396 20298)
Jarndyce	=	Jarndyce, Antiquarian Booksellers, 46 Great Russell Street, Bloomsbury, London WC1B 3PE, England (01 631 4220)
Jenkins	=	The Jenkins Company, Box 2085, Austin, Texas 78768, U.S.A. (512 444 6616)
C.R. Johnson	=	C.R. Johnson, 21 Charlton Place, London N1 8AQ, England (01 354 1077)
Michael Johnson	=	Michael Johnson Books, Oak Lodge, Kingsway, Portishead, Bristol BS20 8HW, Scotland (2 848764)
Karmiole	=	Kenneth Karmiole, Bookseller, 1225 Santa Monica Mall, Santa Monica, California 90401, U.S.A. (213 451 4342)
Lamb	=	R.W. & C.R. Lamb, Talbot House, 158 Denmark Rd., Lowestoft, Suffolk NR32 2EL, England (0502 564306)
Lewton	=	L.J. Lewton, Old Station House, Freshford, Bath BA3 6EQ, England (022 122 3351)
Limestone Hills	=	Limestone Hills Book Shop, P.O. Box 1125, Glen Rose, Texas 76043, U.S.A. (817 897 4991)
Lyon	=	Richard Lyon, 17 Old High Street, Hurstpierpoint, West Sussex BN6 9TT, England (0273 832255)
Marlborough	=	Marlborough Rare Books Ltd., 144-146 New Bond Street, London W1Y 9FD, England (01 493 6993)
Marlborough B'Shop	=	Marlborough Bookshop, 6 Kingsbury Street, Marlborough, Wiltshire, England (0672 54074)

Mendelsohn	=	H.L. Mendelsohn, Fine European Books, P.O. Box 317, Belmont, Massachusetts 02178, U.S.A. (617 484 7362)
Minkoff	=	George Robert Minkoff Inc., Rare Books, R.F.D., Box 147, Great Barrington, Mass 01230, U.S.A. (413 528 4575)
Minster Gate	=	The Minster Gate Bookshop, 8 Minster Gates, York YO1 2HL, England (0904 621812)
Monmouth	=	Monmouth House Books, Llanfapley, Abergavenny, Gwent NP7 8SN, Wales (060 085 236)
Moon	=	Michael Moon, Antiquarian, Booksellers & Publishers, 41, 42 & 43 Roper Street, Whitehaven, Cumbria CA28 7BS, England (0946 62936)
Moorhead	=	Moorhead Books, Suffield Cottage, Moorhead, Gildersome, Leeds LS27 7BA, England
Mordida	=	Mordida Books, P.O. Box 79322, Houston, Texas 77279, U.S.A. (713 467 4280)
M & S Rare Books	=	M & S Rare Books, Inc., Box 311, Weston, Massachusetts 02193, U.S.A. (617 891 5650)
Nouveau	=	Nouveau Rare Books, Steve Silberman, P.O. Box 12471, 5005 Meadow Oaks Park Drive, Jackson, Mississippi 39211, U.S.A. (601 956 9950)
Oak Knoll	=	Oak Knoll Books, 414 Delaware Street, New Castle, Delaware 19720, U.S.A. (302 328 7232)
Offenbacher	=	Emile Offenbacher, 84-50 Austin Street, P.O. Box 96, Kew Gardens, New York 11415, U.S.A. (718 849 5834)
Old Cathay	=	Old Cathay Fine Books, 106 Park Meadow, Old Hatfield, Hertfordshire AL9 5HE, England (07072 71006)
Parmer	=	J. Parmer, Booksellers, 7644 Forrestal Road, San Diego, CA 92120, U.S.A. (619 287 0693)
Patterson	=	Ian Patterson, 21 Bateman Street, Cambridge CB2 1NB, England (0223 321658)
Piccadilly	=	Piccadilly Rare Books Ltd., Old Knowle, Frant, Kent TN3 9EJ, England (089 275 340)
Pickering	=	Pickering & Chatto Ltd., 17 Pall Mall, London SW1Y 5NB, England (01 930 8627)
Polyanthos	=	Polyanthos Park Avenue Books, 600 Park Avenue, Huntington, NY 11743, U.S.A. (516 673 9232)
Ramer	=	Richard C. Ramer, 225 East 70th Street, New York, N.Y. 10021, U.S.A. (212 737 0222)
Rankin	=	Alan Rankin, 72 Dundas Street, Edinburgh EH3 6QZ, Scotland, Scotland (031 556 3705)
Rayfield	=	Tom Rayfield, The Blacksmiths, Radnage Common, Buckinghamshire HP14 4DH, England (024 026 3986)
Reese	=	William Reese Company, 409 Temple Street, New Haven, Connecticut 06511, U.S.A. (203 789 8081)
David Rees	=	David Rees, 22 Wanley Road, London SE5 8AT, England (01 737 4557)
Reference Works	=	Reference Works, Barry Lamb & David Hollister, 12 Commercial Road, Dorset BH19 1DF, England (0929 424423)
Respess	=	L & T Respess Books, PO Box 236, Bristol, RI 02809, U.S.A.
Rittenhouse	=	Rittenhouse Book Store, 1706 Rittenhouse Square, Philadelphia, Pennsylvania 19103, U.S.A. (215 545 6072)
Roberts	=	Graeme Roberts, 57 Queens Road, Leytonstone, London E11 1BA, England (01 539 7095)
Robertshaw	=	John Robertshaw, 5 Fellowes Drive, Ramsey, Huntingdon, Cambridgeshire PE17 1BE, England (0487 813330)

Rootenberg	=	B & L. Rootenberg, P.O. Box 5049, Sherman Oaks, California 91403-5049, U.S.A. (818 788 7765)
Rostenberg & Stern	=	Leona Rostenberg and Madeleine, Stern, Rare Books, 40 East 88 Street, New York, N.Y. 10128., U.S.A. (212 831 6628)
Sanders	=	Sanders of Oxford Ltd., 104 High Street, Oxford OX1 4BW, England (0865 242590)
Savona	=	Savona Books, 9 Wilton Road, Hornsea, North Humberside HU13 1QU, England (0964 535195)
Schoyer	=	Schoyer's Books, 1404 S. Negley Avenue, Pittsburgh, PA 15217, U.S.A. (412 521 8464)
Sklaroff	=	L.J. Sklaroff, The Bookshop, The Broadway, Totland, Isle of Wight PO39 0BW, England (0983 754960)
John Smith	=	John Smith & Son (Glasgow), 57-61 St. Vincent Street, Glasgow G2 5TB, Scotland (041 221 7472)
Sotheran	=	Henry Sotheran Ltd., 2 Sackville Street, Piccadilly, London W1X 2DP, England (01 439 6151)
Spelman	=	Ken Spelman, 70 Micklegate, York YO1 1LF, England (0904 624414)
Stewart	=	Andrew Stewart, 11 High Street, Helpringham, Sleaford, Lincolnshire NG34 9RA, England (052 921 617)
Sumner & Stillman	=	Summer & Stillman, P.O. Box 225, Yarmouth, ME 04096, U.S.A. (207 846 6070)
W. Thomas Taylor	=	W. Thomas Taylor, 1906 Miriam, Austin, Texas 78722, U.S.A. (512 478 7628)
Temple	=	Robert Temple, 65 Mildmay Road, London N1 4PU, England (01 254 3674)
Terramedia	=	Terramedia Books, 19 Homestead Road, Wellesley, MA 02181, U.S.A. (617 237 6485)
Thornton's	=	Thornton's of Oxford, 11 Broad Street, Oxford OX1 3AR, England (0865 242939)
Tiger Books	=	Tiger Books, Yew Tree Cottage, Westbere, Canterbury Kent CT2 0HH, England (0227 710030)
Transition	=	Transition Books, 209 Post Street, Suite 614, San Francisco, CA 94108, U.S.A. (415 391 5161)
Trebizond	=	Trebizond Rare Books, P.O. Box 2430 - Main Street, New Preston, CT 0677, U.S.A. (203 868 2621)
Trophy Room Books	=	Trophy Room Books, Box 3041, Agoura, CA 91301, U.S.A. (818 889 2469)
Upcroft	=	Upcroft Books Ltd., 66 St. Cross Road, Winchester, Hampshire SO23 9PS, England (0962 52679)
Virgo	=	Virgo Books, Mrs. Q.V. Mason, Little Court, South Wraxall, Bradford-on-Avon, Wiltshire BA15 2SE, England (02216 2040)
Washton	=	Andrew D. Washton, 411 East 83rd Street, New York, New York 10028, U.S.A. (212 751 7027)
Waterfield's	=	Waterfield's, 36 Park End Street, Oxford OX1 1HJ, England (0865 721809)
Waterland	=	Waterland Books, 28 North Street, March, Cambridgeshire PE15 8LS, England (0354 52160)
West Side	=	West Side Books, 113 W. Liberty, Ann Arbor, MI 48103, U.S.A. (313 995 1891)
Wheldon & Wesley	=	Wheldon & Wesley Ltd., Lytton Lodge, Codicote, Hitchin, Hertfordshire SG4 8TE, England (0438 820370)
Whitehart	=	F.E. Whitehart, Rare Books, 40 Priestfield Road, Forest Hill, London SE23 2RS, England (01 699 3225)
David White	=	David White, 17 High Street, Bassingbourne, Royston, Hertfordshire SG8 5NE, England (0763 243986)

Charles B. Wood	=	Charles B. Wood III, Inc., 116 Commonwealth Avenue, Post Office Box 310, Boston, Massachusetts 02117, U.S.A. (617 247 7844)
Susan Wood	=	Susan Wood, 24 Leasowe Road, Rubery, Rednal, Worcestershire B45 9TD, England (021 453 7169)
Woolmer	=	J. Howard Woolmer, Revere, Pennsylvania 18953, U.S.A. (215 847 5074)
Words Etcetera	=	Words Etcetera, Julian Nangle, Hod House, Child Okeford, Dorset DT11 8EH, England (0258 860539)
Worldwide Antiqu'n	=	Worldwide Antiquarian, Post Office Box 391, Cambridge, MA 02141, U.S.A. (617 876 6220)
Wreden	=	William P. Wreden, 206 Hamilton Avenue, P.O. Box 56, Palo Alto, CA 94302-0056, U.S.A. (415 325 6851)
Xerxes	=	Xerxes, Fine & Rare Books & Documents, Box 428, Glen Head, New York 11545, U.S.A. (516 671 6235)
Ximenes	=	Ximenes: Rare Books, Inc., 19 East 69th Street, New York, NY 10021, U.S.A. (212 744 10021)
Young's	=	Young's Antiquarian Books, Tillingham, Essex CM0 7ST, England (062187 8187)
Zeno	=	Zeno, 6 Denmark Street, London WC2H 8LP, England (01 836 2522)